Crack
GATE/IES/PSUs & other Competitive Exams *with*

arihant's
Handbook *Series*

Mechanical Engineering

Highly focuses *on*
Key Terms and their Definitions
Well Illustrated Formulae
Theoretical Analysis with Suitable Diagrams
Coverage of Key Points for Additional Info

Gaurav Shukla

arihant
ARIHANT PUBLICATIONS (INDIA) LIMITED

arihant

Arihant Publications (India) Ltd.

卐 **ADMINISTRATIVE & PRODUCTION OFFICES**

Regd. Office

'Ramchhaya' 4577/15, Agarwal Road, Darya Ganj, New Delhi -110002
Tele: 011- 47630600, 43518550

Head Office

Kalindi, TP Nagar, Meerut (UP) - 250002
Tele: 0121-7156203, 7156204

卐 **SALES & SUPPORT OFFICES**

Agra, Ahmedabad, Bengaluru, Bareilly, Chennai, Delhi, Guwahati, Hyderabad, Jaipur, Jhansi, Kolkata, Lucknow, Nagpur & Pune

卐 **ISBN** : 978-93-5094-584-1

卐 **Price** : ₹ 250.00

Published by Arihant Publications (India) Ltd.

For further information about the books published by Arihant log on to www.arihantbooks.com or email to info@arihantbooks.com

Follow us on

Preface

The development of nation is directly proportional to the requirement of engineers. India being a developing country, absorbs huge number of engineers every year, and their demand in coming days cannot be overlooked. There is a lot of craze among the aspirants who like to crack GATE/IES and PSUs through many competitive exams. An aspirant would like to learn lot of things in short duration but in less volume, so keeping above point of view ,I have come up with **Handbook Mechanical Engineering** .This handbook is meant for an exhaustive and precise collection of all subjects that come under **Mechanical Engineering.** It encompasses the topics of leading exams in engineering cadre i.e.,GATE,IES and PSUs.

The key features of this book are

- Each topic is summarized in an exhaustive manner in the form of key points and notes.
- Every topic is taken up separately along with key points and notes.
- Focused material in entirety to prevent ambiguity in concepts.

I am thankful to Arihant Publication (India) Limited for giving me this opportunity to write such a book which covers almost 100% syllabus of GATE,IES and PSUs and thus enlightens the path to your success.

I would also like to thank **Er.Akash Shukla** (Project Coordinator) for giving me full support during this project. Valuable suggestions are always welcome for further improvement.

Gaurav Shukla

Contents

Mechanics 1-16

- Force System 1
- Truss 1
- Friction Force 3
- Plane Motion 4
- Rocket Propulsion 8
- Collision 9
- Moment of Inertia 11

Strength of Materials 17-47

- Engineering Mechanics 17
- Elongation of Bodies 21
- Volumetric Strain 25
- Analysis of Stress and Strain 26
- Mohr's Circle 30
- Shear Force and Bending Moment Diagram 33
- Deflection of Beam 37
- Columns and Struts 42
- Theories of Failure 46

Theory of Machine 48-110

- Basic Concept of Theory of Machines 48
- Degree of Freedom 49
- Velocity Analysis in Mechanism 51
- Instantaneous Centre of Velocity 54
- Acceleration Analysis in Mechanism 56
- Kinematic Synthesis of Mechanism 59
- Cams 60
- Cam with Specified Contours 63
- Gears and Gear Trains 65
- Static and Dynamic Force Analysis 75
- Turning Moment Diagram and Flywheel 80
- Balancing 84
- Governors 91
- Vibrations 99

Machine Design 111-155

- Cotter and Knuckle Joints 111
- Knuckle Joint 113
- Welded Joints 114
- Clutch 124
- Brake 128
- Design of Friction Drives 133
- Gears 143
- Bearing 150

Fluid Mechanics 156-207

- Basic Concept of Fluid Mechanics 156
- Kinematics and Dynamics of Fluids 163
- Flow Measurement 170
- Boundary Layer Theory 176
- Dimensional and Model Analysis 188
- Hydraulic Turbine and Pumps 190
- Centrifugal Pump 201
- Fluid System 206

Heat and Mass Transfer 208-232

- Steady and Unsteady State Conductions 208
- Convection 217
- Heat Exchangers 224
- Radiation 227
- Electrical Network Approach for Radiation Heat Exchange 230

Thermodynamics 233-273

- Basic Concepts of Thermodynamics 233
- Laws of Thermodynamics 238
- Entropy 248
- Availability and Irreversibility 253
- Properties of Pure Substance 257
- Thermodynamics Relations and Joule Thomson Coefficient 270

Power Plant Engineering 274-308

- Analysis of Steam Cycles 274
- Combined Cycle Power Generation 278
- Turbine 281
- Gas Turbine 288
- Component of Gas Turbine Power Plant 291
- Isentropic Flow and Shock Waves 299

Refrigeration and Air Conditioning 309-325

- Refrigeration Cycle 309
- Refrigerant 316
- Psychrometric Charts and Its Applications 317
- Comfort Chart 325

Internal Combustion Engine 326-357

- Basic Concepts Related to Engine 326
- Cycles and Their Analysis 330
- Fuels 339
- Carburetion, Fuel Injections and Ignition 342
- Combustion 347
- Engine Emission 350
- Measurements and Performance Parameters 352

Material Science and Production Engineering 358-425

- Structure of Materials 358
- Metal Cutting and Tools 367
- Metal Forging 382
- Powder Metallurgy 396
- Metal Casting 399
- Welding 406
- Metrology 420

Industrial Engineering 426-480

- Forecasting 426
- Inventory 432
- Linear Programming 439
- Transportation and Assignment Model 443
- Break Even Analysis 447
- Line Balancing 450
- PERT and CPM 452
- Quality Control 457
- Queueing Theory 465
- Work Study and Value Engineering 469
- Sequencing 474
- ABC Analysis and MRP 476
- Plant Layout and PPC 478

Element of Computation 481-486

- Numerical Control & Computer Numerical Control 481
- Part Programming 482
- Robotics 484
- Computer 485

Appendix 487-495

1

Mechanics

Force System

When a member of forces simultaneously acting on the body, it is known as force system. A force system is a collection of forces acting at specified locations. Thus, the set of forces can be shown on any free body diagram makes-up a force system.

Truss

It is a rigid structure composed of number of straight members pin jointed to each other. It can sustain static or dynamic load without any relative motion to each other.

Types of Truss

1. **Plane Truss** It is defined as a truss in which members are essentially lies in a single plane.
2. **Rigid Truss** Rigid means there is no deformation take place due to internal strain in members.
3. **Simple Truss** This type of trusses built a basic triangle by adding different members are known as **simple truss**.

Classification of a Truss (based on joints)

Truss can be classified on the basis of joints (*j*) and members (*m*) in the structure. It can be easily understood with the help of following hierarchical approach.

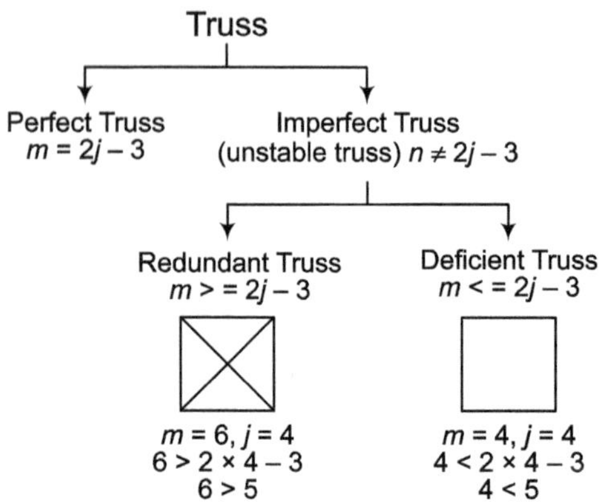

Classification of truss

where m = number of members and j = number of joints

Key Points

- When truss collapse under loading, then truss is known as unstable or imperfect truss.
- When truss is not collapse under the loading, then truss is known perfect truss.

Analysis of a Framed Structure (Section Method)

1. This method is used when the forces in the few members of a truss is required to found out in a truss structure.

To find out force in AC, BC and BD

- First cut a section which passes through AC, BC and BD members.
- Find out reaction at point A and B

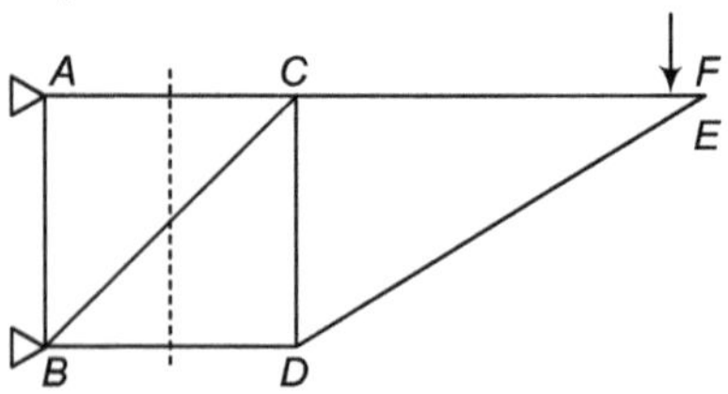

Framed structure using section method

- Find out forces F_{CA}, F_{CB} and F_{DB} in members CA, CB and DB respectively by taking moment about A and B.

2. **Analytical Method** In this method, the free body diagram of each joint is separately analysed to find magnitude of stresses in the truss members.

Lami's Theorem

If a body is in equilibrium under three concurrent forces, the each force is proportional to the sine of the angle between other two.

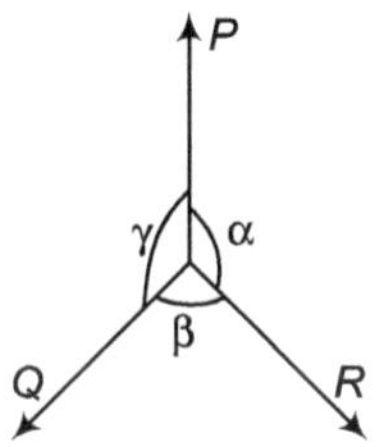

Three concurrent forces P, Q and R

$$\frac{P}{\sin\beta} = \frac{Q}{\sin\alpha} = \frac{R}{\sin\gamma}$$

Friction Force

It is resistant force which acts in opposite direction at the surface in body which tend to move or its move.

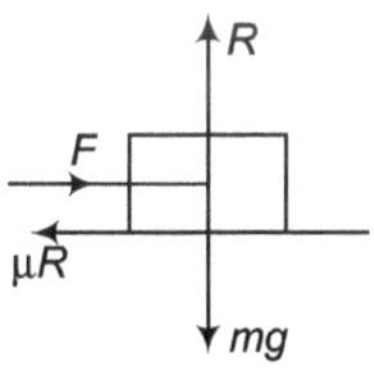

Friction force on a body

Normal force $R = mg$

If $\mu\, mg > F$, the body will not move.

$\mu\, mg = F$ the body will tend to move.

$\mu\, mg < F$ the body will move.

Angle of Friction

It is defined as the angle between normal reaction and resultant reaction when the body is in condition of just sliding.

$$\tan\phi = \frac{\mu\, mg}{R}$$

$$= \frac{\mu\, mg}{mg} = \mu \qquad (\because R = mg)$$

$$\phi = \tan^{-1}\mu$$

μ = coefficient of friction

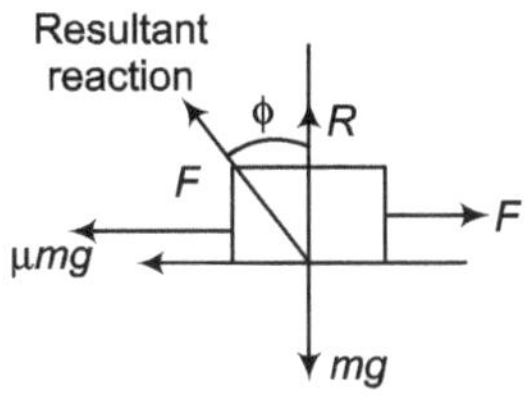

Angle of friction due to resultant reaction

Angle of Repose (α)

It is defined as angle of inclined plane with horizontal at which body is in condition of just sliding.

$$\alpha = \phi$$

Angle of friction is equal to angle of repose.

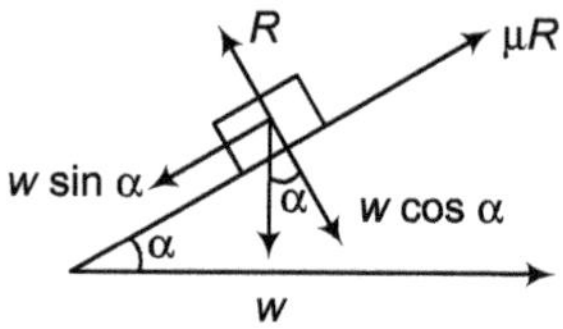

Angle of Repose (α)

Plane Motion

When all parts of the body move in a parallel planes then a rigid body said to perform plane motion.

Key Points

- The motion of rigid body is said to be translation, if every line in the body remains parallel to its original position at all times.
- In translation motion, all the particles forming a rigid body move along parallel paths.
- If all particles forming a rigid body move along parallel straight line, it is known as rectilinear translation.
- If all particles forming a rigid body does not move along a parallel straight line but they move along a curve path, then it is known as curvilinear translation.

Straight Line Motion

It defines the three equations with the relationship between velocity, acceleration, time and distance travelled by the body. In straight line motion, acceleration is constant.

$$v = u + at$$

$$s = ut + \frac{1}{2}at^2$$

$$v^2 = u^2 + 2as$$

where, u = initial velocity
v = final velocity
a = acceleration of body
t = time
s = distance travelled by body

Distance travelled in nth second

$$s_n = u + \frac{1}{2}a\,(2n - 1)$$

Projectile Motion

Projectile motion defines that motion in which velocity has two components, one in horizontal direction and other one in vertical direction. Horizontal component of velocity is constant during the flight of the body as no acceleration in horizontal direction.

Let the block of mass is projected at angle θ from horizontal direction

$$\text{Maximum height } h_{max} = \frac{u^2 \sin^2 \theta}{2g}$$

$$\text{Time of flight } T = \frac{2u \sin \theta}{g}$$

$$\text{Range } R = \frac{u^2 \sin 2\theta}{g}$$

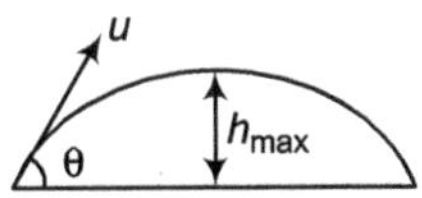

Projectile motion

where, u = initial velocity

Key Points

- At maximum height vertical component of velocity becomes zero.
- When a rigid body move in circular paths centered on the same fixed axis, then the particle located on axis of rotation have zero velocity and zero acceleration.
- Projectile motion describe the motion of a body, when the air resistance is negligible.

Rotational Motion with Uniform Acceleration

Uniform acceleration occurs when the speed of an object changes at a constant rate. The acceleration is the same over time. So, the rotation motion with uniform acceleration can be defined as the motion of a body with the same acceleration over time. Let the rod of block rotated about a point in horizontal plane with ω angular velocity.

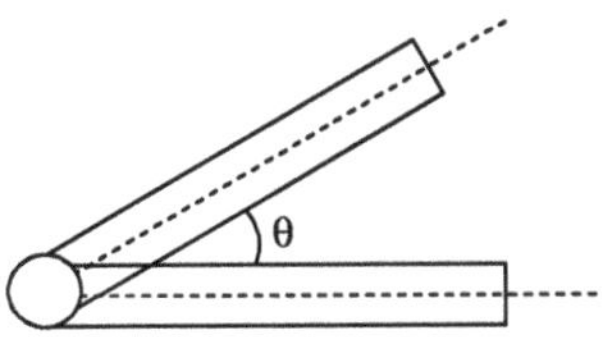

Rotational motion

$$\text{Angular velocity } \omega = \frac{d\theta}{dt} \text{ (change in angular displacement per unit time)}$$

$$\text{Angular acceleration } \alpha = \frac{d\omega}{dt} \Rightarrow \alpha = \frac{d^2\theta}{dt^2}$$

where θ = angle between displacement.

In case of angular velocity, the various equations with the relationships between velocity, displacement and acceleration are as follows.

$$\theta = \omega t$$
$$\alpha = 0$$
$$\omega = \omega_0 + \alpha t$$
$$\theta = \omega_0 t + \frac{1}{2}\alpha t^2$$
$$\omega^2 = \omega_0^2 + 2\alpha\theta$$

where ω_0 = initial angular velocity
ω = final angular velocity
α = angular acceleration
θ = angular displacement

Angular displacement in nth second

$$\theta_n = \omega_0 + \frac{1}{2}\alpha\,(2n-1)$$

Relation between Linear and Angular Quantities

There are following relations between linear and angular quantities in rotational motion.

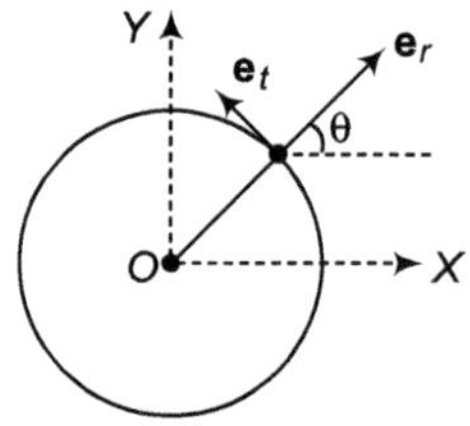

Position of radial and tangential vectors

$$|\mathbf{e}_r| = |\mathbf{e}_t| = 1$$

$\mathbf{e}_r$ and $\mathbf{e}_t$ are radial and tangential unit vector.

Linear velocity $v = r\omega\,\mathbf{e}_t$

Linear acceleration (Net)

$$a = -\,\omega^2 r\,\mathbf{e}_r + \frac{dv}{dt}\,\mathbf{e}_t$$

Tangential acceleration $a_t = \frac{dv}{dt}$ (rate of change of speed)

Centripetal acceleration $a_r = \omega^2 r = \frac{v^2}{r}$ $(\because\ v = r\omega)$

Net acceleration $a = \sqrt{a_r^2 + a_t^2}$

$$= \sqrt{\left(\frac{v^2}{r}\right)^2 + \left(\frac{dv}{dt}\right)^2}$$

where a_r = centripetal acceleration
a_t = tangential acceleration

Centre of Mass of Continuous Body

Centre of mass of continuous body can be defined as

- Centre of mass about x, $x_{CM} = \dfrac{\int x\,dm}{\int dm} = \dfrac{\int x\,dm}{M}$
- Centre of mass about y, $y_{CM} = \dfrac{\int y\,dm}{\int dm} = \dfrac{\int y\,dm}{M}$
- Centre of mass about z, $z_{CM} = \dfrac{\int z\,dm}{\int dm} = \dfrac{\int z\,dm}{M}$

- CM of uniform rectangular, square or circular plate lies at its centre.
- CM of semicircular ring

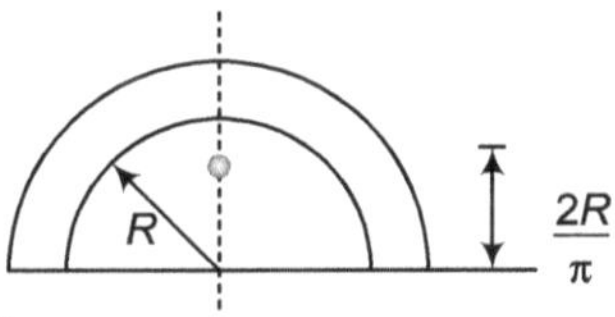

- CM of semicircular disc

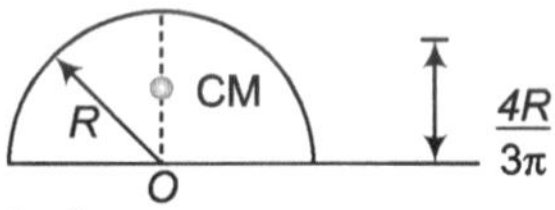

- CM of hemispherical shell

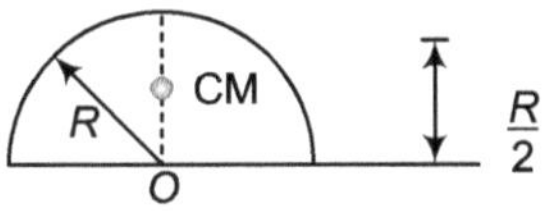

- CM of solid hemisphere

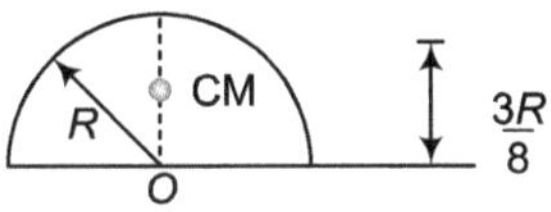

Law of Conservation of Linear Momentum

The product of mass and velocity of a particle is defined as its linear momentum (p).

$$p = mv$$
$$p = \sqrt{2Km}$$

$$F = \frac{dp}{dt}$$

where, K = kinetic energy of the particle
F = net external force applied to body
P = momentum

Rocket Propulsion

Let m_0 be the mass of the rocket at time $t = 0$, m its mass at any time t and v its velocity at that moment. Initially, let us suppose that the velocity of the rocket is u.

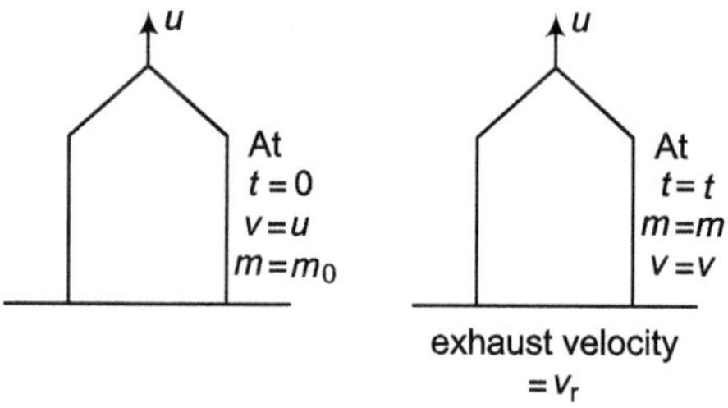

Rocket propulsion

- Thrust force on the rocket $F_t = v_r\left(-\frac{dm}{dt}\right)$

  where, $-\frac{dm}{dt}$ = rate at which mass is ejecting
  v_r = relative velocity of ejecting mass (exhaust velocity)
- Weight of the rocket $w = mg$
- Net force on the rocket $F_{net} = F_t - w = v_r\left(\frac{-dm}{dt}\right) - mg$
- Net acceleration of the rocket

$$a = \frac{F}{m}$$

$$\frac{dv}{dt} = \frac{v_r}{m}\left(\frac{-dm}{dt}\right) - g$$

$$v = u - gt + v_r \ln\frac{m_0}{m}$$

where, m_0 = mass of rocket at time $t = 0$
m = mass of rocket at time t

Impulse

The product of constant force F and time t for which it acts is called the impulse (J) of the force and this is equal to the change in linear momentum which it produces.

$$\text{Impulse } J = Ft$$

$$\Rightarrow \quad \Delta p = p_f - p_i$$

where, F = constant force

P = linear momentum

Instantaneous Impulse *e.g.*, bat and ball contact

$$J = \int F \cdot dt \Rightarrow \Delta p = p_f - p_i$$

Key Points

- The relation between impulse and linear momentum can be understood by the following equation.

$$Ft = m(v - u)$$

where, F = force, t = time, m = mass, v = initial velocity, u = final velocity

- Rotation about a fixed point gives the three dimensional motion of a rigid body attached at a fixed point.

Collision

A Collision is an isolated event in which two or more moving bodies exert forces on each other for a relatively short time.

Collision between two bodies may be classified in two ways

- Head-on collision
- Oblique collision.

Head-on Collision

Let the two balls of masses m_1 and m_2 collide directly with each other with velocities v_1 and v_2 in direction as shown in figure. After collision the velocity become v_1' and v_2' along the same line.

$$v_1' = \left(\frac{m_1 - em_2}{m_1 + m_2}\right) v_1 + \left(\frac{m_2 + em_2}{m_1 + m_2}\right) v_2$$

$$v_2' = \left(\frac{m_2 - em_2}{m_1 + m_2}\right)v_2 + \left(\frac{m_1 + em_1}{m_1 + m_2}\right)v_1$$

where, m_1 = mass of body 1
m_2 = mass of body 2
v_1 = velocity of body 1
v_2 = velocity of body 2
v_1' = velocity of body 1 after collision
v_2' = velocity of body 2 after collision

where e = coefficient restitution

$$e = \frac{\text{Separation speed}}{\text{Approach speed}}$$

$$e = \frac{v_1' - v_2'}{v_2 - v_1}$$

- In case of head-on elastic collision

$$e = 1$$

- In case of head-on inelastic collision

$$0 < e < 1$$

- In case of head-on perfectly inelastic collision

$$e = 0$$

If e is coefficient of restitution between ball and ground, then after nth collision with the floor, the speed of ball will remain $e^n v_0$ and it will go upto a height e^{2n} h.

$$v_n = e^n v_0 = e^n \sqrt{2gh}$$

$$h_0 = e^{2n} h$$

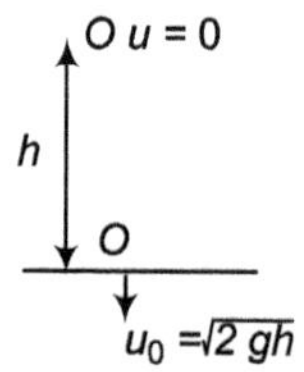

Collision of a ball with floor

Oblique Collision

In case of oblique collision linear momentum of individual particle do change along the common normal direction. No component of impulse act along common tangent direction. So, linear momentum or linear velocity remains unchanged along tangential direction. Net momentum of both the particle remain conserved before and after collision in any direction.

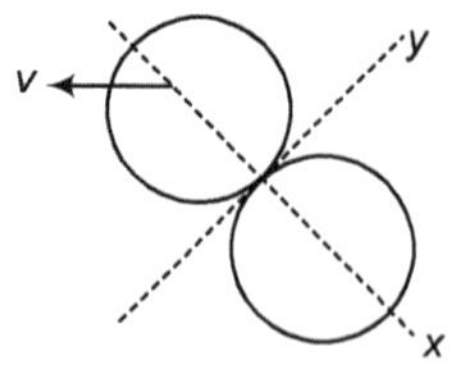

Oblique collision

Moment of Inertia

Momentum of inertia can be defined as

r = distance of the body of mass, m from centre of axis.

$$I = \sum_i m_i r_i^2$$

$$I = \int r^2\, dm$$

- Very thin circular loop (ring)

$$I = MR^2$$

where, M = mass of the body
R = radius of the ring
I = moment of inertia

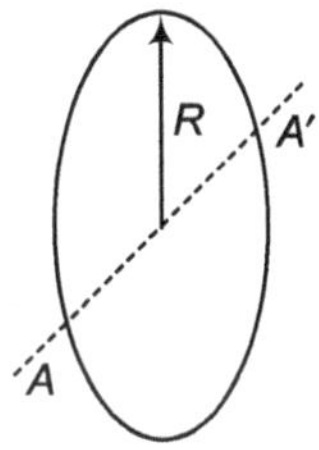

Thin circular ring

- Uniform circular loop $I = M\left(\frac{R_1^2 + R_2^2}{2}\right)$

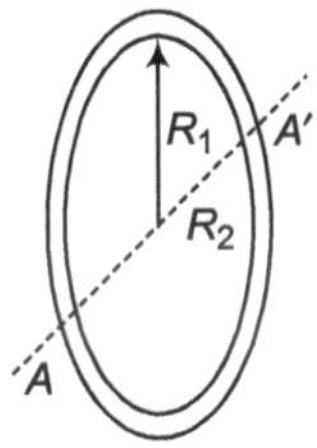

Uniform circular loop

- Uniform solid cylinder $I = \frac{MR^2}{2}$

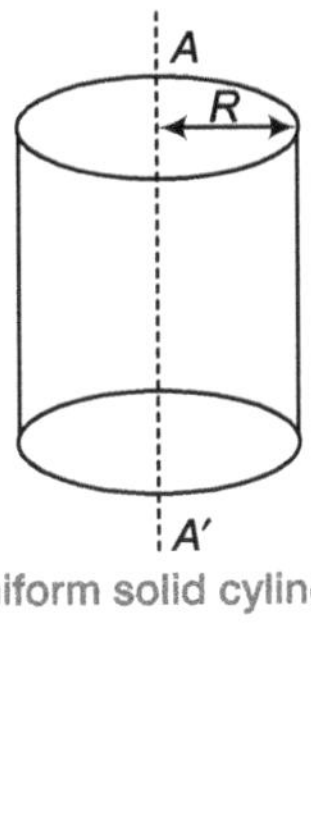

Uniform solid cylinder

- Uniform solid sphere

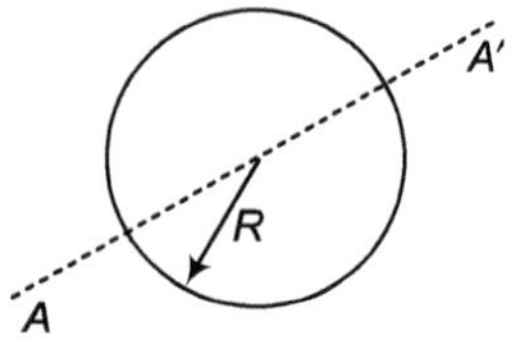

Uniform solid sphere

$$I = \frac{2}{5} MR^2$$

- Uniform thin rod

(AA') moment of inertia about the centre and perpendicular axis to the rod moment of inertia about the one corner point and perpendicular (BB') axis to the rod.

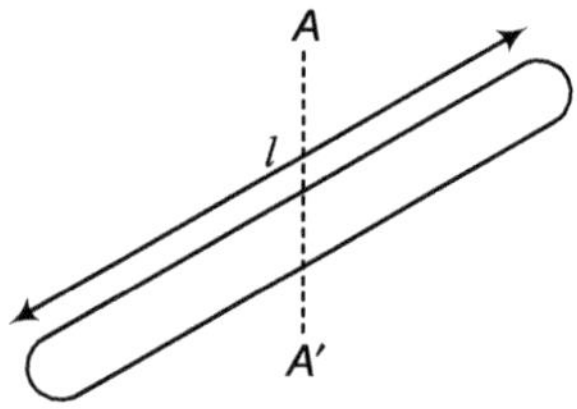

Uniform thin rod

$$I = \frac{Ml^2}{12} \quad \Rightarrow \quad I = \frac{1}{3} Ml^2$$

- Very thin spherical shell

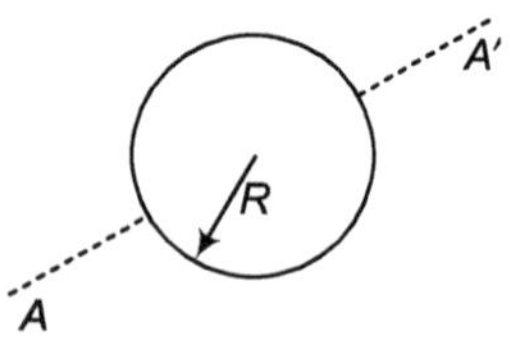

Thin sperical shell

$$I = \frac{2}{3} MR^2$$

- Thin circular sheet

$$I = \frac{MR^2}{4}$$

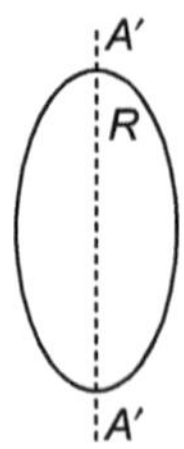

Thin circular sheet

- Thin rectangular sheet

$$I = M\left(\frac{a^2 + b^2}{12}\right)$$

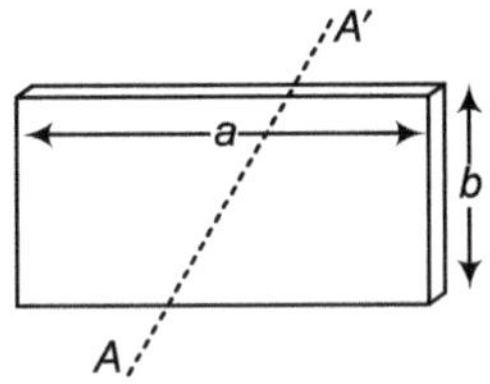

Thin rectangular sheet

- Uniform right cone

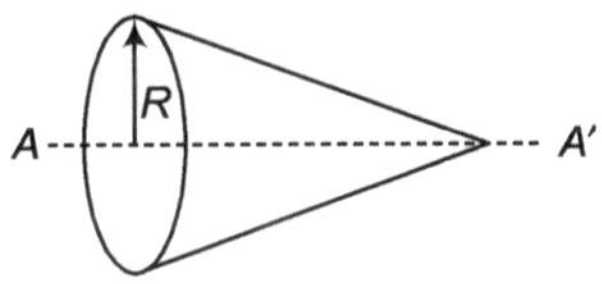

Uniform right cone

$$I = \frac{3}{10} MR^2$$

- Uniform cone as a disc

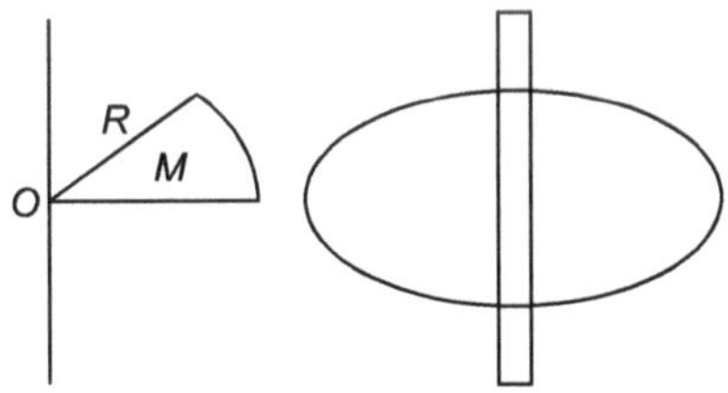

A part of uniform cone as a disc

Suppose the given section is $\frac{1}{n}$th part of the disc, then mass of disc will be nM.

Inertia of the disc,

$$I_{\text{disc}} = \frac{1}{2}(nM) R^2$$

Inertia of the section,

$$I_{\text{section}} = \frac{1}{n} I_{\text{disc}} = \frac{1}{2} MR^2$$

Torque and Angular Acceleration of a Rigid Body

For a rigid body, net torque acting

$$\tau = I\alpha$$

where, α = angular acceleration of rigid body
I = moment of inertia about axis of rotation

- Kinetic energy of a rigid body rotating about fixed axis

$$KE = \frac{1}{2} I\omega^2 \qquad (\omega = \text{angular velocity})$$

- Angular moment of a particle about same point

$$\mathbf{L} = \mathbf{r} \times \mathbf{p}$$
$$\mathbf{L} = m(\mathbf{r} \times \mathbf{v})$$

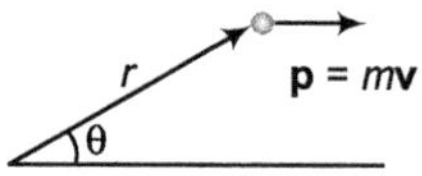

Angular moment

where L = angular displacement

- Angular moment of a rigid body rotating about a fixed axis. $L = I\omega$

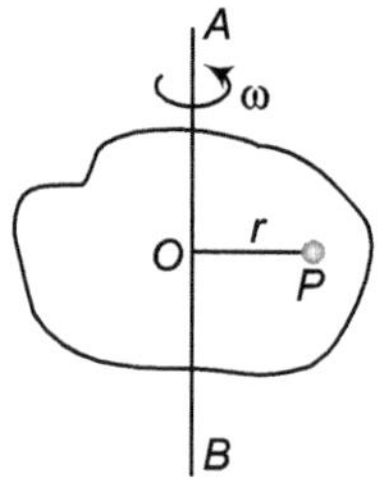

Angular moment of a rigid body

- Angular moment of a rigid body in combined rotation and translation

$$\mathbf{L} = \mathbf{L}_{CM} + M(\mathbf{r}_0 \times \mathbf{v}_0)$$

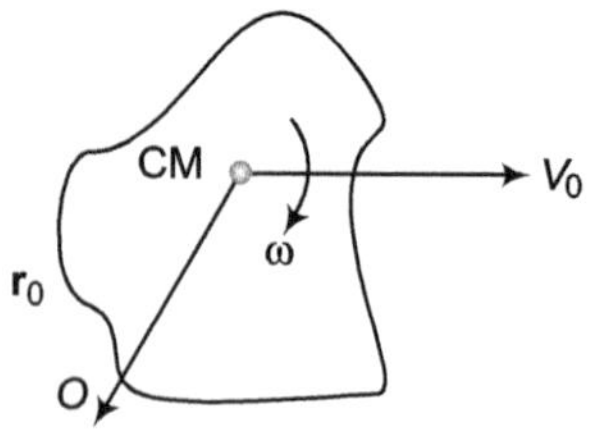

Combined rotation and translation in a rigid body

- Conservation of angular momentum

$$\tau = \frac{d\mathbf{L}}{dt}$$

$$\frac{d\mathbf{L}}{d\mathbf{t}} = \mathbf{r} \times \mathbf{F} + \mathbf{v} \times \mathbf{p}$$

- Kinetic energy of rigid body in combined translational and rotational motion

$$K = \frac{1}{2} m v_{CM}^2 + \frac{1}{2} I_{CM} \omega^2$$

Uniform Pure Rolling

Pure rolling means no relative motion or no slipping at point of contact between two bodies.

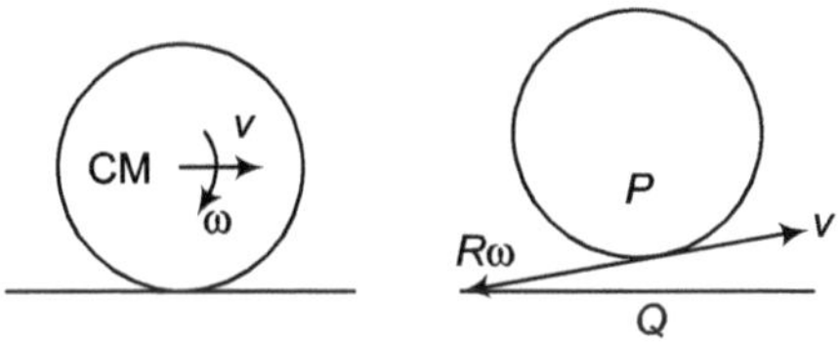

Uniform Pure Rolling

If $v_P = v_Q \Rightarrow$ no slipping
$v = R\omega$

if $v_p > v_Q \Rightarrow$ forward slipping
$v > R\omega$

if $v_P < v_Q \Rightarrow$ backward slipping
$v < R\omega$

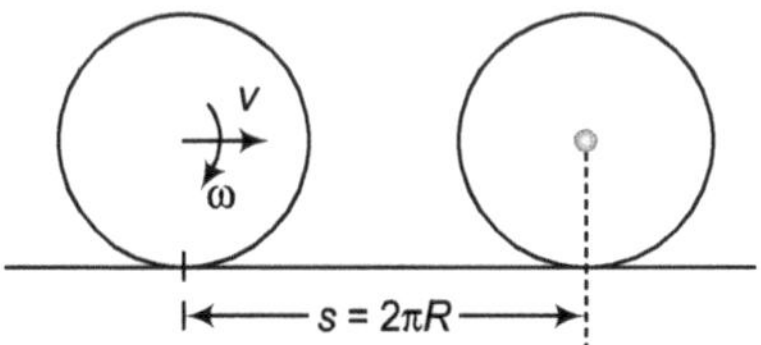

Pure Rolling

No slipping $s = 2\pi R$
Forward slipping $s > 2\pi R$
Backward slipping $s < 2\pi R$

Accelerated Pure Rolling

A pure rolling is equivalent to pure translation and pure rotation. It follows a uniform rolling and accelerated pure rolling can be defined as

$$F + f = Ma$$
$$(F - f) \cdot R = I\alpha$$

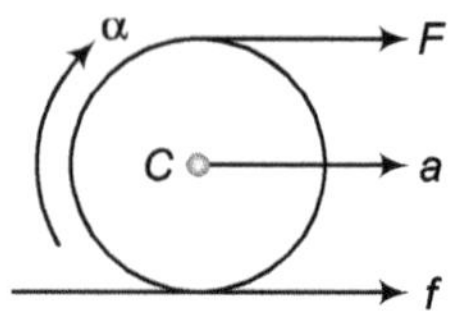

Accelerated pure rolling

F = force acting on a body, f = friction on that body

Angular Impulse

The angular impulse of a torque in a given time interval is defined as

$$\int_{t_1}^{t_2} \tau \cdot dt$$

$$\int_{t_1}^{t_2} \tau \cdot dt = L_2 - L_1$$

where, L_2 and L_1 are the angular momentum at time t_2 and t_1 respectively.

Key Points

- A force, whose line of action does not pass through centre of mass, works as force to produce translational acceleration.
- Different types of collisions are examined, whether they possess kinetic energy or not.
- The radial component of the force, which goes through the axis of rotation, has no contribution to torque.

2

Strength of Materials

Engineering Mechanics

The branch of physical science that deals with the state of rest or the state of motion is termed as **Mechanics.** Starting from the analysis of rigid bodies gravitational force and simple applied forces, the mechanics has grown to the analysis of robotics, air crafts etc. is known is **Engineering Mechanics.**

Stress

When a material in subjected to an external force, a resisting force is set up within the component. The internal resistance force per unit area acting on a material is called the stress at a point.It is a scalar quantity having unit.

$$\text{Stress} = \frac{\text{Force}}{\text{Area}}$$

Strain

It is the deformation produced in the material due to simple stress. It usually represents the displacement between particles in the body relative to a reference length.

$$\text{Tension strain } (e_t) = \frac{\Delta L}{L}$$

$$= \frac{\text{Change in length}}{\text{Initial length}}$$

Stress-Strain Relationship

The stress-strain diagram is shown in the figure. In brittle materials there is no appreciable change in rate of strain. There is no yield point and no necking takes place.

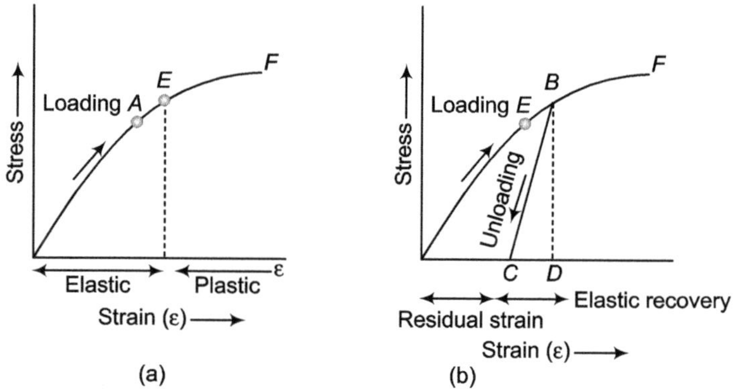

Graph between stress-strain

In figure (a), the specimen is loaded only upto point *A*, is gradually removed the curve follows the same path *AO* and strain completely disappears. Such a behaviour is known as the **elastic behaviour.**

In figure (b), the specimen is loaded upto point *B* beyond the elastic limit *E*. When the specimen is gradually loaded the curve follows path *BC*, resulting in a residual strain *OC* or permanent strain.

Properties of Materials

Some properties of materials which judge the strength of materials are given below

Elasticity

Elasticity is the property by virtue of which a material is deformed under the load and is enabled to return to its original dimension when the load is removed.

Plasticity

Plasticity is the converse of elasticity. A material in plastic state is permanently deformed by the application of load and it has no tendency to recover. The characteristic of the material by which it undergoes inelastic strains beyond those at elastic limit is known as plasticity.

Ductility

Ductility is the characteristic which permits a material to be drawn out longitudinally to a reduced section, under the action of a tensile force (large deformation).

Brittleness

Brittleness implies lack of ductility. A material is said to be brittle when it cannot be drawn out by tension to smaller section.

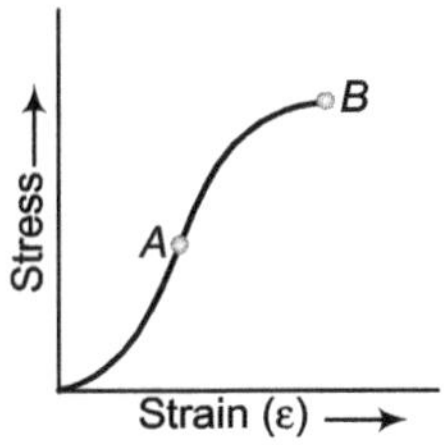

Stress-strain relation

Malleability

Malleability is a property of a material which permits the material to be extended in all directions without rapture. A malleable material possess a high degree of plasticity, but not necessarily great strength.

Toughness

Toughness is the property of a material which enable it to absorb energy without fracture.

Hardness

Hardness is the ability of a material to resist indentation or surface abrasion. Brinell hardness test is used to check hardness.

$$\text{Brinell Hardness Number (BHN)} = \frac{P}{\frac{\pi D}{2}\left(D - \sqrt{D^2 - d^2}\right)}$$

where, P = Standard load, D = Diameter of steel ball
d = Diameter of the indent.

Strength

The strength of a material enables it to resist fracture under load.

Engineering Stress-Strain Curve

The stress-strain diagram is shown in figure. The curve start from origin. Showing thereby that there is no initial stress of strain in the specimen.

The stress-strain curve diagram for a ductile material like mild steel is shown in figure below

- Upto point A, Hooke's Law is obeyed and stress is proportional to strain. Point A is called **limit of proportionality.**

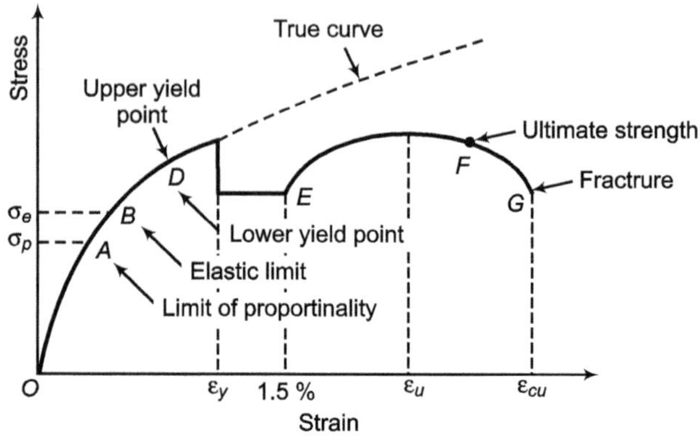

Stress-strain diagram for mild steel

- Point B is called the **elastic limit point.**
- At point B the cross-sectional area of the material starts decreasing and the stress decreases to a lower value to point D, called the **lower yield point.**
- The apparent stress decreases but the actual or true stress goes on increasing until the specimen breaks at point C, called the **point of fracture.**
- From point E ownward, the strain hardening phenomena becomes predominant and the strength of the material increases thereby requiring more stress for deformation, until point F is reached. Point F is called the **ultimate point.**

Hooke's Law (Linear elasticity)

Hooke's Law stated that within elastic limit, the linear relationship between simple stress and strain for a bar is expressed by equations.

$$\sigma \propto \varepsilon, \sigma = E\varepsilon$$

$$\frac{P}{A} = E\frac{\Delta l}{l}$$

where, E = Young's modulus of elasticity
P = Applied load across a cross-sectional area
Δl = Change in length
l = Original length.

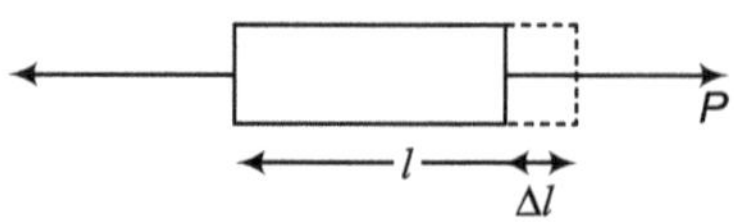

Elastic deformation

Elongation of Bodies

Elongation of a body is defined as the transformation of a body from a reference configuration to a current configuration. A configuration is a set containing the positions of all particles of the body. *The following cases will be considered*

Bars of Varying Sections of Different Materials

It can be shown below

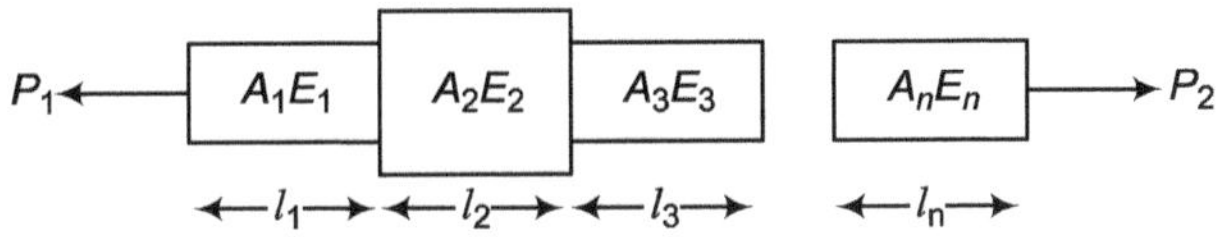

Diagram for different materials

The total deformation for such a bar is given by

$$\Delta = \frac{P_1\, l_1}{A_1\, E_1} + \frac{P_2\, l_2}{A_2\, E_2} \ldots + \frac{P_n\, l_n}{A_n\, E_n}$$

Uniformly Tapering Circular Bar

Let us consider a uniformly tapering circular bar subjected to an axial force P. The bar of length l has diameter d_1 at one end and d_2 at other end.

$$dx = d_1 + \frac{d_2 - d_1}{l} \cdot x$$

Change in length $\Delta\, l = \dfrac{4Pl}{\pi\, E\, d_1\, d_2}$

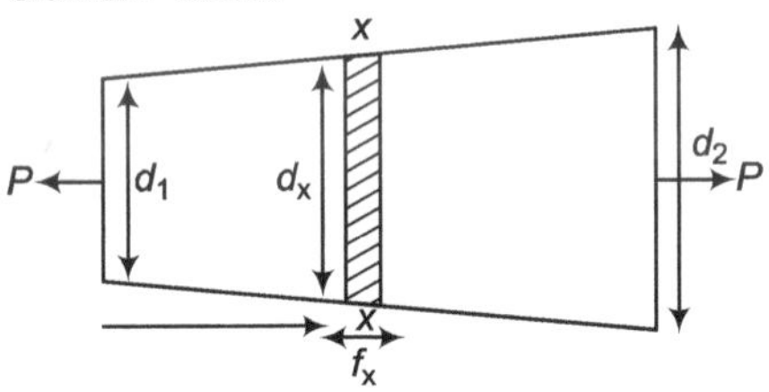

Tapering bars

Elongation of Bar of Uniform Section due to Self Weight

Small deformation of small cross section length dx

$$f_x = \frac{w_x}{A}\frac{f_x}{E}$$

where, w_x = Weight of portion below the section.
Total change in length

$$\Delta\, l = \frac{wL}{2AE}$$

where, w = Total weight of bar.

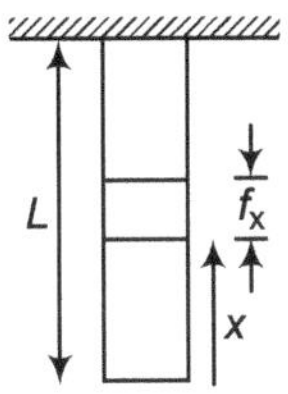

Elongation of bar

Elongation of Bar of Uniform Tapering Section

Total change in length

$$\Delta l = \frac{\rho g L^2}{\sigma E}$$

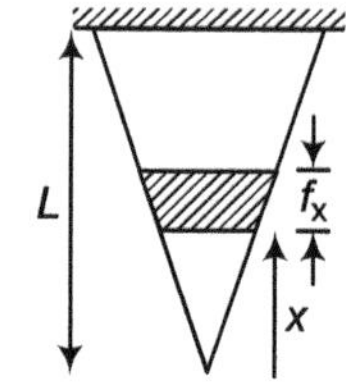

Tapering section

Compounded Bars

Consider a solid box enclosed in the hollow tube and subjected to a compression force P through rigid collars as shown in the figure.

Let P_1 and P_2 are force applied on rod and tube respectively.

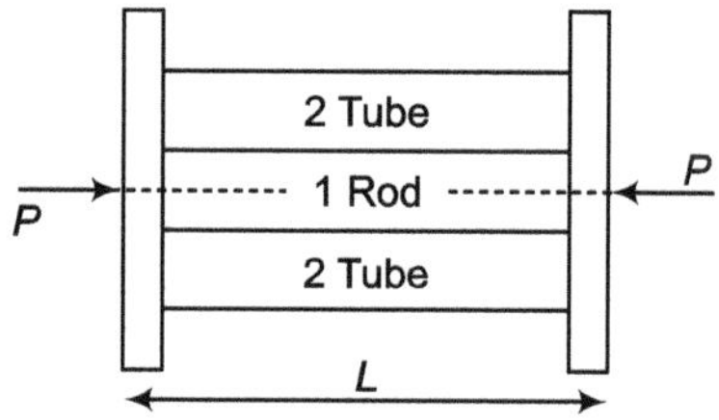

Compound bars

Total force $\quad P = P_1 + P_2$

Change in length for rod = change in length for tube

$$\Delta l_1 = \Delta l_2$$

$$\frac{P_1 L_1}{A_1 E_1} = \frac{P_2 L_2}{A_2 E_2}$$

Stress in Bolts and Nuts

Stress in bolts and nuts are shown as in the figure.

Let stress in steel bolt is σ_s across the cross-sectional area A_s and stress in copper tube is σ_c across cross-sectional area A_c.

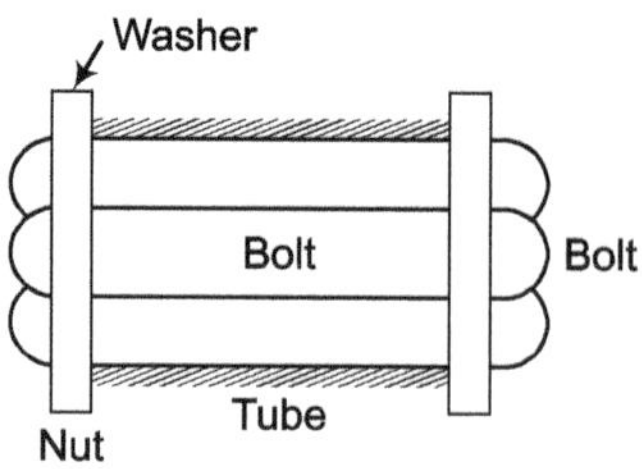

Stress in bolt and nut

$$\sigma_c A_c = \sigma_s A_s$$

where, σ_c = Stress in copper

σ_c = Stress in steel

Since the final length of bolt and tube is same.

Total extension of bolt + Total compression of tube = Moment of nut

$$\frac{\sigma_s L}{E_s} + \frac{\sigma_c L}{E_c} = \text{moment of nut} \qquad \text{(Pitch of thread} \times \text{total turn)}$$

Temperature Stresses in Uniform Bars

Consider a bar of length L subjected to uniform temperature Δt increase.

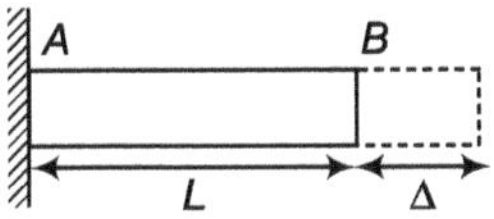

Bar free to expand

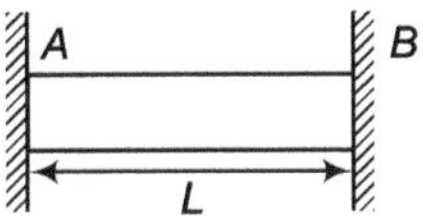

Bar constrained

Increase in length of bar when bar free to expand

$$\Delta l = \varepsilon_t L = \alpha \Delta tL$$

$$\varepsilon_t = \text{thermal strain} = \alpha\ \Delta t$$

where, α = coefficient of the thermal expansion.

If the bar is constrained, net thermal stress

$$\sigma_t = \frac{\Delta l E}{L} = L\alpha\ \Delta t \frac{E}{L} \quad \Rightarrow \quad \sigma_t = E\alpha\ \Delta t$$

Suppose in one of supports yields by an amount a. The total amount of expansion checked will be $(\Delta l - a)$.

$$\sigma_t = (\Delta l - a)\frac{E}{L}$$

$$\sigma_t = (L\alpha\ \Delta t - a)\frac{E}{L}$$

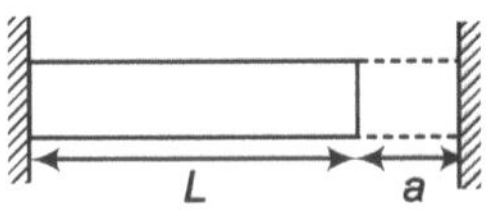

Expansion in rod

Key Points

- The stress will be compressive when the change in temperature is positive.
- The stress will be tensile when the change in temperature is negative.

Temperature Stresses in Composite Bar

Suppose composite bar made of two materials with different coefficient of thermal expansion assume $\alpha_1 < \alpha_2$.

Tensile force in 1 = tensile force in 2

$$P_1 = P_2 = P_1$$

Final extension of 1 = final extension of 2

$$\Delta l_1 = \Delta l_2 = \Delta l$$

From geometry,

$$\Delta l_1 = \Delta l_1' + \Delta l_1''$$

$$\Delta l_2 = \Delta l_2' + \Delta l_2''$$

$\Delta l_1'$ = Free expansion of 1 due to temperature rise
$\Delta l_2'$ = Free expansion of 2 due to temperature rise
$\Delta l_1''$ = Expansion of 1 due to temperature stress (tensile)
$\Delta l_2''$ = Compression of 2 due to temperature stress (compression)

$$\Delta l_1' + \Delta l_1'' = \Delta l_2' - \Delta l_2''$$

$$\Delta l_1'' + \Delta l_2'' = \Delta l_2' - \Delta l_1'$$

$$\frac{PL}{A_1 E_1} + \frac{PL}{A_2 E_2} = L\Delta t\,(\alpha_2 - \alpha_1) \quad \ldots\text{(i)}$$

or

$$P = \frac{\Delta t\,(\alpha_2 - \alpha_1)}{\dfrac{1}{A_s E_s} + \dfrac{1}{A_c E_c}}$$

$$\Delta l = \frac{(\alpha_1 A_1 E_1 + \alpha_2 A_2 E_2)\, L\Delta t}{A_1 E_1 + A_2 E_2}$$

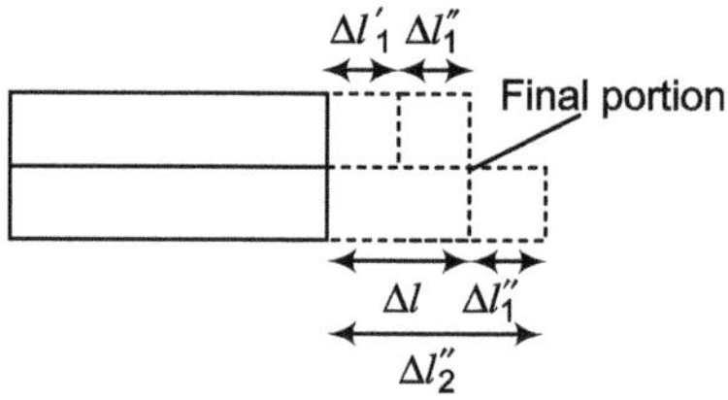

Expansion in composite bar

Complementry Shear Stress

A shear stress in a given direction cannot exist without a balancing shear stress of equal intensity in a direction at right angles to it.

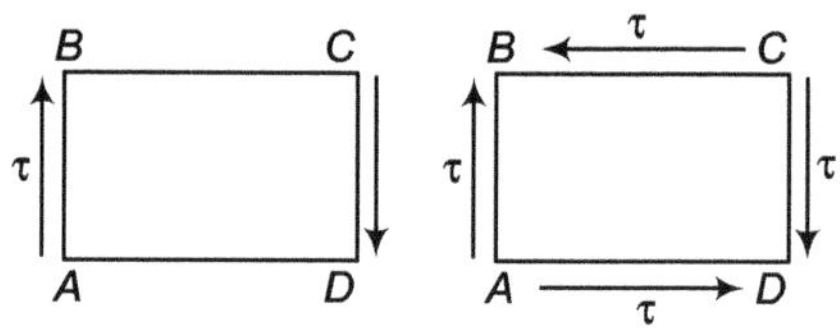

Complementary shear stress

A state of simple shear produces pure tensile and compressive stresses across planes inclined at 45° to those of pure shear and intensities of these direct stresses are each equal to the intensity of the pure shear stress. Linear strain of the diagonal is equal to half the shear strain (ϕ).

$$e = \frac{\phi}{2}$$

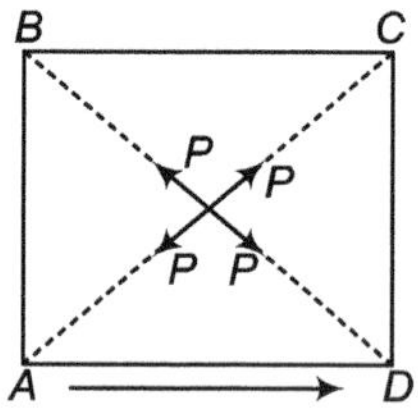

Compressive stress

Poisson's Ratio

When an axial force is applied along the longitudinal axis of a bar, the length of a bar will increase but at the same time its lateral dimension (width) will be decreased so, it is called as Poisson' ratio.

$$\mu = \frac{\text{Lateral strain}}{\text{Longitudinal strain}}$$

Value of Poisson's ratio is same in tension and compression.

Volumetric Strain

It is defined as the ratio of change in volume to the initial volume. Mathematically,

$$\text{Volumetric strain, } e_v = \frac{\text{Change in volume}}{\text{Initial volume}} = \frac{\Delta V}{V}$$

Volumetric Strain Due to Single Direct Stress

The ratio of change in volume to original volume is called **volumetric strain**.

$$e_V = e_1 + e_2 + e_3$$

$$\frac{\Delta V}{V} = \frac{fL}{L} + \frac{fb}{b} + \frac{ft}{t}$$

$$e_1 = \frac{P}{E}$$

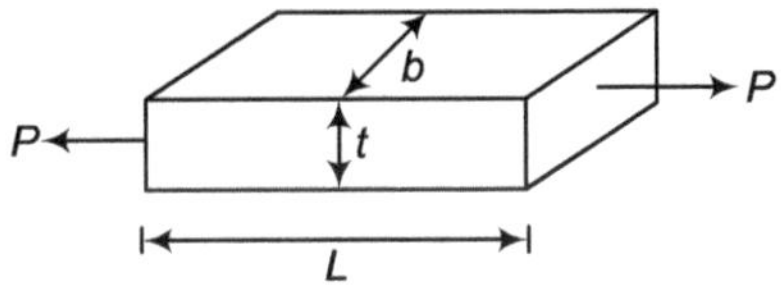

Volumetric strain

$$e_2 = -\frac{P}{E}\mu, e_3 = -\mu\frac{P}{E}$$

$$\text{Volumetric strain,} \quad e_V = \frac{\Delta V}{V}, \; e_V = \frac{P}{E}(1-2\mu)$$

For the circular bar of diameter d, $V = \frac{\pi}{4}d^2L$

$$\frac{fV}{V} = \frac{fL}{L} + \frac{2fd}{d}, \frac{\Delta L}{L} = e_1 = \frac{P}{E}$$

$$\frac{\Delta d}{d} = e_2 = -\mu\frac{P}{E}$$

$$e_V = \frac{P}{E}(1-2\mu), \; e = \frac{P}{E}\left(1-\frac{2}{m}\right), \qquad \left(\because \mu = \frac{1}{m}\right)$$

Volumetric Strain due to Three Mutually Perpendicular Stress System

When a body is subjected to identical pressure in three mutually perpendicular direction, then the body undergoes uniform changes in three directions without undergoing distortion of shape.

$$e_1 = \frac{P_1}{E} - \mu\frac{P_2 + P_3}{E}$$

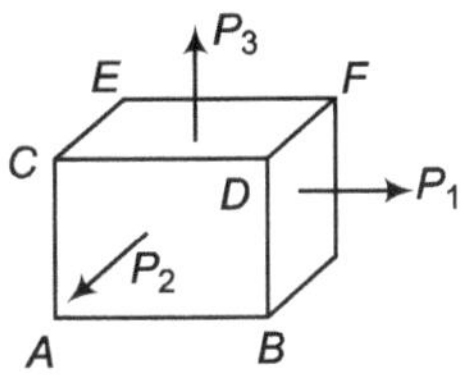

Three stress system

$$e_2 = \frac{P_2}{E} - \mu \frac{P_3 + P_1}{E}$$

$$e_3 = \frac{P_3}{E} - \mu \frac{P_1 + P_2}{E}$$

$$e_V = e_V + e_2 + e_3$$

$$e_V = (1 - 2\mu)\left(\frac{P_1 + P_2 + P_3}{E}\right)$$

or $$e_V = \left(1 - \frac{2}{m}\right)\left(\frac{P_1 + P_2 + P_3}{E}\right) \qquad \left(\because \mu = \frac{1}{m}\right)$$

Shear Modulus or Modulus of Rigidity

Modulus of rigidity $$G = \frac{\text{shear stress}}{\text{shear strain}} = \frac{\tau}{\phi}$$

- At principal planes, shear stress is always zero.
- Planes of maximum shear stress also contains normal stress.

Relationship between *E*, *G*, *K* and μ

Modulus of rigidity $$G = \frac{E}{2(1 + \mu)}$$

Bulk modulus $$K = \frac{E}{3(1 - 2\mu)} \quad \text{or} \quad E = \frac{9KG}{3K + G}$$

$$\mu = \frac{3K - 2G}{6K + 2G}$$

Analysis of Stress *and* Strain

We will derive some mathematical expressions for plains stresses and will study their graphical significance in 2 *D* and 3 *D*.

Stress on Inclined Section *PQ* due to Uniaxial Stress

Consider a rectangular cross-section and we have to calcutate the stress on an inclined section as shown in figure.

Normal stress $$\sigma_n = \sigma \cos^2 \theta$$

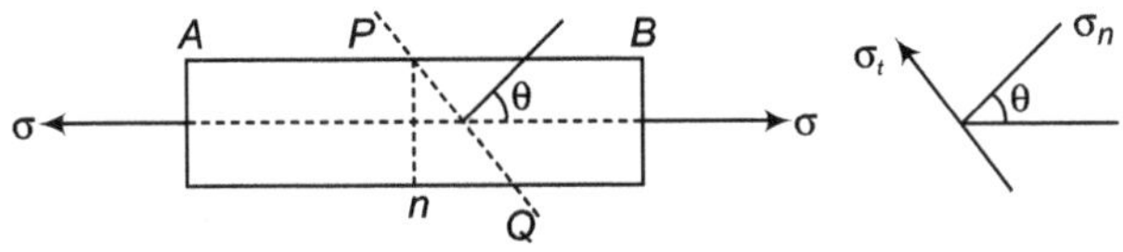

Stress on an inclined section

Tangential stress $\sigma_t = -\frac{\sigma}{2}\sin 2\theta$

Resultant stress $\sigma_r = P\cos\theta$

Stress Induced by State Simple Shear

Induced stress is divided into two components which are given as

$$\text{Normal stress, } \sigma_n = \tau \sin 2\theta$$

$$\text{Tangential stress, } \sigma_t = \tau \cos 2\theta$$

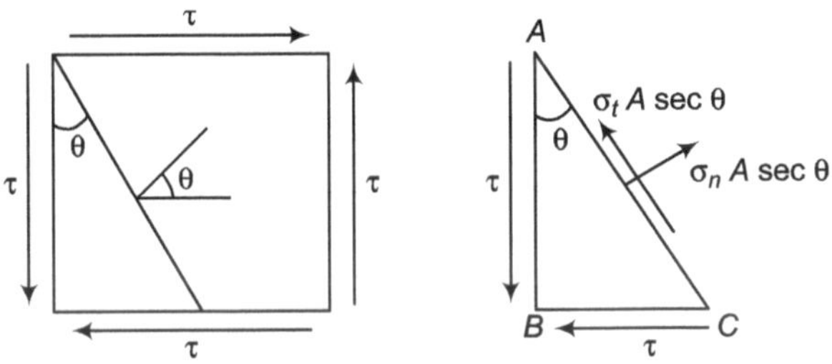

Stress simple shear

Stress Induced by Axial Stress and Simple Shear

Normal stress

$$\sigma_n = \sigma_1 \cos^2\theta + \sigma_2 \sin^2\theta + \tau \sin 2\theta$$

or $$\sigma_n = \frac{\sigma_1 + \sigma_2}{2} + \frac{\sigma_1 - \sigma_2}{2}$$

$$\cos 2\theta + \tau \sin 2\theta$$

Tangential stress

$$\sigma_t = -\left(\frac{\sigma_1 + \sigma_2}{2}\right)\sin 2\theta + \tau\cos 2\theta$$

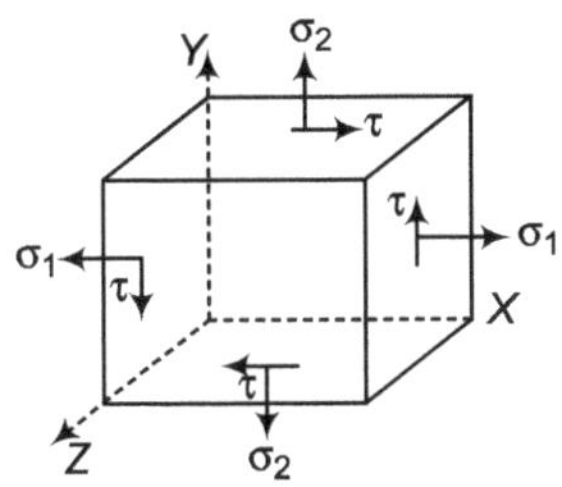

Induced stress body diagram

Principal Stresses and Principal Planes

The plane carrying the maximum normal stress is called the major principal plane and normal stress is called major principal stress. The plane carrying the minimum normal stress is known as minor principal stress.

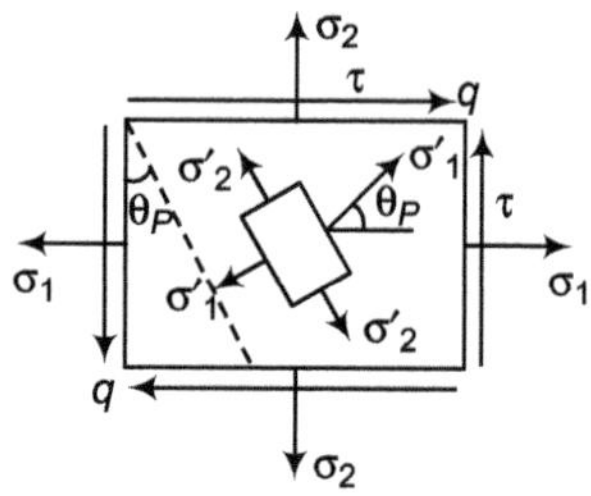

Principal stress and planes

Major principal stress (σ_1'),

$$\sigma_1' = \frac{\sigma_1 + \sigma_2}{2} + \sqrt{\frac{(\sigma_1 - \sigma_2)^2}{2} + \tau^2}$$

Minor principal stress (σ_2'), $\sigma_2' = \frac{\sigma_1 + \sigma_2}{2} - \sqrt{\left(\frac{\sigma_1 - \sigma_2}{2}\right)^2 + \tau^2}$

$$\tan 2\theta_P = \frac{2\tau}{\sigma_1 - \sigma_2}$$

$$\sigma_1' + \sigma_2' = \sigma_1 + \sigma_2$$

When $2\,\theta_P = 0$

$\Rightarrow$ $\sigma_1' = \sigma_1$ and $\sigma_2' = \sigma_2$

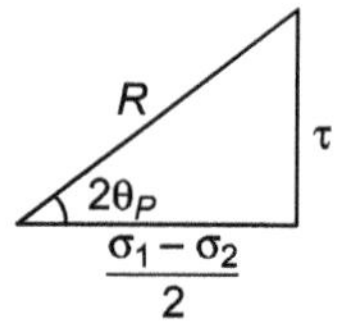

2 θ_P triangle

Across maximum normal stresses acting in plane shear stresses are zero.

Computation of principal Stress from Principal Strain

The three stresses normal to shear principal planes are called principal stress, while a plane at which shear strain is zero is called principal strain.

For two dimensional stress system, $\sigma_3 = 0$

$$\sigma_1 = \frac{E_1(\varepsilon_1 + \mu\varepsilon_2)}{1 - \mu^2}, \quad \sigma_2 = \frac{E(\mu\varepsilon_1 + \varepsilon_2)}{1 - \mu^2}$$

Maximum Shear Stress

The maximum shear stress or maximum principal stress is equal of one half the difference between the largest and smallest principal stresses and acts on the plane that bisects the angle between the directions of the largest and smallest principal stress, *i.e.*, the plane of the maximum shear stress is oriented 45° from the principal stress planes.

$$\tau' = \sqrt{\frac{(\sigma_1 - \sigma_2)^2}{2} + \tau^2}$$

$$\tau' = \frac{\sigma_1' - \sigma_2'}{2}$$

$$\tan 2\theta_s = -\frac{\sigma_1 - \sigma_2}{2\tau} = -\frac{1}{\tan 2\theta_P}$$

$$2\theta_s = 2\theta_P \pm 90^\circ,\ \theta_s = \theta_P \pm 45^\circ$$

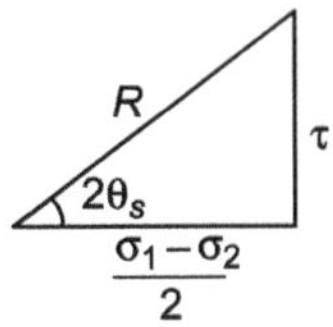

$2\theta_s$ triangle

Principal Strain

For two dimensional strain system,

$$\varepsilon_{1,2} = \frac{e_1 + e_2}{2} \pm \sqrt{\frac{e_1 - e_2}{2} + \left(\frac{\phi}{2}\right)^2}$$

where, e_1 = Strain in *x*-direction

e_2 = Strain in *y*-direction

ϕ = Shearing strain relative to *OX* and *OY*.

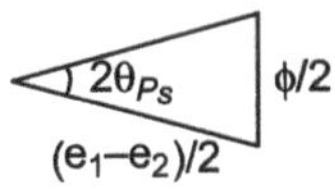

$2\theta_{P_S}$ triangle

$$\tan 2\theta_{P_S} = \frac{\phi/2}{\frac{e_1 - e_2}{2}}$$

$$\tan 2\theta_{P_S} = \frac{\phi}{e_1 - e_2}$$

Maximum Shear Strain

The maximum shear strain also contains normal strain which is given as

$$\frac{\phi_{max}}{2} = \sqrt{\left(\frac{e_1 - e_2}{2}\right)^2 + \left(\frac{\phi}{2}\right)^2}$$

45° Strain Rosette or Rectangular Strain Rosette

Rectangular strains Rosette are inclined 45° to each other.

$$e_a = \frac{1}{2}(\varepsilon_1 + \varepsilon_2) + \frac{1}{2}(\varepsilon_1 - \varepsilon_2)\cos 2\theta$$

$$e_b = \frac{1}{2}(\varepsilon_1 + \varepsilon_2) - \frac{1}{2}(\varepsilon_1 - \varepsilon_2)\sin 2\theta$$

$$e_c = \frac{1}{2}(\varepsilon_1 + \varepsilon_2) - \frac{1}{2}(\varepsilon_1 - \varepsilon_2)\cos 2\theta$$

Principal strains $\varepsilon_{12} = \frac{e_a + e_b}{2} \pm \frac{1}{\sqrt{2}}$

$$= \sqrt{(e_a - e_b)^2 + (e_b - e_c)^2}$$

$$\tan 2q_{pe} = \frac{2e_b - e_a - e_c}{e_c - e_a}$$

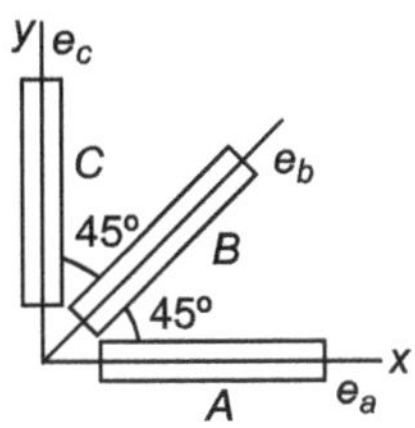

45° strain Rosette

Mohr's Circle

Graphically, variation of normal stress and shear stress are studied with the help of Mohr's circle. A two dimensional Mohr's circle can be constructed, if the normal streses σ_1 and σ_2.

$$\text{Normal stress } \sigma_n = \frac{\sigma_1 + \sigma_2}{2} + \frac{\sigma_1 - \sigma_2}{2}\cos 2\theta$$

$$\text{Shear stress } \tau = -\frac{\sigma_1 - \sigma_2}{2}\sin 2\theta$$

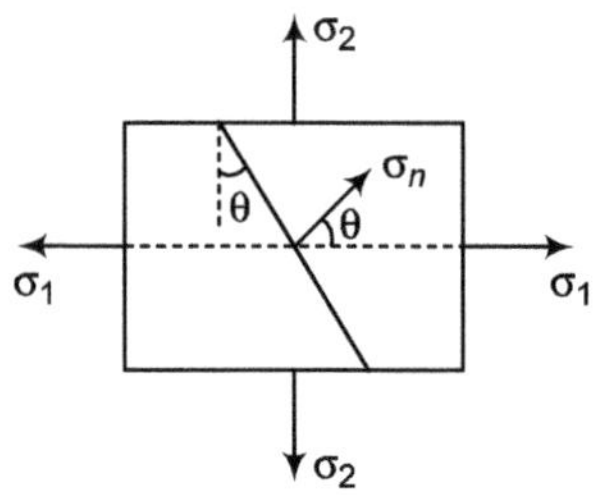

Different stress diagram

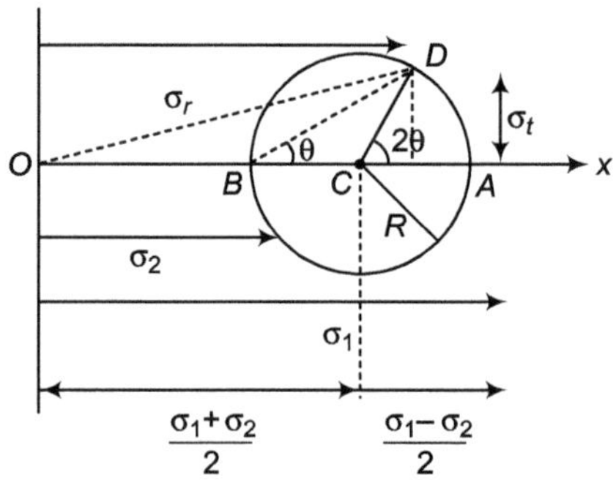

Mohr's circle for plane stress and strain

General State of Stress at an Element

The following procedure is used to draw a Mohr's circle and to find the magnitude and direction of maximum stresses from it.

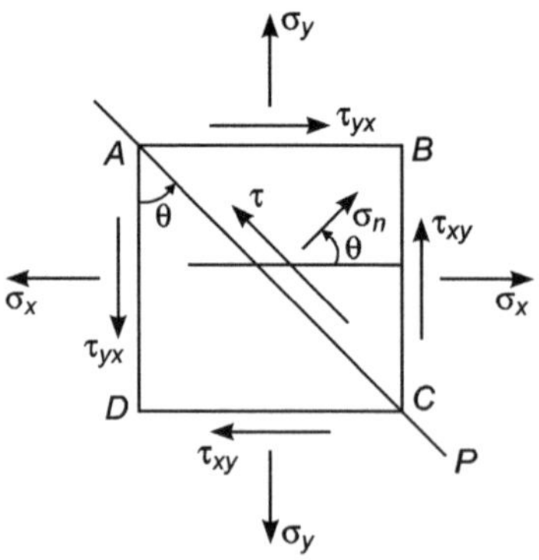

Mohr's circle approach as state of stress

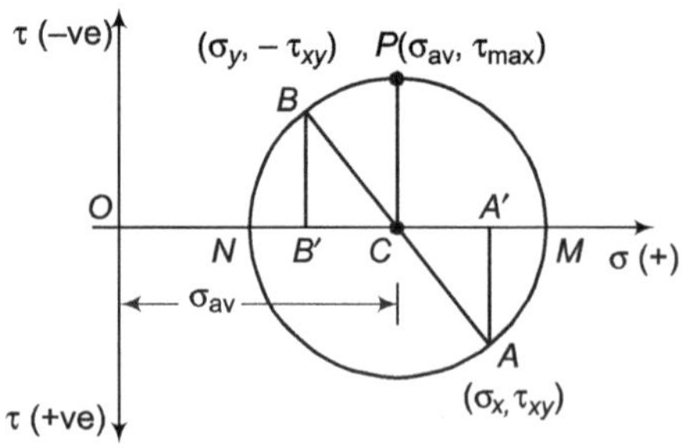

Mohr's circle for plane stressed and plane strain

Observations from Mohr's Circle

The followings are the observations of Mohr's circle as

- At Point M on Circle σ_n is maximum and shear stress is zero.
  $\therefore$ Maximum principal stress $\equiv$ coordinate of M
- At Point N on Circle σ_n is minimum and shear stress τ is zero.
  $\therefore$ Minimum principal stress $\equiv$ coordinate of N
- At Point P on Circle τ is maximum.
  $\therefore$ Maximum shear stress $\equiv$ ordinate of P (*i.e.*, radius of circle)
  Also, normal stress on plane of maximum shear stress

$$\equiv \text{abscissa of } P\left(i.e.,\ \sigma_n = \sigma_{av} = \frac{\sigma_x + \sigma_y}{2}\right)$$

where, $\sigma_n \equiv$ Average stress

- Mohr's circle becomes zero at a point if radius of circle has the following consideration.

$$\text{Radius of circle} = \sqrt{\left(\frac{\sigma_x - \sigma_y}{2}\right)^2 + \tau^2{}_{xy}}$$

- If $\sigma_x = \sigma_y$, then radius of Mohr's circle is zero and $\tau_{xy} = 0$

Centroid and Moment of Inertia of Different Sections

Plane Areas	Properties
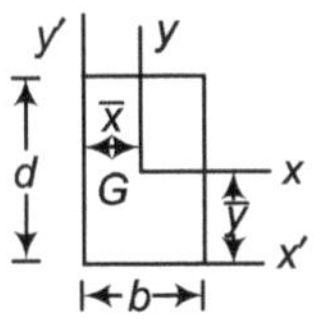	**Rectangle** $A = b \cdot d\,;\ \bar{x} = b/2\,;\ \bar{y} = d/2$ $I_x = \frac{1}{12}bd^3\,;\ I_y = \frac{1}{12}db^3\,;\ I_{xy} = 0$ $I'_x = \frac{1}{3}bd^3\,;\ I'_y = \frac{1}{3}db^3\,;\ I_{x'y'} = \frac{b^2d^2}{4}$
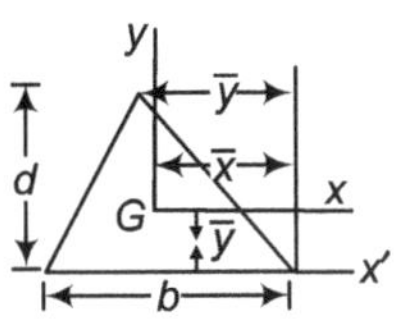	**Triangle** $A = \frac{1}{2}bd\,;\ \bar{x} = \frac{(b+c)}{3}\,;\ \bar{y} = \frac{d}{3}$ $I_x = \frac{1}{36}bd^3\,;\ I_y = \frac{bd}{36}(b^2 - bc + c^2)$ $I_{xy} = \frac{bd^2}{72}(b - 2c)$ $I'_x = \frac{1}{12}bd^3\,;\ I_{y'} = \frac{bd}{12}(3b^2 - 3bc + c^2)$ $I_{x'y'} = \frac{bd^2}{24}(3b - 2c);$

Plane Areas	Properties

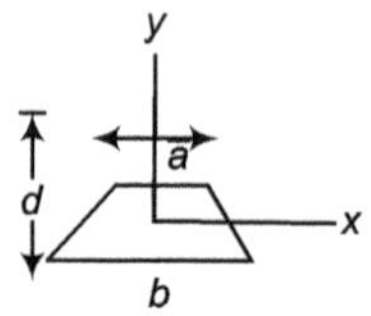

Trapezoid $A = d\,\dfrac{(a+b)}{2}$; $\bar{y} = \dfrac{d}{3}\left(\dfrac{2a+b}{a+b}\right)$

$$I_x = \frac{d^3(a^2+4ab+b^2)}{36(a+b)};\ I'_x = \frac{d^3(3a+b)}{12}$$

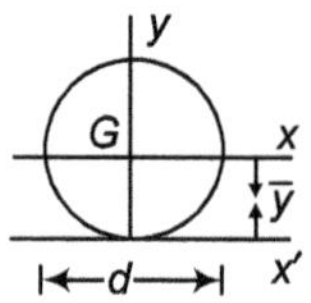

Circle $A = \dfrac{\pi}{4}d^2$; $\bar{y} = \dfrac{d}{2}$; $\bar{x} = 0$

$$I_x = \frac{\pi}{64}d^4 = I_y\ ;\ I_{xy} = 0\ ;\ I'_x = \frac{5\pi}{64}d^4$$

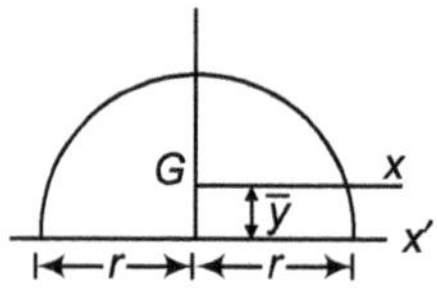

Half circle $A = \dfrac{\pi r^2}{2}$; $\bar{y} = \dfrac{4r}{3\pi}$

$$I_x = \frac{(9\pi^2-64)\,r^4}{72\pi} = 0.1098r^4;\ I_y = \frac{\pi r^4}{8};\ I_{xy} = 0;$$

$$I'_x = \frac{\pi r^4}{8}$$

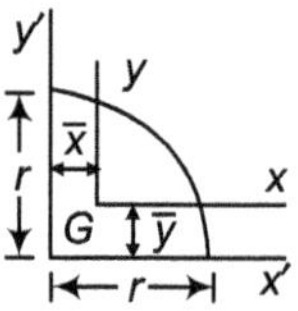

Quarter circle $A = \dfrac{\pi r^2}{4}$; $\bar{x} = \bar{y} = \dfrac{4r}{3\pi}$

$$I_x = I_y = \frac{(9\pi^2-64)\,r^4}{144\,\pi} = 0.05488\,r^4;\ I'_x = I'_y = \frac{\pi r^4}{16}$$

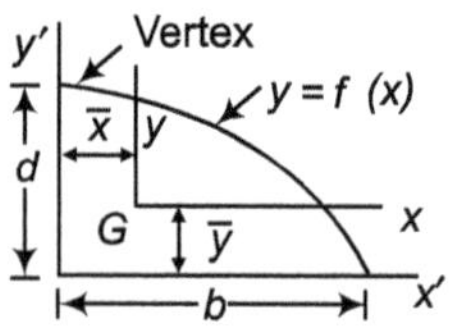

Parabolic semi-segment $y = f(x) = h(1 - x^2/b^2)$;

$$A = \frac{2}{3}bd\ ;\ \bar{x} = \frac{3}{8}b;\ \bar{y} = \frac{2}{5}d$$

$$I_x = \frac{16}{105}bd^3\ ;\ I_y = \frac{2}{15}db^3\ ;\ I_{xy} = \frac{1}{12}b^2d^2$$

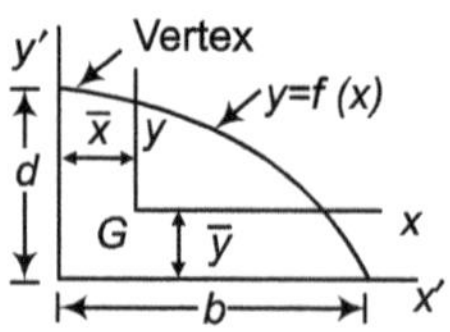

Semi segment of *n*th degree $y = f(x) = h(1 - x^n/b^n)$,

$$n > 0,\ A = bd\left(\frac{n}{n+1}\right),\ \bar{x} = \frac{b\,(n+1)}{2\,(n+2)};\ \bar{y} = \frac{dn}{2n+1}$$

$$I_x = \frac{2bd^3n^3}{(n+1)(2n+1)(3n+1)};\ I_y = \frac{db^3n}{3\,(n+3)}$$

Plane Areas	Properties
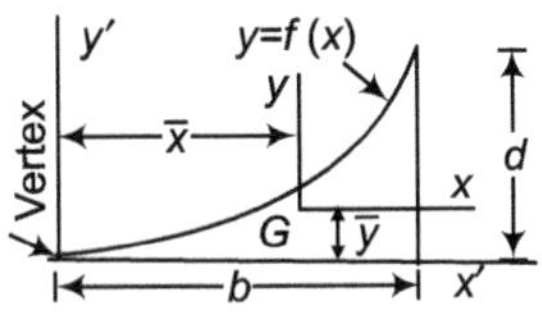	Parabolic spandrel $y = f(x) = d\,(x^2/b^2)$ $A = (b \cdot d)/3\,;\ \bar{x} = 3b/4\,;\ \bar{y} = 3d/10$ $I_x = \dfrac{bd^3}{21}\,;\ I_y = \dfrac{db^3}{5}\,;\ I_{xy} = \dfrac{b^2d^2}{12}$
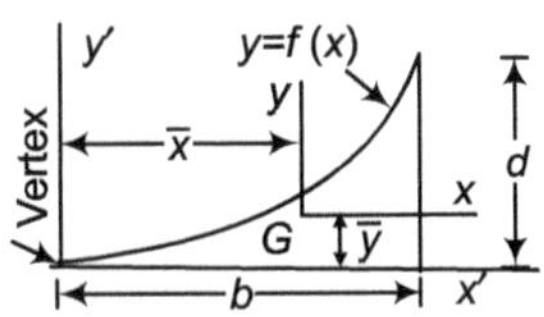	Spandrel of nth degree $y = f(x) = d\,(x^n/b^n),\ n > 0$ $A = \dfrac{bd}{n+1}\,;\ \bar{x} = \dfrac{b\,(n+1)}{(n+2)}\,;\ \bar{y} = \dfrac{d\,(n+1)}{2\,(2n+1)}$ $I_x = \dfrac{bd^3}{3\,(3n+1)}\,;\ I_y = \dfrac{db^3}{(n+3)}\,;\ I_{xy} = \dfrac{b^2d^2}{4\,(n+1)}$

Shear Force *and* Bending Moment Diagram

A Shear Force Diagram (SFD) indicates how a force applied perpendicular to the axis (*i.e.*, parallel to cross-section) of a beam is transmitted along the length of that beam. A Bending Moment Diagram (BMD) will show how the applied loads to a beam create a moment variation along the length of the beam.

Statically Determinate Beam

A beam is said to be statically determinate if all its reaction components can be calculated by applying three conditions of static equilibrium *i.e.*,

$$\Sigma V = 0,\ \Sigma H = 0 \quad \text{and} \quad \Sigma M = 0$$

Statically Indeterminate Beam

When the number of unknown reaction components exceeds the static conditions of equilibrium, the beam is said to be statically indeterminate.

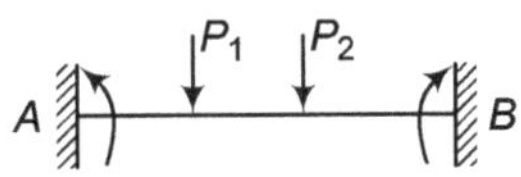

Indeterminate beam

Shear Force

Shear force has a tendency to slide the surface, it acts parallel to surface.

$$\Sigma F_{\text{vert}} = 0$$

$$V - q\,dx - (V + dV) = 0$$

$$\frac{dV}{dx} = -q \quad \Rightarrow \quad \int_A^B dV = -\int_A^B q\,dx$$

$$V_B - V_A = -\int_A^B q\,dx$$

only for distributed load not for point load.

Bending Moment

Any moment produced by forces acting on the beam must be balance by an equal opposite moment produced by internal forces acting in beam at the section. This moment is called **bending moment.**

$$\Sigma M = 0$$

$$-M - qdx\left(\frac{dx}{z}\right) - (V + dV)\,dx + M + dm = 0$$

$$\frac{dM}{dx} = V \quad \Rightarrow \quad M_B - M_A = \int V\,dx$$

only for distributed and concentrated load not for couple.

Bending Moments and Shear Stress Distribution

Bending stress and shear stress distribution are classified in the following groups

Bending Moment in Beam

When beam is subjected to a bending moment or bent there are induced longitudinal or bending stress in cross-section.

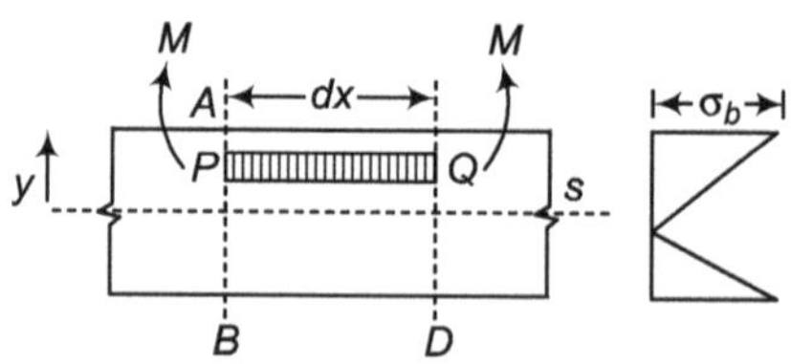

Bending stress in beam

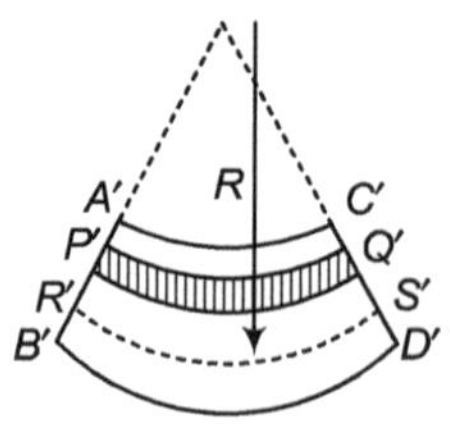

Inertia about neutral axis

I = Moment of inertia about neutral axis.

$$\frac{\sigma_a}{y} = \frac{M}{I} = \frac{E}{R}$$

At the neutral axis, there is no stress of any kind. At one side of the neutral axis, there are compressive stresses, whereas on the other side there are tensile stresses.

Modulus of Section

Section modulus is a geometric property for a given cross-section used in the design of beams or flexural members.

$$z = \frac{I}{y_{max}} \quad \Rightarrow \quad \frac{M}{I} = \frac{\sigma}{y}$$

$$M = \sigma_{max} \frac{I}{y_{max}} \quad \Rightarrow \quad M = \sigma_{max} \times z$$

Rectangular section $I = \frac{bd^3}{12}$

Modulus of section $z = \frac{bd^2}{\sigma}$

Circular section $I = \frac{\pi}{64} d^4$

Modulus of section $z = \frac{\pi}{32} d^3$

Shearing Stress

Shearing stress on a layer JK of beam at distance y from neutral axis.

$$\tau = \frac{VA\bar{y}}{Ib}$$

where,

V = Shearing force

$A\bar{y}$ = First moment of area $\tau = \frac{VQ}{Ib}$.

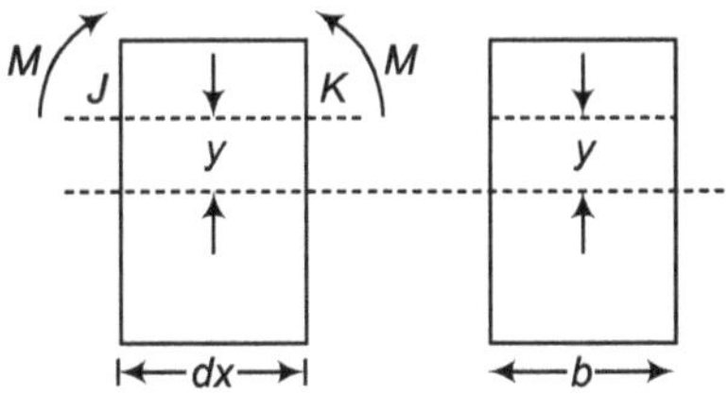

Shearing stress on a beam

Shear Stress in Rectangular Beam

Suppose, we have to determine the shear stress at the longitudinal layer having y distance from neutral axis.

$$Q = b\left(\frac{h}{2} - y_1\right)\left(y_1 + \frac{\frac{h}{2} - y_1}{2}\right)$$

$$Q = \frac{b}{2}\left(\frac{h^2}{4} - y_1^2\right)$$

$$\tau = \frac{V}{2I}\left(\frac{h^2}{4} - y_1^2\right)$$

$$\tau_{max} = \frac{3}{2}\frac{V}{A}$$

$$\tau_{max} = 1.5\,\tau_{av}$$

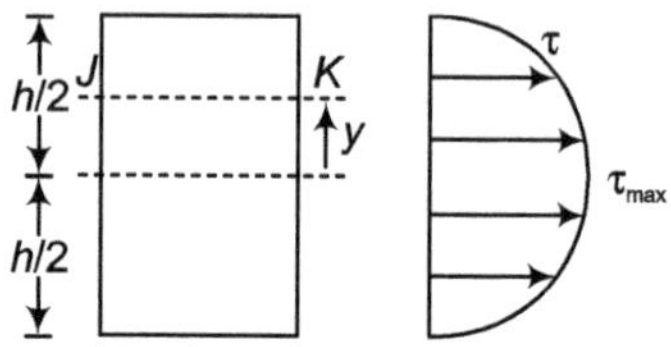

Rectangular beam

Circular Beam

Centre of gravity of semi-circle lies at $\frac{4r}{3\pi}$ distance from centre or base line.

As it is symmetrical above neutral axis, hence at neutral axis shear stress will be maximum.

$$\tau = V\frac{(r^2 - y^2)}{3I}$$

$$Q = A\bar{y} = \frac{\pi r^2}{2}\left(\frac{4r}{3\pi}\right) = \frac{2r^3}{3}$$

$$b = 2r$$

$$\tau_{max} = \frac{V\frac{2r^3}{3}}{\frac{\pi r^4}{4}(2r)} = \frac{4}{3}\frac{V}{A}$$

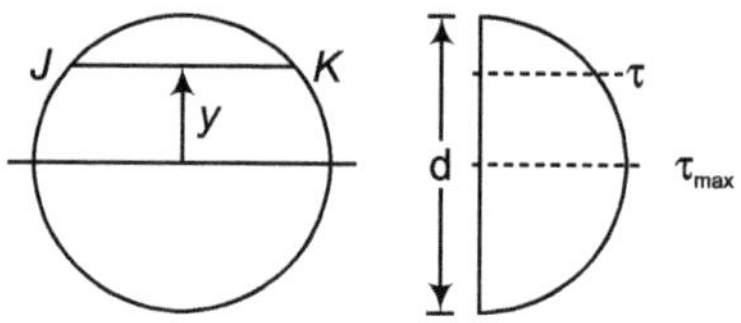

Circular beam

For τ_{max} substituting $y = 0$

$$I = \frac{\pi d^4}{64}$$

$$\tau_{max} = \frac{4V}{3A} = \frac{4}{3}\tau_{av}$$

Shears Stress in Hollow Circular Cross-Section

In hollow circular cross-section, if we have to calculate τ at neutral axis by the formula

$$\tau_{max} = \frac{4V}{3A}\left(\frac{r_2^2 + r_1 r_2 + r_1^2}{r_2^4 - r_1^4}\right)$$

Shear Stress in Triangular Section

In a triangular cross-section, if we have to calculate τ at neutral axis, then in formula

$$\tau = \frac{V}{3I}(hx - x^2) \Rightarrow \tau_{max} = \frac{3}{2}\tau_{av}$$

Shear Stress in I-section

$$\tau = \frac{VX}{Ib}\left[\frac{B(D^2 - d^2)}{8} + \frac{b}{2}\left(\frac{d^2}{4} - y^2\right)\right]$$

$$\tau_{max} = \frac{V}{Ib}\left[\frac{B}{8}(D^2 - d^2) + \frac{bd^2}{8}\right]$$

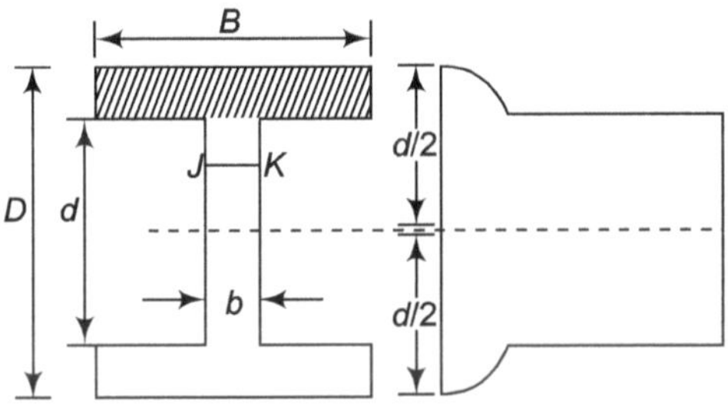

I-Section diagram

Deflection of Beam

Deflection is defined as the vertical displacement of a point on a loaded beam. There are many methods to find out the slope and deflection at a section in a loaded beam.

Double Integration Method

This is most suitable when concentrated or udl over entire length is acting on the beam.

$$EI\frac{d^2y}{dx^2} = -M$$

Integrating one time $\quad EI\frac{dy}{dx} = -\int M$

Integrating again $\quad EIy = -\iint M$

where, M = Bending moment
I = Moment of inertia of the beam section
y = Deflection of the beam
E = Modulus of elasticity of beam material.

Torsion of Shaft and Combined Stresses

If τ_r be the intensity of shear stress, on any layer at a distance r from the centre of shaft, then

$$\frac{\tau_r}{r} = \frac{T}{J} = \frac{G\theta}{l}$$

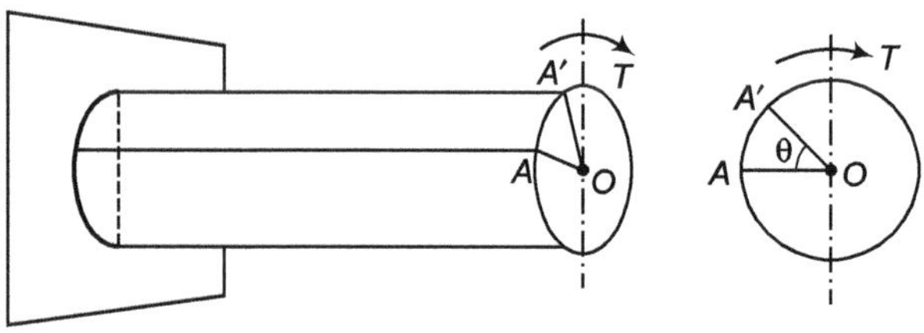

Torsion in shaft and combined stresses

Rate of twist $\left(\frac{\theta}{l}\right)$ $\quad \frac{\theta}{l} = \frac{T}{GJ}$

Total angle of twist $\quad \theta = \frac{Tl}{GJ}$

where, T = Torque, J = Polar moment of inertia
G = Modulus of rigidity, θ = Angle of twist
L = Length of shaft, GJ = Torsional rigidity

$\frac{GJ}{l} \rightarrow$ Torsional stiffness; $\frac{l}{GJ} \rightarrow$ Torsional flexibility

$\frac{EA}{l} \rightarrow$ Axial stiffness; $\frac{l}{EA} \rightarrow$ Axial flexibility

- For solid circular shaft, $J = \frac{\pi d^4}{32}$, $\tau_{max} = \frac{16T}{\pi d^3}$
- For hollow circular shaft, $J = \frac{\pi}{32}(d_o^4 - d_i^4)$
- Power transmitted by shaft, $P = \frac{2\pi NT}{60000}$ kW

where, N = Rotation per minute.

Compound Shaft

An improved type of compound coupling for connecting in series and parallel are given below

1. **Series connection** Series connection of compound shaft as shown in figure. Due to series connection the torque on shaft 1 will be equal to shaft 2 and the total angular deformation will be equal to the sum of deformation of 1st shaft and 2nd shaft.

$$\theta = \theta_1 + \theta_2$$
$$T = T_1 = T_2$$

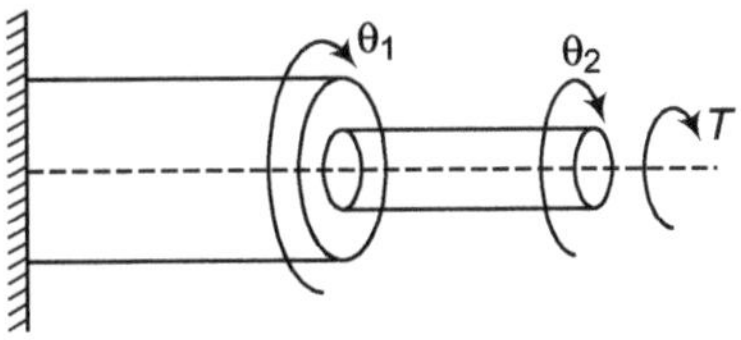

Series connection

where,
θ_1 = Angular deformation of 1st shaft
θ_2 = Angular deformation of 2nd shaft

2. **Parallel connection** Parallel connection of compound shaft as shown in figure. Due to parallel connection of compound shaft the total torque will be equal to the sum of torque of shaft 1 and torque of shaft 2 and the deflection will be same in both the shafts.

$$\theta_1 = \theta_2$$
$$T = T_1 + T_2$$

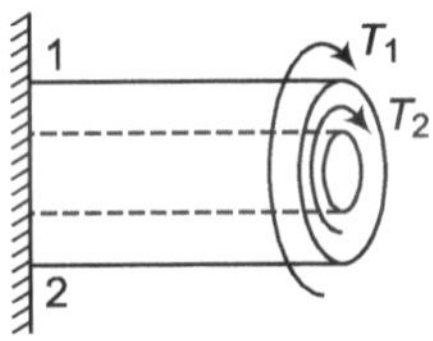

Parallel connection

Effect of Pure Bending on Shaft

The effect of pure bending on shaft can be defined by the relation

$$\sigma = \frac{32\,M}{\pi D^3}$$

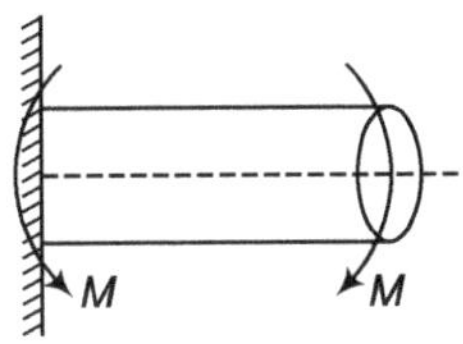

Pure bending on shaft

where, σ = Principal stress
D = Diameter of shaft
M = Bending moment

Effect of Pure Torsion on Shaft

It can be calculated by the formula, which are given below

$$\tau_{max} = \frac{16\,T}{\pi D^3}$$

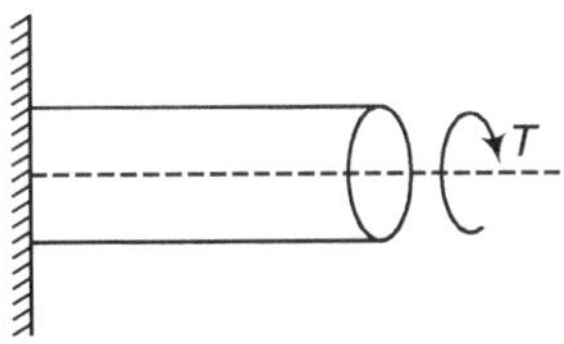

Pure torsion on shaft

where, τ = Torsion
D = Diameter of shaft

Combined effect of bending and torsion

Principal stress $= \frac{16}{\pi D^3}[M \pm \sqrt{M^2 + T^2}]$

Maximum shear stress $= \frac{16}{\pi D^3}\sqrt{M^2 + T^2}$

Equivalent bending moment

$$M_{eq} = \frac{1}{2}[M + \sqrt{M^2 + T^2}]$$

Equivalent torque $T_{eq} = \sqrt{T^2 + M^2}$

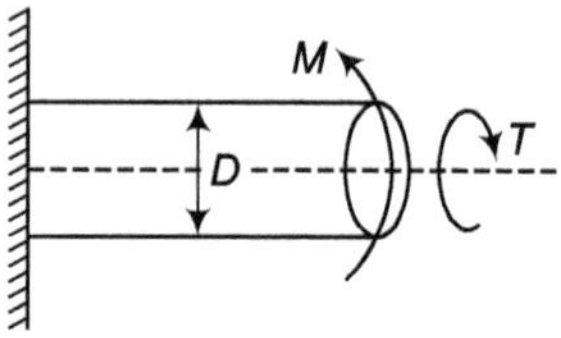

Bending and torsion effect

Thin Shell

If the thickness of the wall of a shell is less than $\frac{1}{10}$th of $\frac{1}{15}$th of its diameter, it is known as a thin shell.

Stresses in Thin Cylindrical Shell

(a) Circumferential stress (hoop stress)

$$\sigma_c = \frac{pd}{2t} \Rightarrow \sigma_c = \frac{pd}{2t\eta}$$

where, p = Intensity of internal pressure
d = Diameter of the shell
t = Thickness of shell
η = Efficiency of joint

(b) Longitudinal stress $\sigma_l = \frac{pd}{4t} \Rightarrow \sigma_l = \frac{pd}{4t\,\eta}$

Change in Dimension of a Thin Cylindrical Shell due to an Internal Pressure

Change in diameter $\delta d = \varepsilon_1 d = \frac{pd}{2tE}\left(1 - \frac{1}{2m}\right) \times d$

ε_1 = Circumferential strain $\varepsilon_1 = \frac{\sigma_c}{E} - \frac{\sigma_l}{mE} \Rightarrow \frac{1}{m} = \mu$

Change in length $f l = \varepsilon_2 l = \frac{pd}{2tE}\left(\frac{1}{2} - \frac{1}{m}\right) \times l$

ε_2 = Longitudinal strain $\varepsilon_2 = \frac{\sigma l}{E} - \frac{\sigma_c}{mE}$

Change in volume $\delta V = (2\varepsilon_1 + \varepsilon_2)V$

Thin Spherical Shell

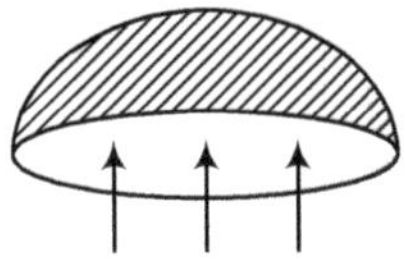
Stresses on spherical shell

1. Stresses in shell material,

$$\sigma = \frac{pd}{4t} \Rightarrow \sigma = \frac{pd}{4t\,\eta}$$

2. Change in diameter $\delta d = \varepsilon \times d$,

$$\text{where } \varepsilon = \left(\frac{\sigma}{E} - \frac{\sigma}{mE}\right)$$

$$= \frac{pd}{4t\varepsilon}\left(1 - \frac{1}{m}\right) \times d$$

Change in volume $\delta V = V \cdot 3\,\varepsilon = V \cdot \frac{pd}{4tE}\left(1 - \frac{1}{m}\right)$

Lake's Theory

Lake's theory is based on the following assumptions

Assumptions

1. Homogeneous material.
2. Plane section of cylinder, perpendicular to longitudinal axis remains under plane and pressure.

   Hoop stress at any section

$$\sigma_r = \frac{b}{r^2} + a$$

Radial pressure $p_r = \frac{b}{r^2} - a$

- **Subjected to Internal Pressure (p)**

1. At $r = r_i$, $\sigma_{r_i} = p\left(\frac{r_0^2 + r_i^2}{r_0^2 - r_i^2}\right)$

2. At $r = r_0$, $\sigma_{r_0} = \frac{2pr_i^2}{r_0^2 - r_i^2}$

- **Subjected to External Pressure (p)**

1. At $r = r_i$, $\sigma_{r_i} = \frac{-2\,pr_0^2}{r_0^2 - r_i^2}$

2. At $r = r_0$, $\sigma_{r_0} = -\,p\left(\frac{r_0^2 + r_i^2}{r_0^2 - r_i^2}\right)$

Note Radial and hoop stresses vary hyperbolically.

Columns *and* Struts

A structural member subjected to an axial compressive force is called **strut.** As per definition strut may be horizontal, inclined or even vertical. Vertical strut is called a **column.**

Euler's Column Theory

This theory has the following assumptions

- Perfectly straight column and axial load apply.
- Uniform cross-section of the column throughout its length.
- Perfectly elastic, homogeneous and isotropic material.
- Length of column is large as compared to its cross-sectional dimensions.
- The shortening of column due to direct compression is neglected.
- The failure of column occurs due to buckling alone.

Euler's Buckling (or crippling load)

The maximum load at which the column tends to have lateral displacement or tends to buckle is known as buckling or crippling load. *Load columns can be analysed with the Euler' column formulas can be given as*

$$p_E = \frac{n^2\pi^2 EI}{l^2} \qquad (n = 1, 2, 3, \dots)$$

or

$$p_E = \frac{\pi^2 EI}{l_{eff}^2}$$

where, E = Modulus of elasticity

l = Length of column

I = Moment of inertia of column section.

Euler's Buckling for Different Structures

1.	For both end hinged $n = 1,$ $l_{\text{eff}} = l$ $p_E = \frac{\pi^2 EI}{l^2}$	 Both end hinged
2.	For one end fixed and other free $n = \frac{1}{2}$ $l_{\text{eff}} = 2l$ $p_E = \frac{\pi^2 EI}{4l^2}$	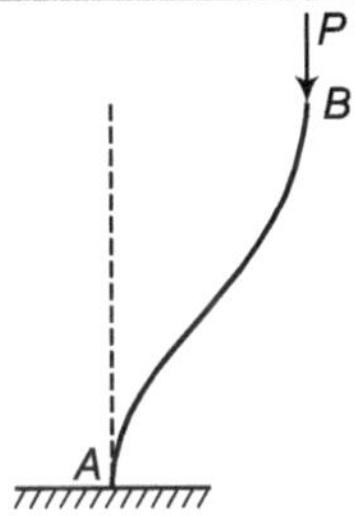 One end fixed
3.	For both end fixed $n = 2,$ $l_{\text{eff}} = \frac{l}{2}$ $p_E = \frac{4\pi^2 EI}{EI}$	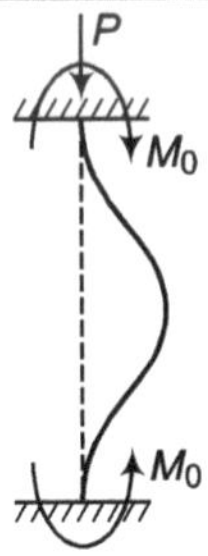 Both end fixed
4.	For one end fixed and other hinged $n = \sqrt{2}$ $l_{\text{eff}} = \frac{l}{\sqrt{2}}$ $p_E = \frac{2\pi^2 EI}{l^2}$	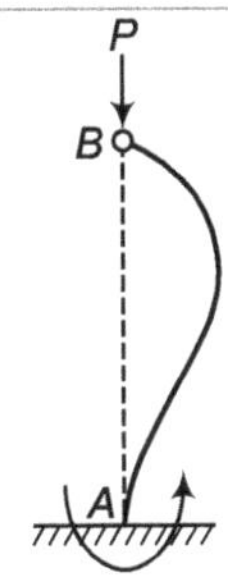 One end fixed and other hinged

Rankine's Formula for Columns

It is an empirical formula, takes into both crushing P_{CS} and Euler critical load (P_R).

$$\frac{1}{P_R} = \frac{1}{P_{CS}} + \frac{1}{P_E}$$

P_R = Crippling load by Rankine's formula

$P_{CS} = \sigma_{CS} A$ = Ultimate crushing load for column

$P_E = \dfrac{\pi^2 EI}{l_{eff}^2}$ = Crippling load obtained by Euler's formula

$$P_R = \frac{\sigma_{CS} A}{1 + a\left(\frac{l_e}{k}\right)^2}, \; I = Ak^2$$

where, A = Cross-section area of column

k = Least radius of gyration

a = Rankine's constant.

Strain Energy

The work done by the load in straining the body is stored within the strained material in the form of **strain energy.**

Strain energy, $U = \dfrac{1}{2} P(\Delta l)$

$$U = \frac{P^2 L}{2AE}$$

Put $P = \dfrac{AE\,\Delta l}{L}$

$$U = \frac{EA\,(\Delta l)^2}{2L}$$

or $U = \dfrac{\sigma^2}{2E} \times V$

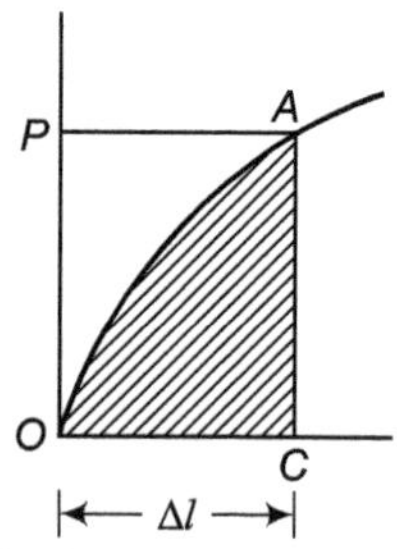

Strain energy diagram

Proof Resilence

The maximum strain energy that can be stored in a material is known as proof resilence.

$$u = \frac{1}{V} \times \frac{P^2 L}{2AE}$$

$$u = \frac{\sigma^2}{2E}, \qquad \text{where } \sigma = \frac{P}{A}$$

$$u = \frac{1}{2} E e^2$$

Strain energy of prismatic bar with varying sections

$$U = \Sigma \frac{P^2 L}{2AE}$$

$$= \frac{P^2}{2E}\left[\frac{L_1}{A_1} + \frac{L_2}{A_2} \dots + \frac{L_n}{A_n}\right]$$

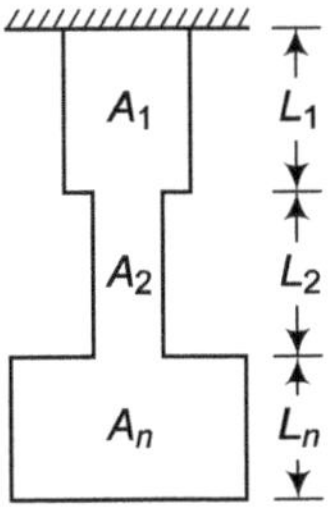

Prismatic bar

Strain energy of non-prismatic bar with varying axial force

$$U = \int_0^L \frac{P_x^2}{2EA_x} dx$$

A_x = Cross-section of differential section.

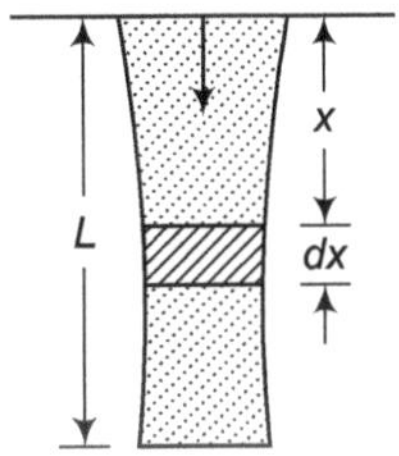

Non-prismatic bar

Strain Energy in Torsion

$$U = \frac{1}{2} T\theta = \frac{1}{2} \frac{T^2 L}{GJ}$$

For solid shaft, $U = \dfrac{\tau^2}{2G} \times$ volume of shaft

For hollow shaft, $U = \dfrac{\tau^2}{2G}\left(\dfrac{D^2 + d^2}{D^2}\right) \times$ volume of shaft

Castigliono's First Theorem

If the strain energy of an elastic structure can be expressed as a function of generalised displacement, then the partial derivative of the strain energy δU with respect to generalised displacement δW, gives the generalised force δ_1.

$$U = \int_0^l \frac{M^2 dx}{2EI} \qquad \text{[where } M \text{ is function of } W \text{ (load)]}$$

Deflection $\delta_1 = \dfrac{\delta U}{\delta W_1} = \displaystyle\int_0^l M\left(\frac{\partial M}{\partial W_1}\right)\frac{dx}{EI}$

Slope $\theta = \dfrac{\partial U}{\partial M_1} = \displaystyle\int_0^l M\left(\frac{\partial M}{\partial M_1}\right)\frac{dx}{EI}$

Theories of Failure

Theories of failure are defined as following groups

Maximum Principal Stress Theory (Rankine theory)

According to this theory, permanent set takes place under a state of complex stress, when the value of maximum principal stress is equal to that of yield point stress as found in a simple tensile test.

For design critical maximum principal stress (σ_1) must not exceed the working stress (τ) for the material.

$$\sigma_{12} \le \tau\sigma_y$$

Note For brittle material, it give satisfactory result.

Yield criteria for 3*D* stress system,

$$\sigma_1 = \sigma_y \quad \text{or} \quad |\sigma_3| = \sigma'_y$$

where, σ_y = Yield stress point in simple tension

σ'_y = Yield stress point in simple compression.

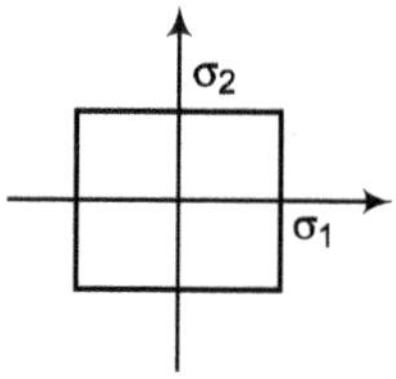

Stresses on rectangular section

Maximum Principal Strain Theory

(ST Venant's theory)

According to this theory, a ductile material begins to yield when the maximum principal strain reaches the strain at which yielding occurs in simple tension.

$$e_{1,2,3} \le \varepsilon_{yp}$$

For 3*D* stress system,

$$e_1 = \frac{\sigma_1}{\varepsilon} - \frac{1}{m\varepsilon}(\sigma_2 + \sigma_3)$$

$$e_2 = \frac{\sigma_2}{\varepsilon} - \frac{1}{m\varepsilon}(\sigma_3 + \sigma_1)$$

$$e_3 = \frac{\sigma_3}{\varepsilon} - \frac{1}{m\varepsilon}(\sigma_1 + \sigma_2)$$

If e_y = Yield point strain tensile $\frac{\sigma_y}{E}$

e'_y = Yield point strain compressive $\frac{\sigma'_y}{E}$

According to theory, $e_1 = e_y$

Yield criteria $\frac{\sigma_1}{\varepsilon} - \frac{1}{m\varepsilon}(\sigma_2 + \sigma_3) = \frac{\sigma_y}{E}$

and $e_3 = e_{y'}$

$$\left| \frac{\sigma_3}{E} - \frac{1}{m}\frac{(\sigma_1 + \sigma_2)}{E} \right| = \frac{\sigma_y'}{E}$$

For $2D$ system, $\frac{\sigma_1}{E} - \frac{\sigma_2}{mE} \le \frac{\sigma_y}{E}$

$$\sigma_1 - \frac{\sigma_2}{m} \le \sigma_y$$

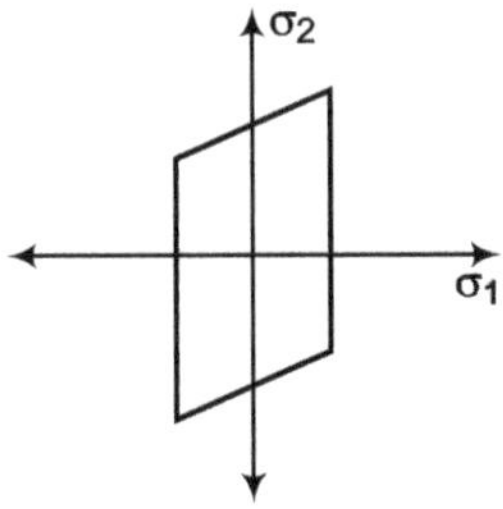

Rhombus

Note This theory over estimate the elastic strength of ductile material.

Maximum Shear Stress Theory (Guest & Tresca's theory)

According to this theory, failure of specimen subjected to any combination of loads when the maximum shearing stress at any point reaches the failure value equal to that developed at the yielding in an axial tensile or compressive test of the same material. For $3D$ system

Yielding criteria, $\tau_{max} = \frac{1}{2}(\sigma_1 - \sigma_3) = \frac{\sigma_y}{2}$

$$\sigma_1 - \sigma_3 = \sigma_y$$

In case of $2D$
Yielding criteria, $\sigma_1 - \sigma_2 = \sigma_y$

Note This theory give well estimation for ductile material.

Maximum Strain Energy Theory (Haigh's theory)

According to this theory, a body under complex stress fails when the total strain energy on the body is equal to the strain energy at elastic limit in simple tension. For $3D$ stress system yield criteria,

$$\sigma_1^2 + \sigma_2^2 + \sigma_3^3 - \frac{2(\sigma_1\sigma_2 + \sigma_2\sigma_3 + \sigma_3\sigma_1)}{m} = \sigma_y^2$$

For $2D$ stress system, $\sigma_1^2 + \sigma_2^2 - \frac{2\sigma_1\sigma_2}{m} = \sigma_y^2$

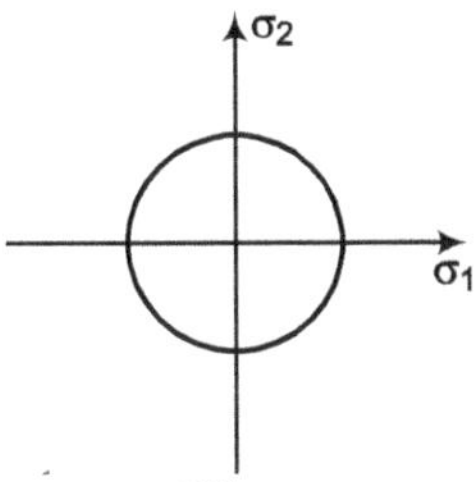

Ellipse

Note This theory does not apply to brittle material for which elastic limit stress in tension and in compression are different.

3

Theory of Machines

Basic Concepts of Theory of Machines

Theory of machine deals with the study of relative motion between the various parts of a machine and forces which act on them.

Mechanism

A combination of rigid or restraining bodies so shaped and connected that they move upon each other with definite relative motion is known as mechanism.

e.g., slider crank mechanism where the rotary movement of the crank is converted through the connecting rod into the reciprocating motion of the slider or *vice-versa*.

Machine

A mechanism or a combination of mechanism which transmit force from the source of power to the resistance (load) to be overcome and thus performs useful mechanical work. *e.g.*, In IC engine, force F available at the piston is transmitted to the crank in the form of driving torque M, to overcome the resistance M_R at the crank shaft.

Link

A link is defined as a member or a combination of members, connecting other members and having motion relative to them.

Joint

Joint is the physical realization of a kinematic pair.

Kinematic Pair

A joint of two links having relative motion between them is known as **kinematic pair**

Grashof's Law for four Bar Chain Mechanism

It states that in a planar four bar chain mechanism, the sum of the length of shortest and longest links should not be greater than the sum of length of remaining two links, if there is to be a continuous relative motion between two links.

Hence, $(S+l) < (P+q)$

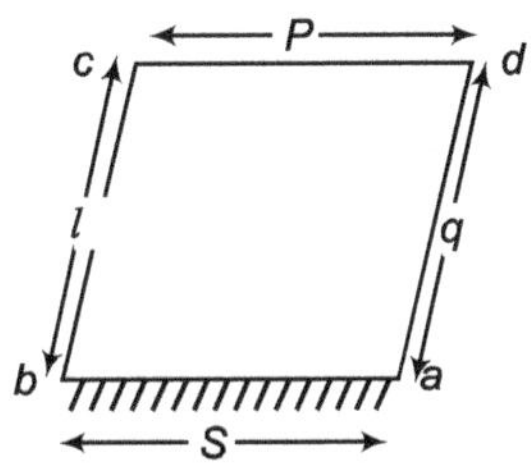

Degree of Freedom

Number of independent coordinates needed to describe the relative portions of pairing elements.

Degree of freedom = 6 – Number of restraints

Degree of Freedom of the Mechanism (3D)

$$F = 6(N-1) - 5P_1 - 4P_2 - 3P_3 - 2P_4 - P_5$$

N = Total number of links in a mechanism

P_n = Number of pairs having n degree of freedom

Degree of Freedom of a Plane Mechanism (2D)

Grubler Criteria for Planar Mechanism

$$F = 3(N-1) - 2P_1 - 1P_2$$

Kutzback's Criteria

Linkages with single degree of freedom $P_2 = 0$

$$F = 3(N-1) - 2P_1$$

Transmission Angle

The angle μ between the output link and the coupler is known as transmission angle.

- If link BC and DC become coincident, the transmission angle is zero and the mechanism would lock or jam.
- For double crank or crank-rocker mechanism. Transmission angle μ will be maximum.

$\Rightarrow$ When $\theta = 180°$; μ minimum $\Rightarrow$ When $\theta = 0°$

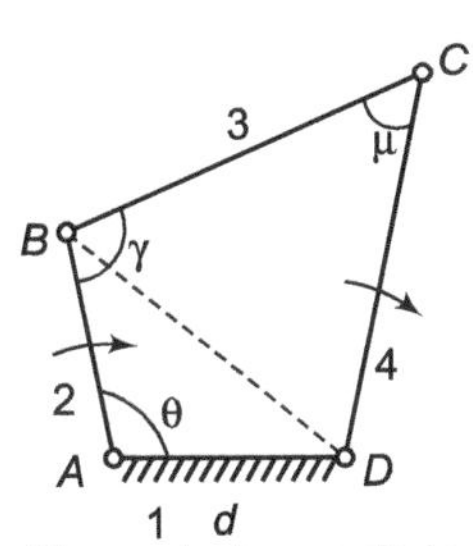

Transmission angle in four bar mechanism

Inversion of Mechanism

The method of obtaining different mechanism by fixing different links of the same kinematic chain, is known as inversion of the mechanism. In the process of inversion, relative motions of the links of the mechanism produced remain unchanged.

Types of kinematic chains and their inversions.

The Four-Bar Chain

It consists of four rigid links which are connected in the form of quadrilateral by four pin joints. It also consists of four turning pair

A link that makes complete revolution is known as crank. The fixed link is known as frame of the mechanism.

Key Points

- Number of degree of freedom of a planar mechanism with n links, j lower kinematic pairs and h higher kinematic pairs can also be expressed as

$$F = 3\,(N - 1) - 2j - h$$

- For constrained motion

(Single degree of freedom and zero higher pair)

$$F = 1$$

$$3\,(N - 1) - 2j - h = 1$$

$$2j - 3N + h + 4 = 0$$

- j can also expressed as

$$j = \frac{1}{2}(2n_2 + 3n_3 + 1n_4)$$

- For planar mechanism degree of freedom can also be expressed as

$$F = N - (2L + 1),\ L = \text{Number of loops in a linkage}$$

$$P_1 = N + (L - 1)$$

Four-Bar Mechanism

a, b = Adjacent link to shortest link

S = Shortest link

L = Longest link

$S + L \leq a + b$ (Grashaf's law)

Case I If $S + L \leq a + b$

- If S fix ⇒ Double-crank mechanism
- If a or b fix ⇒ Crank-rocker mechanism
- If L fix ⇒ Rocker -rocker mechanism

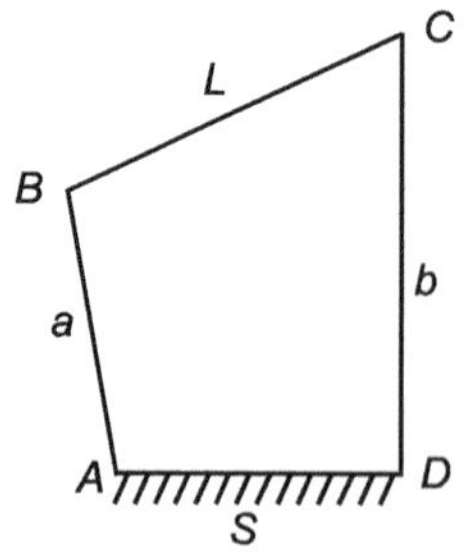

Four-bar mechanism

Case II $S + L > a + b$

Grashaf's law do not satisfy and mechanism is Rocker-rocker mechanism

Case III $S + L = a + b$

- If length of $S \neq L \neq a \neq b$

$\Rightarrow$ Mechanism is same as case I

- If length of $S = a$ and $L = b$

$\Rightarrow$ (Parallelogram linkage) crank-crank mechanism

If S is fixed

$\Rightarrow$ Double-crank mechanism

or L is fixed.

$\Rightarrow$ Double-crank mechanism

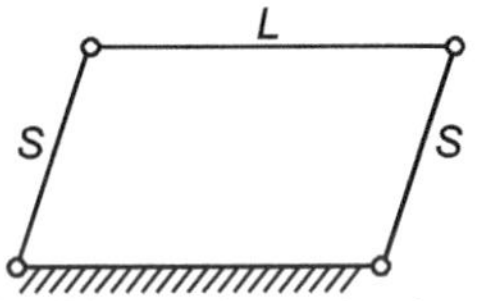

Double-crank mechanism

The Slider-Crank Chain

When one of the turning pairs of a four-bar chain is replaced by a sliding pair, it becomes a single slider-crank chain or simply a slider-crank chain.

Double-Slider Crank Chain

It consists of four pairs out of which two are turning and two others are sliding.Also the two pairs of same kinds are adjacent.

Velocity Analysis in Mechanism

Let a rigid link OA, of length r rotate about a fixed point O with a uniform angular velocity ω rad/s in a counter-clockwise direction OA turns through a small angle $\delta\theta$ in a small interval of time δt. Then, A will travel along the arc AA' as shown in figure.

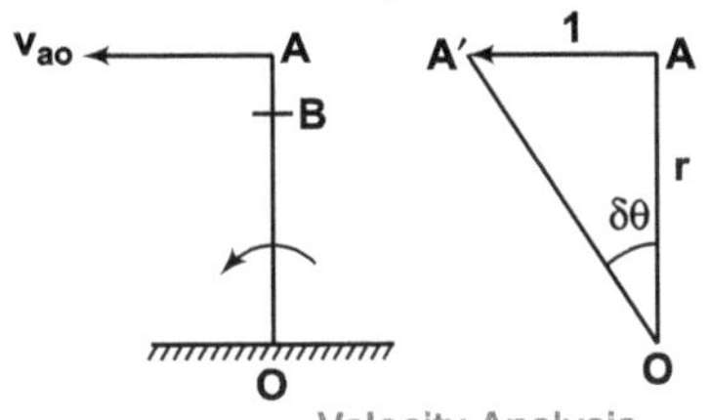

Velocity Analysis

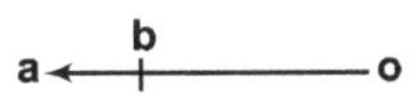

Intermediate points *a* and *b*

$\therefore$ Velocity of A relative to O

$$v_{ao} = \frac{\text{arc } AA'}{\delta\theta} = \frac{r\,\delta\theta}{\delta t}$$

∴ In the limits, when $\delta t \to 0$

$$v_{ao} = r\frac{d\theta}{dt} = r\omega$$

Thus, velocity of *A* is ωr and is perpendicular to *OA*.

Velocity of Intermediate Point

If v_{bo} represent the velocity of *B* with respect to *O*, then

$$\frac{v_{bo}}{v_{ao}} = \frac{\omega OB}{\omega OA} = \frac{OB}{AO}$$

Intermediate points *b*

i.e., *b* divides the velocity vector in the same ratio as *B* divides the link.

The magnitude of the linear velocity of a point on the rotating body at a particular instant is proportional to its distance from the axis of rotation.

Velocity Images of Four Link Mechanism

Figure shows a four link mechanism (quadric cycle mechanism) *ABCD* in which *AD* is fixed link and *BC* is the coupler. *AB* is the driver rotating at an angular speed of ω rad/s in the clockwise direction if it is a crank or moving at angular velocity ω at this instant if it is rocker.

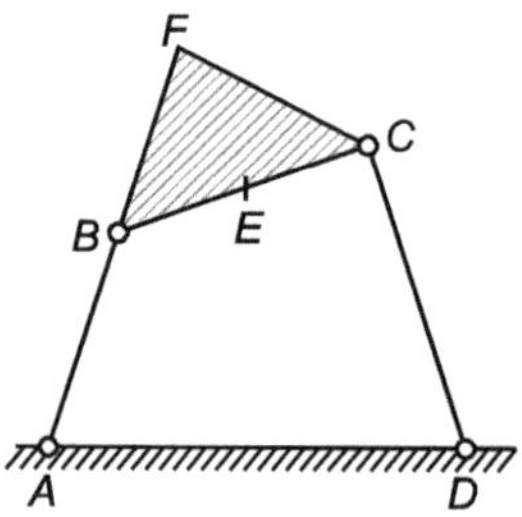

Four link mechanism

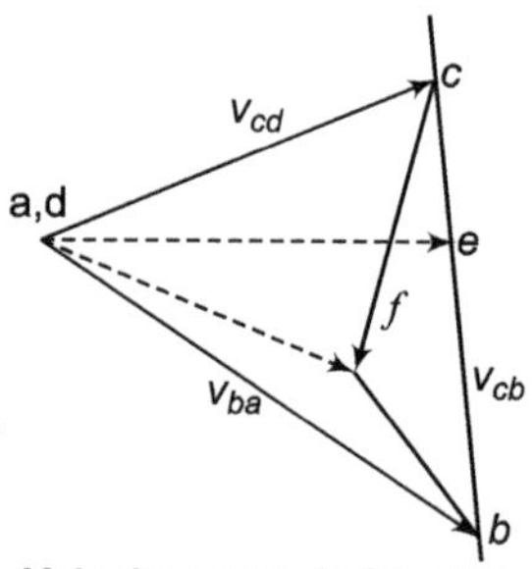

Velocity vector in four link mechanism

Velocity Images of Slider-Crank Mechanism

Consider a slider-crank mechanism in which *OA* is the crank moving with uniform angular velocity ω rad/s in the clockwise direction. At point *B*, a slider moves on the fixed guide *G*.

From the given configuration, the coupler *AB* has angular velocity in the counter-clockwise direction. The magnitude being $\frac{v_{ba}}{BA}$.

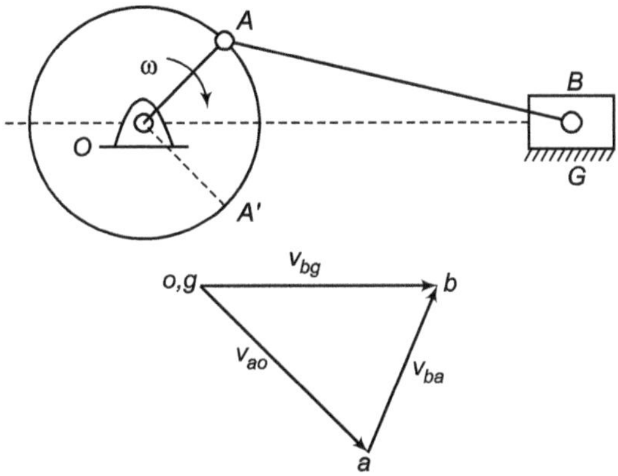

Velocity images of slider-crank mechanism

Velocity of Rubbing

Let us take two links of a turning pair, *a* pin is fixed to one of the links whereas a hole is provided in the other to fit the pin. When joined the surface of the hole of one link will rub on the surface of pin of the other link. The velocity of rubbing of the two surfaces will depend upon the angular velocity of a link relative to the other.

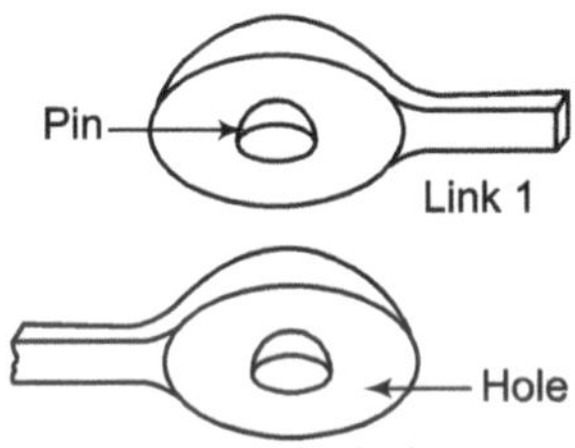

Velocity of rubbing

Pin at *A*

The pin at *A* joins links *AD* and *AB*. *AD* being fixed, the velocity of rubbing will depend upon the angular velocity of *AB* only.

$$\text{Velocity of rubbing} = r_a\omega$$

where, r_a = radius of pin at *A*

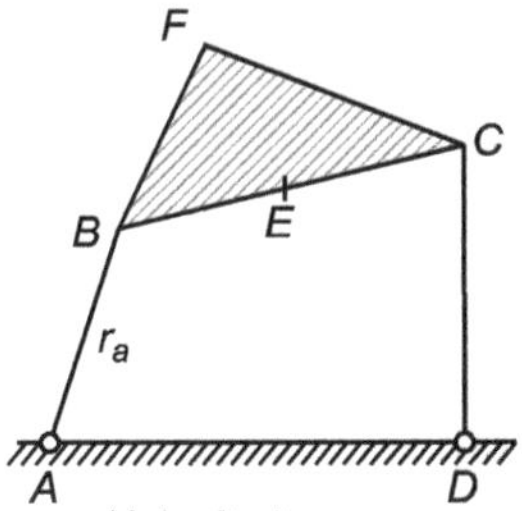

Velocity images mechanism

Pin at *B*

$$\omega_{ba} = \omega_{ab} = \omega \text{ (clockwise)}$$

$$\omega_{bc} = \omega_{cb} = \frac{v_{cd}}{BC} \text{ (counter-clockwise)}$$

$$r_b = \text{Radius of pin at } B$$

$$\text{Velocity of rubbing} = r_b\,(\omega_{ab} + \omega_{bc})$$

Pin at *C*

$$\omega_{bc} = \omega_{cb} \text{ (counter-clockwise)}$$
$$\omega_{dc} = \omega_{cd} \text{ (clockwise)}$$
$$r_c = \text{Radius of the pin at } C$$
$$\text{Velocity of rubbing} = r_c(\omega_{bc} + \omega_{dc})$$

Pin at *D*

where, r_d = radius of pin at D

$$\text{Velocity of rubbing} = r_d\ \omega_{cd}$$

Instantaneous Centre of Velocity (I-centre)

The instantaneous centre of velocity can be defined as a point which has no velocity with respect to the fixed link.

Suppose there are two link 1 *and link* 2

- Link 1 may or may not be fixed. Rigid body 2 is shown to be in plane motion with respect to the link 1.
- In case of fixed link, (link 2) velocity of the point A and B are proportional to PA and PB respectively. So, instantaneously, the rigid body can be thought of as being momentarily in pure rotation about the point P. The velocity of any point C on the body at this instant is given by $v_C = P_C \dfrac{v_B}{r_2}$ in a direction perpendicular to P_C.This point P is called the **instantaneously centre of velocity** and its instantaneously velocity is zero.

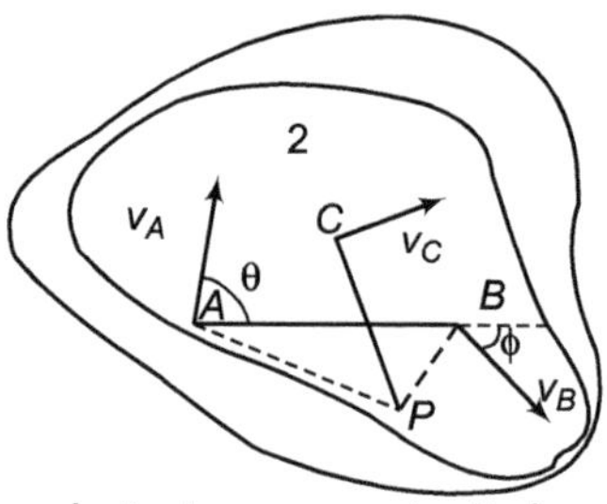

Instantaneous centre P of velocity v_A and v_B

- If both links 1 and 2 are in motion, we can define a relative instantaneous centre P_{12} to be a point on 2 having zero relative velocity with respect to a coincident point on 1. Consequently, the relative motion of 2 with respect to 1 be appears to be pure rotation about P_{12}. So, P_{21} and P_{12} are identical.

Centro

Instantaneous centre is also called centro. So, two coincident points belonging to two rigid bodies having relative motion with the properties.

- They have the same velocities.

- They form a point in one of the rigid bodies about which the other rotates and *vice-versa*. Which is perhaps true for only an instant.

Primary Centro One which can be easily located by a mere observation of the mechanism.

Secondary Centro Centros that cannot be easily located.

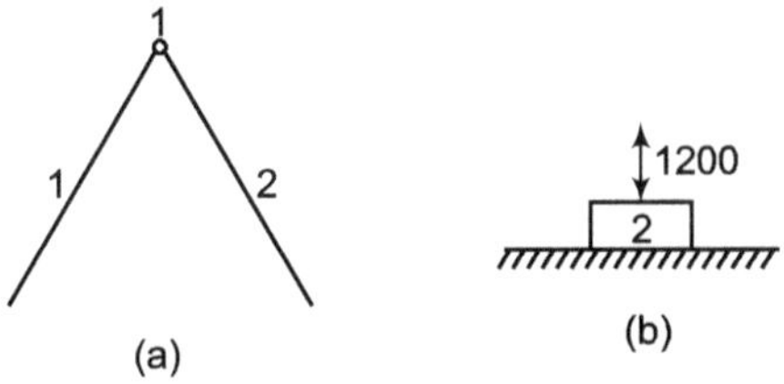

Primary centro Secondary centro

Instantaneous Centre of Acceleration

It is defined as a point on a link having zero relative acceleration with respect to a coincident point on the other link and is different from the instantaneous centre of velocity.

Aronhold-Kennedy Theorem of Three Centre

It state that if three bodies are in relative motion with respect to one another, the three relative instantaneous centers of velocity are collinear.

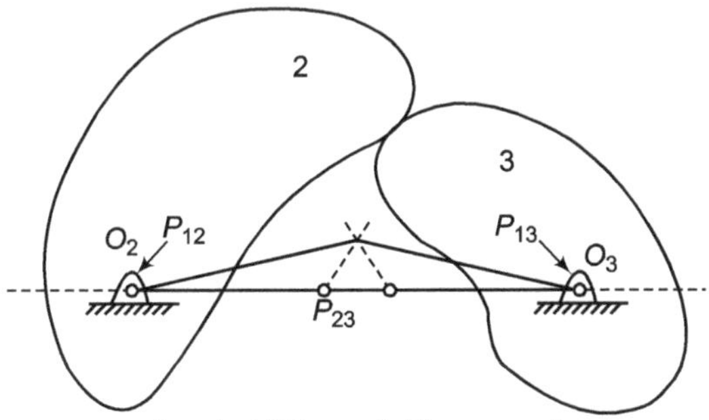

Aronhold-Kennedy Theorem of three centre 1, 2 and 3

- P_{12}-Instantaneous centre of fixed ground 1 and body 2.
- P_{13}-Instantaneous centre of fixed ground 1 and body 3.
- P_{23}-Instantaneous centre of body 2 and body 3.

Number of Centros in a Mechanism

For a mechanism of n links, the number of centros (Instantaneous centre) N is

$$N = \frac{1}{2} n (n - 1)$$

Number of Lines of Centros

The number of lines of centros L for a mechanism with n links is

$$L = \frac{1}{6} n (n - 1) (n - 2)$$

Angular Velocity Ratio Theorem for I centre

When the angular velocity of a link is known and it is required to find the angular velocity of another link,then locate their common I-centre.

The velocity of this I-centre relative to a fixed 3rd link is the same whether the I-centre is considered on the first or the second link. First consider the I-centre to be on the first link and obtain the velocity of the I-centre. Then, consider the I-centre to be on the second link and find its angular velocity.

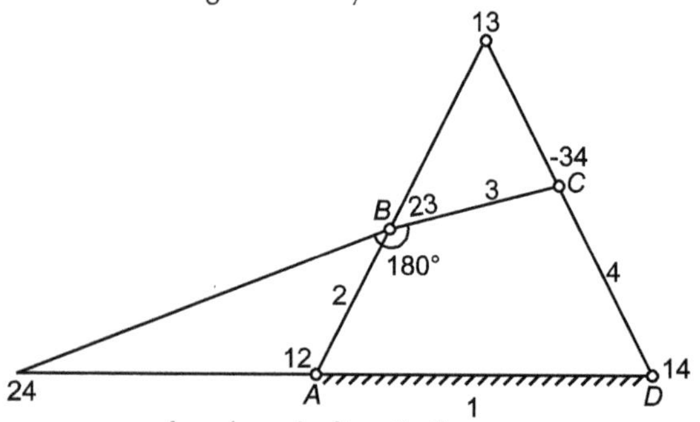

Angular velocity ratio theorem

First imaginary link 2 to be in the form of a flat disc containing point 24 and revolving about 12 or A. Then, $v_{24} = \omega_2 (T_{24} - T_{12})$.

Second imaginary link 4 to be large enough to contain point 24 and revolving about 14 or D.

$$v_{24} = \omega_4 (I_{24} - I_{14}); \quad \frac{\omega_4}{\omega_2} = \frac{I_{24} - I_{12}}{I_{24} - I_{14}}$$

Acceleration Analysis in Mechanism

The rate of change of velocity with respect to time is known as acceleration and acts in the direction of the change in velocity. Velocity can changed by only changing its magnitude or its direction. Let a link OA, of length r, rotate in a circular path in the clockwise direction as shown in figure. It has an instantaneously angular velocity ω and an angular acceleration α in the same direction *i.e.*, the angular velocity increases in the clockwise direction.

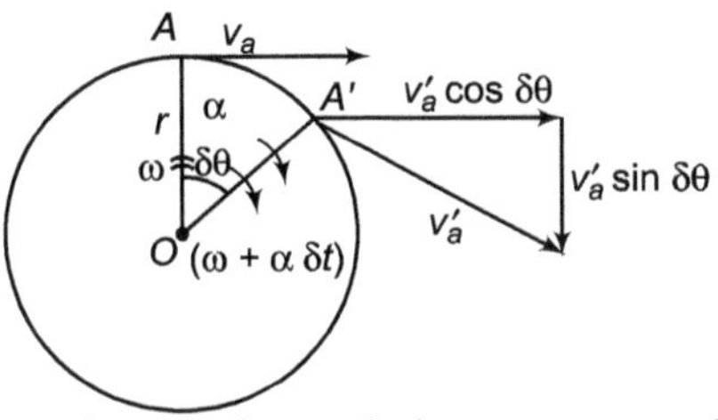

Acceleration analysis

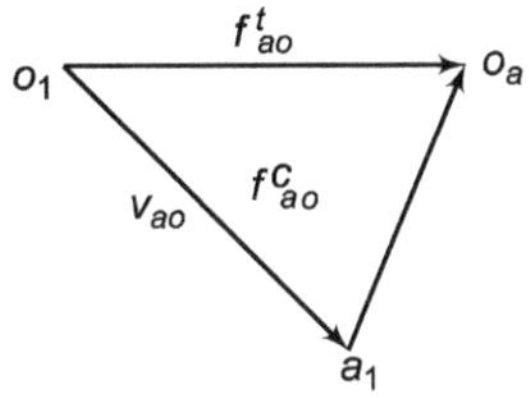

Tangential and radial acceleration analysis

- **Tangential acceleration** of A relative O is defined as

$$f^t_{ao} = \alpha\, r = \left(\frac{d\omega}{dt}\right) r = \frac{dv}{dt}$$

- **Centripetal or radial acceleration** of A relative to O is defined as

$$f^c_{ao} = \omega^2 r = \frac{v^2}{r}$$

- **Total acceleration** (net acceleration) $= f^t_{ao} + f^c_{ao}$

There are three cases occured in the net acceleration as given below

Case I When $\alpha = 0 \Rightarrow \omega = \text{constant}$

$$\Rightarrow \quad f'_{ao} = 0$$

So, net acceleration $= f^c_{ao} = \dfrac{v^2}{r}$

Case II When $\omega = 0 \Rightarrow A$ has linear motion as

$$f^c_{ao} = 0$$

Net acceleration $= a^t_{ao} = \dfrac{dv}{dt}$

Case III When α is negative or the link OA decelerates, tangential acceleration will be negative or its direction will be as shown in figure.

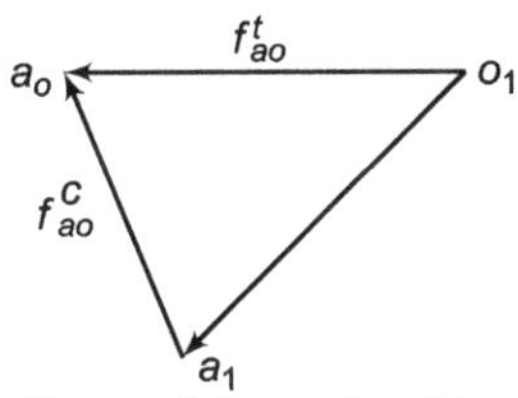

Tangential acceleration as negative

Coriol's Acceleration Component

Consider a link AR rotates about a fixed point A on it. P is a point on a slider on the link.

Here, ω = Angular velocity of the link
α = Angular acceleration of the link
v = Linear velocity of the slider on the link
f = Linear acceleration of the slider on the link
r = Radial distance of point P on the slider.

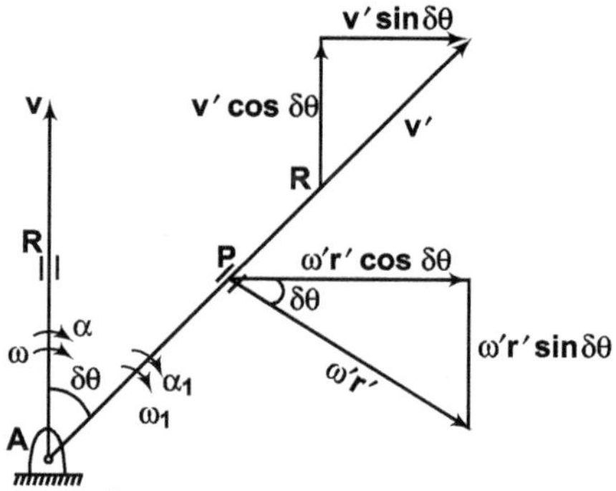

Coriol's acceleration analysis

Key Points

Direction of Coriol's acceleration component $(2\omega \cdot v)$ is perpendicular to AR
Coriol's component is positive if

- The link AR rotate clockwise and the slider moves radially outward.
- The link rotate counter clockwise and the slider moves radially inwards.

Acceleration of slider (f) is positive if

- Slider has a deceleration while moving in the inward direction.
- Slider has acceleration while moving in the outward direction.

- Acceleration of $P \parallel$ to AR

$$f_{P(\parallel AR)} = f - \omega^2 r$$
= acceleration of slider – centripetal acceleration

- Acceleration of $P \perp$ to AR.

$$f_{P(\perp AR)} = 2\omega v + \alpha \cdot r$$
= Coriol's acceleration + tangential acceleration

Let Q be a point on the link AR immediately beneath the point P at the instant, then

- Acceleration of P = acceleration of $P||$ to AR + acceleration of $P\perp$ to AR

$$f_{Pa} = (f - \omega^2 r) + (2\omega v + r\alpha)$$
$$= f + (r\alpha - \omega^2 r) + 2\omega v$$

= acceleration of P relative to Q + Acceleration of Q relative to A + Coriols acceleration component

$$= f_{Pq} + f_{qa} + f_{cr}$$

Kinematic Synthesis of Mechanism

A mechanism is said to be kinematically simple if its acceleration analysis can be carried out completely by applying successively the acceleration equation $A_Q = A_p + A_{QP}$ combined, if necessary, with acceleration-image principle.

Coupler Curve

Coupler curve is a curve traced by a point on one of the coupler links in a mechanism.

Exact Straight Line Motion Mechanism and Condition

Exact straight line mechanism contain a link in which a point describes a coupler curve as an exact straight line.

Condition for straight line mechanism

$$OX = \frac{OS \times OT}{OH}$$

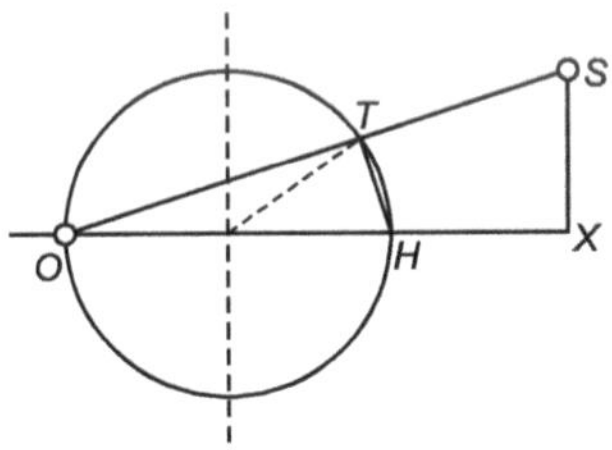

OS, a exact straight line motion

- **Paucelier Mechanism** A paucelier mechanism consists of eight links.
- **Hart Mechanism** It consists of 6 links.
- **Scott-Russel Mechanism** It consists of 4 links.

$$\frac{OQ}{QS} = \frac{QS}{QP}$$

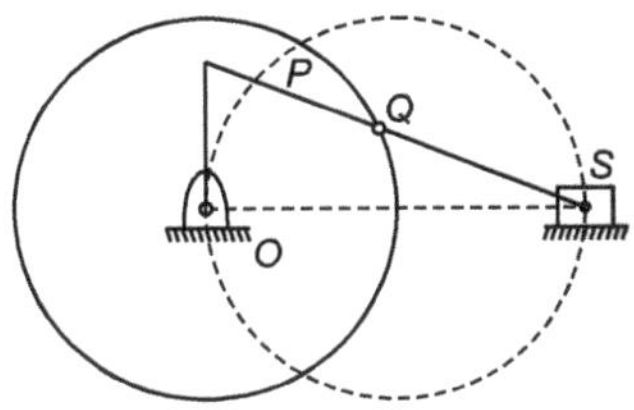

Scott-Russel mechanism

Approximate Straight Line Motion Mechanism

The approximate straight line motion mechanisms are as follows.

Modified Scott-Russel Mechanism

In this, P will approximately traverse a straight line perpendicular to OS.

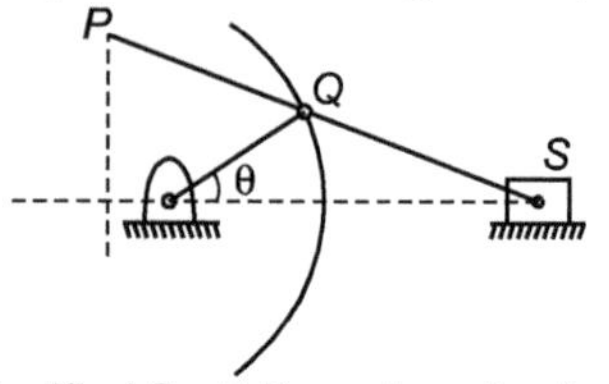

Modified Scott-Russel mechanism

Gross-Hopper Mechanism

This mechanism is a derivation of the modified Scott-Russel mechanism in which the sliding pair at S is replaced by a turning pair.

Pantograph Mechanism

A pantograph is a four bar linkage used to produce paths exactly similar to the ones traced out by a point on the linkage. The paths so produced may be straight or cured ones. It consists of 4 links.

$$\frac{OP}{OQ} = \frac{OA}{OB}$$

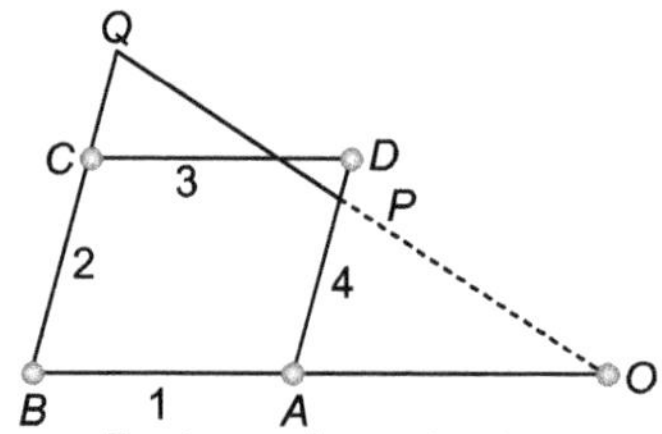

Pantograph mechanism

Cams

A cam is a mechanical member used to impart desired motion (displacement) to a follower by direct contact (either point or line contact). A cam may be rotating or reciprocating whereas the follower may be rotating, reciprocating or oscillating.

- Cam mechanism belong to higher pair mechanism.
- A driver member known as cam.
- A driven member called the follower.
- A frame is one which supports the cam and guides the follower.

Definitions Related to Cam

The following terminologies are used to draw cam profile

Base Circle

The base circle is the smallest circle (with its centre at the cam centre) that can be drawn tangential to the cam profile.

Trace Point

A trace point is a theoretical point on the follower, its motion describing the movement of the follower. *e.g.*, for a knife edge follower, the trace point is at the knife edge.

Pitch Curve

It is the curve drawn by the trace point assuming that the cam is fixed and the trace point of the follower rotates around the cam.

Pressure Angle

It represents the steepness of the cam profile. The angle between the direction of the follower movement and the normal to the pitch curve at any point is referred to as the pressure angle.

Key Points

- For a roller follower, the trace point is at the roller centre.
- For a flat-face follower, it is at the point of contact between the follower and cam surface when the contact is along the base circle of the cam.
- During a complete rotation, the pressure angle varies from its maximum to its minimum value.
- The greater the pressure angle, the higher will be side thrust and consequently the chances of the translating follower jamming in its guide will increase.
- It is not desirable to increase the pressure angle.

Pitch Point and Circle

A pitch point correspond to the point of maximum pressure angle and a circle drawn with its centre at the cam centre, to pass through the pitch point is known as the **pitch circle**.

Prime Circle The prime circle is the smallest circle that can be drawn so as to be tangential to the pitch curve.

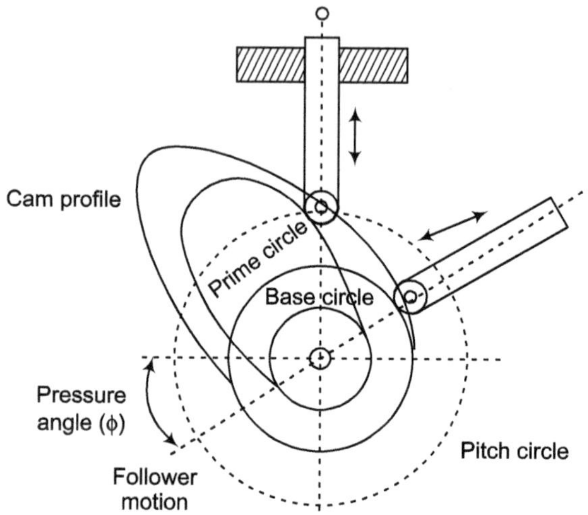

A schematic diagramatic view of pitch circle

Follower Displacement Diagram

The following terms are used with reference to the angular motion of the cam

- **Angle of Ascent (ϕ_a)** It is the angle through which the cam turns during the time the follower rises.
- **Angle of Dwell (*f*)** Angle of dwell is the angle through which the cam turns while the follower remains stationary at the highest or the lowest position.

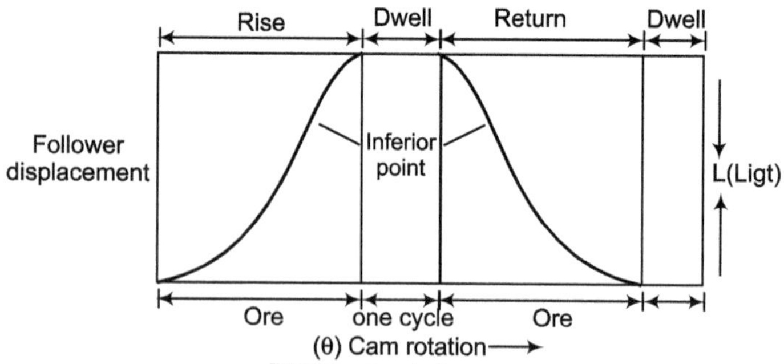

Displacement diagram

- **Angle of Descent (ϕ_d)** Angle of descent is the angle through which the cam turns while follower returns to the initial position.
- **Angle of Action** Angle of action is the total angle moved by the cam during the time between the beginning of rise and the end of return of the follower.

- In a specific interval of time, acceleration must be given to consider the velocity.
- The dynamic effect of acceleration limit the speed and effect of jerk (rate of change of acceleration) produce vibrations of the system.
- The angle made by follower to the surface of cam at the point of contact cannot be reduced from a certain value. So, it exert minimum lateral pressure on the bearing.
- The size of the base circle controls the pressure angle.

Cam with Specified Contours

Followings are the various types of cams, which are symmetrical cams or specified contour cams.

Tangent Cam (with roller follower)

A tangent cam is symmetrical about the centre line.

where, r_c = Least radius of cam, r_n = Radius of nose, r_r = Radius of roller
r = Distance between the cam and nose centre.

Roller on the Flank

The maximum and minimum velocity can be calculated. when the location of roller on the flank as

where, r_c = Least radius of cam
r_n = Radius of nose
r_r = Radius of roller
r = Distance between the cam and the nose centre.

Let θ be angle turned by the cam from the beginning of the follower motion.

$$x = (r_c + r_r)\left(\frac{1}{\cos\theta} - 1\right)$$

$$v = \frac{dx}{dt} = \frac{dx}{d\theta}\frac{d\theta}{dt}$$

$$v = \omega(r_c + r_c)\frac{\sin\theta}{\cos^2\theta}$$

$$v_{max} = \omega(r_c + r_r)\frac{\sin\beta}{\cos^2\beta}$$

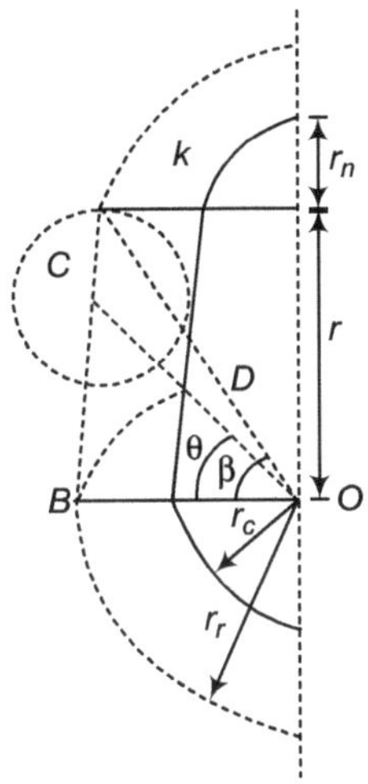

Cam contours

At $\theta = \beta$, (β be the angle turned by the cam when the roller loss the flank)

$$v_{min} = 0 \text{ At } \theta = 0$$

Acceleration

$$f = \frac{dv}{dt} = \frac{dv}{d\theta} \times \frac{d\theta}{dt}$$

$$f = \omega^2 (r_c + r) \frac{(2 - \cos^2 \theta)}{\cos^3 \theta}$$

$$f_{min} = \omega^2 (r_c + r_r) \qquad \text{at } \theta = 0°$$

Roller on the Nose

$$x = r \cos(\alpha - \theta) + \sqrt{l^2 - r^2 \sin^2(\alpha - \theta)} - n$$

$\therefore$ $$l = r_n + r_r$$

and $$n = r_c + r_r$$

$$v = \frac{dx}{dt} = \frac{dx}{d\theta} \frac{d\theta}{dt}$$

$$f = \frac{dv}{dt} = \frac{dv}{d\theta} \frac{d\theta}{dt}$$

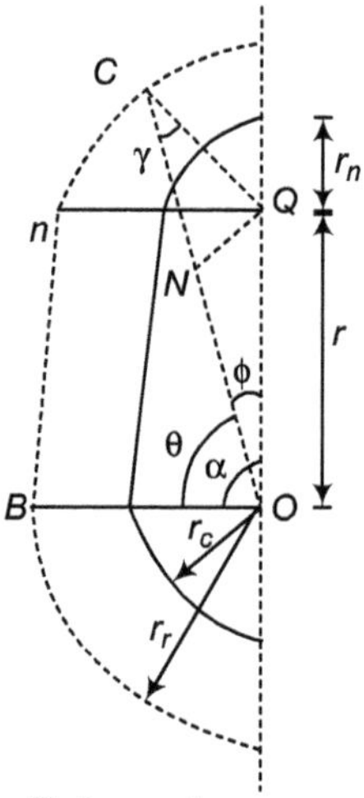

Roller on the nose in cam contours

Circular Arc (convex) Cam (with flat-faced follower)

Follower Touching Circular Flank

$$x = (r_f - r_c)(1 - \cos\theta)$$

$$v = \frac{dx}{dt} = \frac{dx}{d\theta} \frac{d\theta}{dt}$$

$$v = \omega (r_f - r_c) \sin\theta$$

$$v_{min} = 0 \text{ at } \theta = 0$$

$$v_{max} = \omega (r_f - r_c) \sin\beta \text{ at } \theta = \beta$$

Acceleration

$$f = \omega^2 (r_f - r_c) \cos\theta$$

$$f_{max} = \omega^2 (r_f - r_c) \text{ at } \theta = 0$$

$$f_{min} = \omega^2 (r_f - r_c) \cos\beta \text{ at } \theta = \beta$$

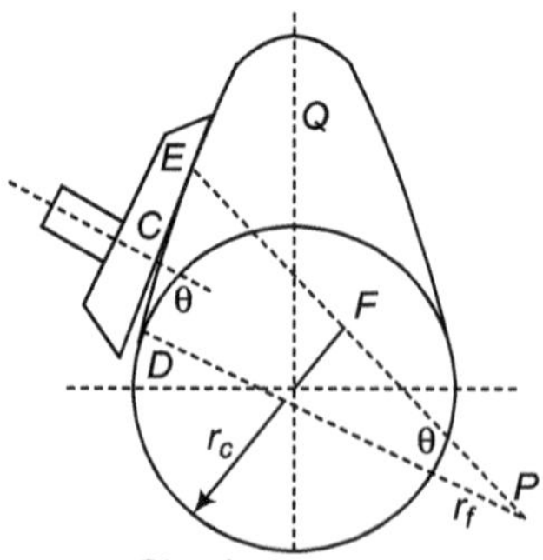

Circular arc cam

Follower on the Nose

$$x = r_n - r_c + r\cos(\alpha - \theta)$$

$$v = \omega r \sin(\alpha - \theta)$$

$$v_{min} = 0 \text{ at } \theta = \alpha$$

v is maximum when the follower just touches the nose of the cam.

$$f = -\omega^2 r \cos(\alpha - \theta)$$

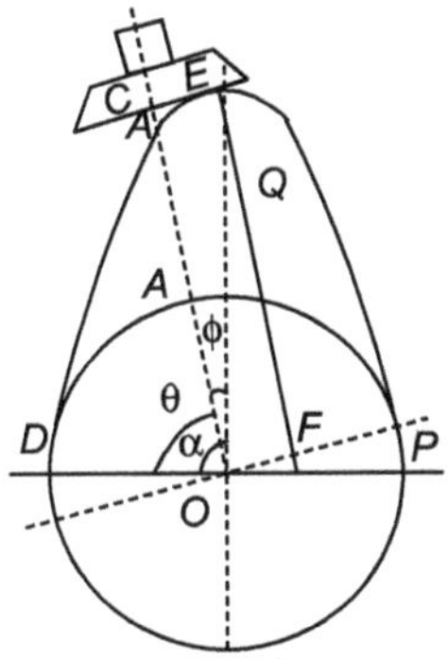

Follower on the nose of the cam

Circular Arc (Convex) Cam (with roller follower)

Follower on the Flank

$$x = (r_f + r_c)\cos\phi - (r_f + r_c)\cos\theta - (r_c + r_c)$$

where, $\cos\phi = \sqrt{\left[1 - \frac{(r_c - r_c)\sin\theta}{(r_f + r_r)}\right]^2}$

or $\quad x = \sqrt{B^2 - A^2\sin^2\theta} - A\cos\theta - (r_c + r_c)$

where, $A = r_f - r_c$ and $B = r_f + r_r$

$$v = \omega A\left[\sin\theta - \frac{A\sin 2\theta}{2\sqrt{B^2 - A^2\sin^2\theta}}\right]$$

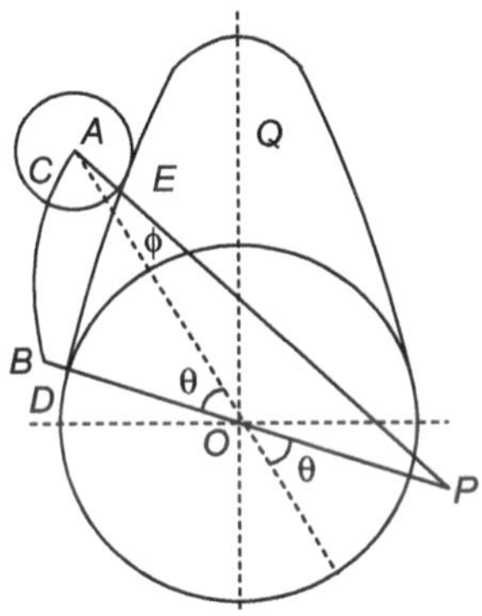

Circular arc cam with follower on the flank

$$f = \omega^2 A\left[\cos\theta - \frac{A\cos 2\theta}{\sqrt{B^2 - A^2\sin^2\theta}} - \frac{A^3\sin 2\theta}{4(B^2 - A^2\sin^2\theta)^{3/2}}\right]$$

Gears *and* Gear Trains

Gear is a positive drive which transmits exact velocity ratio from one shaft to another shaft, gear drive is used when the distance between driver and follower is very less, when two or more gears are made to mesh with each other to transmit power from one shaft to another, the combination is called **gear trains**.

Gears

Gears are used to transmit motion from one shaft to another or between a shaft and slide. It is a higher pair mechanism. Gear use no intermediate link or connector and transmit the motion by direct contact.

Gear Terminology

The basic terms used in gears are

Circular Pitch (p)

It is a distance measured along the circumference of the pitch circle from a point on one tooth to the corresponding point on the adjacent tooth.

$$p = \frac{\pi d}{T}$$

where, p = circular pitch; d = pitch diameter; T = number of teeth.

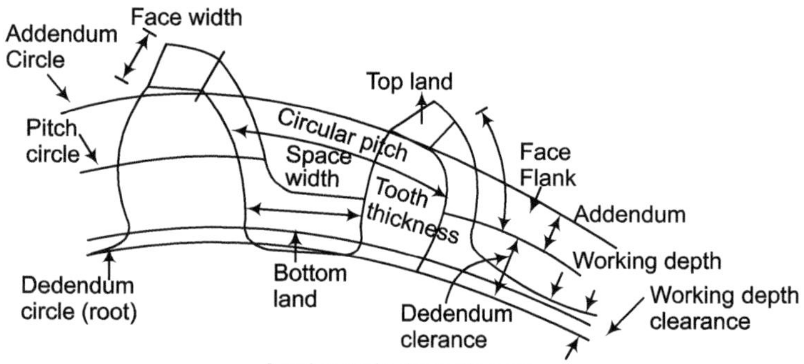

A schematic gear diagram

Diametrical Pitch (P)

It is the number of teeth per unit length of the pitch circle diameter in inches.

$$P = \frac{T}{d}$$

Module (m)

It is the ratio of pitch diameter in mm to the number of teeth. The term is used in SI units in place of diametrical pitch.

$$m = \frac{d}{T}$$

$$\Rightarrow \quad p = \pi m$$

Gear Ratio (G)

It is the ratio of the number of teeth on the gear to that on the pinion.

$$G = \frac{T}{t}$$

wheres, T = number of teeth on the gear; t = number of teeth on the pinion.

Velocity Ratio

The velocity ratio is defined as the ratio of the angular velocity of the follower to the angular velocity of the driver gear.

$$VR = \frac{\omega_2}{\omega_1} = \frac{N_2}{N_1} = \frac{d_1}{d_2} = \frac{T_1}{T_2}$$

Addendum Circle

A circle bounding the ends of the teeth,in a normal section of the gear.

Dedendum circle

The circle bounding the spaces between the teeth, in a normal section of the gear.

Addendum

The radial distance between the pitch circle and the addendum circle.

Dedendum

The radial distance between the pitch circle and the root circle.

Standard Proportions for 20° Full Depth System

The following standard proportions are used in 20° full depth system

Standard Proportions

Addendum $h_a = m$	Dedendum h_f = 1.25 m
Clearance C = 0.25 m	Working depth h_k = 2 m
Whole depth h = 2.25 m	Tooth thickness (s) = 1.5708 m
Tooth space = 1.5700 m	Fillet radius = 0.4 m

Form of Teeth

The form of teeth in a gear system can be as follows.

Conjugate Teeth

Two curves of any arbitrary shape that fulfill the law of gearing can be used as the profiles of teeth. Such a gear are said to have **conjugate teeth.**

The profiles of tooths which are generally used given as

Cycloidal Profile Teeth

As cycloid is the locus of a point on the circumference of a circle that rolls without slipping on a fixed straight line.

In this type, the faces of the teeth are epicycloids and the flanks the hypocycloids.

Epicycloid Teeth

An epicycloid is the locus of point on the circumference of a circle that rolls without slipping on a fixed straight line.

$$\text{Arc } PK = \text{arc } KJG$$

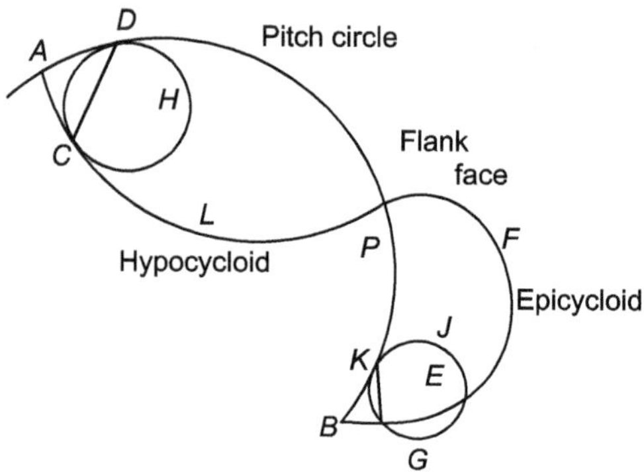

Standardize gear teeth

Hypocycloid Teeth

A hypocycloid is the locus of a point on the circumference of a circle that rolls without slipping inside the circumference of another circle.

$$\text{Arc } AC = \text{Arc } CD$$

Involute Profile Teeth

An involute is defined as the locus of a point on a straight line which rolls without slipping on the circumference of a circle. From figure at the start the tracing point is at *A* as the line rolls on the circumference of the circle, the path *ABC* traced out by the point *A* is the involute.

D is regarded as the instantaneous centre of rotation of *B*. The motion of *B* is perpendicular to *BD*. Since, *BD* is tangent to the base circle, the normal to the involute is a tangent to the base circle.

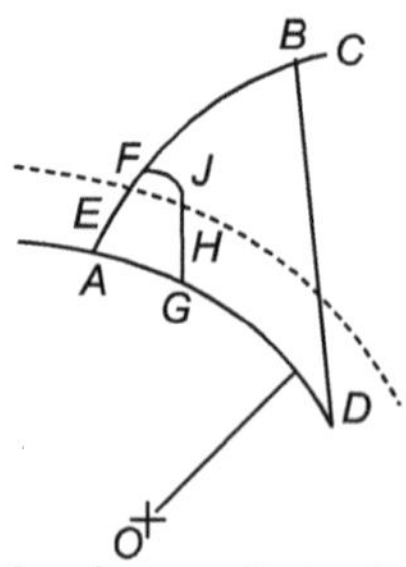

Involute profile teeth

Comparison between Cycloidal Teeth and Involute Teeth

Cycloidal Teeth	Involute Teeth
Pressure angle maximum at beginning of engagement, then reduce to zero at pitch point and again maximum at end of engagement.	Pressure angle is constant throughout the engagement of teeth.
Less smooth running of gears.	Smooth running of gears.
Involve double curve, epicycloid and hypocycloid for the teeth	Involve single curve
Costier manufacturing	Cheaper manufacturing
Exact centre distance is required to transmit a constant velocity ratio.	A little variation in centre distance does not effect the velocity ratio
Phenomenon of interference does not occur at all.	Interference can occur if the condition of minimum number of teeth on a gear is not followed.
Spreading flank teeth and strong.	Radial flank teeth and weaker compare to cycloidal.
Convex flank always has contact with a concave face resulting in less wear.	Two convex surfaces are in contact and thus there is more wear.

Methods of Gear Manufacturing

1. Casting, blanking, machining.
2. For power transmitting gears (made of steel), the following methods generally used are milling, rack generation, hobbing, fellow gear shaper method.
3. For good accuracy and surface finish following finishing process are used grinding, lapping, sharing, barnishing.

The followings are the parameters used in manufacturing of gears as

Path of Contact (*CD*)

According to the figure, the path of the contact is given by the relation

$$
\begin{aligned}
CD &= CP + PD \\
&= (CF - PF) + (DE - PE) \\
&= \left[\sqrt{R_a^2 - R^2\cos^2\phi} - R\sin\phi\right] + \left[\sqrt{r_a^2 - r^2\cos^2\phi} - r\sin\phi\right]
\end{aligned}
$$

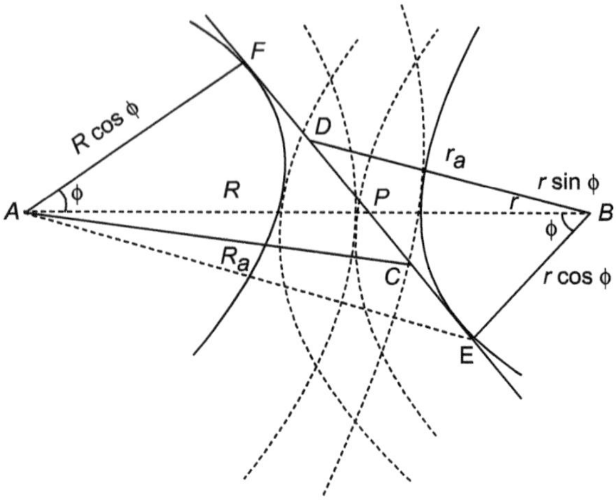

Path of contact diagram

where, r = pitch circle radius of pinion
R = pitch circle radius of wheel
r_a = addendum circle radius of pinion
R_a = addendum circle radius of wheel

Arc of Contact

From the figure, the arc of the contact as given by

- Arc of contact $= \dfrac{\text{path of contact } (CD)}{\cos \phi}$
- Arc of approach $= \dfrac{CP}{\cos \phi}$
- Arc of recess $= \dfrac{PD}{\cos \phi}$
- Number of teeth within the arc $= \dfrac{CD}{\cos \phi} \dfrac{1}{p}$ (p = circular pitch)
- Angle of action δ_p = arc of contact $\times \dfrac{360}{2pr}$

Interference in Involute Gear

- Meshing of two non-conjugate (non-involute) teeth is known as **interference** because the two teeth do not slide properly and thus rough action and binding occurs.
- For equal addenda of the wheel and pinion, the addendum radius of the wheel decides whether the interference will occur or not.

Minimum Number of Teeth

- Maximum value of the addendum radius of wheel to avoid interference can be upto BE

$$(BE)^2 = (BF)^2 + (FE)^2$$

$$BE = R\sqrt{1+\left(\frac{r}{R}\right)\left(\frac{r}{R}+2\right)-\sin^2\phi}$$

Addendum value of wheel $\leq BE - R$

$$T \geq \frac{2a_\omega}{\sqrt{1+\frac{1}{G}\left(\frac{1}{G}+2\right)\sin^2\theta}-1}$$

where, G = gear ratio $\left(\frac{T}{t}\right)$

T and t are number of teeth on the wheel and pinion respectively.

When addendum is equal to one module $a_\omega = 1$

$$T \geq \frac{2}{\sqrt{1+\frac{1}{G}\left(\frac{1}{G}+2\right)\sin^2\phi}-1}$$

Interference between Rack and Pinion

The interference between rack and pinion is shown in the figure.

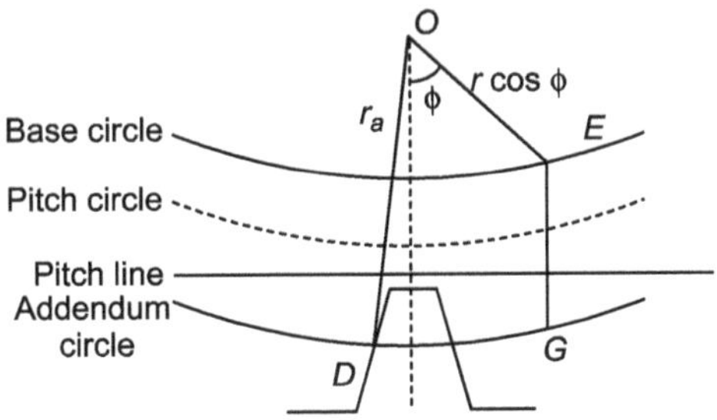

Rack and pinion interference

To avoid interference $\quad GE \geq a_r m$

$$t \geq \frac{2a_r}{\sin^2\phi}$$

For standard addendum, $a_r = 1; t \geq \frac{2}{\sin^2\phi} \Rightarrow t_{min} = \frac{2}{\sin^2\phi}$

To eliminate interference, *there are following methods*

- By using more teeth on gears (however, it can make gear noiser).
- By larger pressure angle.
- By using gears with shorter teeth.
- To eliminate the undercutting or interference, the pinion addendum may be enlarged until the addendum circle of the pinion passes through the interference point of gear.

Undercutting

When gear teeth are produced such that interference is eliminated then this effect is called undercutting.

Long and Short Addendum System

In the long and short addendum system, there is no change in pitch circle and consequently not in pressure angle. The pinion addendum enlarge until the addendum circle of the pinion passes through the point of gear.

In long and short addendum system, recess action is obtained more than approach action.

Gear Train

A gear train is a combination of gears used to transmit motion from one shaft to another. Gear trains are used to speed up or stepped down the speed of driven shaft. *The following are main types of gear trains*

Simple Gear Train

Series of gears, capable of receiving and transmitting motion from one gear to another is called a **simple gear train**.

- **Train value** $= \dfrac{N_3}{N_1} = \dfrac{T_1}{T_3}$

  $= \dfrac{\text{Number of teeth on driving gears}}{\text{Number of teeth on driver gear}}$

- **Speed ratio** $= \dfrac{1}{\text{Train value}}$

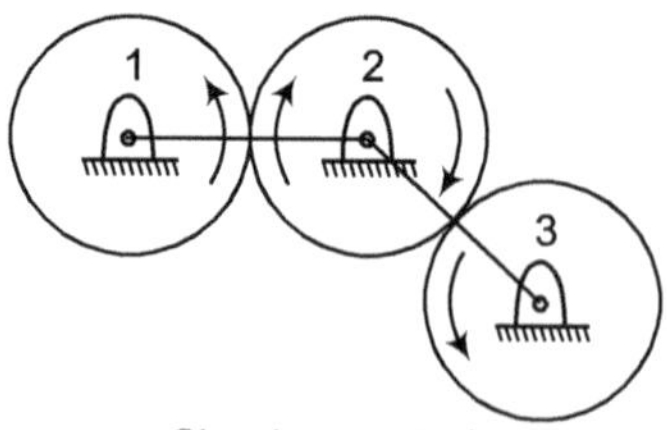

Simple gear train

- The intermediate gears have no effect on the speed ratio and therefore they are known as **idlers**.

Compound Gear Train

When a series of gears are connected in such a way that two or more gears rotate about an axis with the same angular velocity.

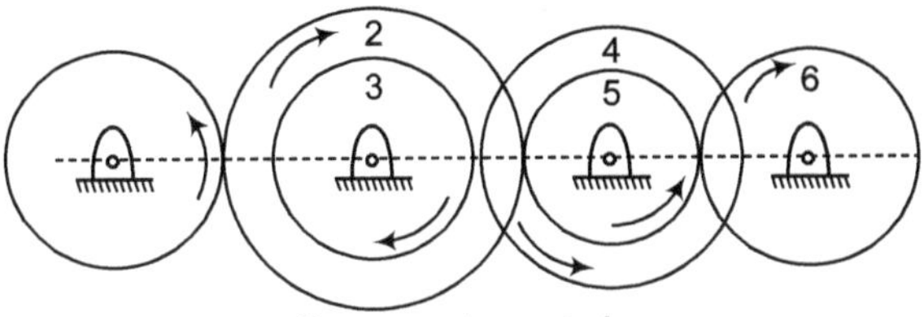

Compound gear train

- Train value = $\dfrac{\text{Product of number of teeth on driving gears}}{\text{Product of number of teeth on driven gears}}$

$$\frac{N_6}{N_1} = \frac{T_1}{T_2}\frac{T_3}{T_4}\frac{T_5}{T_6}$$

Reverted Gear Train

If the axis of first and last wheel of a compound gear coincide, it is called as **reverted gear train.**

- Train valve, $\dfrac{N_4}{N_1} = \dfrac{T_1}{T_2}\dfrac{T_3}{T_4}$
- If r is the pitch circle radius of a gear

$$r_1 + r_2 = r_3 + r_4$$

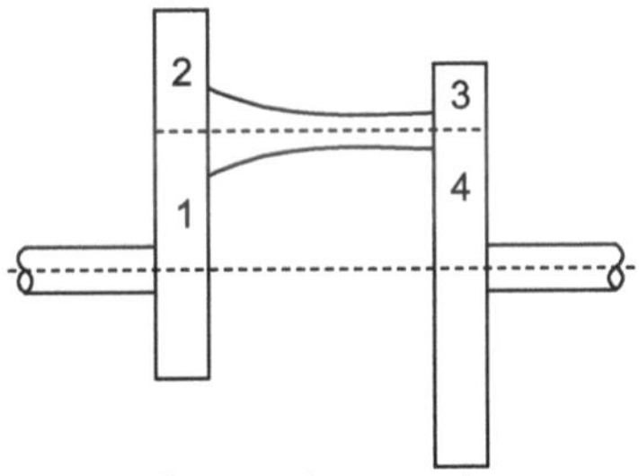

Reverted gear train

Planetary or Epicyclic Gear Train

A gear train having a relative motion of axes is called a planetary or an epicyclic gear train. In an epicyclic train, the axis of at least one of the gears also moves relative to the frame.

If the arm a is fixed the wheels S and P constitute a simple train. However if the wheel S is fixed so that arm a can rotate about the axis of S. The P would be moved around S therefore it is an epicyclic train.

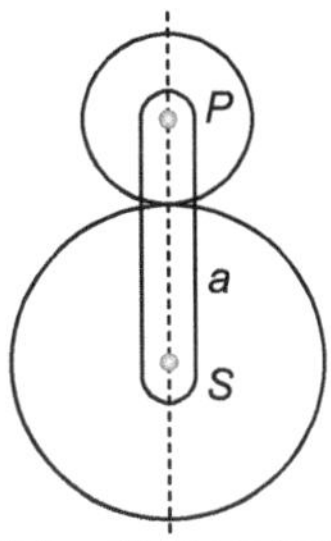

Epicyclic gear train

- In general, gear trains have 2 degree of freedom.

$$\theta'_{5/2}\,(\theta_{L/R}) = \left(-\frac{T_2}{T_4}\right)\cdot\left(\frac{-T_4}{T_5}\right)$$

$$= \frac{\omega_{5/3}}{\omega_{2/3}}$$

$$\left(\frac{-T_2}{T_4}\right)\left(\frac{-T_4}{T_5}\right) = \frac{\omega_5 - \omega_3}{\omega_2 - \omega_3}$$

So, $$\theta'_{L/F} = \frac{\omega_L - \omega_a}{\omega_F - \omega_a}$$

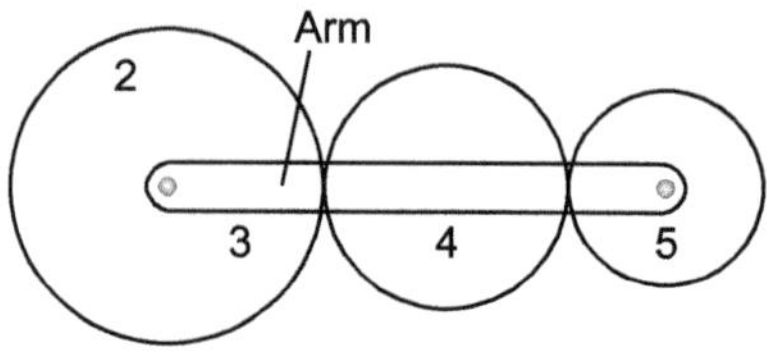

First order kinematic coefficient

Torque in Epicyclic Trains

Let N_S, N_a, N_P, and N_A be the speeds and T_S, T_a, T_P and T_A the external transmitted by S, a, P and A

$$\Sigma T = 0$$

$$T_S + T_a + T_P + T_A = 0$$

Since planet P can rotate on its our pinfixed to a but is not connected to any thing outside.

$$T_P = 0$$

$$T_S + T_a + T_A = 0$$

Assuming no loss in power transistor

$$\Sigma T\omega = 0$$

$$T_S N_S + T_a N_a + T_A N_A = 0$$

If efficiency η to deliver power to S.

$$T_a = \frac{T'_a}{\eta}\left(\frac{\text{actual torque at output}}{\eta}\right)$$

$$T_S = \frac{\text{power input}}{2\pi N/60}$$

So, $$\begin{cases} T_S N_S + \dfrac{T'_a}{h} N_a + T_A N_A = 0 \\ \text{and } T_S + T'_a + T_A = 0 \end{cases}$$

Tabular Method

In tabular method, we apply following procedure

Assume gear makes m revolution and arm makes n revolution. Consider clockwise rotation is positive and anticlockwise rotation is negative.

Analysis of Tabular Method

Step	Operation	Revolution of elements (N)		
		Arm C	Gear A	Gear B
1.	Fix the arm and give + 1 revolution to gear A clockwise.	0	+1	$\frac{N_B}{N_A} = \frac{-t_A}{t_B}$ $N_B = \frac{-t_A}{t_B} \times 1$
2.	Now, multiplying by m (It means m rotation is given to gear A in clockwise.)	0	$+m$	$-m.\frac{t_A}{t_B}$
3.	Add n (It means $+n$ rotations are given to all elements of gear train in clock wise)	n	n	n
4.	Total motion	n	$m+n$	$-m\frac{t_A}{t_B}+n$

On the basis of given condition, we can determine the values m and n.

For example, if gear A is at rust, then

$$N_A = 0$$

$$\therefore \quad m + n = 0 \quad \ldots(i)$$

and arm is given +k revolution, then

$$n = k \quad \ldots(ii)$$

By solving Eqs. (i) and (ii), we can find the values of m and n. Once we find out m and n then number of revolutions and direction of gear B can be determined.

as $$N_B = -m\frac{t_A}{t_B} + n$$

Static *and* **Dynamic Force Analysis**

If components of a machine accelerate, inertia forces are produced due to their masses. However, if the magnitudes of these forces are small compared to the externally applied loads, they can be neglected while analysing the mechanism. Such an analysis is known as static force analysis. When the inertia effect due to the mass of the components is also considered, it is called dynamic-force analysis.

The static and dynamic force analysis define by the following forces.

Constraint Force

A pair of action and reaction forces which depend upon the nature of connection and constraint two connected bodies to behave in a particular manner are known as **constraint forces**.

Applied Forces

Forces which acting from outside on a system of bodies are called **applied forces**.

Static Equilibrium

A body is in static equilibrium if it remains in its state of rest or motion. If the body is at rest, it tends to remain at rest and if in motion, it tends to keep the motion.

In Static Equilibrium

The vector sum of all the forces acting on the body is zero.

$$\Sigma F = 0$$

The vector sum of all the moments about any arbitrary point is zero.

$$\Sigma T = 0$$

Principle of Virtual Work

It can be stated as the work done during a virtual displacement from the equilibrium is equal to zero.

Virtual displacement may be defined as an imaginary infinitesimal displacement of the system.

The following symbols in the figure is defined as

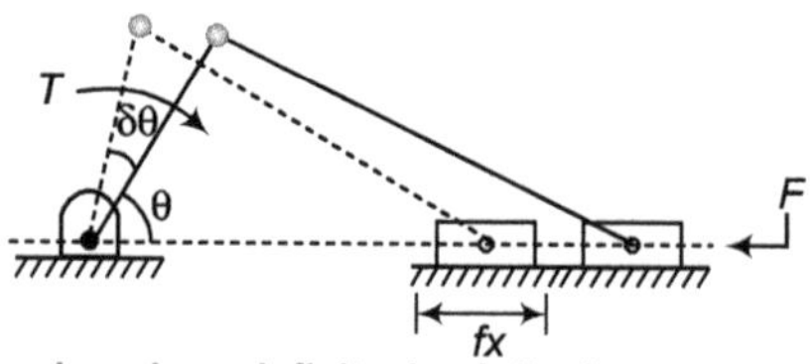

Imaginary infinitesimal displacement system

T = torque
ω = angular velocity of the crank
v = linear velocity of the force
F = applied force

$$T\omega + Fv = 0$$

or

$$T = \frac{-F}{\omega} v$$

D'Alembert's Principle and Dynamic Equilibrium

The vector sum of all external force and inertia forces acting on a rigid body is zero and that the vector sum of all external moments and inertia torques acting on a rigid body is also separately equal to zero.

Here, a_G = rectilinear acceleration of the centre of gravity

α = angular acceleration

R = externally applied force

I = moment of inertia of the about the same axis

$Re = I\alpha$

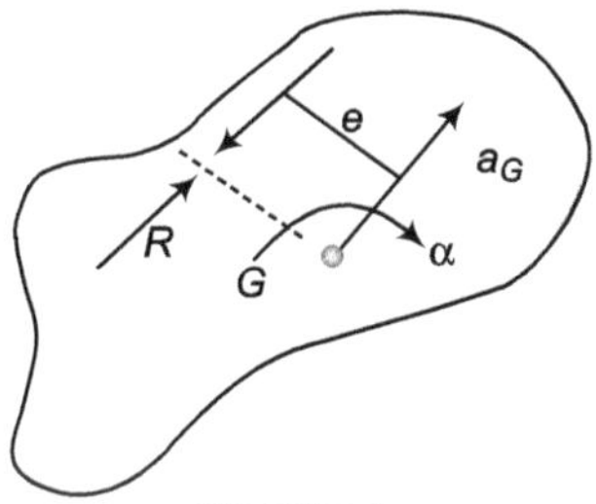

Rigid body

According to D'Alembert principle the vector sum of forces and torques (or couples) has to be zero *i.e.*,

$$\Sigma F + F_i = 0$$

$$\Sigma T + C_i = 0$$

Dynamic Analysis of Slider-Crank Mechanism

The dynamic analysis of slider-crank mechanism describe by the diagram as given below

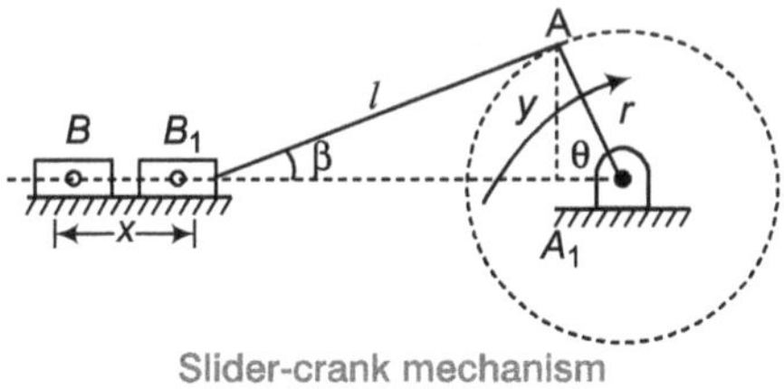

Slider-crank mechanism

Let, x = displacement of piston from inner dead centre

l = length of the connecting rod

r = crank radius

$\frac{l}{r} = n$

The various parameters used in dynamic analysis of slider-crank mechanism is given below

- **Displacement of Piston** It can given by

$$x = r[(1 - \cos\theta) + (n - \sqrt{n^2 - \sin^2\theta})]$$

$$\cos\beta = \frac{1}{n}\sqrt{n^2 - \sin^2\theta}$$

$$x = r\,(1 - \cos\theta) \text{ (assume } n \text{ large } \Rightarrow \sqrt{n^2 - \sin^2\theta} = n)$$

- **Velocity of Piston** It may be given by

$$v = \frac{dx}{dt} = \frac{dx}{d\theta}\frac{d\theta}{dt}$$

$$v = r\,\omega\left[\sin\theta + \frac{\sin 2\theta}{2n}\right]$$

$$v = r\,\omega\sin\theta \ (n \text{ large } \Rightarrow \frac{\sin 2\theta}{2n} \text{ tend to zero})$$

- **Acceleration of Piston** It may be given by

$$f = \frac{dv}{dt} = \frac{dv}{d\theta}\frac{d\theta}{dt}$$

$$f = r\,\omega^2\left(\cos\theta + \frac{\cos 2\theta}{n}\right)$$

$$f = r\,\omega^2\cos\theta \ (n \rightarrow \text{large})$$

- **Angular Velocity of Connecting Rod** (ω_c) It may be defined as

$$l\sin\beta = r\sin\theta \quad \Rightarrow \quad \sin\beta = \frac{\sin\theta}{n}$$

$$\omega_c = \omega\frac{\cos\theta}{\sqrt{n^2 - \sin^2\theta}}$$

- **Angular Acceleration of Connecting Rod** It can be defined as

$$\alpha_c = \frac{d\omega_c}{dt}$$

$$\alpha_c = -\,\omega^2\sin\theta\left[\frac{n^2 - 1}{(n^2 - \sin^2\theta)^{3/2}}\right]$$

The negative sign indicates that the sense of angular acceleration of the rod is such that it tends to reduce the angle β.

- **Piston Effort** (Effective driving force) Piston effort is termed as the net or effective force applied on the piston.

Let, A_1 = area of the corner end
A_2 = area of the piston rod
p_1 = pressure on the corner end
p_2 = pressure on the rod end
m = mass of the reciprocating parts

- **Force on piston due to gas pressure,**

$$= F_p = p_1 A_1 - p_2 A_2$$

Inertia force $$F_b = mr\omega^2\left(\cos\theta + \frac{\cos 2\theta}{n}\right)$$

Friction resistance $= F_f$
Weight of piston $= mg$
Net force on piston $F = F_P + mg - F_b - F_f$

- **Force** (thrust) **Along the Connecting Rod** (F_c) Thrust force along the connecting rod is given by

$$F_c \cos\beta = F \quad \Rightarrow \quad F_c = \frac{F}{\cos\beta}$$

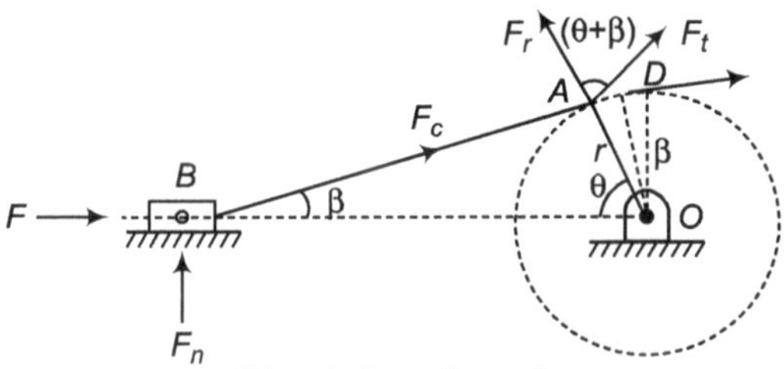

Thrust along the rod

- **Thrust on the Sides of the Cylinder** (F_n) The normal reaction on the cylinder walls are given below

$$F_n = F_c \sin\beta$$

$$\Rightarrow \quad F_n = F \tan\beta$$

- **Crank Effort** (F_t) It is the net effort applied at the crank pin perpendicular to the crank.

$$F_t \times r = F_c \, r \sin(\theta + \beta)$$

$$F_t = \frac{F}{\cos\beta}\sin(\theta + \beta)$$

- **Thrust on the Bearings** (F_r) The force acting along the crank which applied a thrust on the bearings.

  It is known as **thrust on the bearings.**

  The thrusting on the bearings is given by

$$F_r = F_c \cos(\theta + \beta)$$

$$\Rightarrow \quad F_r = \frac{F}{\cos\beta}\cos(\theta + \beta)$$

- **Turning Moment on Crank Shaft** The turning moment on the crank shaft is defined by the relation as given below

$$T = F_t \times r$$

$$T = \frac{F}{\cos\beta} r \sin(\theta + \beta)$$

$$T = F \times OD$$

$$T = F_r \left(\sin\theta + \frac{\sin 2\theta}{2\sqrt{n^2 - \sin^2\theta}} \right)$$

- **Correction Couple of Connecting Rod** (ΔT) The correction couple of the connecting rod is given by the relation.

$$\Delta T = mb\,\alpha_c(l - L) \qquad \text{(taking } b + d = L\text{)}$$

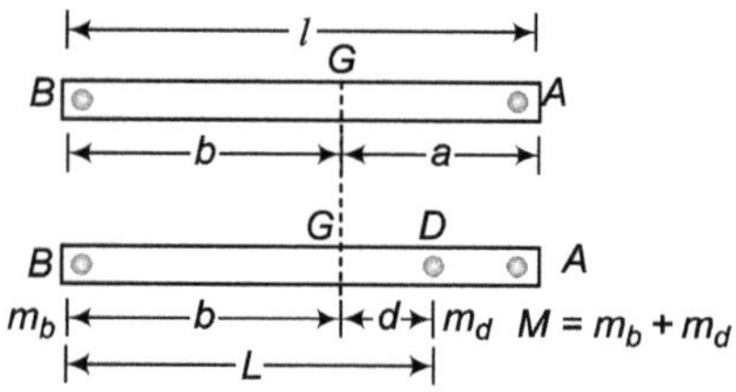

- **Correction Torque at Crank Shaft** (T_c) It may be defined by the formula

$$T_c = \Delta T \frac{\cos\theta}{\sqrt{n^2 - \sin^2\theta}} \qquad \left(\begin{aligned} \Delta T &= F_y\, l \cos\beta \\ &= mb\,\alpha_c(l - L) \end{aligned} \right)$$

Turning Moment Diagram *and* Flywheel

Turning moment diagram is a diagram which shows the variation of turning moment (or torque) on a crank for various position of the crank. Hence turning moment diagram is a graph between M and θ. The turning moment (M) is taken along y axis and crank angle (θ) along x-axis.

During one revolution of the crank shaft of a steam engine or IC engine the torque on it varies and is given by

$$M = F_t \times r = F_r \left(\sin\theta + \frac{\sin 2\theta}{2\sqrt{n^2 - \sin^2\theta}} \right)$$

where, f_t = Tangential force
r = Crank radius
f_r = Net force acting on the system

θ = Angle turned by crank.
n = Ratio of the length of connecting rod to crank radius

Note $M = F_t \times r$ a plot of F_t versus θ is identical to a turning moment (M versus θ) diagram.

Turning Moment Diagram for Single Cylinder Double Acting Steam Engine

For a complete revolution of a crank-shaft, the variation of turning moment (or torque T) for various value of θ is plotted

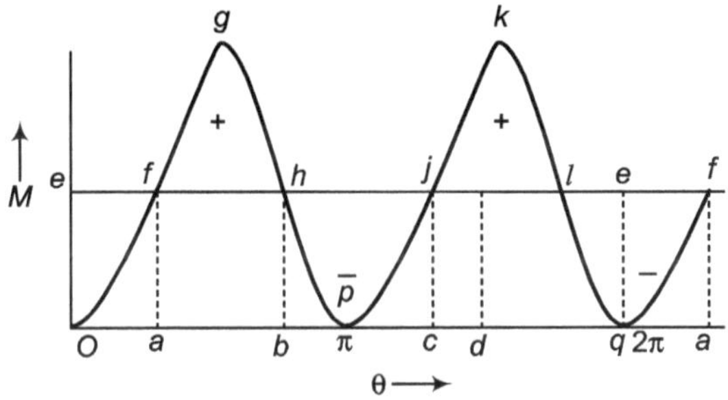

Graph between M and θ

Mean torque against which the engine works, is given by

$$T_{\text{mean}} = \frac{\text{Area } ogpkq}{2\pi}$$

The magniated of the Turning moment is given by

$$M = F_P \times r \times (\sin\theta + \frac{\sin 2\theta}{2n})$$

where, F_b = net force exerted by steam on piston

TMD for Single Cylinder Four Strokes Engine

Four strokes engine with single cylinder graph vary in the following manner

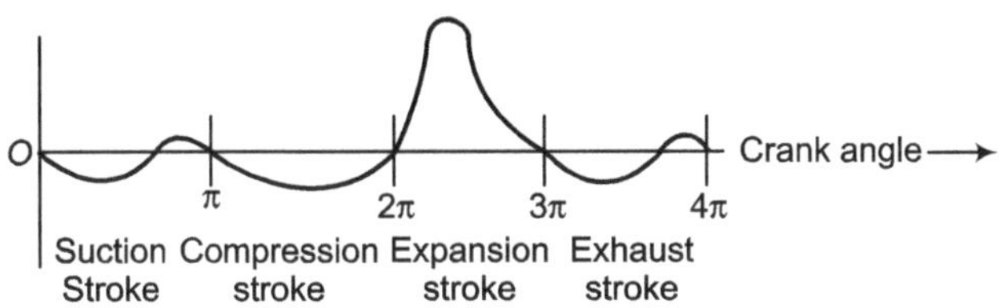

Single cylinder four strokes engine

Multi-cylinder Engine

For multi cylinder engine, the turning moment diagram is obtained by combining the turning moment diagram for each cylinder. *The multi-cylinder engine's displacement diagram is given below*

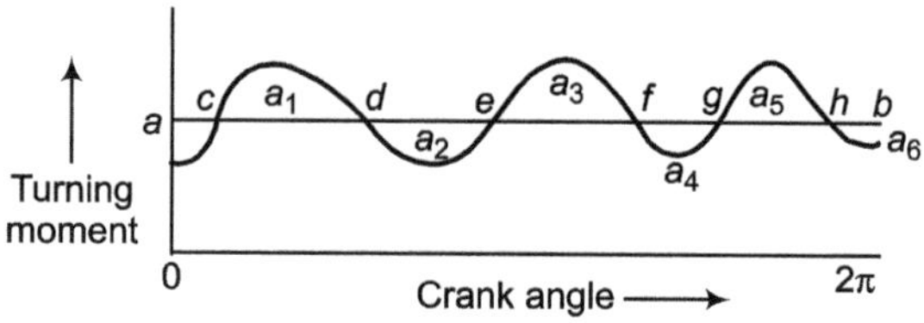

Multi-cylinder engine

Note As the number of cycle increases the variation of turning moment decreased.

Fluctuation of Crank Shaft Speed

For most engines, the load torque or the resisting moment M_R (which opposes the motion of the crank shaft) remains constant over cycle where the turning moment M varies.

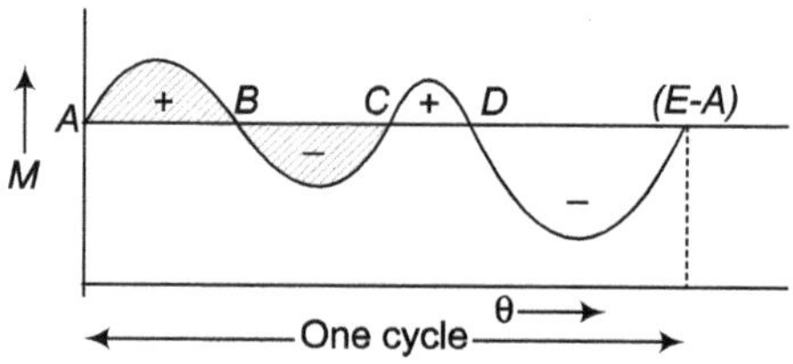

Crank shaft speed

- Energy output of the engine in one cycle $E = \oint M d\theta$
- Average turning moment in this cycle $M_{av} = \frac{E}{\theta} = \frac{1}{\theta}\int_0^{\theta} M d\theta$
- The work done in one cycle against the load is $M_R\theta$.
- If $M_R - M_{av} = \frac{E}{\theta}$

Then condition refer to stable operation and there is no net energy input in the crank shaft and so the speed remains the same at the beginning and at the end of the cycle.

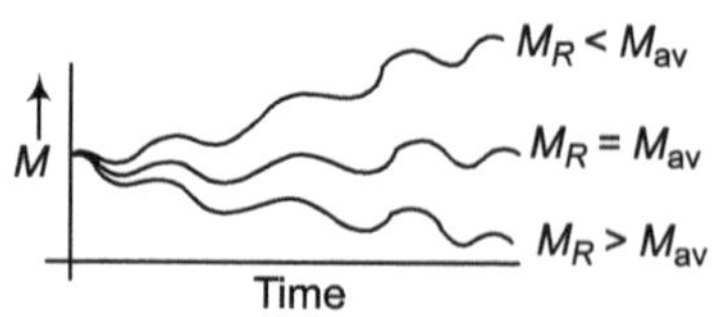

Speed from cycle to cycle

- If $M_R > M_{av} \Rightarrow$ Speed decreases from cycle to cycle.
- If $M_R < M_{av} \Rightarrow$ Speed increases from cycle to cycle.

The difference between the maximum and minimum kinetic energy is known as **maximum fluctuation of energy**.

$$(\Delta KE)_{max} = \int_{\theta_2}^{\theta_1} (M - M_R)\, d\theta$$

- The **coefficient of fluctuation of energy** in a cycle is defined as

$$k_e = \frac{(\Delta KE)_{max}}{E}$$

- The **coefficient of fluctuation of speed** in a cycle

$$k_s = \frac{\omega_{max} - \omega_{min}}{\omega_{av}}$$

Flywheel

A flywheel is used to control the variations in speed during each cycle of an operation. A flywheel acts as a reservoir of energy which stores energy during the period when the supply of energy is more than the requirement and releases the energy during the period when the supply energy is less than the requirement.

- **Maximum fluctuation of energy** (e),

$$e = (\Delta KE) = \frac{1}{2} I (\omega_{max}^2 - \omega_{min}^2)$$

$$e = I\omega(\omega_1 - \omega_2) \quad \left(\omega = \frac{\omega_1 + \omega_2}{2}\right)$$

$$e = I\omega^2 k_s \qquad \left[k_s = \frac{\omega_1 - \omega_2}{\omega}\right]$$

$$k_s = \frac{e}{\frac{1}{2} I\omega^2 \times 2}, \quad k_s = \frac{e}{2E}$$

where, ω_{max} and ω_{min} are the maximum and minimum angular speed respectively.

E = kinematic energy of the flywheel at mean speed.

The equation of motion for the angular oscillation of the flywheel,

$$I \frac{d^2\theta}{dt^2} = M - M_R$$

where, M_R is resistsing moment

Flywheel in Punching Press

Generally, flywheel is used to reduce fluctuation of speed where the load on the crank shaft constant while the applied torque varies.

However, the flywheel can also be used to reduce fluctuation of speed when the torque is constant but load varies during the cycle *e.g.*, in punching press in rivetting machine.

Let E be the energy required for one punch energy supplied to crank shaft from the motor during punching $= E\left(\frac{\theta_2 - \theta_1}{2\pi}\right)$

- The valance energy require for punching

$$= E_1 - E_1\left(\frac{\theta_2 - \theta_1}{2\pi}\right)$$

If, t = thickness of job

v = velocity of tool

s = stroke of the punch ($2r$)

- Maximum fluctuation of energy,

$$\frac{\theta_2 - \theta_1}{2\pi} \approx \frac{t}{2s} = \frac{t}{4r}$$

$$e = (\Delta KE)_{max} = E\left(1 - \frac{\theta_2 - \theta_1}{2\pi}\right)$$

$$= \frac{1}{2} I\,(\omega_{max}^2 - \omega_{min}^2) = I\omega^2 k_s$$

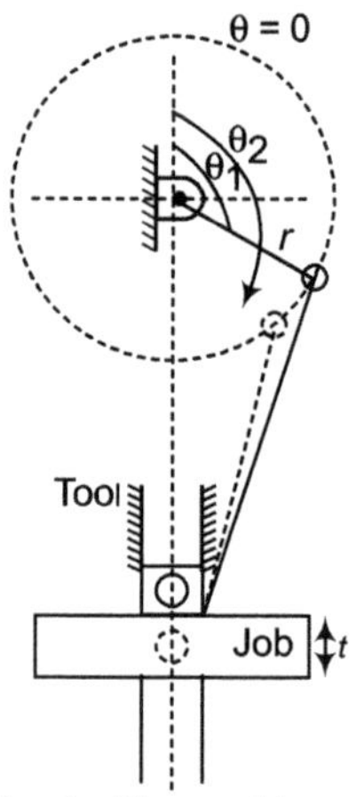

Flywheel with punching press

Balancing

Balancing is defined as the process of designing a machine in which unbalance force is minimum. The rotating and reciprocating parts of a high speed engine if are not properly balanced, the dynamic forces will be setup.

Balancing of Rotating Masses

If the centre of mass of rotating machines does not lie on the axis of rotation, the inertia force is given by $F_I = m\omega^2 e$

where, m = mass of the machine

ω = angular speed of the machine

e = eccentricity *i.e.*, the distance from the centre of mass to the axis of rotation.

Internal and External Balancing

Let a shaft carry an unbalanced mass m_1 with the centre of mass located at distance e from the axis of rotation. The shaft can be completely balancing by adding a mass m_1 at a distance e_1 from the axis of rotation diametrically opposite to m so that,

$$m\omega^2 e = m_1\,\omega^2 e_1$$

(where, ω = speed of rotation of the shaft)

$$me = m_1 e_1$$

In this case $R_A = 0, R_B = 0$

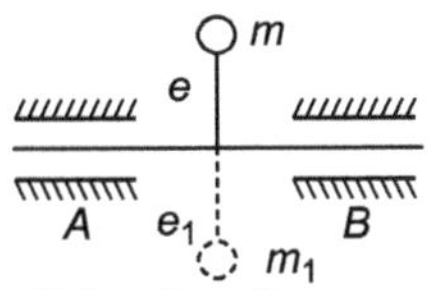

Balancing diagram

The shaft will also be free from dynamic bending stress. This situation is referred to as internal balancing.

The dynamic reactions R_A and R_B can be reduced to zero, by adding two balancing masses m_1 and m_2 at distances e_1 and e_2 respectively from the shaft in the same axial plane as m but on the opposite side of the axis of rotation.

$$m_e = m_1 e_1 + m_2 e_2$$

$$m e a = m_1 e_1 a_1 + m_2 e_2 a_2$$

This situation is referred to as external balancing where the entire length of the shaft is not free from dynamic bending stresses.

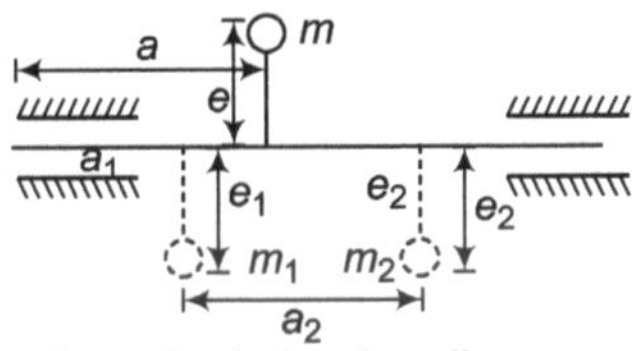

Describe balancing diagram

Static Balancing

If a shaft carries a number of unbalanced masses such that the centre of mass of the system lies on the axis of rotation, the system is said to be statically balanced.

Statically balancing is defined as below

$$m_1 r_1 + m_2 r_2 + m_3 r_3 + m_c r_c = 0$$

$$\Sigma m r + m_c r_c = 0$$

$$\Sigma m r \cos\theta + m_c r_c \cos\theta = 0$$

$$\Sigma m r \sin\theta + m_c r_c \sin\theta = 0$$

$$\tan\theta_c = \frac{-\Sigma m r \sin\theta}{-\Sigma m r \cos\theta}$$

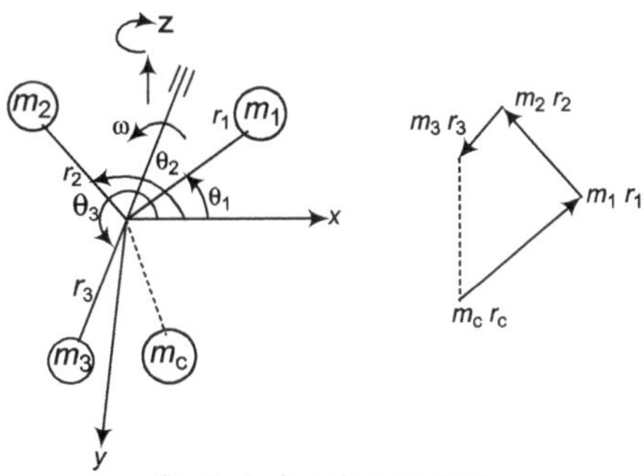

Static balancing system

Dynamic Balancing

A system of rotating masses in dynamic balance when there does not exist any resultant centrifugal force as well as resultant couple.

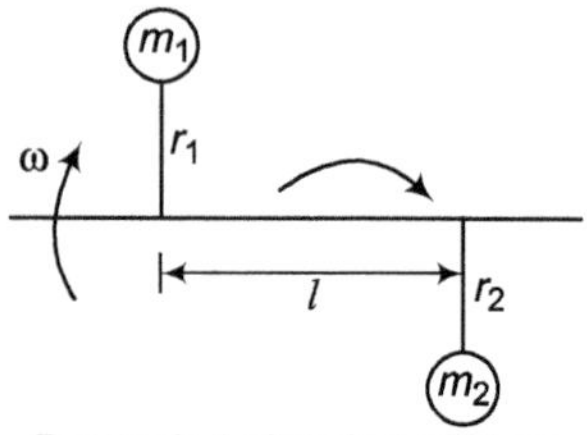

Dynamic balancing system

$$m_1 r_1 = m_2 r_2$$

$$m_1 r_1 l_1 = m_2 r_2 l_2$$

$$\Sigma m r l + m_c r_c l_c = 0$$

Balancing of Several Masses in Different Planes

Let there be a rotor revolving with a uniform angular velocity ω, m_1, m_2 and m_3 are the masses attached to the rotor at radii r_1, r_2 and r_3 respectively. The masses m_1, m_2 and m_3 rotate in planes 1, 2, 3 respectively. In order to complete balancing, Let introduce two counter masses m_{c_1} and m_{c_2} at radii r_{c_1} and r_{c_2} respectively.

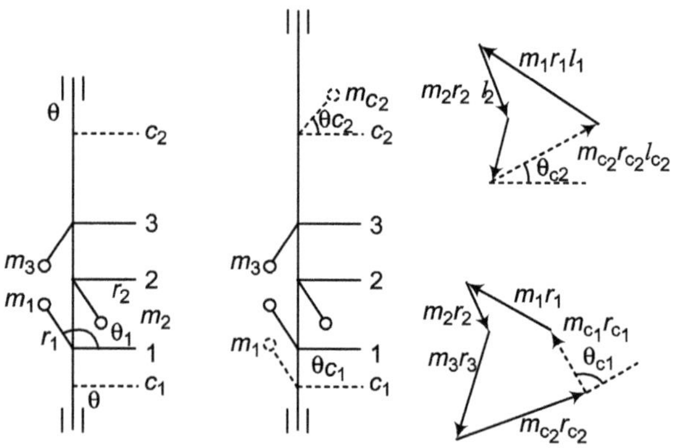
Balancing in different planes with masses

- **Balancing Forces** According to figure, the balancing forces are defined as

$$m_1 r_1 \omega^2 + m_2 r_2 \omega^2 + m_3 r_3 \omega^2 + m_{c_1} r_{c_1} \omega^2 + m_{c_2} r_{c_2} \omega^2 = 0$$

or

$$m_1 r_1 + m_2 r_2 + m_3 r_3 + m_{c_1} r_{c_1} + m_{c_2} r_{c_2} = 0$$

$$\Sigma mr + m_{c_1} r_{c_1} + m_{c_2} r_{c_2} = 0$$

- **Balancing Couples** Let m_{c_1} be placed in reference plane and m_{c_2} be l_2 from reference plane

$$m_1 r_1 l_1 \omega^2 + m_2 r_2 l_2 \omega^2 + m_3 r_3 l_3 \omega^2 + m_{c_2} r_{c_2} l_{c_2} \omega^2 = 0 \quad (l_{c_1} = 0)$$

or $$\Sigma m r l + m_{c_2} r_{c_2} l_{c_2} = 0$$

$$\Sigma m r l \cos\theta + m_{c_2} r_{c_2} l_{c_2} \cos\theta_{c_2} = 0$$

$$\Sigma m r l \sin\theta + m_{c_2} r_{c_2} l_{c_2} \sin\theta_{c_2} = 0$$

$$\tan\theta_{c_2} = -\frac{\Sigma m r l \sin\theta}{-\Sigma m r l \cos\theta}$$

$$\tan\theta_{c_1} = \frac{-(\Sigma m r l \sin\theta + m_{c_2} r_{c_2} \sin\theta_{c_2})}{-(\Sigma m r \cos\theta + m_c r_{c_2} \cos\theta_{c_2})}$$

By complex method,

$$m_1 r_1 l_1 \angle\theta_1 + m_2 r_2 l_2 \angle\theta_2 + m_3 r_3 l_3 \angle\theta_3 + m_{c_2} r_{c_2} l_{c_2} \angle\theta_{c_1} = 0$$

$$\Sigma m r l \angle\theta + m_{c_2} r_{c_2} l_{c_2} \angle\theta_{c_2} = 0$$

and $m_1 r_1 \angle\theta_1 + m_2 r_2 \angle\theta_2 + m_3 r_3 \angle\theta_3 + m_{c_1} r_{c_1} \angle\theta_{c_1} + m_{c_2} r_{c_2} \angle\theta_{c_2} = 0$

$$\Sigma m r < \theta + \Sigma m_c r_c < \theta = 0$$

Balancing of Reciprocating Mass

This diagram shows the condition of the balancing of reciprocating mass

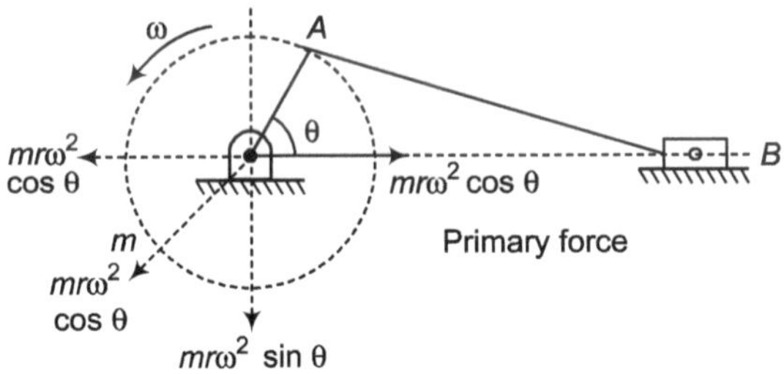

Balancing of reciprocating mass

Force required to accelerate mass is

$$F = \underbrace{m r \omega^2 \cos\theta}_{\text{Primary accelerating force}} + \underbrace{m r \omega^2 \frac{\cos 2\theta}{n}}_{\text{Secondary accelerating force}}$$

- **Primary Balancing of Reciprocating Mass** If c is the fraction of the reciprocating mass, primary force balanced by the mass

$$= c\, m r \omega^2 \cos\theta$$

Primary force unbalanced by the mass

$$= (1 - c)\, m r \omega^2 \cos\theta$$

Vertical component of centrifugal force which remains unbalanced

$$= c\, m r \omega^2 - \sin\theta$$

Resultant unbalanced force at any instant

$$= \sqrt{[(1-c)\, m\, r\, \omega^2 \cos\theta)]^2 + (c\, m\, r\, \omega^2 \sin\theta)^2}$$

The resultant unbalanced force is minimum when $c = \frac{1}{2}$

Secondary Balancing of Reciprocating Mass

The secondary balancing of reciprocating mass is defined as the given below.

$$\text{Secondary force} = mr\omega^2 \frac{\cos 2\theta}{\eta} = mr\,(2\omega)^2 \frac{\cos 2\theta}{4\eta}$$

Key Points

- Primary forces must balance *i.e.*, primary force polygon is enclosed.
- Primary couples must balance *i.e.*, primary couple polygon is enclosed.
- Secondary force must balance *i.e.*, secondary forces polygon is enclosed.
- Secondary couples must balance *i.e.*, secondary couple polygon is enclosed.

Balancing of Inline Engine

The in-line engine is a multi-cylinder engine in which the line of stroke of all reciprocating parts are placed parallel to each other.The net force acting on inline is equal to zero. For this, the line of action of the forces must be same, or in other words, the centre of mass of the system lie on the line.

- **Primary force** $= m\, r\, \omega^2 [\cos\theta + \cos(180° + \theta)] = 0$
- **Primary couple** $= mr\, \omega^2\, l \cos\theta$

Maximum value at $\theta = 0°$ and $180°$

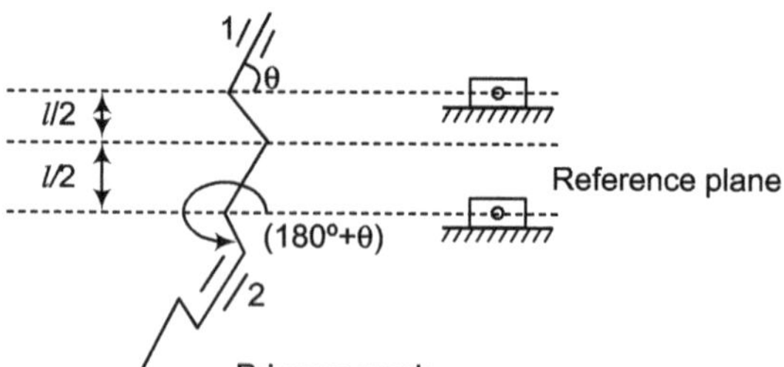

Primary crank

- Secondary force $= \frac{2\,m\,r\,\omega^2}{n}\cos\theta$

  Maximum value at

  $$\theta = 0°, 90°, 180°, 270°$$

- Secondary couple

  $$= \frac{m\,r\,\omega^2}{\theta n}\left[\frac{l}{2}\cos 2\theta + \left(\frac{-l}{2}\right)\cos(360° + 2\theta\right]$$

  $$= 0$$

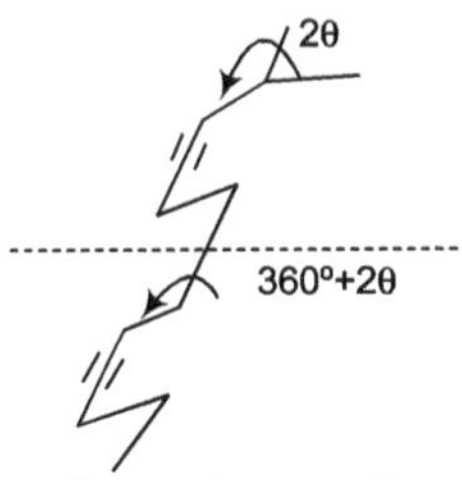

Secondary crank

Balancing of *V*-Engines

Consider a symmetrical two cylinder *V*- engine as shown in figure.

Primary Force

- Total primary force along **x**-axis

  $$= 2m\,r\,\omega^2\cos^2\alpha\cos\theta$$

- Total primary force along *z*-axis

  $$= 2m\,r\,\omega^2\sin^2\alpha\,\sin\theta$$

- Resultant primary force

  $$= \sqrt{(\cos^2\alpha\cos\theta)^2 + (\sin^2\alpha\,\sin\theta)^2}\;2\,m\,r\,\omega^2$$

  $$\tan\beta = \frac{\sin^2\alpha\,\sin\theta}{\cos^2\alpha\cos\theta}$$

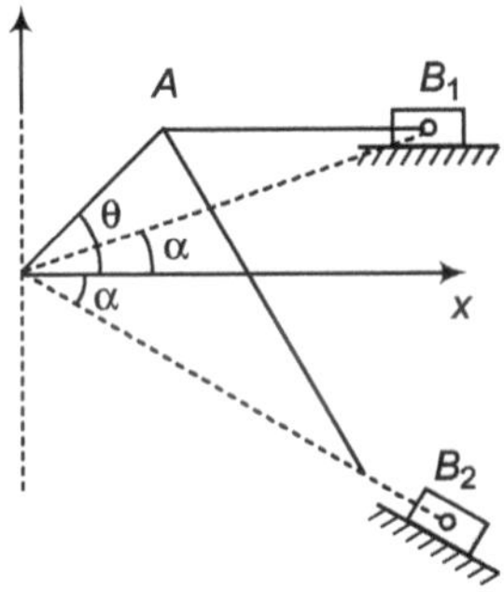

Balancing of V-engines

In case of $2\alpha = 90°$

$$\tan\beta = \tan\theta$$

and resultant force $= m\,r\,\omega^2$ *i.e.*, $\beta = \theta$ or it acts along the crank and it can be complete balance volume by diametrically opposite mass such that $m_r r_r = mr$

For a given value of α, the resultant primary force is maximum when θ is 0°.

Secondary Force

- Total secondary force along *x*-axis $= \frac{2mr\,\omega^2}{n}\cos\alpha\,\cos 2\theta\,\cos 2\alpha$

- Total secondary force along **z**-axis $= \frac{2mr\,\omega^2}{n}\sin\alpha\,\sin 2\theta\,\sin^2 2\alpha$

- Resultant secondary force

$$= \frac{2mr\omega^2}{n}\sqrt{(\cos\alpha\cos\theta\cos 2\alpha)^2 + (\sin\alpha\sin\theta\sin 2\alpha)^2}$$

$$\tan\beta' = \frac{\sin\alpha\sin 2\theta\sin 2\alpha}{\cos\alpha\cos 2\theta\cos 2\alpha}$$

- If $2\alpha = 90°$ or $\alpha = 45°$

Secondary force $= \dfrac{\sqrt{2}\,mr\omega^2}{n}\sin 2\theta$

$$\tan\beta' = \infty \quad \Rightarrow \quad \beta' = 90$$

This means that at $\alpha = 45°$, the force out along z-axis and is a harmonic force and special methods are needed to balance it.

Balancing of Radial Engine

Crank rotating in the direction of engine rotation is known as the direct crank and imaginary crank rotating in the opposite direction is known as reverse crank.

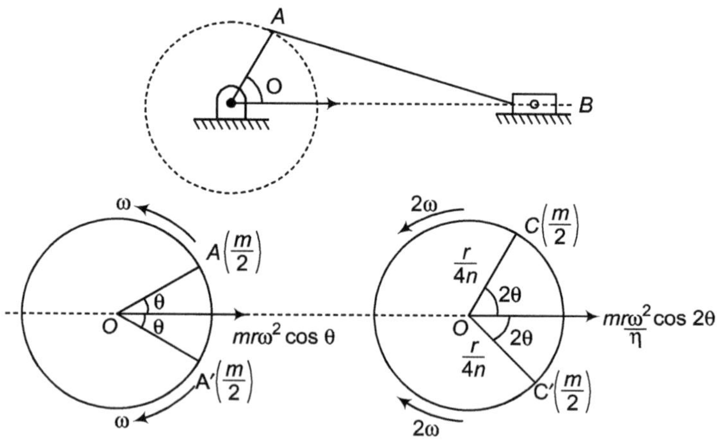

Balancing of radial engine system

OA = primary direct crank
OA' = primary reverse crank
OC = secondary direct crank
OC' = secondary reverse crank

Component of centrifugal force due to rotating mass along the line of stroke are given below

Due to mass at $A = \dfrac{m}{2}r\omega^2\cos\theta$

Due to mass at $A' = \frac{m}{2} r\,\omega^2 \cos\theta$

Thus total force along line of stroke $= m\,r\,\omega^2 \cos\theta$

Which is equal to the primary force.

- Secondary accelerating force $\left(m \frac{r}{4n} (2\omega)^2 \cos 2\theta\right)$

Component of centrifugal force due to rotating mass along the line of stroke are given below

- Due to mass at $c = \frac{m}{2}\left(\frac{r}{4n}\right)(2\omega)^2 \cos 2\theta = \frac{m\,r\,\omega^2}{2n} \cos 2\theta$
- Due to mass at $c' = \frac{m}{2}\left(\frac{r}{4n}\right)(2\omega)^2 \cos 2\theta = \frac{m\,r\,\omega^2}{2n} \cos 2\theta$
- Total secondary force along line of stroke

$$= 2 \times \frac{m}{2}\left(\frac{r}{4n}\right)(2\omega)^2 \cos 2\theta = \frac{m\,r\,\omega^2}{n} \cos 2\theta$$

Governors

The function of a governor is to maintain or regulate the speed of an engine within specified limits whenever there is variation of load.

Flywheel and governor both are used to maintain the speed of an engine but the difference between governor and flywheel is that governor maintain the speed due to variation over a number of revolution while flywheel maintain the speed of engine due to variation each revolution (cyclic variation).

Types of Governors

The broadly classification of the governors are given below.

Centrifugal Governor

- In this type of governor, the action of governor depends upon the centrifugal effects produced by the masses of two balls.
- The balls are operated by actual change of engine speed in case of centrifugal governor.

Inertia Governor

- In this type of governor, positions of the balls are effected by the forces set up by an angular acceleration or deceleration of the given spindle in addition to centrifugal forces on the balls.
- The balls are operated by the rate of change of speed.

  So, response of inertial governors is faster than that of centrifugal types.

Types of Centrifugal Governor

The centrifugal governors are the following types as given below

Pendulum Type Watt Governor (simple conical governor)

Assume sleeve is frictionless and link is massless.

m = mass of each ball
w = weight of each mass (mg)
T = tension in the arm
r = radial distance of ball centre and spindle axis.

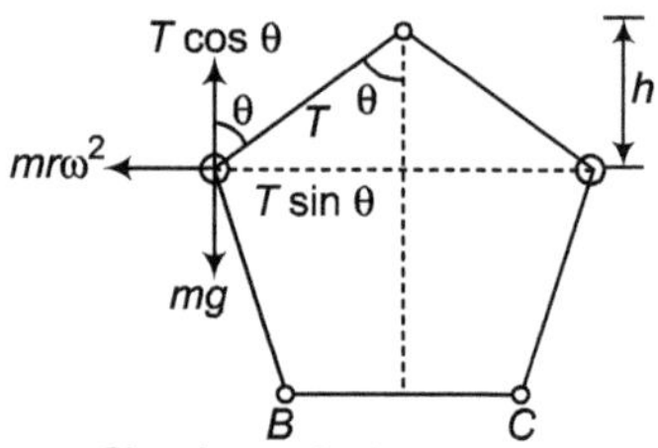

Simple conical governor

So, height of each ball

$$h = \frac{g}{w^2}$$

$$h = \frac{895}{N^2} \text{ (metre)}$$

N = speed in rpm

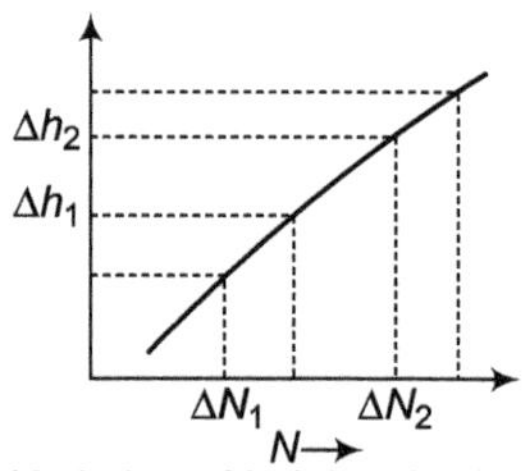

Variation of height Δh with speed

Height h is independent of mass of ball.

As the speed increases, sensitivity decreases.

Porter Type Governor

Porter governor can be shown as

$$h = \frac{895}{N^2}\left(\frac{2mg + (Mg \pm f)(1+k)}{2\ mg}\right)$$

where, $g = 9.8\ \text{m/s}^2$, N = rpm, h = rpm
M = mass of sleeve
m = mass of each ball
f = force of friction at the sleeve

From figure, $\tan\theta = \frac{c}{a}$, $\tan\beta = \frac{b}{a}$

$$k = \frac{\tan\beta}{\tan\theta}$$

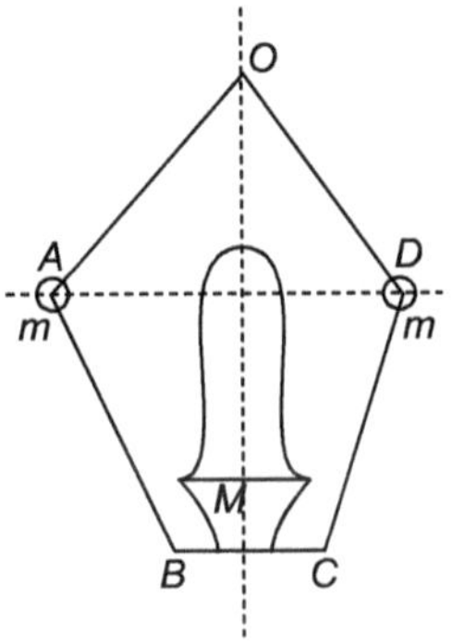

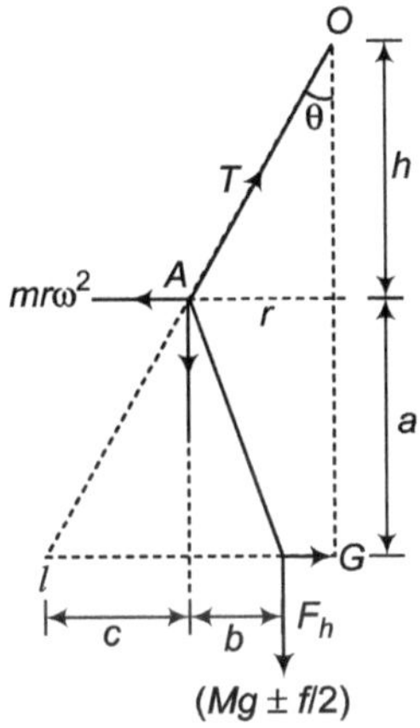

Porter governor   Variation of height with speed

Proell Type Governor

Proell gavernor can be shown below

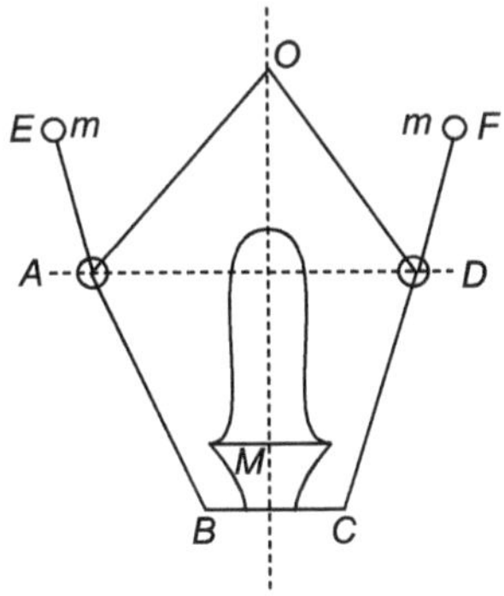

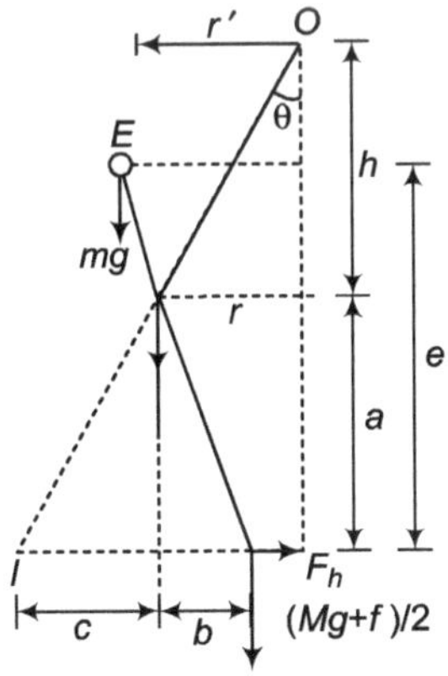

Proell governor   Variation of height with speed

Taking $r' = r$

$$h = \frac{895}{N^2}\frac{a}{e}\left[\frac{mg + (Mg \pm f)}{mg}\right]$$

Hartness Governor

In this type, balls are controlled by a spring as shown in figure.

By neglecting obliquity effects of the arms of the ball crank levers (assume θ small)

$$a_1 = a_2 = a,\ b_1 = b_2 = b,\ c_1 = c_2 = 0$$

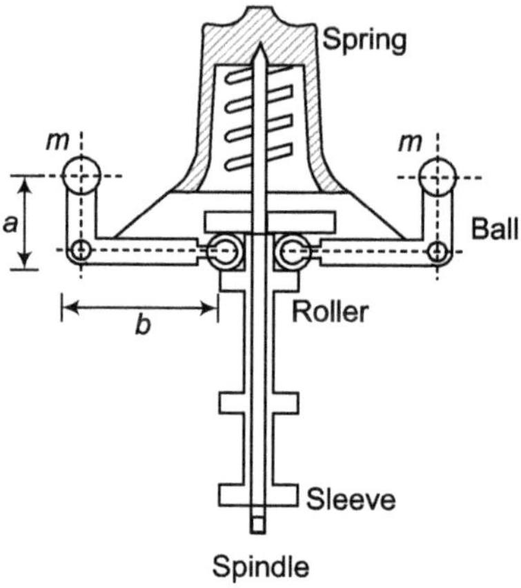

Hartness governor

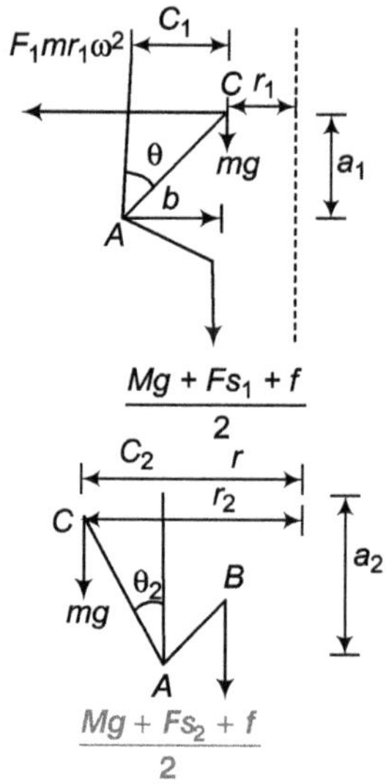

Force analysis in hartness governor

Centrifugal force $\quad F = mr\,\omega^2$

$$\text{Spring force} = F_s$$

$$F_1 a_1 = \frac{1}{2}(Mg + F_{s_1} + f)\,b_1 + mg\,c_1$$

$$F_2 a_2 = \frac{1}{2}(Mg + F_{s_2} + f)\,b_2 + mg\,c_2$$

$$F_{s_2} - F_{s_1} = \frac{2a}{b}(F_2 - F_1)\left\{\begin{matrix} a_1 = a_2 = a \\ b_1 = b_2 = b \\ c_1 = c_2 = 0 \end{matrix}\right\}$$

Let s = Stiffness of the spring, h = Moment of the sleeve

$$F_{s_2} - F_{s_1} = h_1 s = \frac{2a}{b}(F_2 - F_1)$$

$$\frac{r_2 - r_1}{a} \times b\,s = \frac{2a}{b}(F_2 - F_1)$$

$$s = 2 \times \frac{a^2}{b^2}\left(\frac{F_2 - F_1}{r_2 - r_1}\right)$$

Wilson Hartness Governor

This type of governor can be shown below

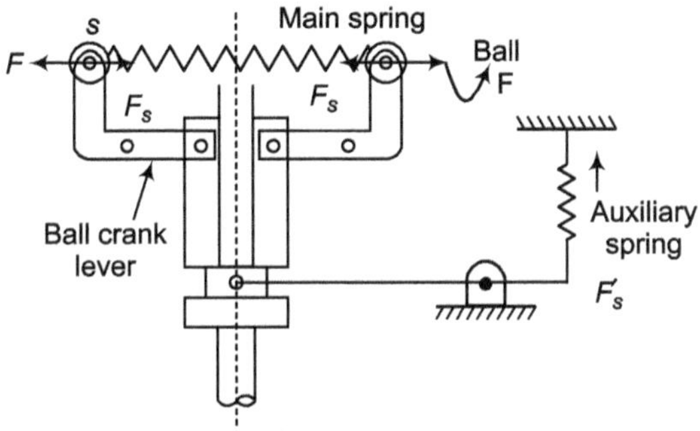

Radial spring governor

Let, s = stiffness of each of main spring
s_a = stiffness of auxiliary spring

- $F_1 a_1 - F_{s_1} a_1 = \frac{1}{2}\left(Mg + F'_{s_1}\frac{y}{x} + f\right) b_1 + mgc_1$

  $F_2 a_2 - F_{s_2} a_{s_2} = \frac{1}{2}\left(Mg + F'_{s_2}\frac{y}{x} + f\right) b_2 + mgc_2$

- Net main spring force

  $$F_{s_2} - F_{s_1} = 4s\,(r_2 - r_1)$$

- Net auxiliary spring force

  $$F'_{s_2} - F'_{s_1} = h_2 - s_a = \left(h_1\frac{y}{x}\right)\frac{s}{a}$$

  $$= (r_2 - r_1)\,\frac{b}{a}\frac{y}{x}\,s_a$$

- If obliquity effects are neglected,

  $$a_1 = a_2 = a,\ b_1 = b_2 = b$$
  $$c_1 = c_2 = 0$$
  $$\frac{F_2 - F_1}{r_2 - r_1} = 4s + \frac{s_a}{2}\left(\frac{b}{a} - \frac{y}{x}\right)^2$$

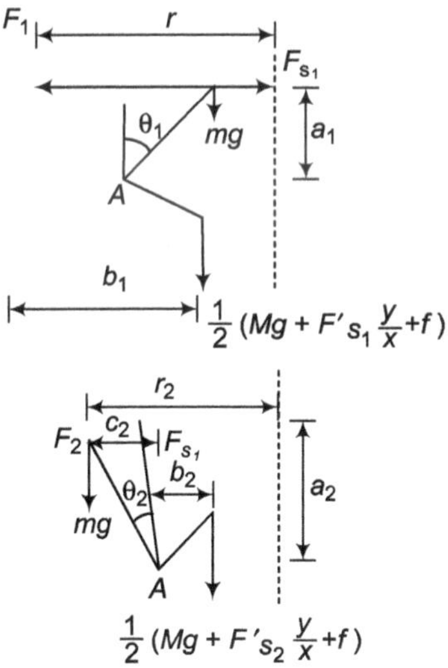

h_1 = movement of sleeve
h_2 = deflection of the auxiliary spring

Force analysis in radial spring governor

Pickering Governor

Let, m = mass fixed to each spring

e = distance between spindle axis and centre of mass when the governor is at rest

ω = angular speed of the sleeve

f = deflection of the centre of the leaf spring for spindle speed ω.

$$\text{Centrifugal force } F = m\,(e + f)\,\omega^2$$

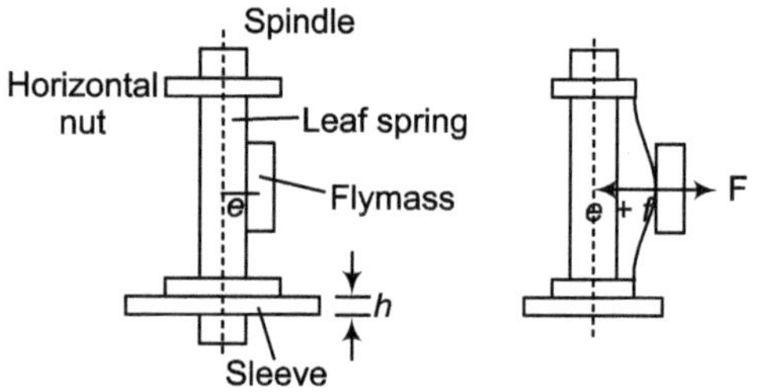

Pickering governor

$$f = \frac{m\,(e + f)\,\omega^2 l^3}{192\,EI}$$

where, E = modulus of elasticity of the spring material

I = moment of inertia of the cross-section of the spring about neutral axis

Inertial Governor

This type of governor is a rotating governor. The inertial governor can be shown in the figure

r = radial distance OG

ω = angular velocity of the disc

v = tangential velocity of G

Centrifugal force of the rotating mass

$$F = m\,r\,\omega_2^2$$

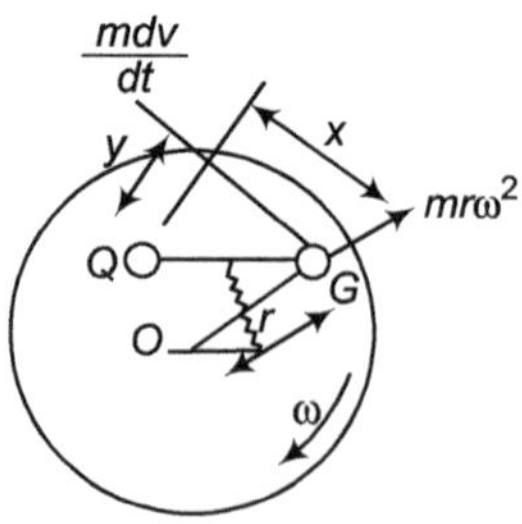

Inertial governor

If the engine shaft is accelerated due to increase in speed, the ball mass does not get accelerated at the same amount on account of its inertia, the inertia force being equal to

$$F_i = mf = m\frac{dv}{dt}$$

$$\text{Torque on the arm } = I_c \frac{d\omega}{dt}$$

Sensitiveness of a Governor

- The governor is said to be sensitive when it readily responds to a small change of speed.
- Sensitiveness of a governor is defined as the ratio of difference between the maximum and minimum speeds to the mean equilibrium speed.

$$\text{Sensitiveness} = \frac{\text{range of speed}}{\text{mean speed}} = 2\,\frac{(N_2 - N_1)}{N_1 + N_2}$$

where, N = mean speed
N_1 = minimum speed corresponding to full load conditions
N_2 = maximum speed corresponding to no load conditions.

Hunting

Sensitiveness of a governor is a desirable quality. However, if a governor is too sensitive, it may fluctuate continuously. This phenomenon of fluctuation is known as **hunting.**

Isochronism

If a governor is at equilibrium only for a particular speed, it is called isochronous governor, for which $\omega_1 = \omega_2 = \omega$ we can say that an isochronous governor is infinitely sensitive.

$$\frac{dF}{dr} = m\,\omega^2$$

Stability

A governor is said to be stable if it brings the speed of the engine to the required value and there is not much hunting. Obviously, the stability and sensitivity are two opposite characteristics.

Effort and Power of a Governor

The effort of the governor is the mean force acting on the sleeve to raise or lower it for a given change of speed capacity.

$$E = 2\int_{r_{min}}^{r_{max}} F \cdot dr = 2\int_{r_{min}}^{r_{max}} F(r)\,dr \quad \Rightarrow \quad \text{Effort} = \frac{E}{2}$$

The power of governor is defined as the work done at the sleeve for a given percentage change of speed.

$$\text{Power} = \text{Mean effort} \times \text{lift of sleeve}$$

Controlling Force

When the balls of a governor rotate in their circular path, the centrifugal force on each ball tends to move in outwards. This is resisted by an equal and opposite forces acting radially inwards and is known as the **controlling force.**

$\therefore$ Controlling force,

$$F_C = m.\omega^2.r$$

The controlling force is provided by the weight of the sleeve and balls as in Porter governor and by the spring and weight as in Hartness governor (or spring controlled governor).

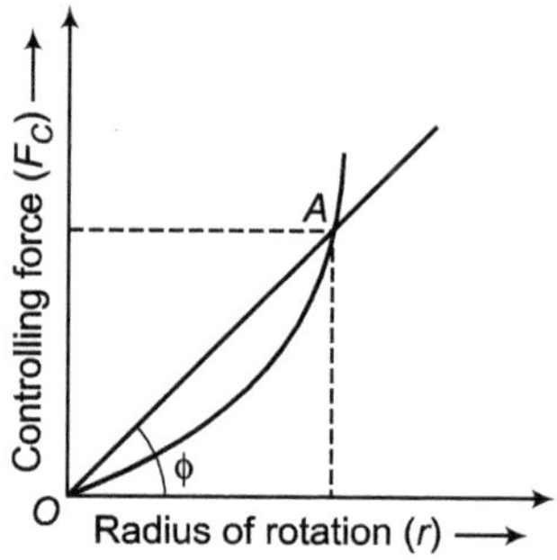

Controlling force diagram

When the graph between the controlling force (F_C) as ordinate and radius of rotation of the balls (r) as abscissa is drawn, then the graph obtained is known as **controlling force diagram.**

Key Points

- Controlling force curve for spring loaded governor as shown below

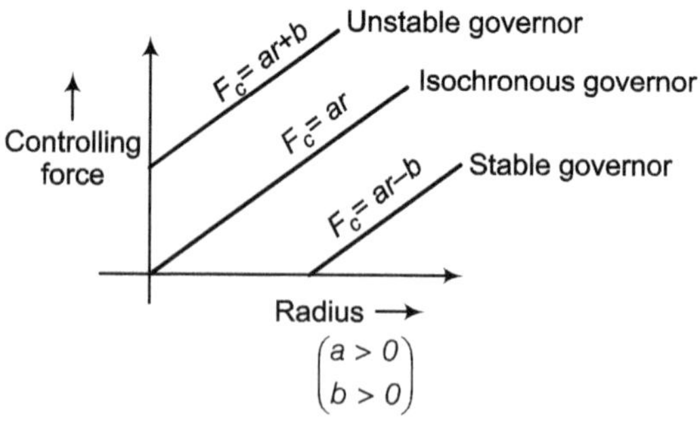

- Controlling force curve is parabolic curve in case of dead weight governor.
- Controlling force curve is straight line in case of spring loaded governor.

Vibrations

Vibration refers to mechanical oscillations about an equilibrium point. In its simplest form, vibration can be considered to be the oscillation or repetitive motion of an object around an equilibrium position.

Vibrations or mechanical oscillations are of many types as given below

- **Free Vibration** (Natural vibration) Vibration over an interval of time during which the system is free from excitation is known as **free vibration**.
- **Damped Vibration** Energy of a vibrating system is gradually dissipated by friction and other resistance.
- **Forced Vibration** When a repeated force continuously acts on a system, the vibrations are said to be **forced.**
- **Harmonic Vibration** Vibration in which the motion is a sinusoidal function of time.
- **Fundamental Vibration** Harmonic component of a vibration with the lowest frequency.
- **Steady State Vibration** When the particles of the body move in steady state condition or continuing period vibration is called **steady state vibration.**
- **Transient Vibration** Vibratory motion of a system other than steady state.
- **Longitudinal Vibration** Vibration parallel to the longitudinal axis of a member.
- **Transverse Vibration** Vibration in a direction perpendicular to the longitudinal axis or central plane of a member.
- **Torsional Vibration** Vibration that involves torsion of a member.
- **Mode of Vibration** Configuration of points of a SHM is called the mode of vibration.
- **Natural Frequency** Frequency of free simple harmonic vibration of an undamped linear system.
- **Time Period** Time taken for one oscillation is called time period.
- **Simple Pendulum** If time period of the pendulum is 1s, then pendulum is called simple pendulum.

$$\left|\frac{\theta}{\alpha}\right| = \frac{l}{g} \qquad \left(\text{where}\ , \alpha = \frac{d^2\theta}{dt^2}\right)$$

Time period is given by

$$T = 2\pi\sqrt{\left|\frac{\theta}{\alpha}\right|}$$

$$T = 2\pi\sqrt{\frac{l}{g}} \quad \text{(for small amplitude } \sin\theta \approx \theta)$$

$$T = 2\pi\sqrt{\frac{l}{g}\left(1+\frac{\theta_0^2}{16}\right)} \quad \text{[large amplitude } (\theta_0)]$$

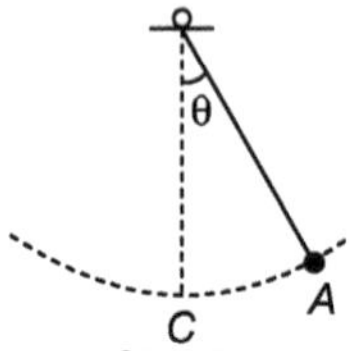

Simple pendulum

Key Points

- If the time period of a simple pendulum is 2s, it is called second pendulum.
- If a simple pendulum is in a carriage which is accelerating with acceleration a, then $g_{\text{eff}} = g - a$.
- If the acceleration **a** is upward

$$g_{\text{eff}} = g + a$$

$$T = 2\pi\sqrt{\frac{l}{g_{\text{eff}}}}$$

- If the acceleration **a** is downward $g_{\text{eff}} = g - a$
- If the pendulum moves on edge which makes angle θ from horizonal.

$$g_{\text{eff}} = g\cos\theta$$

$$T = 2\pi\sqrt{\frac{l}{g\cos\theta}}$$

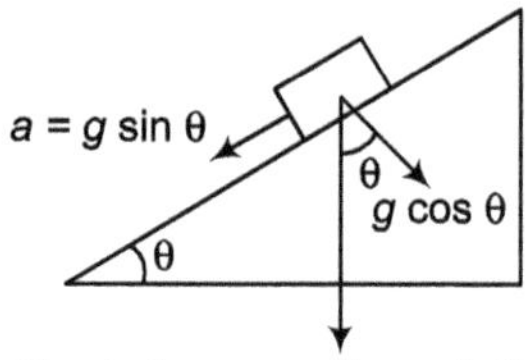

Block diagram with angle θ

- Simple pendulum consists of small sphere of mass m with charge q suspended by a thread of length l

$$g_{\text{eff}} = \frac{w - f_e}{m} = \frac{mg - q\varepsilon}{m}$$

$$T = 2\pi\sqrt{\frac{l}{g - \frac{q\varepsilon}{m}}}$$

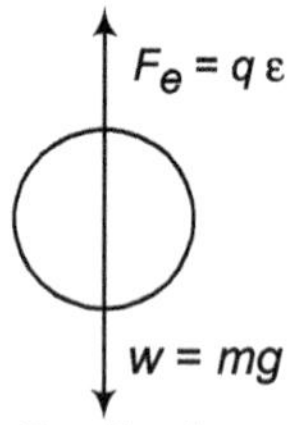

Small sphere

Spring Block System

This type of system can be shown in the figures.

$$ma = -kx$$

$$\left|\frac{x}{a}\right| = \frac{m}{k} \Rightarrow T = 2\pi\sqrt{\frac{m}{k}}$$

Spring block system

- Equivalent force constant (k) is given by

$$\frac{1}{k}=\frac{1}{k_1}+\frac{1}{k_2}$$

$$k=\frac{k_1k_2}{k_1+k_2}$$

$$k=k_1+k_2$$

$$k=k_1+k_2$$

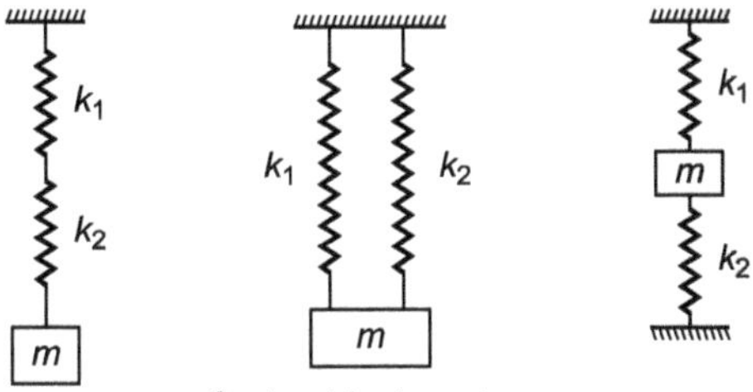

Spring block system

- If spring has a mass m_s and a mass m is suspended from it,

$$T=2\pi\sqrt{\frac{m+\frac{m_s}{3}}{k}}$$

- If block oscillated by massless spring,

$$T=2\pi\sqrt{\frac{m_e}{k}}$$

$$m_e=\frac{m_1 m_2}{m_1+m_2}$$

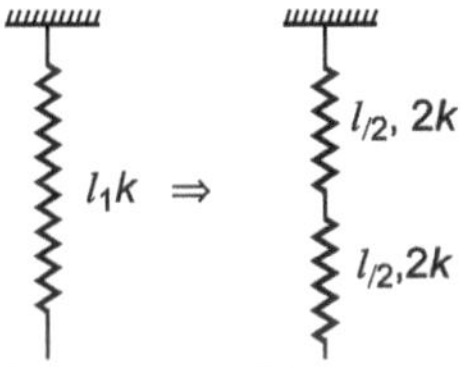

Before cut After cut

- If spring cut new $k_e=2k$

  Oscillation of a fluid column time period is given by

$$T=2\pi\sqrt{\frac{m}{\rho g\, A(\sin\theta_1+\sin\theta_2)}}$$

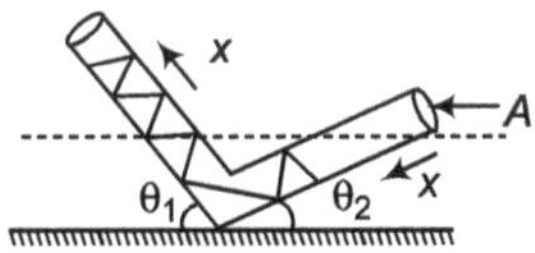

Fluid column

- The time period of the physical pendulum is given by

$$T=2\pi\sqrt{\left|\frac{\theta}{\alpha}\right|}$$

$$T=2\pi\sqrt{\frac{I}{mg\,l}}$$

  and moment of inertia $I=I_G+ml^2$

  $I_G=mk^2, k=$ Radius of gyration

  $I=m\,(k^2+l^2)$

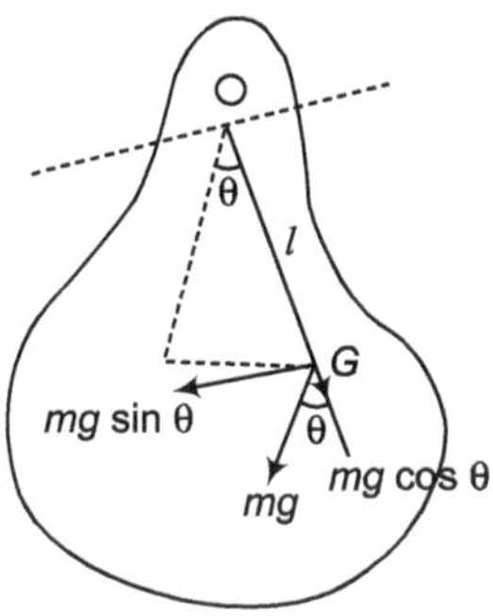

Physical pendulum

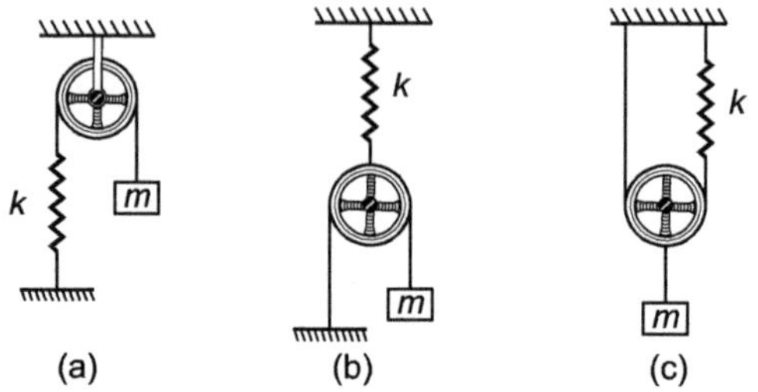

Pulley spring-block system

$$\text{(a) } T = 2\pi\sqrt{\frac{m}{k}} \qquad \text{(b) } T = 2\pi\sqrt{\frac{4m}{k}} \qquad \text{(c) } T = 2\pi\sqrt{\frac{m}{4k}}$$

- $F = kx_0 + mg\sin\theta$

$$2F = mg \quad \Rightarrow \quad T = 2\pi\sqrt{\frac{5m}{4k}}$$

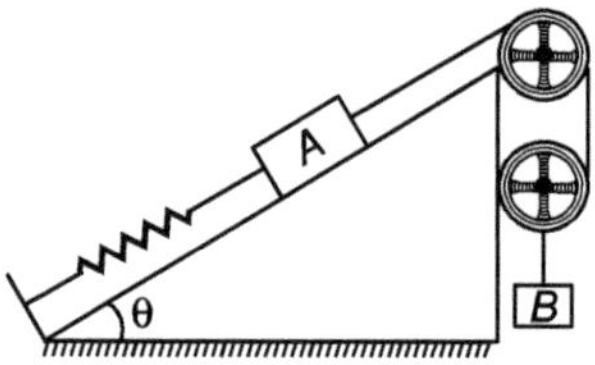

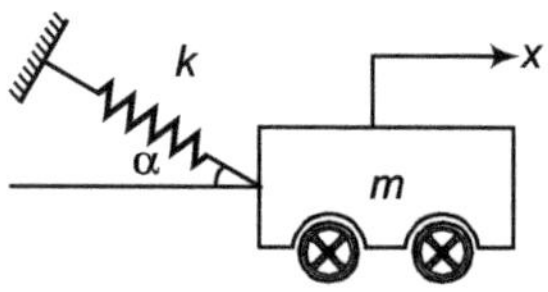

Block spring system

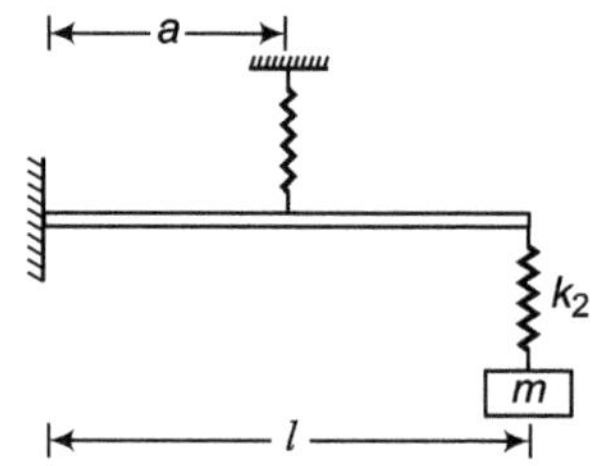

Spring wheel system

- $k_{eq} = k\cos^2\alpha$

$$T = 2\pi\sqrt{\frac{m}{k\cos^2\alpha}} \quad \Rightarrow \quad k_{1e} = k_1\frac{a^2}{l^2}$$

$$\frac{1}{k_e} = \frac{1}{k_{1e}} + \frac{1}{k_2} = \frac{l^2}{k_1 a^2} + \frac{1}{k_2}$$

$$k_e = \frac{k_1 k_2 a^2}{k_1 a^2 + k_2 l^2}$$

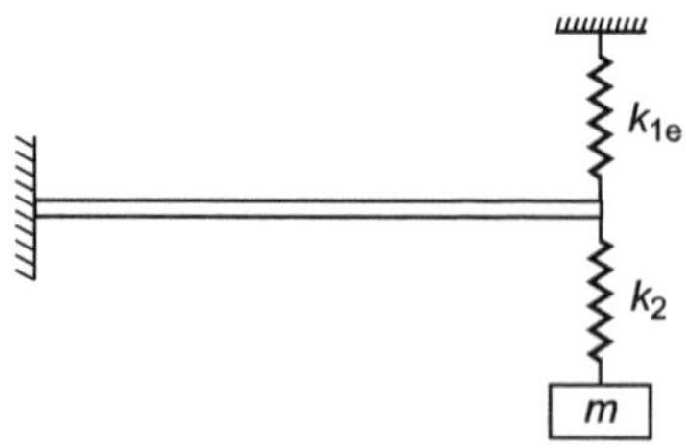

Rod block-spring system

Time period $T = 2\pi\sqrt{\frac{m}{k_e}}$ and natural Frequency $\omega_n = \sqrt{\frac{k_1 k_2 a^2}{m(k_1 a^2 + k_2 l^2)}}$

Geared torsional system can be shown in the figure

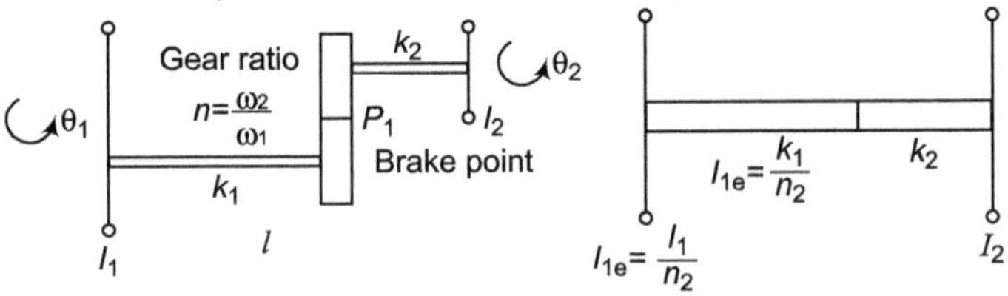

Geared torsional system

$$I_{1e} = \frac{I_1}{n^2}$$

$$I_{1e} = \frac{I_1}{n^2} = I_1 \frac{\omega_1^2}{\omega_2^2}$$

$$k_{1e} = \frac{1}{n^2} k_1 = k_1 \frac{\omega_1^2}{\omega_2^2}$$

Damping

Any influence which tends to dissipate the energy of a system.

Single degree of freedom system with viscous damping can be shown in the figure below

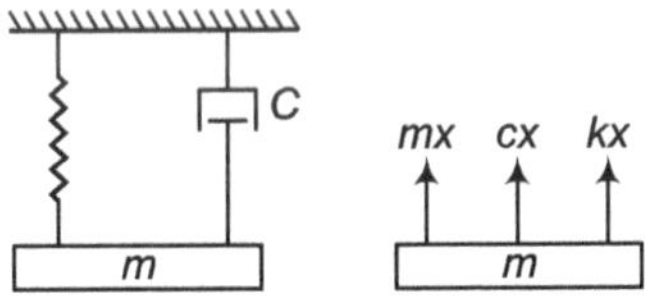

1-Degree of freedom system

$$m\ddot{x} + c\dot{x} + x = 0$$

$$\ddot{x} + \frac{c}{m}\dot{x} + \frac{k}{m}x = 0$$

$$\alpha^2 + \frac{c}{m}\alpha + \frac{k}{m} = 0$$

$$\alpha_{12} = \frac{-c}{2m} \pm \sqrt{\left(\frac{c}{2m}\right)^2 - \left(\frac{k}{m}\right)^2}$$

Damping Factor or Damping Ratio

It is the ratio of actual to critical damping coefficient

$$\xi = \sqrt{\frac{(c/2m)^2}{k/m}} = \frac{c}{2\sqrt{mk}}$$

Damping Coefficient

Coefficient of proportionality between the damping force and relative velocity

$$c = 2\,\xi\sqrt{mk} = 2\,\xi\, m\omega_n,\quad \omega_n = \frac{k}{m} = 2\,\xi\,\frac{l}{\omega_n}$$

- $\xi = 1$, the damping is known as critical under critical damping condition. Critical damping coefficient

$$c_c = 2\sqrt{mk}$$

$$\Rightarrow \quad c_c = 2m\,\omega_n$$

- $\xi > 1$ *i.e.*, the system is over damped

$$\alpha_{1,2} = (-\,\xi \pm \sqrt{(\xi^2 - 1)})\,\omega_n$$

$$x = Ae^{(-\,\xi + \sqrt{\xi^2 - 1})\,\omega_n t} + Be^{(-\,\xi - \sqrt{\xi^2 - 1})\,\omega_n t}$$

Motion is a periodic (non-oscillatory). In practice, no mechanical systems have over damping.

- $\xi < 1$ *i.e.*, system is underdamped.

$$\alpha_{1,2} = (-\,\xi \pm i\sqrt{1 - \xi^2})\omega_n$$

$$x = Ae^{(-\,\xi + i\sqrt{1-\xi^2})\omega_n t} + Be^{(-\,\xi - i\sqrt{1-\xi^2})\,\omega_n t}$$

- If the system oscillates with frequency

Damped frequency

$$\omega_d = \sqrt{1 - \xi^2}\;\omega_n$$

(ω_d is always less than ω_n)

The above equation can be written as

$x = X\,e^{-\xi\omega_n t}\,(\sin\omega_d t + \phi)$

X = constant

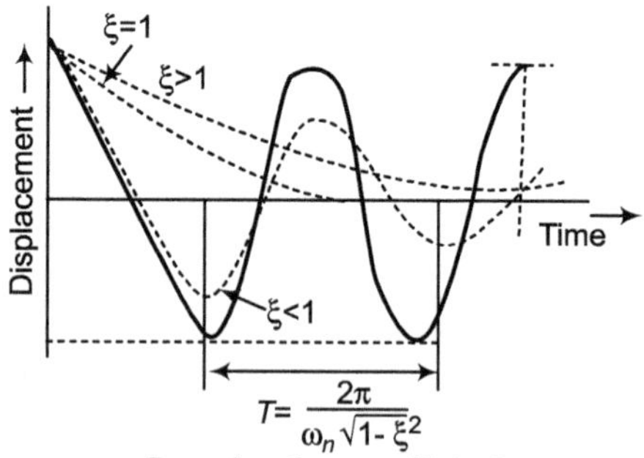

Damping-time coefficient

Linear frequency $f_d = \dfrac{\omega_d}{2\pi}$; Time period $T_d = \dfrac{2\pi}{\omega_d}$

where, C = actual damping coefficient

C_c = critical damping coefficient

ω_d = damped frequency

ξ = damping factor

ω_n = natural frequency

Logarithmic Decrement

In an underdamped system, natural logarithmic ratio of two successive oscillations is called logarithmic decrement (constant).

$$\text{Since, } \frac{X_n}{X_{n+1}} = e^{\xi\omega_n T_d} = \frac{X_0}{X_1} = \frac{X_1}{X_2} = \frac{X_2}{X_3}$$

Logarithmic decrement,

$$\delta = \ln\left(\frac{X_n}{X_{n+1}}\right) = \ln e^{(\xi\,\omega_n T_d)}\ \delta = \xi\omega_n T_d$$

or $$\delta = \frac{2\pi\xi}{\sqrt{1-\xi^2}} \quad \text{or} \quad \delta = \frac{1}{n}\ln\left(\frac{X_0}{X_n}\right)$$

Forced Vibration

Equation of forced vibration can be given as

$$m\ddot{x} + kx = F_0 \sin \omega t$$

$$x = X \sin(\omega_n t + \phi) + \frac{F_0/k}{1-\left(\frac{\omega}{\omega_n}\right)^2} \sin \omega t$$

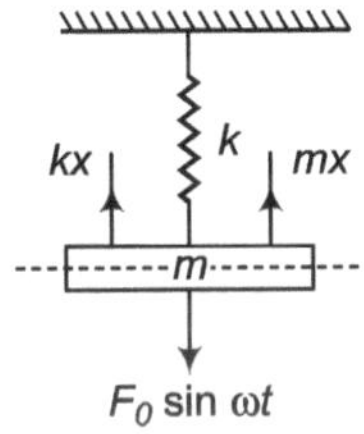

Forced vibration

$$m\ddot{x} + c\dot{x} + kx = F_0 \sin \omega t$$

$$x = Xe^{-\xi\omega_n t} \sin(\omega_d t - \phi_1) + \frac{F_0 \sin(\omega t - \phi)}{\sqrt{(k - m\omega^2)^2 + (c\omega)^2}}$$

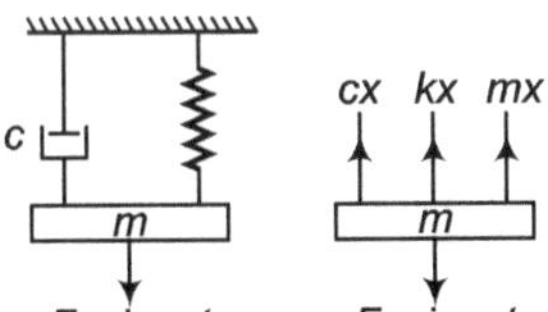

In the case of steady state response first term zero ($e^{-\infty} = 0$).

$$x = \frac{F_0}{\sqrt{(k - m\omega^2)^2 + (c\omega)^2}} \sin(\omega t - \phi)$$

The amplitude of the steady state response is given by

$$A = \frac{F_0}{\sqrt{(k - m\omega^2)^2 + (c\omega)^2}}$$

$$A = \frac{F_0/k}{\sqrt{\left[1-\left(\frac{\omega}{\omega_n}\right)^2\right]^2 + \left(2\,\xi\frac{\omega}{\omega_n}\right)^2}} \quad \tan\phi = \left(\frac{c\omega}{k - m\omega^2}\right)$$

Magnification Factor

Ratio of the amplitude of the steady state response to the static deflection under the action force F_0 is known as magnification factor.

$$MF = \frac{\dfrac{F_0}{\sqrt{(k - m\omega^2)^2 + (c\omega)^2}}}{F_0/k} = \frac{k}{\sqrt{(k - m\omega^2)^2 + (c\omega)^2}}$$

$$= \frac{1}{\sqrt{\left[1 - \left(\dfrac{\omega}{\omega_n}\right)^2\right]^2 + \left(2\xi\dfrac{\omega}{\omega_n}\right)^2}}$$

Let frequency ratio $r = \dfrac{\omega}{\omega_n}$

$$\text{MF [Function } H(\omega)] = \frac{1}{\sqrt{(1-r^2)^2 + (2\xi r)^2}}$$

For small values of damping, the peak can be assumed to be at $\omega = \omega_n$ which define the quality factor.

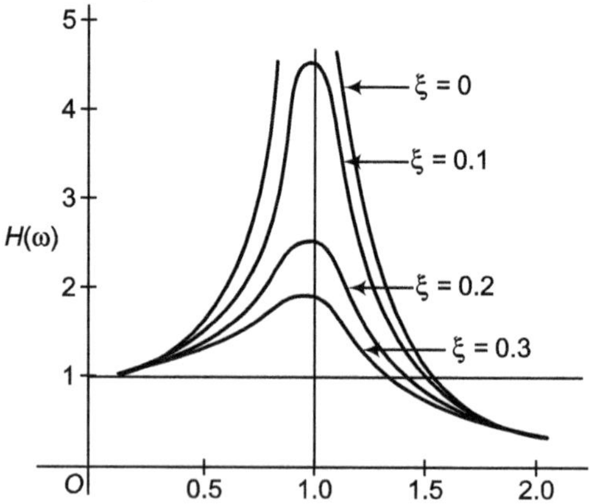

Response of the damping ratio

$$H(\omega)_{max} = \frac{1}{2\xi}$$

For $\xi \neq 0$, the peak amplitudes occur at

$$\frac{d}{dr^2}[(1 - r^2) + (2\xi r)^2] = 0\,,\; r = \sqrt{1 - 2\xi^2}$$

$$r = \frac{\omega}{\omega_n}$$

Transmissibility

Transmissibility is defined as the ratio of the force transmitted to the force applied. It is measure of the effectiveness of the vibration isolating material.

Transmissibility $\varepsilon = \dfrac{F_t}{F_0} = \dfrac{\sqrt{1+(2\xi r)^2}}{\sqrt{(1-r^2)^2+(2\xi r)^2}}$

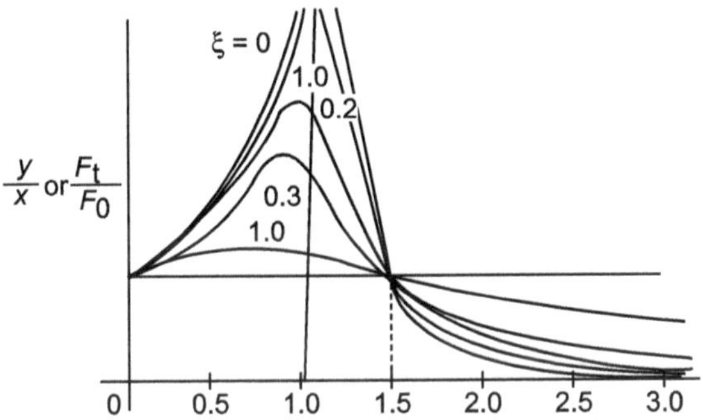

Transmissibility curve using damping ratio

At resonance $r = 1;\ \varepsilon = \dfrac{\sqrt{1+(2\xi)^2}}{2\xi}$

When no damper is used $\xi = 0$

$$\varepsilon = \frac{1}{\pm\left[1-\left(\dfrac{\omega}{\omega_n}\right)^2\right]}$$

Frequency response and phase relationship of single degree of freedom system with base excitation transmissibility and phase relatively of a system.

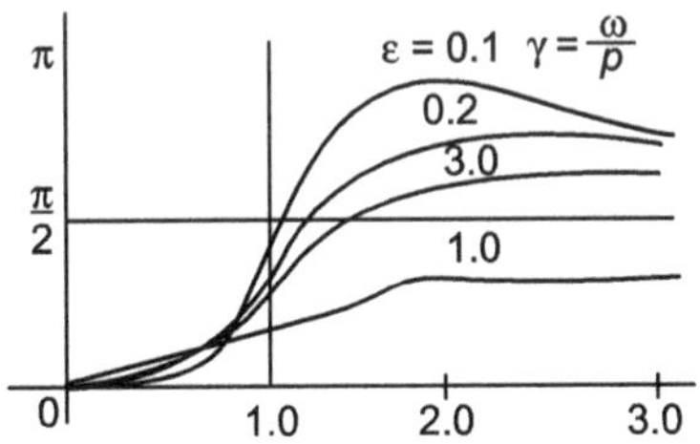

Transmissibility curve with $\xi = 0.1$ to 1

Key Points

- At $\omega = \sqrt{2}\,\omega_n$ $(\varepsilon = 1)$, the response is same as the magnitude of the excitation amplitude for the values of damping.
- For $\omega > \sqrt{2}\omega_n$, $(\varepsilon < 1)$, the maximum experiences lower amplitude of vibration than the base excitation.
- For $\omega < \sqrt{2}\,\omega_n$ $(\varepsilon > 1)$, $F_t > F_0$ damping is not useful, actually it increases the amplitude for a given system.
- At resonance $\omega = \omega_n$, the phase angle is not 90°.

A Single Degree of Freedom System Excited by Support Motion

$$\frac{Y}{X} = H(\omega) = \sqrt{\frac{1+(2\xi r)^2}{(1-r)^2+(2\xi r)^2}}$$

$$\tan\phi = \frac{2\xi r^3}{1-r^2+(2\xi r)^2}$$

$H(\omega)$ is maximum when

$$r = \frac{1}{2\xi}\sqrt{\sqrt{1+8\xi^2}-1}$$

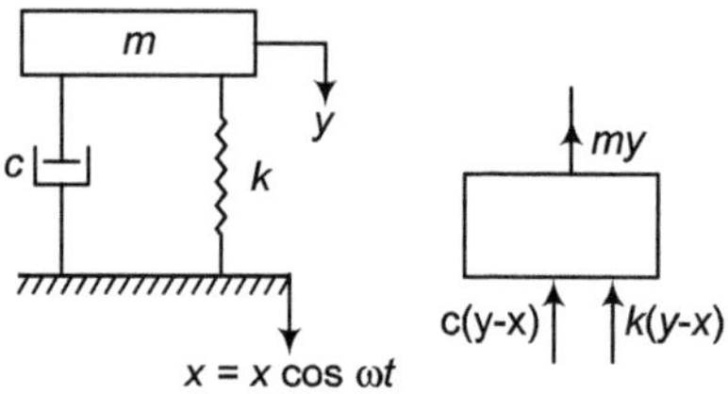

Support motion diagram

Critical Speed

Critical or whirling or whipping speed is the speed at which the shaft tends to vibrate violently in transverse direction.

For equilibrium, $ky = m\,(y+e)\,\omega^2$

$$y = \frac{e}{\left(\frac{\omega_n}{\omega}\right)^2 - 1}$$

$y \to \infty$, when $\omega_n = \omega$

Critical speed,

$$\omega_c = \omega_n = \sqrt{\frac{k}{m}} = \sqrt{\frac{g}{\Delta}}$$

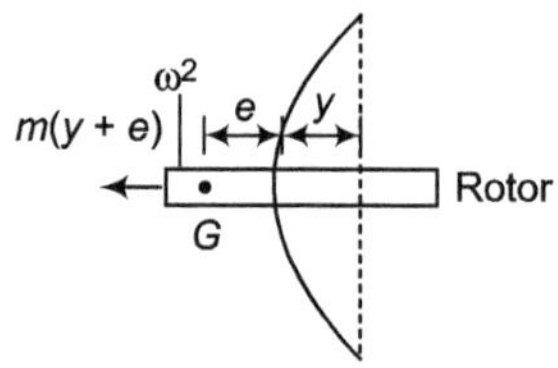

Shaft with vibrate

where, ω = Angular velocity of shaft

k = Stiffness of shaft

e = Initial eccentricity of centre of mass of rotor

m = Mass of rotor

y = Additional of rotor due to centrifugal force

Dynamic force on the bearings, $ky = m\,\omega_n^2\,y$

Critical speed for Simple Shaft

Bending Critical Speed We can also write function as total displacement

$$r_\omega = Re^{i(\omega t - \phi)}$$

where, $R = \dfrac{m\omega^2 a}{(k - m\omega^2) + (c\omega)^2}$

Hence, dynamic magnifier and phase angle.

$$H(\omega) = \frac{R}{a} = \frac{r^2}{\sqrt{(1-r^2)^2 + (2\xi r)^2}}$$

$$\theta = \tan^{-1}\left(\frac{2\xi r}{1-r^2}\right)$$

For an undamped rotor resonance occurs,

when $\omega = \omega_n$

Also at resonance, $\phi = 90^\circ$

$$m\omega^2 a = c\omega R$$

$$\frac{R}{a} = \frac{m\omega}{c} = \frac{1}{2\xi}$$

or $$R_{max} = \frac{a}{2\xi}$$

Distance of the centre of gravity from the bearing axis or whirl amplitude

$$R_G^2 = R^2 + a^2 + 2Ra\cos\phi$$

and $$R_G = a\sqrt{\frac{1+(2\xi r)^2}{(1+r^2)^2 + (2\xi r)^2}}$$

Critical Speed for Multi-mass System

Bending Critical speed

The synchronous whirl frequency increases with the rotational speed linearly and can be represented $1\times$ rev excitation frequency, whenever, this excitation line intersects the natural frequencies, critical speeds occur.

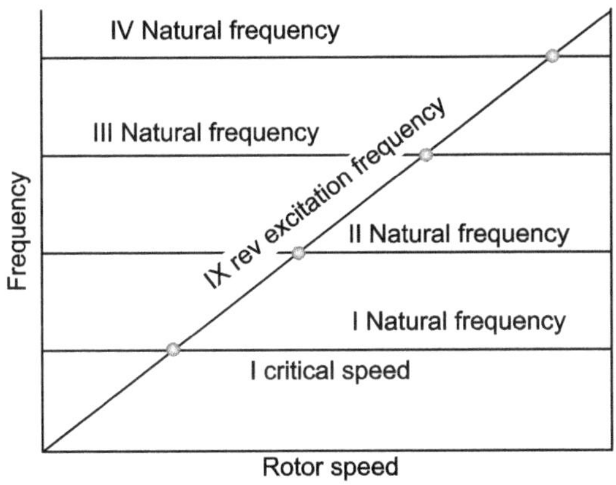

Campbell diagram

Durkerley's lower bound approximation,

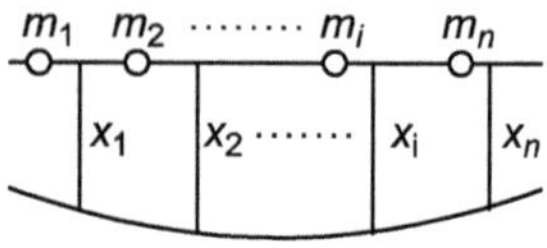

Durkerley's lower bound approximation

Considering n degree of freedom,

$$\frac{1}{\omega_n^2} = \frac{1}{\omega_{n_{11}}^2} + \frac{1}{\omega_{n_{22}}^2} \ldots + \frac{1}{\omega_{nn}^2} = \Sigma \alpha_{ii} m_i$$

where influence coefficient $\alpha_{ii} = \dfrac{1}{k_{ii}}$

Rayleigh's upper bound approximation, $\omega_n^2 = g \dfrac{\Sigma m_i x_i}{\Sigma m_i x_i^2}$

4

Machine Design

Cotter *and* Knuckle Joints

A cotter joint is a temporary fastening and is used to connect rigidly two co-axial rods or bars which are subjected to axial tensile or compressive forces. It is used in steam engine to connect piston rod end crosshead of a steam engine, piston rod and tail or pump rod.

Socket *and* Spigot Cotter Joints

In a socket and spigot cotter joint, one end of the rods is provided with a socket type of end as shown in figure and the other end of the rod is inserted into a socket. The end of the rod which goes into a socket is also called spigot.

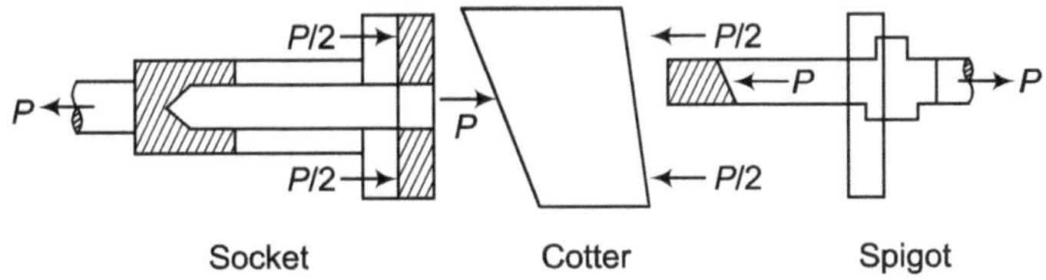

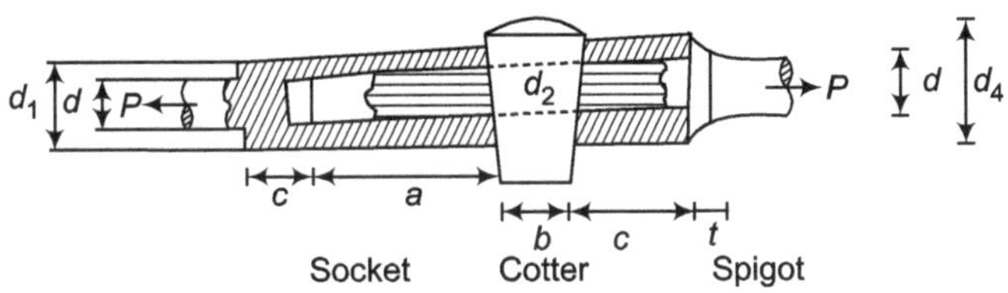

Socket cotter and spigot layout

Failure

In order to design the cotter joint and find out the above dimensions, failure in different parts and different cross-sections are considered.

Design of Socket and Spigot Cotter Joint

Let, P = Load carried by the rods
d = Diameter of the rods
d_1 = Outside diameter of socket
d_2 = Inside diameter of socket
d_3 = Outside diameter of spigot collar
t_1 = Thickness of spigot collar
d_4 = Diameter of socket collar
C = Thickness of socket collar
b = Mean width of cotter
t = Thickness of cotter
l = Length of cotter
a = Distance from the end of the slot to the end of rod
σ_τ = Permissible shear stress for the rods material
t = Permissible shear stress for the cotter material
σ_c = Permissible crushing stress for the cotter material

Failures in Socket and Spigot Cotter Joints

Failure Cases	Tensile Force
Failure of the rod in tension	$P = \frac{\pi}{4} \times d^2 \times \sigma_t$
Failure of spigot in tension across the weakest section	$P = \left[\frac{\pi}{4}(d_2)^2 - d_2 \times t\right]\sigma_t$
Failure of the rod or cotter in crushing	$P = d_2 \times t \times \sigma_c$
Failure of the socket in tension across the slot	$P = \left\{\frac{\pi}{4}[(d_1)^2 - (d_2)^2] - (d_1 - d_2)t\right\}\sigma_t$
Failure of cotter in shear	$P = 2b \times t \times \tau$
Failure of the socket collar in crushing	$P = (d_4 - d_2)t \times \sigma_c$
Failure of socket end in shearing	$P = 2(d_4 - d_2)c \times \tau$
Failure of rod end in shear	$P = 2a \times d_2 \times \tau$
Failure of spigot collar in crushing	$P = \frac{\pi}{4}[(d_3)^2 - (d_2)^2]\sigma_c$
Failure of the spigot collar in shearing	$P = \pi d_2 \times t_1 \times \tau$

Failures in Sleeve and Cotter Joint

Failure Cases	Tensile Force
Failure of the rods in tension	$P = \frac{\pi}{4} \times d^2 \times \sigma_t$
Failure of the rod in tension across the weakest section	$P = \left[\frac{\pi}{4}(d_2)^2 - d_2 \times t\right]\sigma_t$
Failure of the rod or cotter in crushing	$P = d_2 \times t \times \sigma_c$
Failure of sleeve in tension across the slot	$P = \left\{\frac{\pi}{4}[(d_1)^2 - (d_2)^2] - (d_1 - d_2)t\right\}\sigma_t$
Failure of cotter in shear	$P = 2b \times t \times \tau$
Failure of rod end in shear	$P = 2a \times d_2 \times \tau$
Failure of sleeve end in shearing	$P = 2(d_1 - d_2)c \times \tau$

Failures Gib and Cotter Joint

Failure Cases	Tensile Force
Failure of the strap in tension	$P = 2B_1 \times t_1 \times \sigma_t$
Failure of the gib and cotter in shearing	$P = 2B \times t \times \tau$
Failure of the rod in tension	$P = x^2 \times \sigma_t$
Failure of the strap end in tension at the location of gib and cotter	$P = 2[x \times t_1 - t_1 \times t]\sigma_t$
Failure of the strap or grib in crushing	$P = 2t_1 \times t \times \sigma_c$
Failure of the strap end in shearing	$P = 4l_2 \times t_1 \times \tau$

where, P = Load carried by the rod
d = Diameter
B_1 = Width of the strap
B = Total width of gib and cotter
t = Thickness of cotter
t_1 = Thickness of the strap at the thinnest part
σ_t = Permissible tensile stress for the material of the strap
τ = Permissible shear stress for the material.

Knuckle Joint

- It is used to connect two rods whose axis either coincide or intersect and lie in one plane.
- This joint generally found in the link of a cycle chain tie rod joint for roof truss, valve rod joint with eccentric rod tension link in bridge structure, lever and rod connection of various types.
- It is sometimes also called forked pin joint.

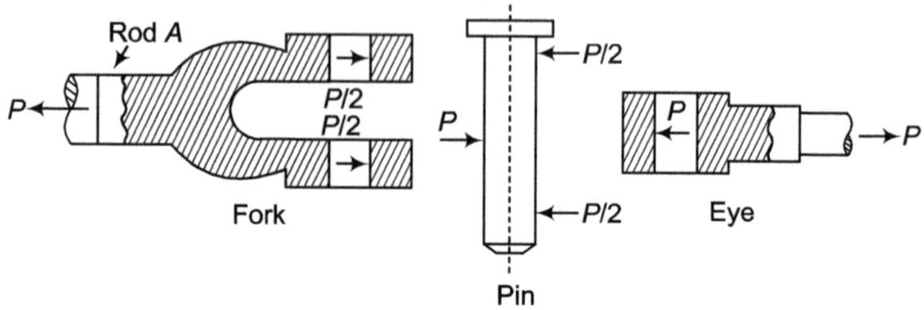

Knuckle joint

Failures in Knuckle Joint

Failure Case	Tensile Force
Failure of the solid rod in tension	$P = \frac{\pi}{4} \times d^2 \times \sigma_t$
Failure of the knuckle pin in shear	$P = 2\ \frac{\pi}{4} (d_1)^2 \tau$
Failure of the single eye or rod end in tension	$P = (d_2 - d_1) t \times \sigma_t$
Failure of the single eye or rod end in shearing	$P = (d_2 - d_1) t \times \tau$
Failure of the forked end in tension	$P = (d_2 - d_1)\, 2t_1 \times \sigma_t$
Failure of the forked end in crushing	$P = d_1 \times 2t_1 \times \sigma_c$

- To connect the transmission shaft to rotating machine elements like pulley, gear, sprocket or flywheel.
- Cotter and knuckle joints are not used for connecting shafts that rotate and transmit torque.

Welded Joints

- It is a **permanent joint**.
- When the two parts are joined by heating to a suitable temperature with or without application of pressure.
- It is lighter than riveted joint, great strength, maximum efficiency.

Welding Processes

If two rods have to joined then we have to adopt the welding process with which two or more rods can be joined together. *Some types of welding are given below*

Fusion Welding

In case of fusion welding, the parts to be jointed are held in position while the molten metal is supplied to the joint. The molten metal may come from the parts themselves (*i.e.,* parent metal) or filler metal which normally have the composition of the parent metal. The joint surface become plastic or even molten because of the heat from the molten filler metal or other source.

Thus, when the molten metal solidifies or fuses, the joint is formed. *The types of fusion welding are given below*

Thermit Welding

In thermit welding, a mixture of iron oxide and aluminium called thermit is ignited and the iron oxide is reduced to molten iron. The molten iron is poured into a mould made around the joint and fuses with the parts to be welded.

Gas Welding

A gas welding is made by applying the flame of an oxy-acetylene or hydrogen gas from a welding torch upon the surfaces of the prepared joint.

Electric Arc Welding

In electric arc welding, the work is prepared in the same manner as for gas welding. In this case the filler metal is supplied by metal welding electrode. The operator, with his eyes and face protected, strikes an arc by touching the work of base metal with the electrode. The base metal in the path of the arc stream is melted, forming a pool of molten metal, which seems to be forced out of the pool by the blast from the arc

Forge Welding

In this type of welding, first the parts are to be heated about a suitable temperature and then they hammered to form a joint.

Types of Welded Joints

There are two types of welded joint

- Lap joint or fillet joint
- Butt joint

Lap Joint or Fillet Joint

In lap joint, overlapping the plate and welding the edge of the plates takes place in welding process. The strength of different types of fillet joint can be given according to their welding process as

Strength of Parallel Fillet Joint

$$t = h\cos 45°, \quad t = 0.707\ h$$

Shear strength in parallel fillet weld,

$$\tau = \frac{P}{0.707\ hl} \quad \text{or } P = 0.707\ hl\tau$$

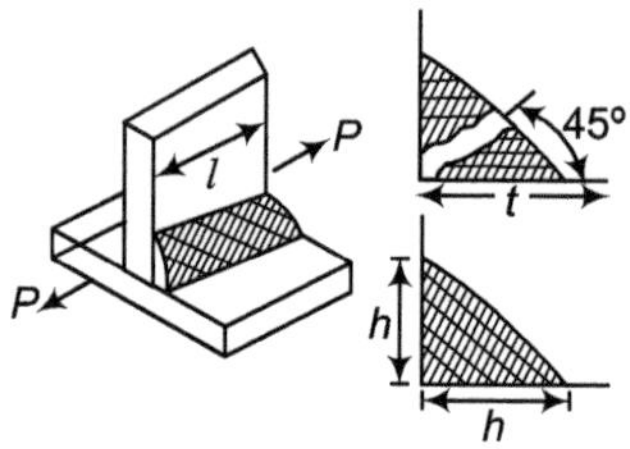

Parallel fillet joint

where, P = Tensile force on the plates

h = Leg of the weld

l = Length of the weld

τ = Permissible shear stress

For double parallel fillet weld, $P = 1.414\ hl\tau$

Strength of Transverse Fillet Weld

p = Throat area × Allowable tensile stress

$= 0.707\ s \times l \times \sigma_t$

For double transverse fillet joint

$P = 1.414\ hl\sigma_t$

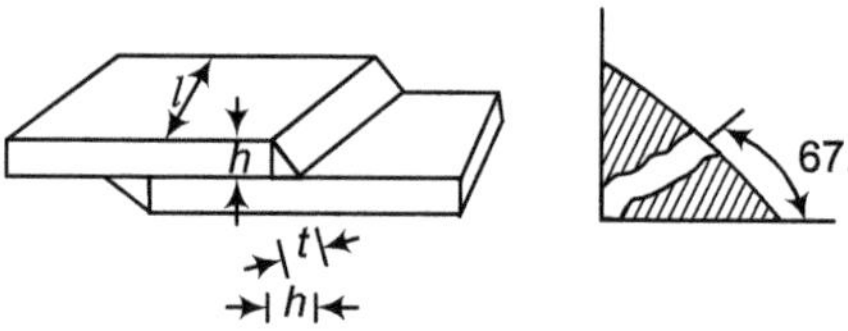

Transverse fillet weld

Special Cases of Fillet Welded Joint

The following are the considerations when the fillet welded joints are subjected to some forces (*i.e.*, torsion and bending)

Circular Fillet Weld Subjected to Torsion

Shear stress $\tau = \dfrac{2T}{\pi t d^2}$

or $\tau_{max} = \dfrac{2.83\ T}{\pi h d^2}$

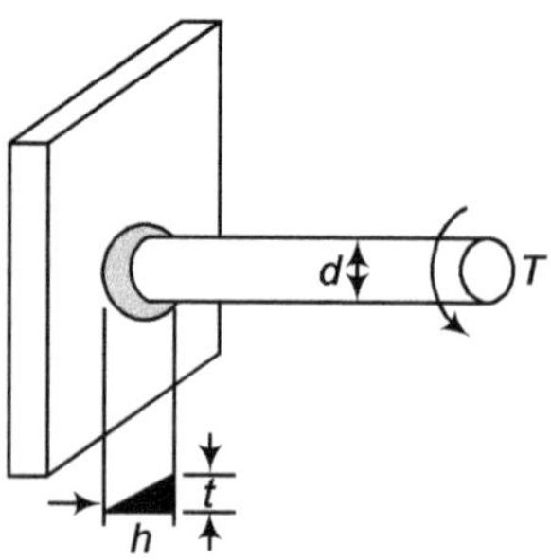

Torsion in circular fillet weld

where, T = Torque acting on rod

h = Size of weld

t = Throat thickness

Circular Fillet Weld Subjected to Bending Moment

Bending stress

$$\sigma_b = \frac{4\,M}{\pi\, t d^2}$$

$$\sigma_{b\,(\max)} = \frac{5.66\,M}{\pi h d^2}$$

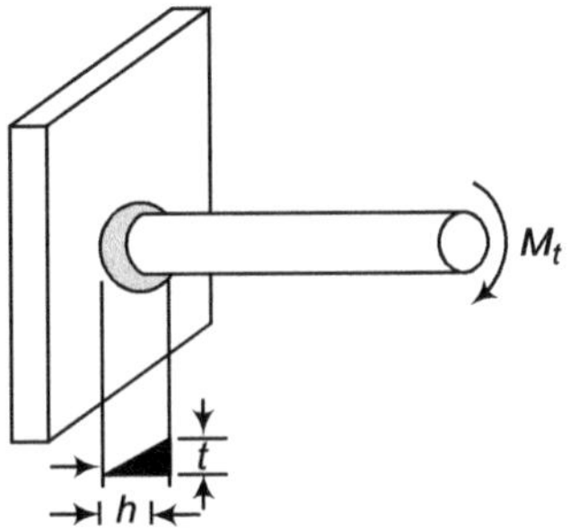

Bending moment in circular fillet weld

Long Fillet Weld Subjected to Torsion

Shear stress $\tau = \dfrac{3T}{tl^2}$

$$\tau_{\max} = \frac{4.242\,T}{hl^2}$$

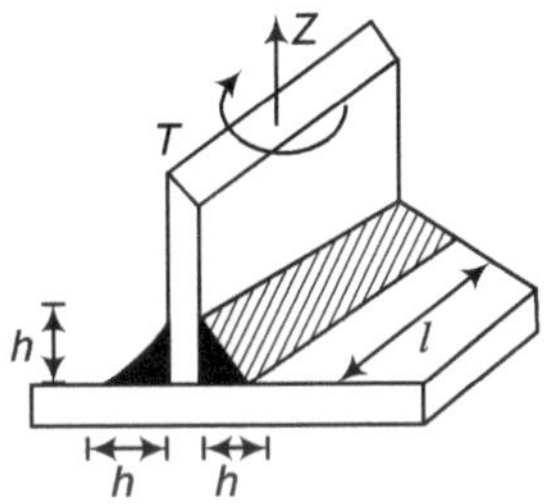

Torsion in long fillet weld

Butt Joint

- The butt joint is the joint between two plates lying on same plane.
- In this type of joint, cutting the members to the appropriate length and butting them together.

Strength of Butt Joint

For single V- butt joint, $P = t \times l \times \sigma_t$

For double V-butt joint, $P = (t_1 + t_2)\, l \times \sigma_t$

where, t_1 = Throat thickness at the top, t_2 = Throat thickness at the bottom

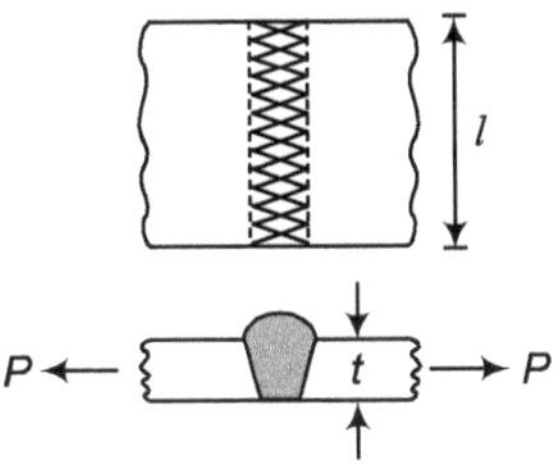

Single V-butt joint

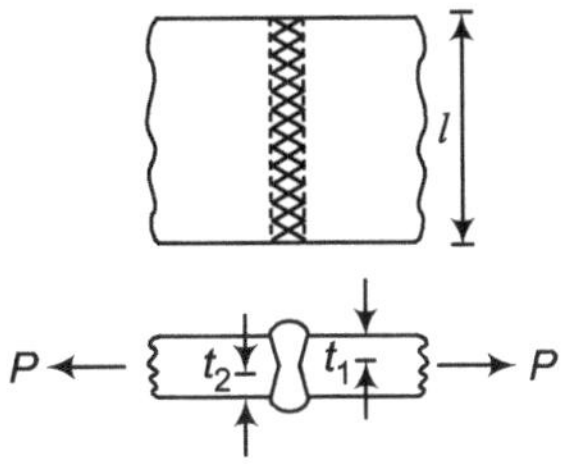

Double V-butt joint

Axial Loaded Unsymmetrical Welded Joints

Figure shows an angle section welded to a vertical beam by means of two parallel fillet weld 1 and 2, G is centre of gravity of the angle section.

$$l = l_1 + l_2$$

$$P = P_1 + P_2$$

$$l_1 y_1 = l_2 y_2$$

$$l_1 = \frac{l \times y_1}{y_1 + y_2} \quad \text{and} \quad l_2 = \frac{l \times y_2}{y_1 + y_2}$$

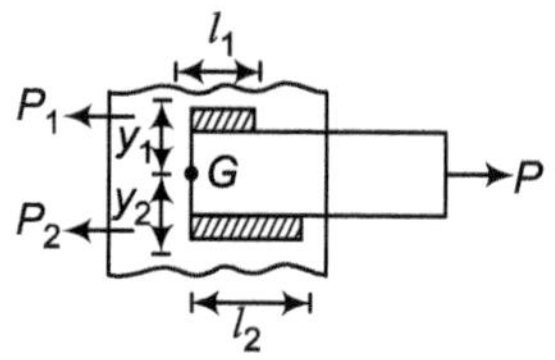

Unsymmetrical welded joints

Eccentric Loaded Welded Joints

When the shear and bending stresses are simultaneously present in a joint.

Maximum normal stress $\sigma_{t_{max}} = \frac{\sigma_b}{2} + \frac{1}{2}\sqrt{(\sigma_b)^2 + 4\tau^2}$

Maximum shear stress $\tau_{max} = \frac{1}{2}\sqrt{(\sigma_b)^2 + 4\tau^2}$

Direct or primary shear stress

$$\tau_1 = \frac{\text{Load}}{\text{Throat area}} = \frac{P}{2tl}$$

$$\tau_1 = \frac{P}{1.414\, hl}$$

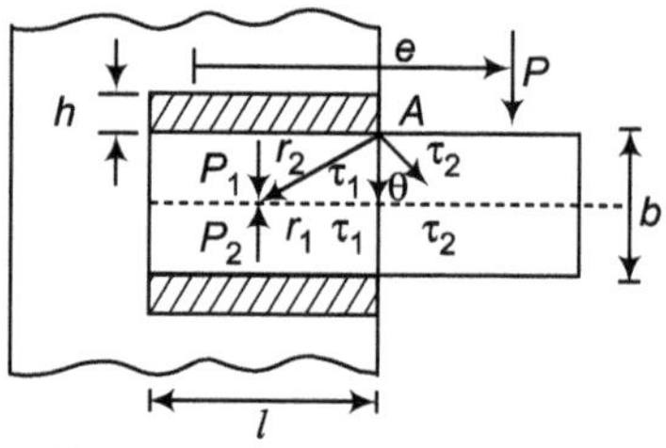

Eccentric loaded welded joints

Secondary shear stress $\tau_2 = \frac{T \times r_2}{J_G}$ and $T = P \times e$

$$\tau_2 = \frac{P \times e \times r_2}{J_G}$$

$$J_G = J_{G_1} + Ar_1^2, \quad J_G = A\left[\frac{l^2}{12} + r_1^2\right]$$

where, G_1 = Centre of gravity of weld

G = Centre of gravity of group of weld

Resultant shear stress at A $\tau_A = \sqrt{\tau_1^2 + \tau_2^2 + 2\tau_1\tau_2 \cos\theta}$

Threaded Joints

Separable joint of two or more machine parts that are held together by means of a threaded fastening such as bolt and nut is called threaded joint.

Bolts

- Bolts are generally subjected to shock and impact loads.
- Resilience (ability to absorb and release energy when loaded and unloaded) of the bolt can be increased by length of shank portion of bolt and reduce the shank diameter to the core diameter of threads.

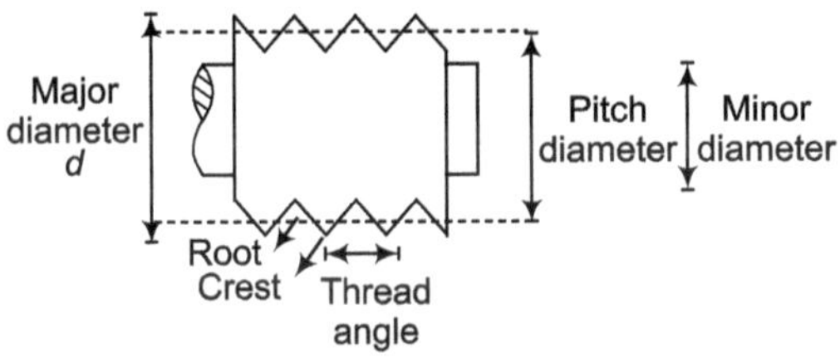

Terminology of thread

Key Points

- Through bolts used in threaded joints are simply a bolt or a nut type bolt.
- Top bolt is turned into threaded hole in one of the parts being connected and not into nut.
- Through bolt is turned into nut form a cap screws.
- Cylindrical rod threaded at both ends forms a stud.
- The locking devices are Jam nut and lock nut.
- Castle nuts used for joints that are subjected to vibrations mostly in automobile industry.
- Spring washers are used to prevent loosing of threads.

Strength of Bolted Joint

Maximum tensile stress in the bolt

$$\sigma_t = \frac{P}{\left(\frac{\pi}{4} d_c^2\right)}$$

where, d_c = Core diameter

$$\frac{\sigma_{yt}}{f_s} = \frac{P}{\frac{\pi}{4} d_c^2}$$

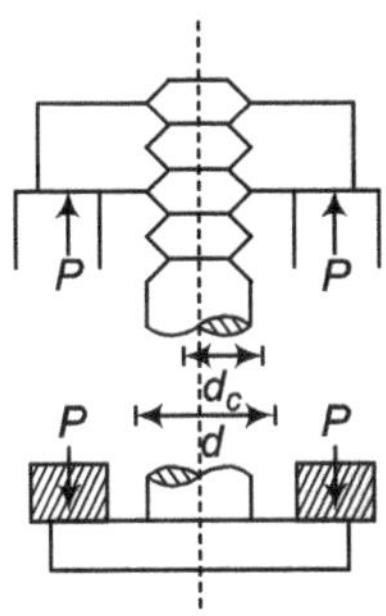

Bolted joint

Acting Force in Bolted Joints

The following are the considerations to find out the forces acting on a bolted joint

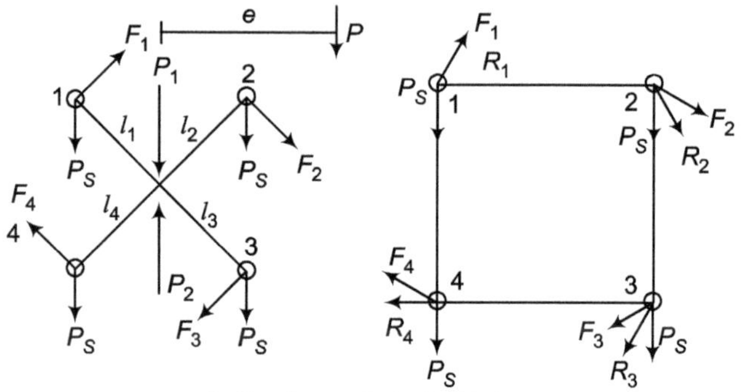

Acting forces in bolted joints

- Direct shear force $P_s = \dfrac{P}{n}$ acting parallel to load P.
- Secondary shear forces $F_1 = \dfrac{P_e \cdot l_1}{l_1^2 + l_2^2 + l_3^2 + l_4^2}$

  and $F_2 = \dfrac{P_e l_2}{l_1^2 + l_2^2 + l_3^2 + l_4^2}$

  where, n = Number of bolts $F_3 = \dfrac{P_e\ l_3}{l_1^2 + l_2^2 + l_3^2 + l_4^2}$

  and $F_4 = \dfrac{P_e\ l_4}{l_1^2 + l_2^2 + l_3^2 + l_4^2}$

- Resultant of forces $R_1 = \sqrt{(P_s)^2 + F^2 + 2\,P_s \times F_1 \cos\theta}$

  θ = Angle between the primary or direct shear load (P_s) and secondary shear load (F).

  Similarly, calculate R_2, R_3, R_4.

Torque Requirement for Bolt Tightening

$$M_t = \frac{P_i\, d_m}{2}\left(\frac{\mu \sec\theta + \tan\alpha}{1 - \mu \sec\theta \tan\alpha}\right)$$

where, P_i = Pretension in bolt, $d_m = 0.9\,d$
d = Nominal diameter

For ISO metric screw thread $\theta = 30°$

For ISO metric $\alpha = 2.5°$

Eccentric Load on Bracket with Circular Base

Let L_1, L_2, L_3, L_4 = Distance of bolt centre from tilting edge
R = Radius of the column flange
r = Radius of bolt pitch circle
w = Load per bolt per unit distance from the tilting edge
L = Distance of the load from the tilting edge

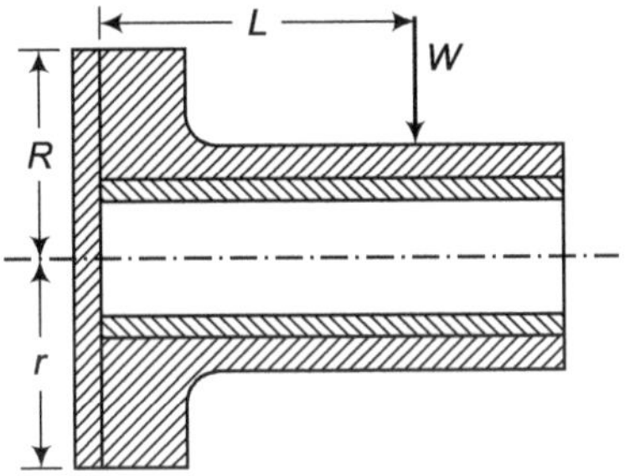

Eccentric Load w on bracket

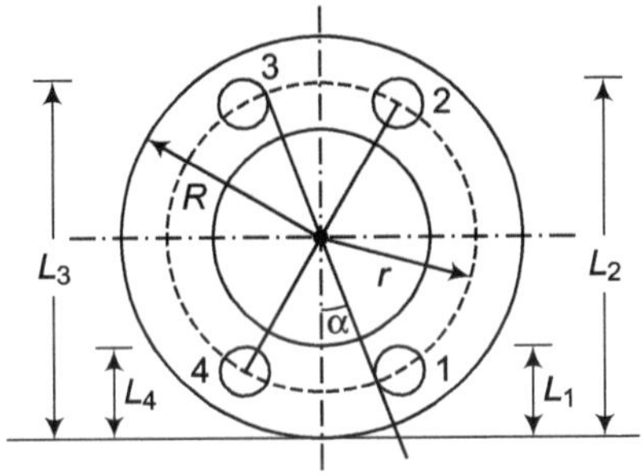

Circular base

$$L_1 = R - r\cos\alpha,\ \ L_2 = R + r\sin\alpha$$
$$L_3 = R + r\cos\alpha,\ L_4 = R - r\sin\alpha$$

If there are n number of bolts, then

$$\text{load in a bolt } w_{b_1} = \frac{2wL\,(R - r\cos\alpha)}{n\,(2R^2 + r^2)}$$

In above case when $n = 4$

$$w_b = \frac{w \cdot L\,(R - \alpha\cos\alpha)}{2\,(2R^2 + r^2)}$$

Maximum load in bolt is given by

$$(w_{b_1})_{\max} = \frac{2wL}{n}\left(\frac{R + r}{2R^2 + r^2}\right)$$

where, $\cos\alpha = -1$

Factor of Safety (FOS) in Bolted Joints

It is defined as the ratio of failure stress to allowable stress.

$$\text{FOS} = \frac{\text{Failure stress}}{\text{Allowable stress}}$$

- For ductile material, $\text{FOS} = \dfrac{S_{yt}}{\sigma}$
- For brittle material, $\text{FOS} = \dfrac{S_{ut}}{\sigma}$

where, S_{yt} = Yield strength of component material
S_{ut} = Ultimate tensile stress of components material
σ = Allowable stress.

Stress Concentration of Joints

It is defined as the localisation of high stresses due to the irregularities present in the component and abrupt changes of the cross-section.

Stress concentration Factor

It is defined as the ratio of highest value of actual stress near discontinuity to nominal stress obtained by elementary equations for minimum cross-section. It is denoted by k_t.

$$k_t = \frac{\text{Highest value of actual stress near discontinuity}}{\text{Nominal stress obtained by elementary equations}}$$

or
$$k_t = \frac{\sigma_{max}}{\sigma_0} = \frac{\tau_{max}}{\tau_0}$$

where, σ_0, τ_0 = Nominal stresses

The magnitude of stress concentration factor depends upon the geometry of the component.

In this case, $k_t = 1 + 2\,(a / b)$

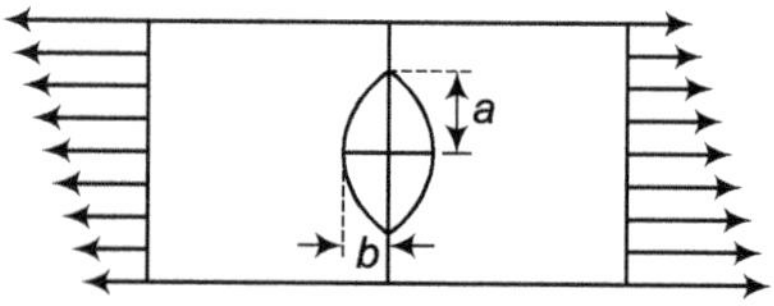

Design against fluctuating load

where, a = Semi-axis of ellipse perpendicular to the direction of load
b = Semi-axis of ellipse parallel to the direction of load

If $b = 0$ then, hole is like as very sharp crack then,
$k_t = \infty$

If $a = b$ then, hole becomes a circular hole then,
$k_t = 1 + 2 = 3$

Fluctuating Load

It is defined as the load, of which magnitude and direction both changes with respect to time. The stresses induced due to these forces are called fluctuating stresses.

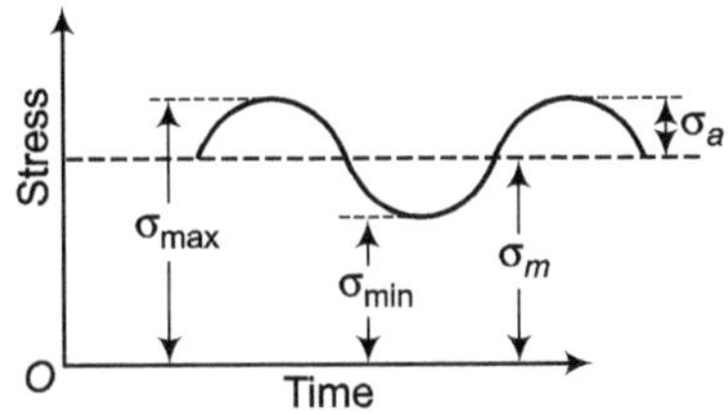

(a) Fluctuating load

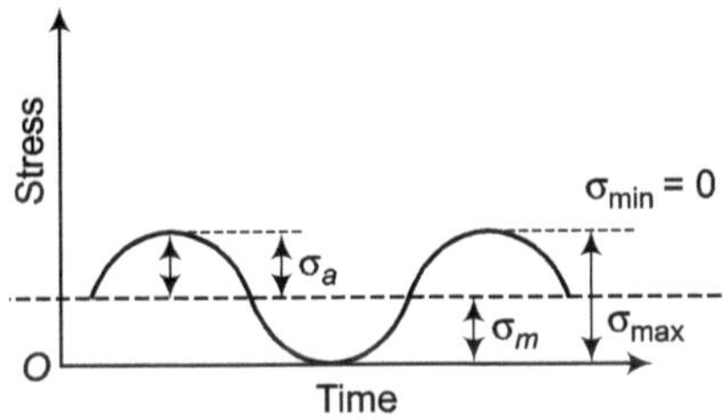

(b) Repeated load

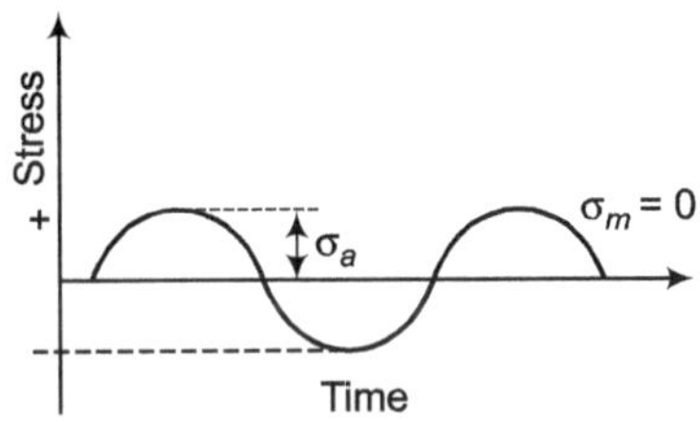

σ_m = Mass stress, σ_a = Stress amplitude

- Mass stress and stress amplitude

$$\sigma_m = \frac{1}{2}(\sigma_{max} + \sigma_{min})$$

$$\sigma_a = \frac{1}{2}(\sigma_{max} - \sigma_{min})$$

- For repeated stress, $\sigma_{min} = 0$, $\sigma_m = \frac{\sigma_{max}}{2}$ and $\sigma_a = \frac{\sigma_{max}}{2}$
- For reversed stress, $\sigma_m = \frac{1}{2}(\sigma_{max} - \sigma_{max}) = 0$

$$\sigma_a = \frac{1}{2}(\sigma_{max} + \sigma_{max}) = \sigma_{max}$$

Clutch

A clutch is a machine member used to connect a driving shaft to a driven shaft so that the driven shaft may be started or stopped at will, without stopping the driving shaft. A clutch is a mechanical device that provides for the transmission of power (and therefore usually motion) from one component (the driving member) to another (the driven member) when engaged, but can be disengaged.

Friction Clutch

The friction clutch is used to transmit power of shafts and machines which must be started and stopped frequently.

Friction surfaces of a clutch remain in contact to each other by applying an axial thrust or load *w*.

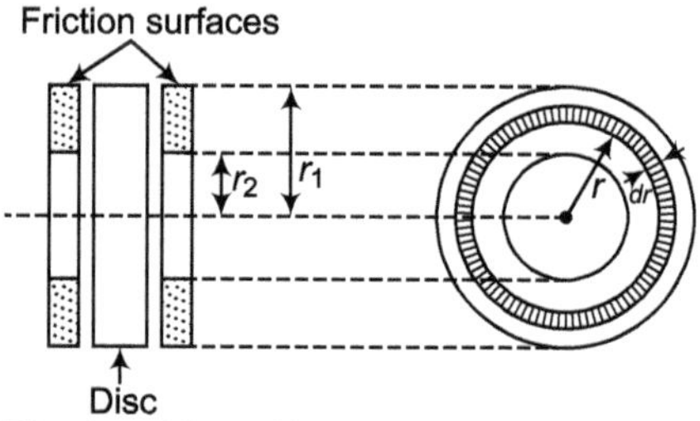

Disc is subjected between friction surfaces

where, r_2 = External radius of friction faces

m = Coefficient of friction

p = Intensity of axial pressure.

Considering Uniform Pressure

Let us consider r_1 and r_2 are the external and internal radius of friction faces respectively. The axial thrust w with which the friction surfaces held together, then the uniform pressure p can be evaluated as.

$$p = \frac{w}{\pi (r_1^2 - r_2^2)}$$

Total frictional torque given in this case,

$$T = \frac{2}{3} \mu w \left(\frac{r_1^3 - r_2^3}{r_1^2 - r_2^2} \right) = mw R_m$$

where, R_m = Mean radius of friction surfaces

$$R_m = \frac{2}{3} \left(\frac{r_1^3 - r_2^3}{r_1^2 - r_2^2} \right)$$

Considering Uniform Wear

Assume, p is the pressure on friction surfaces at a distance r from the axis of the clutch.

According **uniform wear theory**,

$$pr = \text{constant} = c \Rightarrow p = \frac{w}{2\pi r\,(r_1 - r_2)}$$

- Total frictional torque acting on clutch

$$T = \frac{1}{2}\mu w\,(r_1 + r_2) = \mu w R_m$$

where, R_m = Mean radius of friction surfaces = $\dfrac{r_1 + r_2}{2}$

- In uniform wear theory, **Maximum pressure** acts at the inner radius and minimum pressure acts at the outer radius.

$$p_{max} \times r_2 = c,\ p_{min} \times r_1 = c$$

$$\frac{p_{max}}{p_{min}} = \frac{r_1}{r_2}$$

- Average pressure on the friction surfaces

$$p_{av} = \frac{w}{\pi\,(r_1^2 - r_2^2)}$$

Multiple Disc Clutch

In multiple disc clutch, the inside discs are fastened to the driven shaft to permit axial motion. The outside discs are held by bolts and are fastened to the housing which is keyed to the driving shaft.

Number of pairs of contact surfaces

$$n = n_1 + n_2 - 1$$

where, n_1 = Number of discs on the driving shaft
n_2 = Number of discs on the driven shaft

- Total frictional torque acting on the frictional surface

$$T = \eta\mu w R_m$$

where, $R_m = \dfrac{2}{3}\left(\dfrac{r_1^3 - r_2^3}{r_1^2 - r_2^2}\right)$ [in case of uniform pressure]

$= \dfrac{1}{2}(r_1 + r_2)$ [in case of uniform wear]

where, r_1 and r_2 are outer and inner radii of the friction plates.

Cone Clutch

In cone clutch, the driver is keyed to the driving shaft by a sunk key and has an inside conical surface or face which exactly fits into the outside conical surface of the driven.

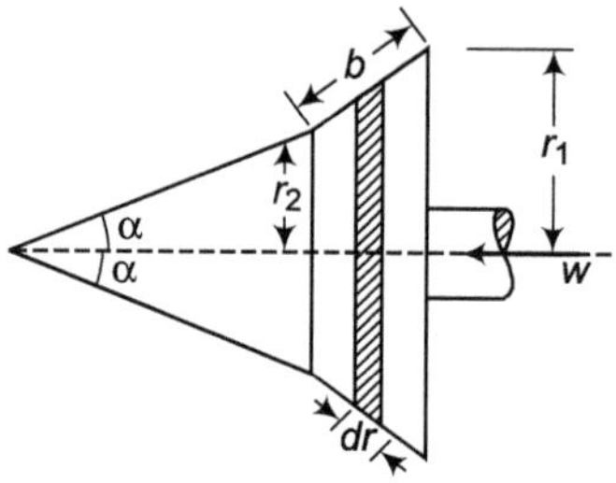

Cone clutch

- Total torque on the clutch,

$$T = \frac{2}{3}\mu w \operatorname{cosec}\left[\frac{r_1^3 - r_2^3}{r_1^2 - r_1^2}\right]$$

(for uniform pressure)

$$= \frac{1}{2}\mu w \operatorname{cosec}\alpha\,(r_1 + r_2)$$

(for uniform wear)

α = Semi angle of cone or face angle of the cone

$$w_n = \frac{w}{\sin\alpha}$$

where, w = Axial load or thrust

- Axial force required for engaging the clutch,

$$= w_n(\sin\alpha + \mu\cos\alpha)$$

- Axial force required to disengaged the clutch

$$w_d = w_n(\mu\cos\alpha - \sin\alpha)$$

- If face width b and mean radius of cone clutch is R_m.

Then, $$R_m = \left(\frac{r_1 + r_2}{2}\right)$$

Then, $$P_n = \frac{w}{2\,\pi R_m\, b\sin\alpha},\quad T = 2\pi\mu\cdot P_n R_m b$$

- Power transmitted by clutch $= \dfrac{2\pi NT}{60}$

where, N = Speed of shaft (rpm), T = Frictional torque

Centrifugal Clutch

It consists of number of shoes on the inside of a rim of the pulley. The outer surface of the shoes are covered with a friction material. These shoes, which can move radially in guides, are held against the boss on the driving shaft by means of spring. The springs exert a radially inward force which is assumed constant.

Let M = Mass of each shoe

r = Distance between centre of gravity of the shoe and the centre of spider

N = Running speed of the pulley

m = Coefficient of friction between the shoe and rim

R = Inner radius of pulley rim

n = Number of shoes in the pulley rim

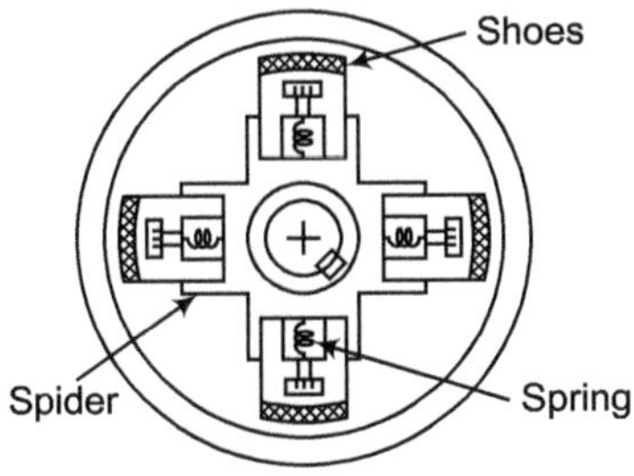

Centrifugal clutch

- When engagement starts then, the spring force equal to the centrifugal force acts at shoe at that time.

$$P_s = m\,(\omega_1)^2 r$$

where, ω_1 is the speed of rim when engagement starts.

- Centrifugal force acting on each shoe at running speed $P_c = m\omega^2 r$

where, $\omega = \left(\dfrac{2\pi N}{60}\right)$

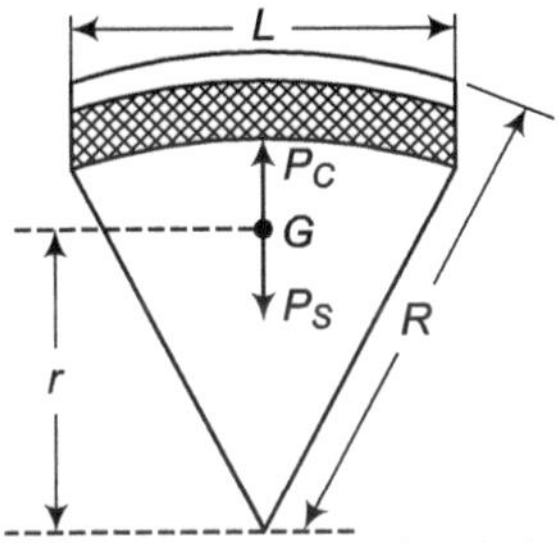

A section view of centrifugal clutch

- Friction force acting on each shoe = $\mu\,(P_c - P_s)$

The direction of force is perpendicular to the radius of the rim pulley.

- Frictional torque on each shoe = $\mu\,(P_c - P_s) \times R$

$$\text{Total torque transmitted} = \text{Number of shoes} \times \mu\,(P_c - P_s)\,R$$

$$= n\mu\,(P_c - P_s)\,R$$

$$\text{Arc} = \text{Angle (in radian)} \times \text{Radius}$$

$$l = \theta R$$

where, area of contact = lb

- Force exerted on each shoe = plb

$\therefore$ $$P_c - P_s = lbp$$

where, l = Contact length of the shoe

b = Width of the shoe

p = Pressure intensity on shoe

q = Angle made by shoe at the centre of spider in radian

R = Contact radius of shoe = inside radius of the rim of the pulley

Brake

A brake is a device by means of which artificial frictional resistance is applied to a moving machine member, in order to retard or stop the motion of a machine. The most commonly brakes use friction to convert kinetic energy into heat, though other methods of energy conversion may be employed.

Single Block or Shoe Brake

It consists of a block or shoe which is passed against the rim of revolving brake wheel drum. The block is made of a softer material than the rim of the wheel.

Let, P = Force applied at the end of the lever

r = Radius of the wheel

R_N = Normal force pressing the block on the wheel

F_t = Tangential braking or frictional force acting at the contact sur face of the block and wheel

2θ = Angle of contact surface of the block

μ = Coefficient of friction

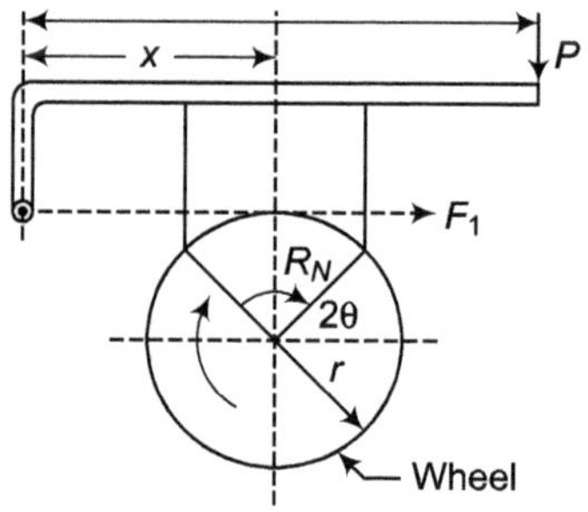

A schematic diagram of shoe brake

If the angle of contact is less than 60° then, it may be assumed that normal pressure or force between the block and the wheel is uniform.

The following are cases to describe the location of line of action of tangential braking force with respect to fulcurm point O.

Case I

When the line of action of tangential braking force passes through the fulcrum O of the lever.

Then, *there are two cases*

- If wheel is rotating in clockwise direction then, Free Body Diagram (FBD) of wheel and block is

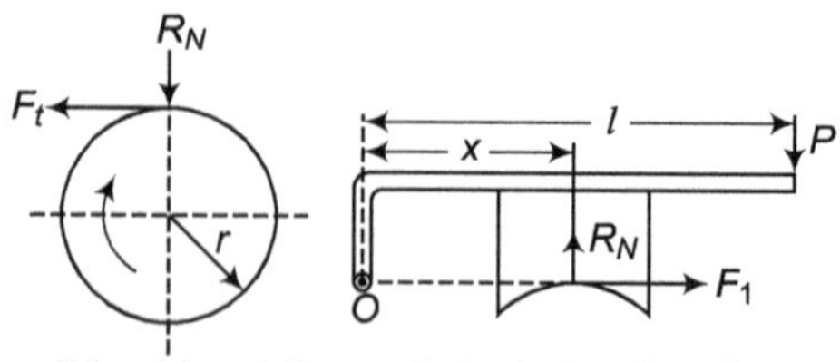

Wheel is rotating anti-clockwise direction

- If wheel is rotating in anticlockwise direction then, FBD of wheel and block is

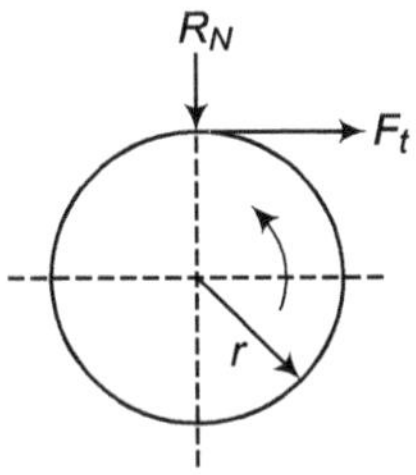

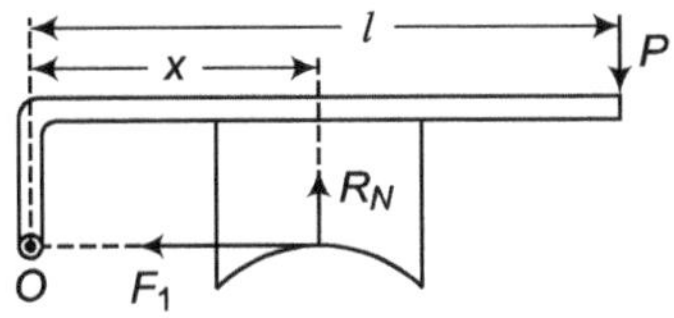

Wheel is rotating anti-clockwise direction

Now, taking moment about fulcrum O of the lever when wheel is rotating in clockwise direction.

$$R_N \times x = P \times l$$

$$R_N = \frac{Pl}{x}$$

- Braking force $= \mu R_N = \frac{\mu Pl}{x}$
- Braking torque $(T_B) = r \times$ Braking force

$$T_B = \frac{\mu Plr}{x}$$

When wheel is rotating in anticlockwise direction then, the braking torque is same as above

$$T_B = \frac{\mu Plr}{x}$$

***Case* II**

When the line of acting of the tangential braking force (F_t) passes through a distance a below the fulcrum O.

Then, *there are two cases*

If wheel rotates in clockwise direction then, the FBD of block is

$$R_N = \frac{Pl}{x + \mu a}$$

- Braking force $F_t = \mu R_N = \frac{\mu Pl}{x + \mu a}$
- Braking torque $T_B = \frac{\mu Plr}{x + \mu a}$

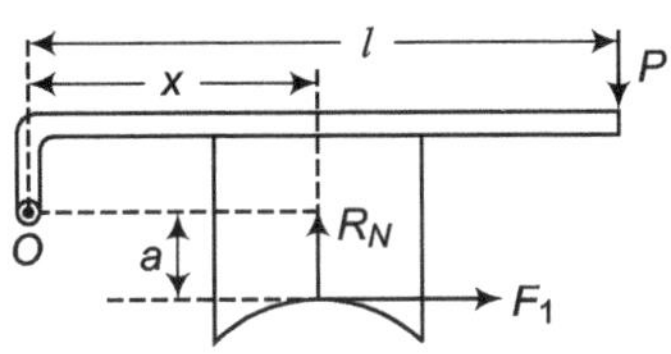

FBD of block

If wheel rotates in anticlockwise direction then, FBD of block is given as

$$\therefore \qquad R_N = \frac{Pl}{x - \mu a}$$

- **Braking force** $F_t = \dfrac{\mu Pl}{x - \mu a}$
- **Braking torque** $T_B = \dfrac{\mu Plr}{x - \mu a}$

(as $T_B = F_t \times r$)

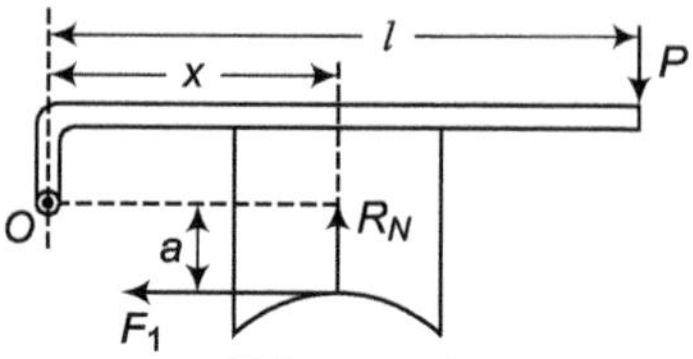

FBD of block

***Case* III**

When the line of action of tangential braking force (F_t) passes through a distance *a* above the fulcrum *O*.

Then, *there are two cases*

When wheel rotates in clockwise direction, then, FBD of block is Taking moment about point O,

we get $\quad R_N = \dfrac{Pl}{x - \mu a}$

- Braking force $F_t = \dfrac{\mu Pl}{x - \mu a}$
- Braking troque $T_B = \dfrac{\mu Plr}{x - \mu a}$

$(T_B = F_t \times r)$

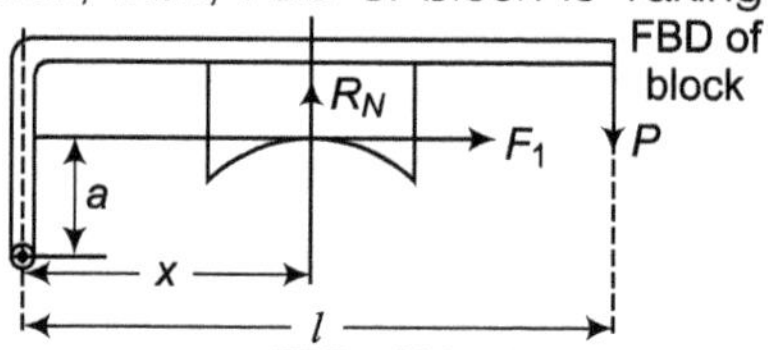

FBD of block

When wheel rotates in anticlockwise direction then, FBD of block is

$$R_N = \frac{Pl}{x - \mu a}$$

- **Braking force** $F_t = \dfrac{\mu Pl}{x + \mu a}$
- **Braking torque** $T_B = \dfrac{\mu Plr}{x + \mu a}$

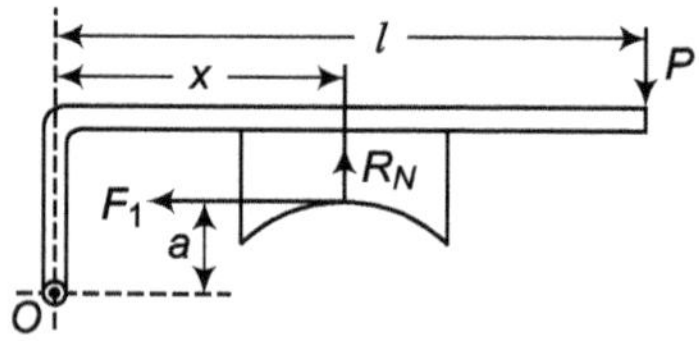

FBD of block

When the frictional force helps to apply the brakes then, such type of brakes are said to self energising brakes.

When *P* is negative or equal to zero then, these are known as self locking brakes.

Bearing pressure on shoe $P_b = \dfrac{R_N}{\text{Projected area}} = \dfrac{R_N}{(2r \sin\theta)\, w}$

where, w = Width of shoe

$2r \sin\theta$ = Projected length of shoe

Pivoted Block or Shoe Brake

When the angle of contact is greater than 60° then, the unit pressure normal to the surface of contact is less at ends than at the centre. In this case, the block or shoe pivoted to the lever gives the uniform wear of the brake lining in the direction of the applied force.

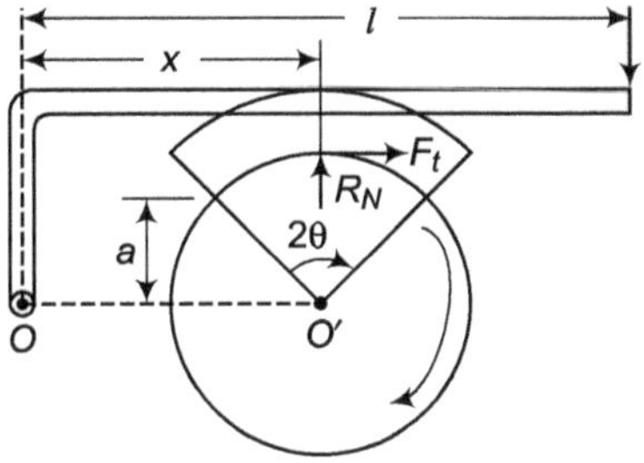

Shoe brake diagram

In this case,

Braking torque $T_B = \mu' R_N r$

$$= \frac{4\mu \sin\theta}{2\theta + \sin 2\theta} \qquad \text{(where, } 2\theta > 60^\circ\text{)}$$

where, μ' = Equivalent coefficient of friction
μ = Actual coefficient of friction

These brakes having more life and provides a higher braking torque.

Simple Band Brake

A band brake consists of a flexible band of leather, one or more ropes, or a steel lined with friction material, which embraces a part of the circumference of the drum is called **simple band brake**.

- When wheel rotates in the clockwise direction then, the end of the band attached to B is in tight side and end of the band attached to fulcrum O is in slack side.
- When wheel rotates in the anticlockwise direction then, the end of the band attached to B is in slack side and end of the band attached to the fulcrum O is in tight side.

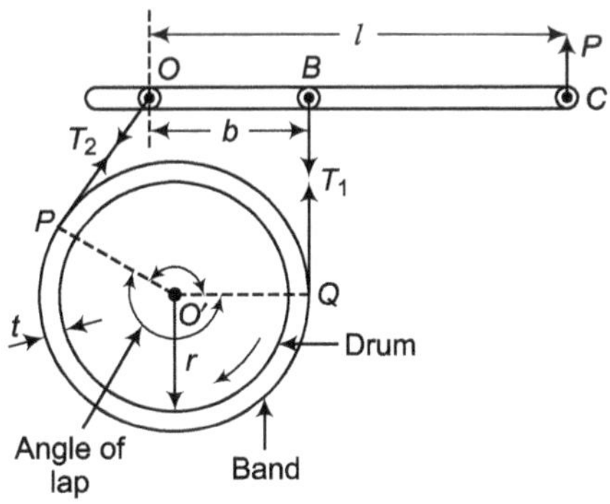

Simple band brake

Let, T_1 = Tension in the tight side of the band
T_2 = Tension in the slack side of the band
t = Thickness of the band
θ = Angle of lap
μ = Coefficient of friction between band and the drum

r = Radius of the drum
$r_e = r + t/2$ = Effective radius of the drum

We know,

$$\frac{T_1}{T_2} = e^{\mu\theta}$$

$$(\theta = 360° - \theta')$$

or $2.3 \log\left(\frac{T_1}{T_2}\right) = \mu\,(360° - \theta')$

$$= \mu\theta$$

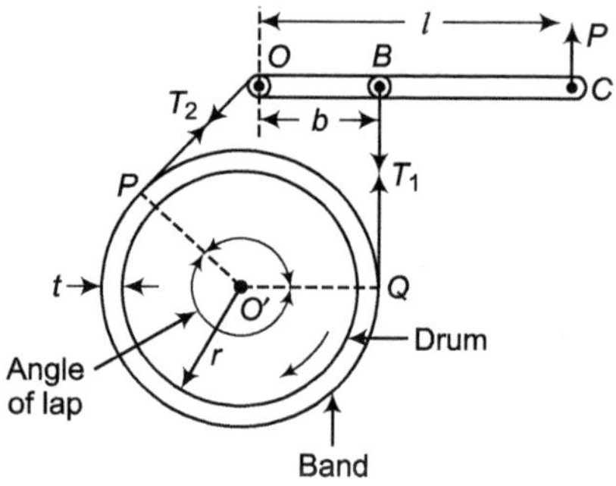

Simple band brake

- **Braking force** on the drum
$$= (T_1 - T_2)$$
- **Braking torque** on the drum
$$(T_B) = (T_1 - T_2) \times r$$
- When wheel rotates in the clockwise direction and taking moment about fulcrum O
$$Pl = T_1 \times b$$
$$T_1 = \frac{Pl}{b}$$
- For anticlockwise rotation of the drum $Pl = T_2 b$
$$T_2 = \frac{Pl}{b}$$

where, b = Perpendicular distance from O to the line of action T_1 or T_2
l = Length of the lever from the fulcrum
$T_1 = \sigma_t\, wt$
w = Width of the band
t = Thickness of the band
σ_t = Permissible stress in the band

Design of Friction Drives

Coupling

Coupling is basically done in shafts to join permanently two rotating shafts to each other.

Types of Coupling

Coupling can be classified as

Rigid Coupling

The rigid coupling used to connect two shafts which are perfectly aligned and these coupling widely use for transmitting large torque. *The rigid coupling are classified as*

- Sleeve or muff coupling
- Clamp or split-muff or compression coupling
- Flange coupling

Flexible Coupling

These couplings are used to connect two shafts having both lateral and angular misalignment. *The flexible coupling is further classified as*

- Bashed pin type coupling
- Universal coupling
- Oldham coupling

Calculations of Parameters in Coupling

Muff Coupling or Sleeve Coupling or Box Coupling

It is the simplest type of rigid coupling, made of cast-iron. It consists of a hollow cylinder whose inner diameter is the same as that of the shaft. It is fitted over the ends of the two shafts by means of a gib head key.

Calculating Diameter

$$M_t = \frac{60 \times 10^6 \text{ (kW)}}{2\pi n}$$

and $$\tau = \frac{16 M_t}{\pi d^3}$$

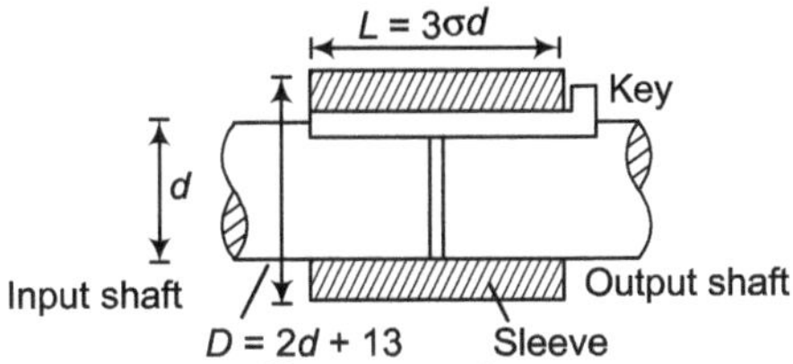

Sleeve coupling

Length of key = half length of sleeve

$$l = \frac{L}{2}$$

- Shear stress in key $\tau = \dfrac{2M_t}{dbl}$
- Compressive stress in key $\sigma_C = \dfrac{4M_t}{dhl}$

Clamp Coupling

It is also known as split muff coupling. In this case, the muff or sleeve is made into two halves and are bolted together.

Clamping force of each bolt $P_1 = \dfrac{\pi}{4} d_1^2 \sigma_t$

where, P_1 = Tensile force on each bolt
d_1 = Core diameter of clamping bolt
σ_t = Permissible tensile stress

For sleeve halves $D = 2.5\,d$, $L = 3.5\,d$

where, D = Outer diameter of sleeve halves
L = Length of sleeve
d = Diameter of shaft
n = Total number of bolt
f = Coefficient of friction

$$M_t = \frac{fdP_1 n}{2}, \quad P_1 = \frac{2M_t}{fdn}$$

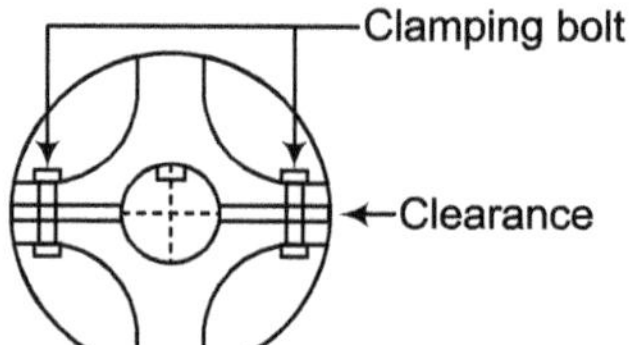

Clamp coupling diagram

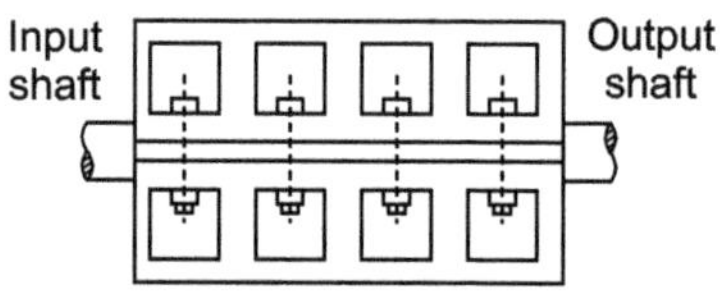

Location of input and output shaft in clamp clamp coupling

- The torque (τ) for a hub is given by

$$\tau = \frac{\pi}{16} \times \tau_c \left(\frac{D^4 - d^4}{D}\right)$$

- The torque (τ) for a flange is given by

$$\tau = \frac{\pi D^2}{2} \times \tau_c \times t_f$$

- The torque (τ) for a bolts is given by

$$\tau = (n \times d_1 \times t_f \times \sigma_{cb}) \frac{D_1}{2}$$

Flange Coupling

A flange coupling usually applies to a coupling having two separate cast iron flanges. Each flange is mounted on the shaft and a keyed to it. The face are turned up at right angle to the axis of the shaft. One of the flanges has a projected portion and the other flange has a corresponding recess.

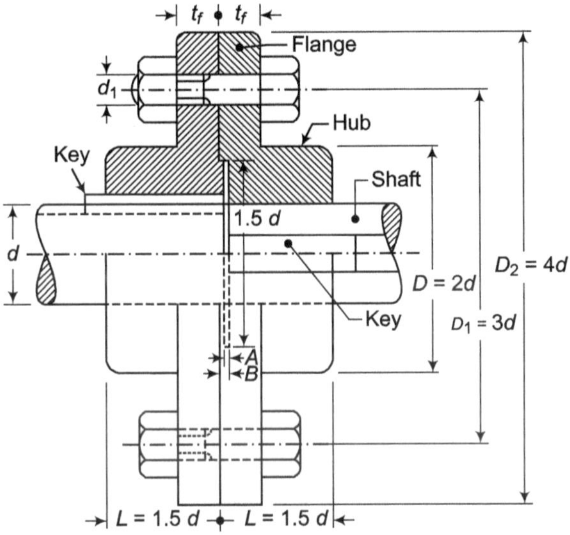

Flange coupling diagram

Belt Drives

The belts or ropes are used to transmit power from one shaft to another by means of pulleys which rotate at the same speed or at different speeds.

Types of Belts

The belts can be classified as

Flat Belt

These belt use when distance between pulleys are not more than 8 m. Mostly use in factories and workshop where moderate power is required.

V-Belt

- Use when two pulleys are very near to each other.
- Use in factories where high power is required.

Circular Belt

- Use when two pulleys are more than 8 m apart.
- Use where high power is required.

Parameters used in Belt Drives

The basic parameters used in belt drives are discussed below

Velocity Ratio of a Belt Drive

It is the ratio between the velocities of the driver and the follower or driven.

Let, d_1 = Diameter of driver, d_2 = Diameter of follower

N_1 = Speed of driver in rpm, N_2 = Speed of follower in rpm

$$\frac{N_2}{N_1} = \frac{d_1}{d_2}$$

When thickness of belt t is considered.

Velocity ratio
$$\frac{N_2}{N_1} = \frac{d_1 + t}{d_2 + t}$$

Slip of the Belt

Sometimes due to insufficient frictional grip driver moves forward without carrying the belt. This is called slip of the belt .It reduces the velocity ratio of the system.

S_1 = % slip between driver and belt

S_2 = % slip between belt and follower

$$\frac{N_2}{N_1} = \left(\frac{d_1 + t}{d_2 + t}\right)\left[1 - \left(\frac{S_1 + S_2}{100}\right)\right] \text{ or } \frac{N_2}{N_1} = \left(\frac{d_1 + t}{d_2 + t}\right)\left[1 - \left(\frac{S}{100}\right)\right]$$

$$S = S_1 + S_2$$

Creep of the Belt

When there is relative motion between the belt and pulley surfaces due to extension and contraction of belt. This relative motion is termed as creep of the belt. It reduces slightly the speed of driven pulley or follower.

$$\frac{N_2}{N_1} = \frac{d_1}{d_2} \times \left[\frac{E + \sqrt{\sigma_2}}{E + \sqrt{\sigma_1}}\right]$$

where, σ_1 and σ_2 = Stress in the belt on tight and slack side respectively.

E = Young modulus for the material of the belt.

Length of an Open Belt Drive

In this open belt drive both the pulleys rotate in the same direction.

An open belt drive is shown in figure below

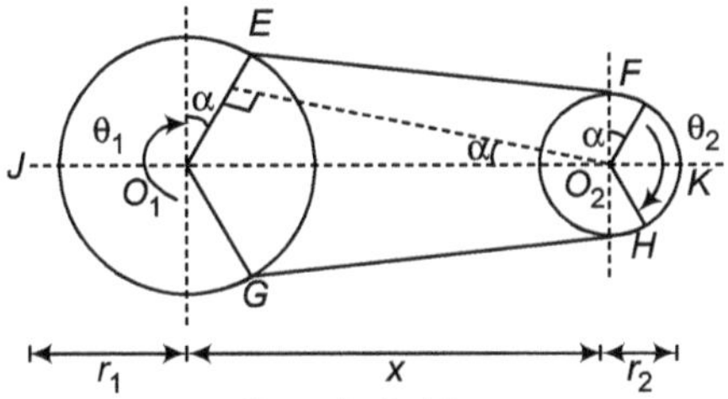

Open belt drive

The dimension of open belt drive are as follow

where, α = Wrap angle for pulley, r_1 = Radius of big pulley
r_2 = Radius of small pulley, r = Centre distance

- Length of belt (L) $L = \pi (r_1 + r_2) + \dfrac{(r_1 - r_2)^2}{x} + 2x$

$$\sin \alpha = \frac{r_1 - r_2}{x} \text{ or } \alpha = \sin^{-1}\left(\frac{r_1 - r_2}{x}\right)$$

Angle of contact on smaller pulley $\theta_2 = 180° - 2\alpha$

Angle of contact on larger pulley $\theta_1 = 180° + 2\alpha$

Length of a Cross Belt Drive

A cross belt drive are shown in figure below

Angle of contact on smaller pulley $\theta_1 = 180° + 2\alpha$

Angle of contact on larger pulley $\theta_2 = 180° + 2\theta$

$$L = \pi (r_1 + r_2) + \frac{(r_1 + r_2)^2}{x} + 2x$$

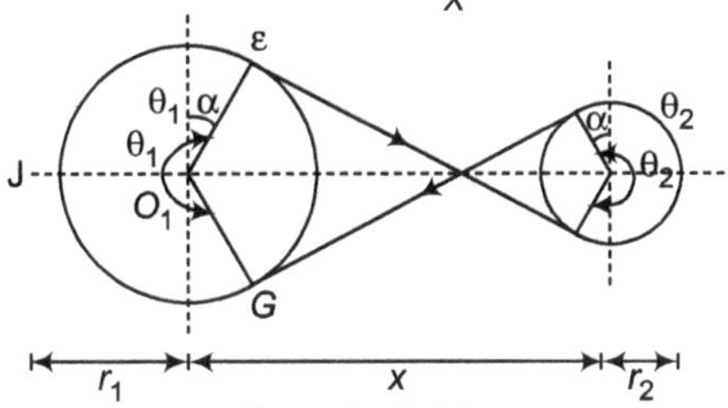

Cross belt drive

$$\alpha = \frac{r_2 + r_1}{x}$$

- Power transmitted by belt $P = (T_1 - T_2)\,v$

  where, v is the velocity of the belt.

Ratio of Driving Tension

- For flat drive, ratio of driving tension $\frac{T_1}{T_2} = e^{\mu\theta}$
- For V-belt, ratio of driving tension $\frac{T_1}{T_2} = e^{\frac{\mu\theta}{\sin\alpha}}$

  where, α = Cone angle
  θ = Angle of contact in radians
  μ = Coefficient of friction
  T_1 = Tension in belt on tight side
  T_2 = Tension in belt on slack side.

Centrifugal Tension

The belt continuously runs over the pulleys. Therefore, some centrifugal force is caused, whose effect is to increase the tension on both the tight as well as the slack sides. The tension caused by centrifugal force is called centrifugal tension.

- Centrifugal tension is given by

$$T_c = mv^2$$

  where, m = Mass of belt per unit length
  v = Belt velocity
- Maximum tension in belt $T = T_1 + T_c$
- Condition for transmission of maximum power $T = 3\,mv^2 = 3T_c$

Initial Tension in Belt

When the pulleys are stationary then the belt which was running over those pulley subjected to some tension. This tension is called initial tension in the belt.

Let the consider initial tension to cut on belt

- Neglecting centrifugal tension $T_0 = \frac{T_1 + T_2}{2}$
- Considering centrifugal tension $T_0 = \frac{T_1 + T_2 + 2T_c}{2}$

Chain Drives

The chains are generally used to transmit motion and power from one shaft to another, when the centre distance between their shafts is short. The chains may also used for long distance of up to 8 m centre.

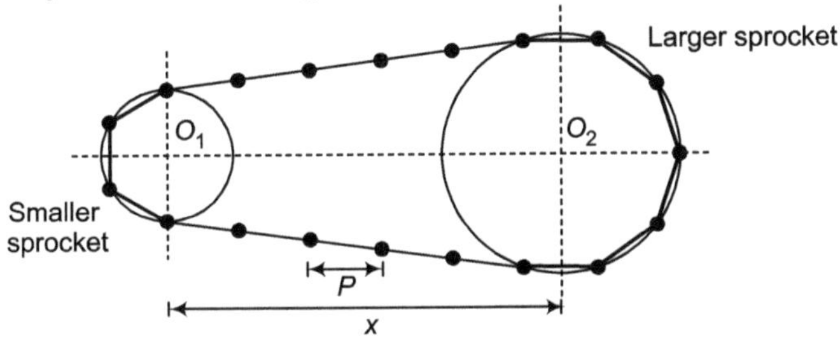

Schematic diagram of chain drives

Let, T_1 = Number of teeth on smaller sprocket
T_2 = Number of teeth on larger sprocket
P = Pitch of chain
x = Centre distance

- Length of chain $L = K \cdot P$

  where, K = Number of chain links
  P = Pitch of chain

- Number of chain links $K = \dfrac{T_1 + T_2}{2} + \dfrac{2x}{P} + \left[\dfrac{T_2 - T_1}{2\pi}\right]^2 \dfrac{P}{x}$

- Centre distance $x = \dfrac{P}{4}\left[K - \dfrac{T_1 + T_2}{2} + \sqrt{\left(K - \dfrac{T_1 + T_2}{2}\right)^2 - 8\left(\dfrac{T_2 - T_1}{2\pi}\right)^2}\right]$

- Power transmitted by chains, $P = \dfrac{w_B \times v}{n \times K_s}$

  where, w_B = Breaking load in Newton
  v = Velocity of chain
  n = Factor of safety
  K_s = Service factor = $K_1K_2K_3$
  K_1 = Multiple strand factor
  K_2 = Tooth correction factor.

Power Screw

- The power screw is used to convert rotary motion into translatory motion.
- Hand press, screw jack, hosting screw, lead screw are applications of screw.

Types of Screw Threads Used for Power Screw

There are three types of screw threads which are given below

Square Thread

The square thread is used for transmission of power in either direction. The thread results in maximum efficiency and minimum radial or bursting pressure on the nut.

Acme or Trapezoidal Thread

- Use where a split nut is required and where provision is made to take up wear as in the lead screw of a lathe.
- It is a modification of square thread with lower efficiency than square thread and more bursting pressure on nut than square thread but increases its area in shear.

Buttress Thread

- Use when large forces act along the screw axis in one direction.
- Higher efficiency, stronger than other thread because of greater thickness at the base of thread.
- Employed as the thread for light jack screw.

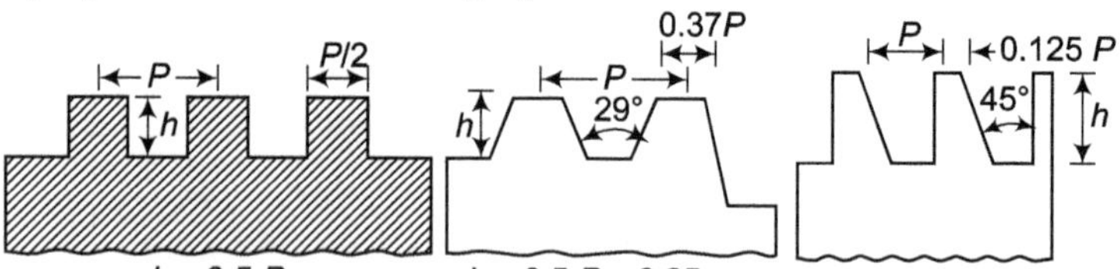

$h = 0.5\,P$ (a) Square thread $h = 0.5\,P + 0.25$ mm (b) Acme thread $h = 0.75\,P$ (c) Buttress thread

Multiple Thread

Power screw with multiple threads such as double, triple etc., are employed when, it is desired to secure a large lead with fine threads or high efficiency. Such types of threads are usually found in high speed actuators.

Torque Required to Raise Load by Square Threaded Screws

- Effort applied at the circumference of the screw to lift the load (P)

$$P = w \tan(\alpha + \phi)$$

where, w = Load to be lifted

μ = Coefficiency of friction between screw and nut
= $\tan \phi$, where ϕ is friction angle
P = Pitch of the screw
α = Helix angle $\tan \alpha = \frac{P}{\pi d}$
d = Mean diameter of the screw

- Torque required to overcome friction between screw and nut

$$T_1 = P \times \frac{d}{2} = w \tan(\alpha + \phi) \frac{d}{2}$$

- Torque required to overcome friction at the collar

$$T_2 = \frac{2}{3} \mu_1 w \left(\frac{R_1^3 - R_2^3}{R_1^2 - R_2^2} \right)$$

$$= \mu_1 w \left(\frac{R_1 + R_2}{2} \right) = \mu_1 w R$$

(Assuming uniform pressure condition)

R_1 and R_2 = Outside and inside radii of collar

R = Mean radius of collar = $\frac{R_1 + R_2}{2}$

$$= \frac{2}{3} \left(\frac{R^3{}_1 - R_2^3}{R^2{}_1 - R_2^2} \right)$$

μ_1 = Coefficient of friction for the collar

- **Total torque** required to overcome friction (*i.e.*, to rotate the screw)

$$T = T_1 + T_2$$

- If an effort P_1 is applied at the end of lever of arm length l then total torque applied to overcome friction must be equal to the torque applied at the end of lever.

$$T = P_1 \times l$$

- **Mechanical Advantages** (MA)

  It is ratio of load lifted to the effort applied P_1 at the end of lever.

$$\text{MA} = \frac{w}{P_1} = \frac{w \times 2l}{P \times d} \qquad \left(\begin{array}{l} T = T_1 = P \times \frac{d}{2}, T_2 = 0 \\ \text{neglect friction of collar} \end{array} \right)$$

$$= \frac{2l}{d \tan(\alpha + \phi)}$$

Note In the case torque required to lower loudly regular threaded screw effort $P = \tan(\phi - \alpha)$ if $\phi > \alpha$ torque required to overcome friction between screw and nut $T_1 = P \times \frac{d}{2}$, $T = T_1 + T_2 = P_1 \times l$

Efficiency of Square Threaded Screw

The efficiency of the square threaded screw may be defined as,

In the case of lifting the load $\eta = \frac{\text{Ideal effort}}{\text{Actual effort}}$

- If only screw friction consider

$$\eta = \frac{P_o}{P} = \frac{w \tan\alpha}{w \tan(\alpha + \phi)} = \frac{\tan\alpha}{w \tan(\alpha + \phi)}$$

(ideal effort $P_o = w \tan\alpha$ $(\phi = 0)$)

- If both screw friction and collar friction consider

$$\eta = \frac{P_o \times d/2}{P \times \frac{d}{2} + \mu_1 wR}$$

Maximum Efficiency of Square Thread

$$2\alpha = 90 - \phi, \quad \alpha = 45^\circ - \frac{\phi}{2}$$

$$\eta_{max} = \frac{1 - \sin\phi}{1 + \sin\phi}$$

where, ϕ = Angle of friction.

Self Locking Screw

In case of self locking $\phi > \alpha$ *i.e.*, the torque required to lower the load will be positive indicating that an effort is applied to lower the load, such a screw is known as self locking screw.

Efficiency of self locking $\eta = \frac{\tan\phi}{\tan(\alpha + \phi)} < \frac{1}{2}$

$\Rightarrow$ efficiency of self locking is less than $\frac{1}{2}$ or 50%. If the efficiency is more than 50% than the screw is said to be over hauling.

Gears

Gear can be defined as the mechanical element used for transmitting power and rotary motion from one shaft to another by means of progressive engagement of projections called teeth.

Classification of Gears

- Spur Gear
- Helical Gear
- Bevel Gear
- Worm Gear

Spur Gear

In spur gears, teeth are cut parallel to axis of the gear.

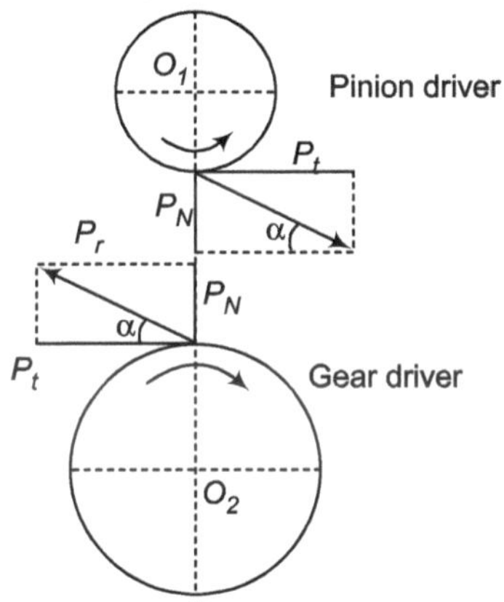

Spur gear

- Circular pitch $P = \dfrac{\pi d}{z}$
- Diametrical pitch $P = \dfrac{z}{d}$
- Module $m = \dfrac{d}{z}$

  where, z = Number of teeth

  d = Pitch diameter
- Torque transmitted by gear

$$M_t = \frac{60 \times 10^6 \text{ (kW)}}{2\pi n}$$

where, M_t = Torque transmitted by gear

n = Speed of rotation

kW = Power transmitted by gear

$$P_t \times \frac{d}{2} = M_t \; P_r = P_t \tan\alpha \; P_N = \frac{P_t}{\cos\alpha}$$

- Beam strength of gear tooth $S_b = mb\sigma_b Y$

  where, σ_b = Permissible bending stress, Y = Lewis form factor

  m = Module.
- Dynamic load or incremental dynamic load

$$P_d = \frac{21v\,(ceb + P_t)}{21v + \sqrt{ceb + P_t}}$$

where, v = Pitch line velocity, c = Deformation factor

b = Face width of tooth, P_t = Tangential force due to rated torque.

e = Sum of errors between two meshing teeth

- Estimation of module based on beam strength

$$m = \left[\frac{60 \times 10^6}{\pi}\left\{\frac{(kw)\,c_s\,(f_s)}{z n c_v \left(\frac{b}{m}\right)\left(\frac{S_W}{3}\right) Y}\right\}\right]^{1/3}$$

where, c_s = Service factor, c_v = Velocity factor

f_s = Factor of safety, n = Speed (rpm)

- **Wear strength** of gear tooth

$$S_w = bQd_pK$$

where, K = Load stress factor

b = Gear width

Q = Ratio factor $= \dfrac{2z_g}{z_g + z_p}$ d_p = Pitch diameter

- Estimation of module based on wear strength

$$m = \left[\frac{60 \times 10^6}{\pi}\left\{\frac{(kw)\,c_s\,(f_s)}{z_p^2 n_p c_v \left(\frac{b}{m}\right)\rho K}\right\}\right]^{1/3}$$

Helical Gear

- The teeth of helical gear cut in the form of helix or an angle on the pitch cylinder.

$$P_n = \frac{P}{\cos \psi}$$

where, P_n = Normal diametrical pitch, P = Transverse diametrical pitch

ψ = Helix angle

$m_n = m \cos \psi$, m_n = Normal module

m = Transverse module

- **Axial pitch** $P_a = \dfrac{P}{\tan \psi}$
- Pitch circular diameter $d = \dfrac{zm_n}{\cos \psi}$
- Tooth proportions

Addendum $h_a = m_n$

Dedendum $h_f = 1.25\ m_n$

Clearance $c = 0.25\ m_n$

- Addendum circle diameter $d_a = d + 2h_a$ or $d_a = \dfrac{z\,m_n}{\cos\psi} + 2\,m_n$
- Dedendum circle diameter $d_f = \dfrac{z\,m_n}{\cos\psi} - 2.5\,m_n$

Force analysis in helical gears is given below

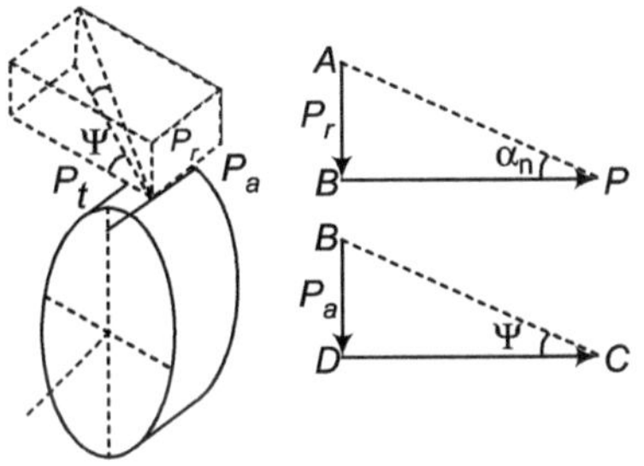

Pitch cylinder

Component of tooth forces $P_t = \dfrac{2M_t}{d_p}$

$$P_r = P_t\left[\frac{\tan\alpha_n}{\cos\psi}\right]$$

$$P_a = P_t \tan\psi$$

- Beam strength of helical gear $S_b = m_n\, b\,\sigma_b Y$

  where, m = Module, σ_b = Permissible bending stress
  y = Lewis form factor

- Dynamic load or incremental dynamic load P_d

$$P_d = \frac{21v\,(ceb\cos^2\psi + P_t)\cos\psi}{21v + \sqrt{(ceb\cos^2\psi + P_t)}}$$

  where, e = Sum of errors, C = Deformation factor

- Wear strength of helical gear $S_w = \dfrac{bQ\,d_pK}{\cos^2\psi}$

Herringbone Gear

In order to avoid an axial thrust on the shaft and the bearings, the double helical gears or Herringbone gears are used.

Bevel Gears

- Use to transmit power between two intersecting shafts.
- High speed high power transmission.

Classification of Bevel Gear

- **Mitre Gear** When two bevel gears are mounted on shafts that are intersecting at right angle.
- **Crown Gear** In pair of bevel gear, when one of the gear has a pitch angle of 90°.
- **Internal Bevel Gear** When the teeth of bevel gear are cut on the inside of the pitch.
- **Skew Bevel Gear** Mounted on non-parallel and non-intersecting shafts. It constant of straight teeth.
- **Hypoid Gear** Similar to skew bevel gear, non-parallel and non-intersecting shafts. It consists of curved teeth.
- **Zerol Gear** Spiral bevel gear with zero spiral angle.
- **Face Gear** Consists of a spur or helical pinion meshing with a conjugate gear or disk form.

Terminologies in Bevel Gear

- Pitch circle radius $r_b = \dfrac{D}{2\cos\gamma}$ where, γ = Pitch angle
- **Formative number of teeth** $z' = \dfrac{z}{\cos\gamma}$,

$$\tan\gamma = \frac{z_p}{z_g}$$

where, z = Actual number of teeth

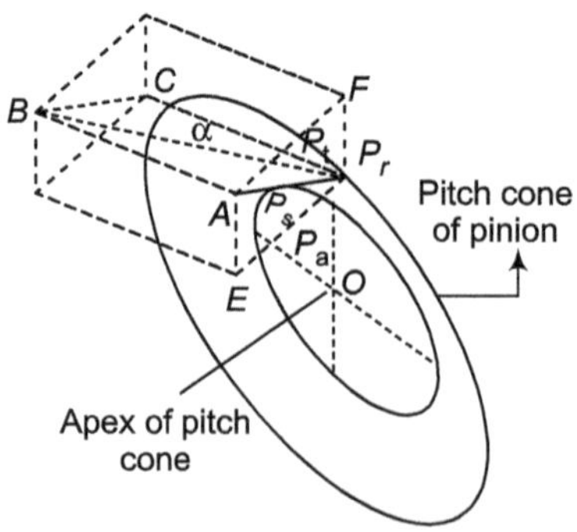

Cone distance in bevel gear section

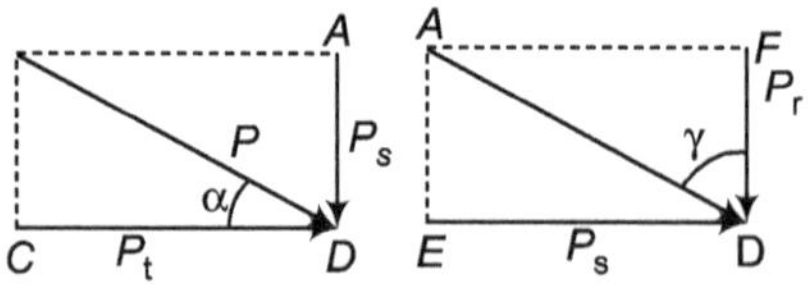

Component of tooth face

- Cone distance $A_o = \sqrt{\left(\frac{D_p}{2}\right)^2 + \left(\frac{D_g}{2}\right)^2}$

- Mean radius $r_m = \left[\frac{D_p}{2} - \frac{b \sin\gamma}{2}\right]$

- $M_t = \frac{60 \times 10^6\ (\text{kW})}{2\pi\, n_p}$
- $P_t = \frac{M_t}{r_m}$
- $P_r = P_t \tan\alpha \cos\gamma$
- $P_a = P_t \tan\alpha \sin\gamma$

  where, Tangential component of force-P_t

  Radial component-P_r, Thrust component-P_a

- Beam strength of bevel gear $S_b = mb\sigma_b Y\left[1 - \frac{b}{A_o}\right]$

  where, $\left[1 - \frac{b}{A_o}\right]$ = bevel factor.

  Wear strength of bevel gears $S_w = \frac{0.75\, b\, Q\, D_p K}{\cos\gamma}$

  where, K = Material constant, $K = 0.16\left(\frac{\text{BHN}}{100}\right)^2$

Worm Gear

Worm gears are used to transmit power between two non-intersecting shafts which are in general at right angle to each other.

- High speed reduction, compact with small size self locking operation.
- Drawback low efficiency.

Terminologies in Worm Gear

There are four terminology in worm gear system as given below

where, z_1 = Number of starts on the worm

z_2 = Number of teeth on worm wheel

q = Diametrical quotient

m = Module (mm) $q = \frac{d_1}{m}$

Lead angle (γ) $\tan\gamma = \frac{z_1}{q}$, Helix angle (ψ) $\gamma + \psi = \frac{\pi}{2}$

Properties of Worm Gear

For involute helicoidal tooth form $h_{a_1} = m$

$$h_{f_1} = (2.2 \cos\gamma - 1)\, m, c = 0.2\, m \cos\gamma$$

h_{a_1} = Addendum h_{f_1} = Dedendum

c = Clearance, $d_{a_1} = d_1 + 2ha_1 = qm + 2m$

Outside diameter of worm $d_{a_1} = m\,(q+2)$

Root diameter of worm $d_{f_1} = m\,(q + 2 - 4.4 \cos\gamma)$

Effective face width $F = 2\, m\sqrt{q+1}$

Length of the root $l_r = (d_{a_1} + 2c)\sin^{-1}\left[\dfrac{F}{d_{a_1} + 2c}\right]$

Force Analysis in Worm Gear

The forces in worm gears are given by

- **Tangential force,** $P_t = P\cos\alpha\sin\gamma$
- **Axial force,** $P_a = P\cos\alpha\cos\gamma$
- **Radial force,** $P_r = P\sin\alpha$

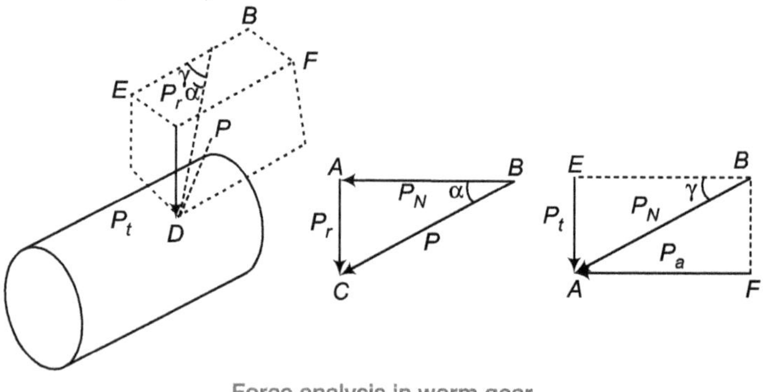

Force analysis in worm gear

A worm gear drive is said to be self locking if the coefficient of friction (μ) is greater than tangent of lead angle.

i.e., $\mu > \tan\gamma$

Efficiency $\eta = \dfrac{\cos\alpha - \mu\tan\gamma}{\cos\alpha + \mu\cot\gamma}$

Strength Rating of Worm Gear

Maximum permissible torque can be calculated by the lower of the following two valves

$$(M_t)_1 = 17.65\, X_{b1}\, S_{b1}\, ml_r\, d_2 \cos\gamma$$

$$(M_t)_2 = 17.65 X_{b2}\, Sb_2\, ml_r d_2 \cos\gamma$$

where, $(M_t)_1\ (M_t)_2$ = Permissible torque on the worm wheel

X_{b_1}, X_{b_2} = Speed factors for strength of worm and worm wheel

S_{b_1}, S_{b_2} = Bending stress factors of worm and worm wheel

m = Module

l_r = Length of root of worm wheel teeth

d_2 = Pitch circle diameter of worm wheel

γ = Lead angle of worm.

Wear Rating of Worm Gear

Maximum permissible torque is given by the lower of the following two valves

$$(M_t)_3 = 18.64\ X_{C_1}\ S_{C_1}\ Y_z\ (d_2)^{1.8} m$$

$$(M_t)_4 = 18.64\ X_{C_2}\ S_{C_2}\ Y_z\ (d_2)^{1.8} m$$

where, $(M_t)_3, (M_t)_4$ = Permissible torque on the worm wheel

X_{C_1}, X_{C_2} = Speed factors for the wear of worm and worm wheel

S_{C_1}, S_{C_2} = Surface stress factors of the worm and worm wheel

Y_z = Zone factor.

Wire Rope

- Wire rope is used in hoisting and materials handling equipments.
- It consists of high strength to weight ratio and greater reliability.

Lay in Wire Rope

- Refer to the manner in which the wires are helically laid into strands and strands into the rope.

Key Points

- **Regular Lay** When the wires in the strands are twisted in a direction opposite to that of strands.
- **Lang's Lay** When the wires in strands are twisted in the same direction as strand.
- Regular lay are more popular due to more structural stability and resistance to crushing and distortion.

Stresses in Wire Rope

Bending stress in one of the individual wire

$$\sigma_b = \frac{M_b y}{I} \quad \text{and} \quad y = \frac{d_w}{2}$$

Therefore, $$\sigma_b = \frac{M_b d_w}{2I}$$

$$\Rightarrow \quad \frac{M_b}{EI} = \frac{2}{D} \quad \Rightarrow \quad \sigma_b = \frac{E\,d_w}{D}$$

where, D = Shear diameter
E = Modulus elasticity of rope
d_w = Diameter of individual wire.

Force Per Unit Area of the Wire Rope

$$FX = \frac{2P}{d_r D}$$

where, P = Tension in the rope
d_r = Nominal diameter of wire rope
D = Shear diameter.

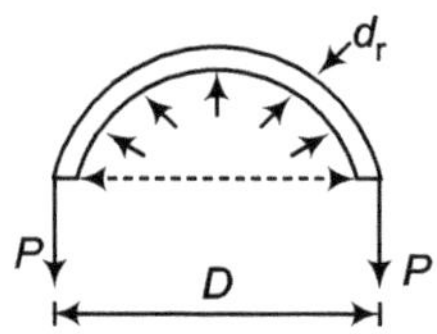

Stress in wire rope

Bearing

A bearing is a mechanical element that permits relative motion between two components or parts, such as the shaft and housing, with minimum friction.

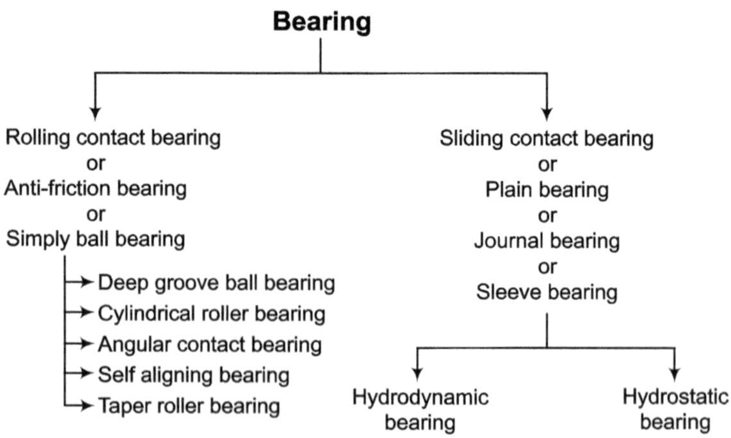

Classification of bearing

Dynamic Load Carrying Capacity

The dynamic load carrying capacity bearing can be defined as the thrust load in thrust bearing or radial load in radial bearing that can be carried for minimum life in one million revolutions. It is denoted by *C*.

Bearing Life

The life of an individual bearing is defined as the total number of revolutions (or the number of hours at a given constant speed) which the bearing can complete before the first evidence of fatigue failure develops on the ball or races.

The bearing life can be defined by either of the following terms

Fatigue Life

It is the number of revolutions that the bearing runs before the first evidence of fatigue crack in balls or races.

Rating Life

It is the number of revolutions that 90% of the bearings will complete or exceed before the first evidence of fatigue crack.

Comparison Between Ball and Roller Bearings

S. No.	Comparison Parameter	Ball Bearing	Roller Bearing
1.	Rolling Element	In ball bearings, spherical balls are used as the rolling elements.	In roller bearings, cylindrical rollers, or spherical rollers are used as the rolling elements.
2.	Nature of contact	In ball bearings, the contact between the inner race and ball or the outer race and ball is point contact.	In roller bearings, the contact between the inner race and roller or the outer race and roller is line contact.
3.	Load Carrying Capacity	Because of point contact, the load carrying capacity of the ball bearing is low.	Because of line contact, the load carrying capacity of the roller bearings is high.
4.	Radial Dimensions	For ball bearings, the radial dimension is more.	For roller bearings., the radial dimension is less. Hence, roller bearings are used where compactness is required in radial direction.
5.	Axial Dimensions	For ball bearings, the axial dimension is less.	For roller bearings, the axial dimension is more.

Load Life Relationship

The bearing life is given by $L_{10} = \left(\frac{C}{P}\right)^n$

where, L_{10} = Rated bearing life (in million revolutions)

C = Dynamic load capacity

n = 3 (for ball bearing) = 10/3 (for roller bearing)

$$L_{10} = \frac{60 n_p L_{10_h}}{10^6}$$

where, L_{10_h} = Rated bearing life (hours)

n_p = Speed of rotation (rpm)

Equivalent Bearing Load

It can be defined as the constant radial load in radial bearing that if applied to the bearing would give same life as which the bearing will attain under actual conditions of forces. It is denoted by P.

Equivalent dynamic load $P = X V F_r + Y F_a$

where, F_r = Radial load

F_a = Axial or thrust load

V = Race rotation factor

X = Radial factor

Y = Thrust factor

- Design for cyclic load $P_c = \sqrt[3]{\dfrac{N_1 P_1^3 + N_2 P_2^3 + \ldots}{N_1 + N_2 + \ldots}}$
- Bearing with a probability of survival other than 90 per cent

$$\frac{L}{L_{10}} = \left(\frac{\log_e\left(\frac{1}{R}\right)}{\log_e \frac{1}{R_{90}}}\right)^{\frac{1}{b}}, \text{ where } R = \text{Reliability}$$

Needle Bearing

- In Needle bearing cylindrical rollers of very small diameter and relatively long length.
- Large load carrying capacity.
- Use in oscillatory motion such as pipson pin bearings, rocker arm and universal joint.

Sliding Contact Bearing

Use lubrication between rubbing surfaces of bodies have relative motion. This bearing supports the load in radial direction. The portion of shaft inside the bearing is know as **journal**.

Hydrodynamic Bearing

In this system, load supporting fluid film is created by shape and relative motion of sliding surface.

Hydrostatic Bearing

In this system, load supporting fluid film separating the two surfaces is created by an external source like pump supplying sufficient fluid under pressure.

Bearing Characteristic Number (BCN)

The transition from thin film lubrication to thick film hydrodynamic lubrication can be understood with the help of curve called $\frac{\mu N}{P}$ curve.

$$\text{BCN} = \frac{\mu N}{P}$$

where, μ = Absolute viscosity of lubricant
N = Speed of journal
p = Unit bearing pressure

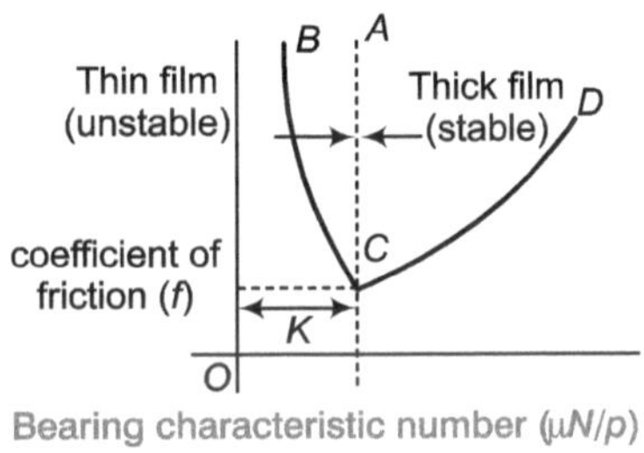

Hydrostatic Step Bearing

The hydrostatic step bearing is shown in figure and following notations are used in analysis

Let, w = Thrust load (N)
R_o = Outer diameter of shaft (mm)
R_i = Radius of the recess or pocket (mm)
p_i = Supply of inlet pressure (N / m^2) (MPa)
P_o = Outlet or atmospheric pressure (N / mm^2)
h_o = Fluid film thickness (mm)
Q = Flow of lubricant (mm^3 / s)
μ = Viscosity of the lubricant (MPa-s)

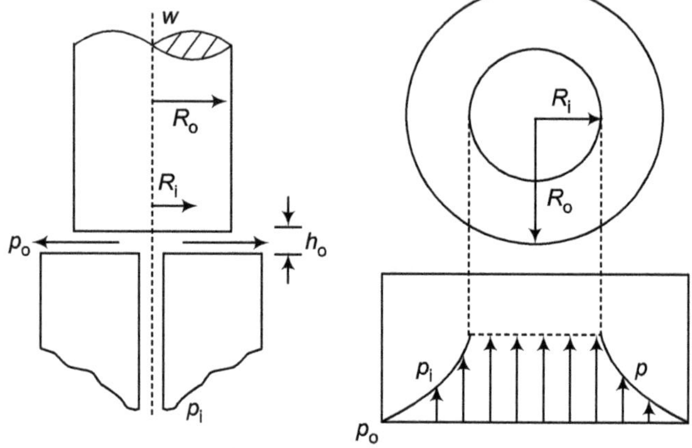

Hydrostatic step bearing

Flow of lubricant
$$Q = \frac{\pi\, p_i\, h_o^3}{6\mu \log_e\left(\frac{R_o}{R_i}\right)}$$

Thrust load
$$w = \frac{\pi p_i}{2}\left[\frac{R_o^2 - R_i^2}{\log_e\left(\frac{R_o}{R_i}\right)}\right]$$

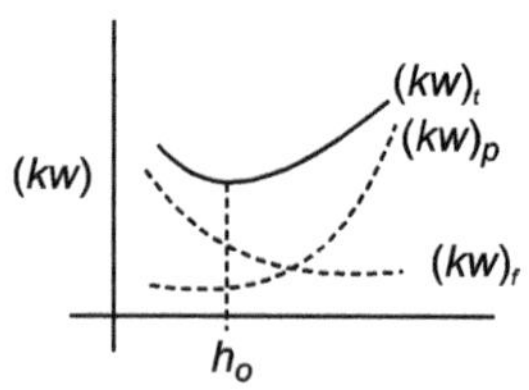

Total power loss in hydrostatic bearing

Energy Losses in Hydrostatic Bearing

- Power Loss in Pumping
$$(kw)_p = Q\,(p_i - p_0) \times 10^{-6}$$
- Frictional Power loss $(kw)_f = \dfrac{1}{58.05 \times 10^6}\,\dfrac{\mu n^2 (R_o^4 - R_i^4)}{h_o}$
- Total power loss $(kw_t) = (kw)_p + (kw)_f$

Raimondi and Boyd Method of Bearing

In Raimondi and Boyd method performance of bearing is expressed in terms of dimensionless parameter.

Radial clearance $c = R - r$

where, R = Radius of bearing

r = Radius of journal

Eccentricity ratio $E = \frac{e}{c}$

where, e = Eccentricity

Minimum film thickness variable = $\frac{h_o}{c}$

h_o = Minimum film thickness

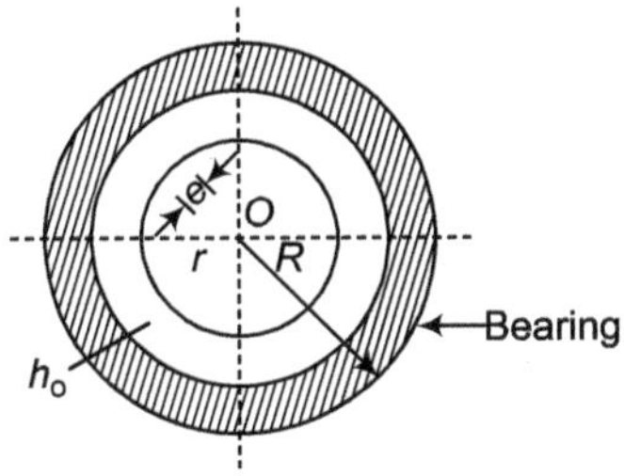

Raimondi and Boyd Method

- **Sommerfield number** $S = \left(\frac{r}{c}\right)^2 \times \frac{\mu n_s}{p}$

where, μ = Viscosity of lubricant

p = Unit bearing pressure (load per unit of projected area)

n_s = Journal speed

- **Coefficient of Flow Variable**

$$\text{CFV} = \left(\frac{r}{c}\right) f, \quad f = \text{Coefficient of friction}$$

- **Frictional power** $(kw)_f = \frac{2\pi n_s fwr}{10^6}$

- **Flow Variable** $\text{FV} = \frac{Q}{r c n_s l}$

where, l = Length of bearing

Q = Flow of lubricant

- **Temperature rise** $\Delta t = \frac{8.3 p(\text{CFV})}{\text{FV}}$

- **Average temperature** $T_{av} = T_i + \left(\frac{\Delta t}{2}\right)$

where, T_i = Inlet temperature

5

Fluid Mechanics

Basic Concept of Fluid Mechanics

Fluid mechanics describes the behaviour of water under various static and dynamic conditions. This gives the information about an ideal liquid, a frictionless inelastic liquid whose particles follow smooth flow paths.

Solid

Its molecules are very closely spaced. Intermolecular cohesive force is high between its molecules. Solid undergoes static deformation.

Fluid

The space between fluid molecules is relatively large *e.g.,* liquid and gas. Their intermolecular force is low. Fluid undergoes continuous deformation.

Ideal Fluid

It has no viscosity, no surface tension and incompressible.

Real Fluid

It possesses the properties like viscosity, surface tension and compressibility.

Properties of Fluid

The material properties of a fluid, which may vary, sometimes sensitively with temperature, pressure and composition; determine its mechanical behaviour.

Some of the properties of fluids are given below

Density

Mass of liquid in unit volume is known as density.

$$\text{Density } \rho = \frac{\text{Mass}}{\text{Volume}}$$

$$\text{Specific weight} = \frac{\text{Weight}}{\text{Volume}} = \frac{m.g}{V} = \rho \cdot g$$

Specific Weight

It is defined as the weight per unit volume.

Specific Gravity or Relative Density

It is the ratio of specific weight of fluid to the specific weight of a standard fluid. Standard fluid is water in case of liquid and H_2 or air in case of gas.

$$\text{Specific gravity} = \frac{\gamma}{\gamma_w} = \frac{\rho \cdot g}{\rho_w \cdot g} = \frac{\rho}{\rho_w}$$

where, γ_w = Specific weight of water

ρ_w = Density of water specific.

Specific Volume

Specific volume of a liquid is defined as volume per unit mass. It is also defined as the reciprocal of specific density.

$$\text{Specific volume} = \frac{V}{m} = \frac{1}{\rho}$$

Fluid Pressure at a Point

If a fluid is **stationary**, then force acting on any surface or area is perpendicular to that surface.

$$p = \frac{dF}{dA}$$

If force is uniformly distributed over surface, then

$$p = \frac{F}{A}$$

Its unit is pascal.

$$1 \text{ Pa} = 1 \text{ N/m}^2$$

and $\quad 1 \text{ bar} = 100 \text{ kPa} = 10^5$

$$\text{Pa} = 10^5 \text{ N/m}^2$$

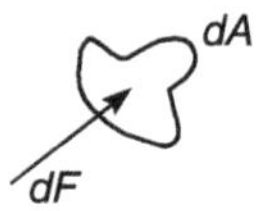

Force over surface

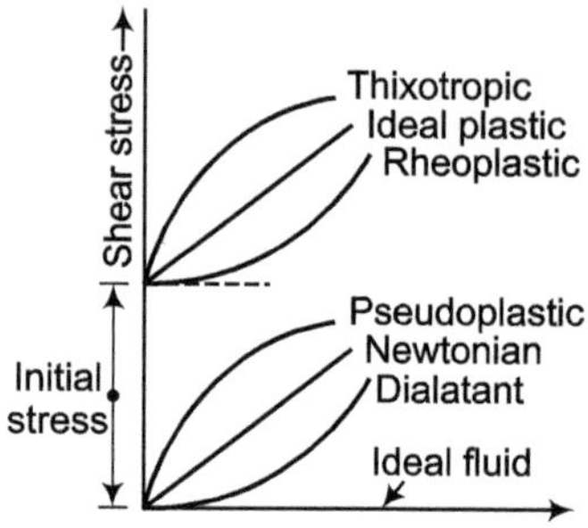

Stress-fluid graph

Pascal's Law

It states that pressure or intensity of pressure at a point in a static fluid (fluid is in rest) is equal in all directions. If fluid is not in motion then according to Pascal's law,

$$p_x = p_y = p_z$$

where, p_x, p_y and p_z are the pressure at point x, y, z respectively.

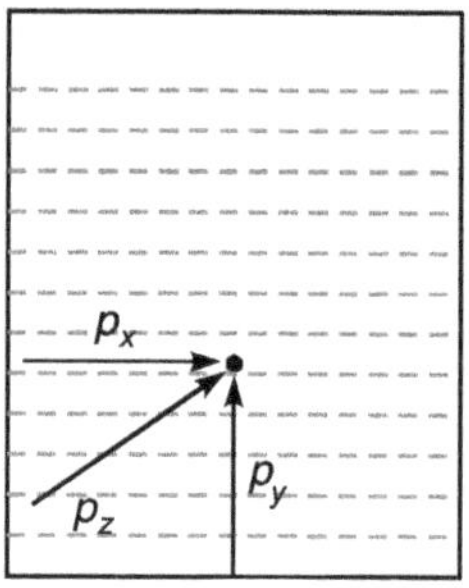

Pressure at a point

Hydrostatic Law

It states that rate of increase of pressure in a vertical direction is equal to weight density of fluid at that point. *e.g.*, if fluid is in rest, then balancing the force acting on small element of fluid is

$$\frac{\partial p}{\partial z} = \rho g \quad \text{or} \quad dp = \rho g \, dz$$

Now, integrating from $z = 0$ to $z = h$ depth, then pressure at h depth from free surface is

$$\int_0^p dp = \rho g \int_0^h dz$$

$$p = \rho g h \quad \text{and} \quad h = \frac{p}{\rho g}$$

where, h is known as pressure head.

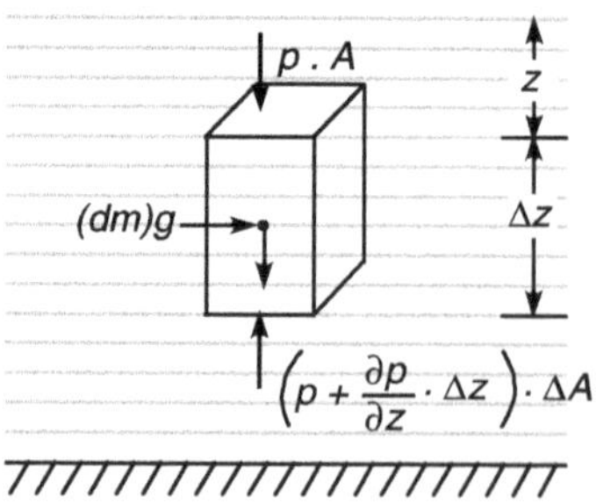

Pressure-density diagram

Archimedes' Principle

It states, 'when a body is immersed completely or partially in a fluid, it is lifted up by a force equal to weight of fluid displaced by the body.

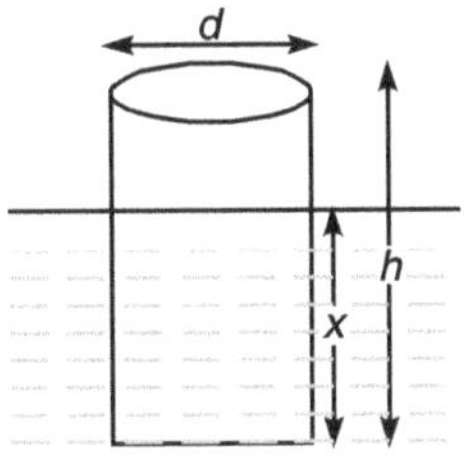

Cylinder-fluid system

Buoyant force = Weight of fluid displaced by body

Buoyant force on cylinder

$$= \text{Weight of fluid displaced by cylinder}$$
$$= V_{S_{in}} \rho_l g$$
$$V_{S_{in}} = \text{Value of immersed part of solid}$$

or
$$F_B = \rho_{water} \times g \times \text{Volume of fluid displaced}$$
$$= \rho_{water} \times g \times \text{Volume of cylinder immersed inside the water}$$
$$= \rho_w g \frac{\pi}{4} d^2 x \qquad (\because w = mg = \rho V g)$$

Concept of Buoyancy

When a body is immersed in a fluid, an upward force is exerted by fluid on the body which is equal to weight of fluid displaced by body. This upward force is known as **buoyancy**.

Principle of Floatation

According to this principle, if weight of body is equal to buoyant force then, body will float.

$$F_B = mg$$
$$V_{S_{in}} \rho_l g = V_S \rho_S g$$
$$\rho_w g \frac{\pi}{4} d^2 x = \rho_{cylinder}\, g \frac{\pi}{4} d^2 h$$
$$\rho_w x = \rho_{cylinder} h$$

Centre of Buoyancy

The point at which force of buoyancy acts is called **centre of buoyancy**. It lies on centre of gravity of volume of fluid displaced or centre of gravity of the part of the body which is inside the water. Point B is the centre of buoyancy.

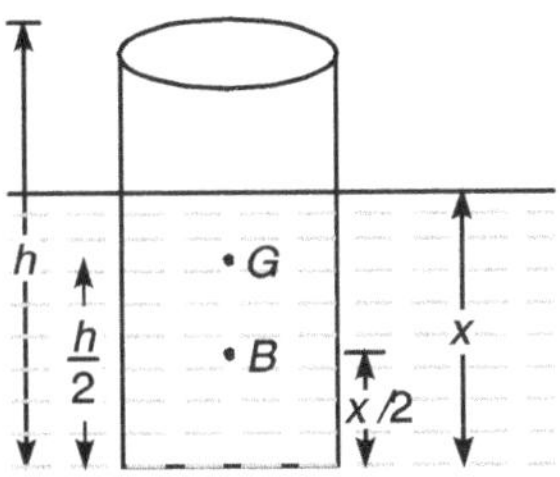

Centre of buoyancy diagram

Metacentre of a Floating Body

If a body which is floating in liquid is given small angular displacement, it starts oscillating about some point *M*. This point is called metacentre.

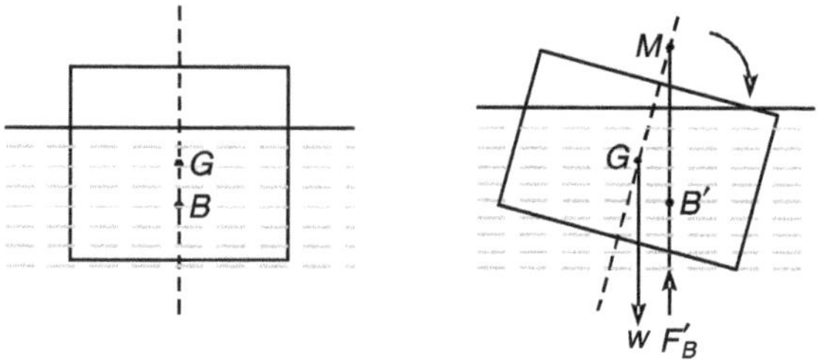

Body floating in liquid

The equilibrium of a submerged body in a liquid requires that the weight of the body acting through its centre of gravity should be colinear with an equal hydrostatic lift acting through the centre of buoyancy. Let us suppose that a body is given a small angular displacement and then released. Then it will be said to be in distance *MG* is called metacentric height (it is the distance between gravity centre and metacentre).

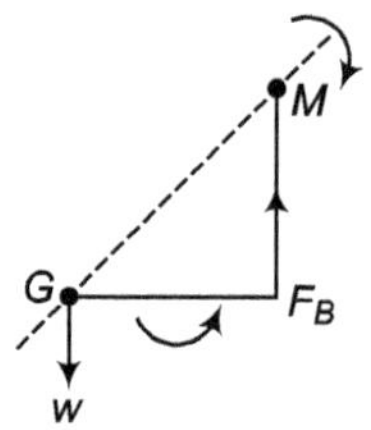

Gravity centre and metacentre

Stability of Submerged Body

It is classified into the three groups, which defined as

Stable Equilibrium

When centre of buoyancy lies above the centre of gravity, submerged body is stable.

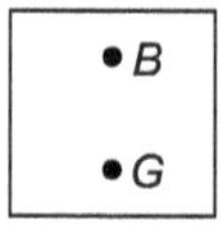

Stable equilibrium

Unstable Equilibrium

When *B* lies below *G*, then body is in unstable equilibrium.

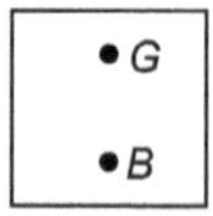

Unstable equilibrium

Neutral Equilibrium

When B and G coincide then, body is in neutral equilibrium.

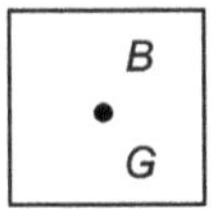

Neutral equilibrium

Stability of Floating Bodies

When the body undergoes an angular displacement about a horizontal axis, the shape of the immersed volume changes and so the centre of buoyancy moves relative to the body.

Stable Equilibrium

When a body is given a small angular displacement by external means and if body comes to its original position due to internal forces then, it is called **stable equilibrium.**

It occurs, when metacentre lies above centre of gravity.

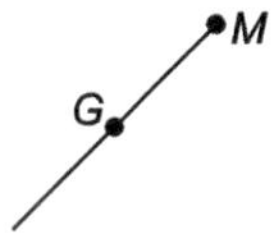

Stable position

Unstable Equilibrium

In the above case, if body does not come in its original position and moves further away then, it is known as **unstable equilibrium.** M lies below centre of gravity.

Unstable position

Neutral Equilibrium

When a body is given a small angular displacement and it sets on new position then, body is called in **neutral equilibrium.** In this, M and G coincide.

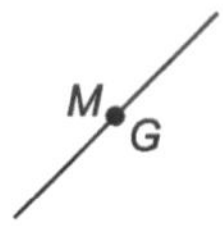

Neutral position

Relation between B, G and M is $GM = \frac{I}{V} - BG$

Here, I = Least moment of inertia of plane of body at water surface
G = Centre of gravity
B = Centre of buoyancy
M = Metacentre

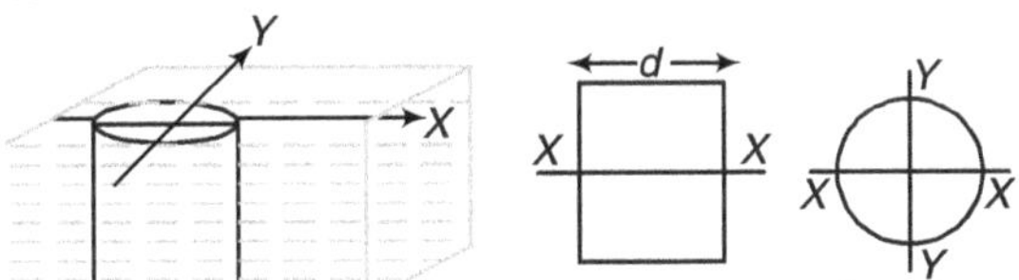

Front view Top view

$$I = \frac{\pi}{64} d^4$$

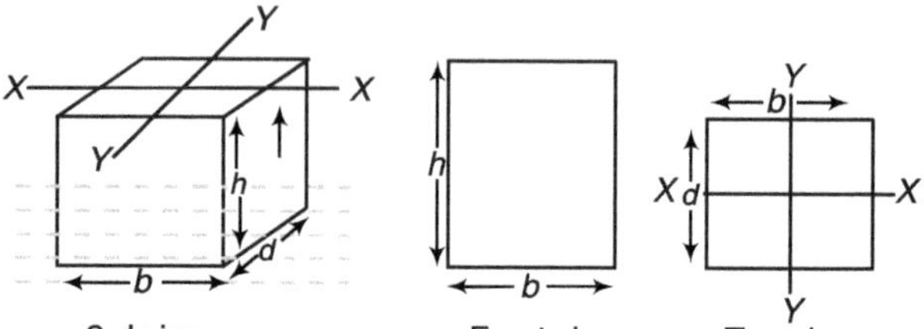

3-d view Front view Top view

Different views of rectangular and circular cross-sections

$$I = \min(I_{xx}, I_{yy}), \quad I_{xx} = \frac{bd^3}{12}, \quad I_{yy} = \frac{db^3}{12}$$

V is volume submerged inside the water can be given as

$$V = bdx$$

where b, d and x are the length, width and depth of the section or body.

BG is distance between centre of gravity and centre of buoyancy. (In other words, BG = distance between centre of gravity of whole body and centre of gravity of submerged part of body)

When we find out GM then, we can determine the status of body as $GM > 0$ (stable equilibrium), $GM < 0$ (unstable equilibrium), $GM = 0$ (neutral equilibrium).

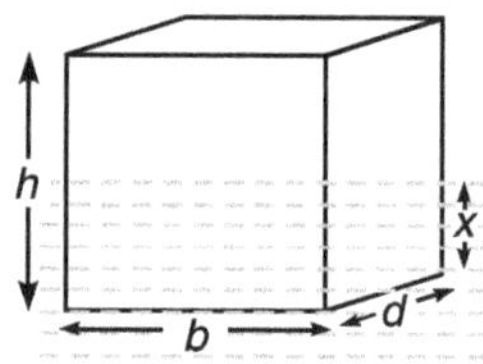

Submerged part of body

Viscosity

It is property of fluid by which it offers resistance to deformation under the action of shear force.

As shown in figure, upper layer moves with higher velocity relative to lower layer. Upper layer tries to move lower layer but lower layer moves with smaller velocity so, it opposes the motion of upper layer. Shear stress between two joining layers is proportional to velocity gradient

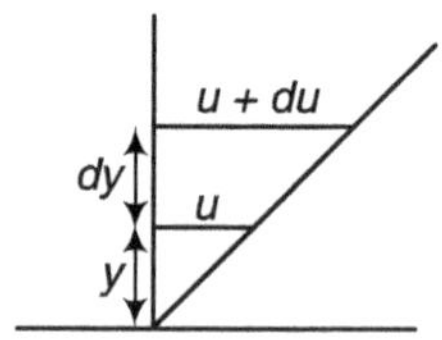

Viscosity at different layers

$$\tau \propto \frac{du}{dy}, \quad \tau = \mu \cdot \frac{du}{dy}$$

μ is the coefficient of viscosity.

Unit of viscosity = pascal-s (Pa-s) = N-s/m^2

1 poise = 0.1 pascal-s (Pa-s) = 0.1 N-s/m^2

Kinematics *and* Dynamics of Fluids

Kinematics is the branch of classical mechanics that describes the motion of bodies and systems without consideration of the forces the cause the motion.

Dynamics is that branch of mechanics which treats the motion of bodies and the action of forces in producing or changing their motion. Dynamics is also closely related to Newton's second law.

Kinematics deals with motion of particles without considering the forces causing the motion.

Steady *versus* Unsteady Flow

Steady Flow	Unsteady Flow
The flow in which characteristics of fluid like velocity, pressure, density etc., at a point, do not change with time is called as steady flow. $\frac{\partial v}{\partial t} = 0, \frac{\partial p}{\partial t} = 0, \frac{\partial p}{\partial t} = 0$	If velocity pressure and density changes with time then flow is unsteady flow. $\frac{\partial v}{\partial t} \neq 0, \frac{\partial p}{\partial t} \neq 0, \frac{\partial p}{\partial t} \neq 0$

Uniform *versus* Non-uniform Flow

Uniform Flow	Non-uniform Flow
The flow in which velocity at any given time does not change with respect to distance. $\left(\frac{\partial v}{\partial s}\right)_{t = c} = 0$	In this flow, velocity at any given time changes with respect to distance. $\left(\frac{\partial v}{\partial s}\right)_{t = c} \neq 0$

Laminar *versus* Turbulent Flow

Laminar Flow	Turbulent Flow
The flow in which the adjacent layers do not cross each other and move along well defined path. 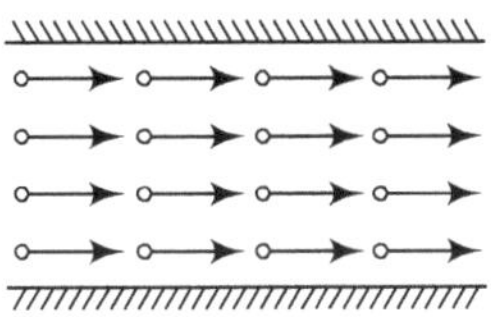	The flow in which adjacent layers cross each other and do not move along well defined path. 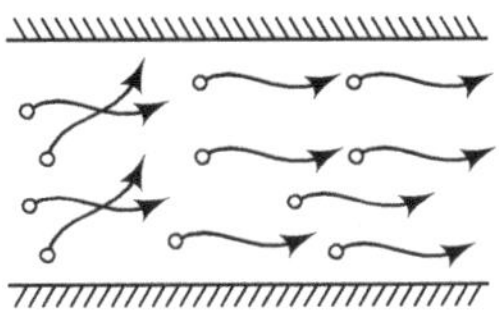

Rotational *versus* Irrotational Flow

Rotational Flow	Irrotational Flow
If the fluid particles flowing along stream lines, also rotate about their own axes, then flow is rotational.	If fluid particles do not rotate about their own axes, then flow is irrotational.

Compressible *versus* Incompressible Flow

Compressible Flow	Incompressible Flow
If density of fluid varies from point to point then fluid flow is compressible flow $\rho \neq$ constant	In incompressible flow, density remains constant ρ = constant

Continuity Equation

This equation is based on the principle of conservation of mass. According to this equation, mass flow rate is constant.

By continuity equation, $\rho_1 A_1 v_1 = \rho_2 A_2 v_2$ or $\frac{A_1 v_1}{V_1} = \frac{A_2 v_2}{V_2}$

where, V_1 and V_2 are the specific volume for inlet and outlet flows respectively.

Here, V is specific volume.

If fluid is incompressible, then

$$\rho_1 = \rho_2$$

$$\therefore \quad A_1 v_1 = A_2 v_2$$

Quantity Av is known as discharge Q.

Continuity Equation in 3-D

Continuity equation in 3-D can be derive from the diagram

$$\frac{\partial}{\partial x}(\rho \cdot u) + \frac{\partial}{\partial y}(\rho \cdot v) + \frac{\partial}{\partial z}(\rho \cdot w) + \frac{\partial \rho}{\partial t} = 0$$

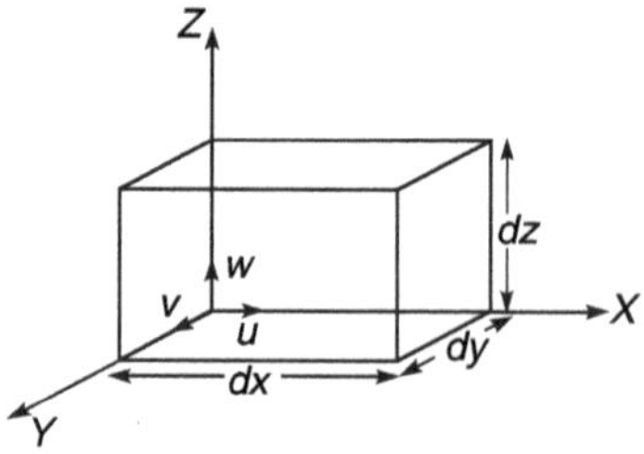

3-D body diagram

If flow in steady then, $\frac{\partial \rho}{\partial t} = 0$

Continuity equation will be

$$\frac{\partial}{\partial x}(\rho \cdot u) + \frac{\partial}{\partial y}(\rho \cdot v) + \frac{\partial}{\partial z}(\rho \cdot w) = 0$$

If fluid is incompressible then, ρ = constant

Continuity equation will be

$$\frac{\partial u}{\partial x}+\frac{\partial v}{\partial y}+\frac{\partial w}{\partial z}=0$$

Velocity of Fluid Particle

Let v_R is resultant velocity at any point in a fluid flow. Let u, v and w are velocity components in X, Y and Z directions.

Resultant velocity $\mathbf{v}=u\mathbf{i}+v\mathbf{j}+w\mathbf{k}$

$$v_R=\sqrt{u^2+v^2+w^2}$$

where u, v and w are the resultant vectors in X, Y and Z directions, respectively.

Acceleration of Fluid Particle

Let a_x, a_y, a_z are the accelerations in X, Y and Z directions, respectively.

For stead flow, $a_x=u\frac{\partial u}{\partial x}+v\frac{\partial u}{\partial y}+w\frac{\partial u}{\partial z}$

$$a_y=u\frac{\partial v}{\partial x}+v\frac{\partial v}{\partial y}+w\frac{\partial v}{\partial z}$$

$$a_z=u\frac{\partial w}{\partial x}+v\frac{\partial w}{\partial y}+w\frac{\partial w}{\partial z}$$

Stream Function

The partial derivative of stream function with respect to any direction gives the velocity component at right angles to that direction. It is denoted by ψ.

$$\frac{\partial\psi}{\partial x}=v,\ \frac{\partial\psi}{\partial\psi}=-u$$

Continuity equation for two-dimensional flow is

$$-\frac{\partial^2\psi}{\partial x\partial y}+\frac{\partial^2\psi}{\partial y\partial x}=0$$

Equations of Rotational Flow

As ψ satisfies the continuity equation hence, if ψ exists then it is a possible case of fluid flow. Rotational components of fluid particles are

$$\omega_x=\frac{1}{2}\left(\frac{\partial\omega}{\partial y}-\frac{\partial v}{\partial z}\right)$$

$$\omega_y = \frac{1}{2}\left(\frac{\partial u}{\partial z} - \frac{\partial \omega}{\partial x}\right)$$

$$\omega_z = \frac{1}{2}\left(\frac{\partial v}{\partial x} - \frac{\partial u}{\partial y}\right)$$

Equations of Irrotational Flow

If $\omega_x = \omega_y = \omega_z$ then, flow is irrotational.

For irrotational flow, $\omega_z = 0$

$$\omega_z = \frac{1}{2}\left(\frac{\partial v}{\partial x} - \frac{\partial u}{\partial y}\right) = 0$$

$$\frac{\partial}{\partial x}\left(\frac{\partial \psi}{\partial x}\right) - \frac{\partial}{\partial y}\left(-\frac{\partial \psi}{\partial y}\right) = 0$$

$$\frac{\partial^2 \psi}{\partial x^2} + \frac{\partial^2 \psi}{\partial y^2} = 0$$

This is Laplace equation for ψ.

It can be concluded that

- If stream function (ψ) exists, it is a possible case of fluid flow. But we can't decide whether flow is rotational or irrotational.
- But if stream function ψ satisfies Laplace equation then, it is a possible case of irrotational flow otherwise it is rotational flow.

Velocity Components in Terms of ψ

The velocity components in cylindrical polar coordinates in terms of stream function in $u_r = \frac{1}{r}\frac{\partial \psi}{\partial \theta}$, $u_\theta = \frac{-\partial \psi}{\partial r}$.

where, u_r = Radial velocity, u_θ = Tangential velocity

Velocity Potential Function

It is a scalar function of space and time such that its negative derivative with respect to any direction gives the fluid velocity in that direction. It is denoted by ϕ.

$$-\frac{\partial \phi}{\partial x} = u, \quad -\frac{\partial \phi}{\partial y} = v, \quad -\frac{\partial \phi}{\partial z} = w$$

We know that continuity equation for steady flow is

$$\frac{\partial u}{\partial x} + \frac{\partial v}{\partial y} = 0$$

$$\frac{\partial}{\partial x}\left(-\frac{\partial \phi}{\partial x}\right) + \frac{\partial}{\partial y}\left(-\frac{\partial \phi}{\partial y}\right) = 0$$

$$\frac{\partial^2 \phi}{\partial x^2} + \frac{\partial^2 \phi}{\partial y^2} = 0$$

If ϕ satisfies the Laplace equation, then it is a possible case of fluid flow. Rotational component ω_z can be given by

$$\omega_z = \frac{1}{2}\left(\frac{\partial v}{\partial x} - \frac{\partial u}{\partial y}\right) = \frac{1}{2}\left[\frac{\partial}{\partial x}\left(\frac{-\partial \phi}{\partial y}\right) - \frac{\partial}{\partial y}\left(-\frac{\partial \phi}{\partial x}\right)\right]$$

$$= \frac{1}{2}\left(\frac{-\partial^2 \phi}{\partial x \cdot \partial y} + \frac{\partial^2 \phi}{\partial y \cdot \partial x}\right) = 0$$

It shows that ϕ exists then, flow will be irrotational.

Relation between Stream Function and Velocity Potential

We know $$-\frac{\partial \phi}{\partial x} = u, -\frac{\partial \phi}{\partial y} = v$$

and $$\frac{\partial \psi}{\partial x} = v, -\frac{\partial \psi}{\partial y} = u$$

∴ $$\frac{\partial \phi}{\partial x} = \frac{\partial \psi}{\partial y} \text{ and } \frac{\partial \phi}{\partial y} = -\frac{\partial \psi}{\partial x}$$

Stream *versus* Velocity Function

Stream Function	Velocity Potential
$\frac{\partial \psi}{\partial x} = v$ $\frac{\partial \psi}{\partial y} = -u$	$-\frac{\partial \phi}{\partial x} = u$ $-\frac{\partial \phi}{\partial y} = v$
If ψ exists then, a possible case of fluid flow.	If ϕ exists then, flow is irrotational.
If ψ satisfies Laplace equation then, flow is irrotational.	If ϕ satisfies the Laplace equation then, a possible case of fluid flow.

Equipotential Line *versus* Stream Line

Equipotential Line	Stream Line
It is the line along which velocity potential ϕ is constant. $d\phi = \frac{d\phi}{dx} \cdot dx + \frac{\partial \phi}{\partial y} \cdot dy = 0$ $\Rightarrow \quad - u \cdot dx - v \cdot dy = 0$ $\Rightarrow \quad \frac{dy}{dx} = \frac{-u}{v}$	It is the line along which stream function ψ is constant. $d\psi = \frac{\partial \psi}{\partial x} \cdot dx + \frac{\partial \psi}{\partial y} \cdot dy = 0$ $\Rightarrow \quad v\ dx - u\ dy = 0$ $\Rightarrow \quad \frac{dy}{dx} = \frac{v}{u}$

Vorticity

Vorticity Ω is two times of value of rotation.

Vortex flow It is defined as the flow of a fluid along a curved path or the flow of rotating mass of fluid is known as **vortex flow.**

Forced Vortex *versus* Free Vortex Flow

Forced Vortex Flow	Free Vortex Flow
In this flow, some external torque is required to rotate the fluid mass. ω = constant or $\frac{v}{r}$ = constant	In this flow external torque is zero. $T = \frac{d}{dt}(mvr)$ As $T = 0$ $\therefore \quad \frac{d}{dt}(mvr) = \theta$ So, mvr = constant As m is constant So, $v \cdot r$ = constant
Examples (a) Flow of liquid inside the impeller of centrifugal pump. (b) A vertical cylinder containing liquid which is rotated about its central axis with constant angular velocity ω.	**Examples** (a) Flow of liquid around a circular bend in pipe. (b) A whirlpool in river.

Dynamics of Fluid Flow

In this, we study the forces acting on the fluid. According to Newton's law of motion acceleration of fluid in a direction is the ratio of net force acting in that direction to mass of fluid.

$$a = \frac{F}{m}$$

Energy Equations

The common fluid mechanics equations used in fluid dynamics are given below

Let, Gravity force F_g Pressure force F_p

Viscous force F_v Compressibility force F_c

Turbulent force F_t

$$F_{net} = F_g + F_p + F_v + F_c + F_t$$

(a) If fluid is incompressible, then $F_c = 0$

$\therefore$ $$F_{net} = F_g + F_p + F_v + F_t$$

This is known as **Reynolds equation of motion.**

(b) If fluid is incompressible and turbulence is negligible, then

$$F_c = 0, F_t = 0 \quad \therefore \; F_{net} = F_g + F_p + F_v$$

This equation is called as **Navier-Stokes equation.**

(c) If fluid flow is considered ideal then, viscous effect will also be negligible. Then,

$$F_{net} = F_g + F_p$$

This equation is known as **Euler's equation.**

Euler's equation can be written as

$$\frac{dp}{\rho} + gdz + vdv = 0$$

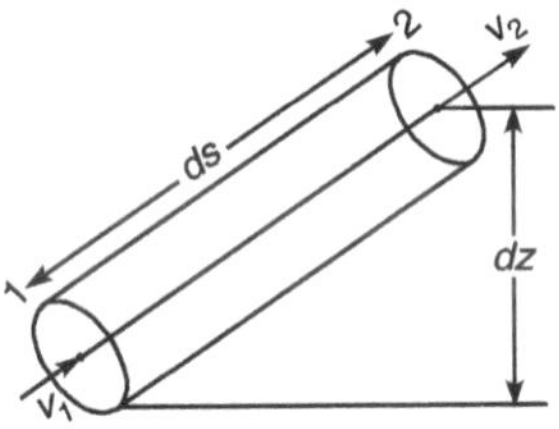

Incompressible flow

Bernoulli's Equation

It is based on law of conservation of energy. *This equation is applicable when it is assumed that*

1. Flow is steady and irrotational
2. Fluid is ideal (non-viscous)
3. Fluid is incompressible

Bernoulli's equation can be obtained by **Euler's equation**

$$\frac{dp}{\rho} + v\,dv + g\,dz = \text{constant}$$

As fluid is incompressible, ρ = constant

$$\int \frac{dp}{\rho} + \int v\,dv + \int g\,dz = \text{constant}$$

or $$\underset{\text{Pressure head}}{\frac{p}{\rho g}} + \underset{\text{Kinetic head}}{\frac{v^2}{2g}} + \underset{\text{Potential head}}{z} = \text{constant}$$

where, $\text{Head} = \dfrac{\text{Energy}}{\text{Weight}}$

But, real fluids are viscous and hence, offer resistance to flow. Bernoulli's equation can be modified by taking head loss into consideration.

Between points 1 and 2,

$$\frac{p_1}{\rho g} + \frac{v_1^2}{2g} + z_1 = \frac{p_2}{\rho g} + \frac{v_2^2}{\rho g} + z_2 + h_L$$

Flow Measurement

Flow measurement is the quantification of bulk fluid movement. Flow can be measured in a variety of ways. Positive displacement flow meters accumulate of a fixed volume of fluid and then count the number of times the volume is filled to measure flow.

Some flow measuring instruments are given as

Venturimeter

It measures the rate of flow of fluid flowing through a pipe. Applying Bernoulli's equation between points 1 and 2.

∴ Venturi meter head, $$h = \frac{v_2^2 - v_1^2}{2g}$$

where, v_1 = Velocity at inlet section

v_2 = Velocity at throat

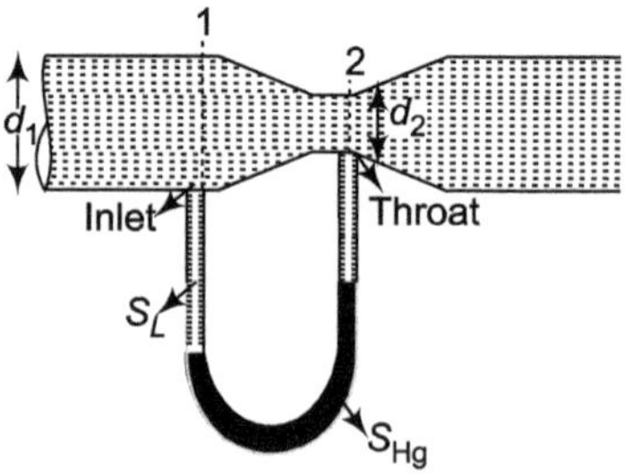

Venturimeter

In term of discharge, venturimeter head,

$$h = \frac{Q^2_{th}}{2g}\left(\frac{1}{A_2^2} - \frac{1}{A_1^2}\right)$$

$$Q_{act} = C_d \cdot Q$$

C_d is coefficient of discharge ($C_d < 1$)

If $S_{Hg} > S_L$ $\quad h = x\left(\dfrac{S_{Hg}}{S_L} - 1\right)$

If $S_L > S_{Hg}$ $\quad h = x\left(1 - \dfrac{S_{Hg}}{S_L}\right)$

Hg is liquid filled in manometric tube while L is liquid flowing inside venturimeter.

Pitot Tube

Pitot tube is used to measure mean velocity of flow.

Now, applying Bernoulli's equation between points 1 and 2,

Stagnation pressure = static pressure + dynamic pressure

$$\frac{p_1}{\rho g} + \frac{v_1^2}{2g} = \frac{p_2}{\rho g}$$

Actual velocity, $v_{actual} = C_v \sqrt{2gh}$

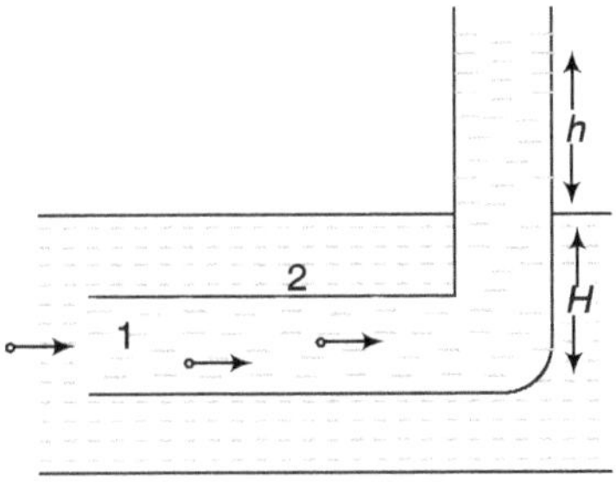

Pitot tube

Force Exerted by Fluid Flow on Bend

The net force exerted on fluid on X-direction is equal to rate of change of momentum in the X-direction *i.e.*, net force on fluid in X-direction = rate of change of momentum in X-direction

$$\Rightarrow \quad F_X = \rho Q(v_2 \cos\theta - v_1) + p_2 A_2 \sin\theta - p_1 A_1$$

Net force in Y-direction = Rate of charge of momentum in Y-direction

$$F_Y = \rho Q\,(v_2 \sin\theta) + P_2 A_2 \sin\theta$$

Net force, $\quad F = \sqrt{F_X^2 + F_Y^2}$

Here F_X and F_Y are forces exerted by fluid flow on the bend.

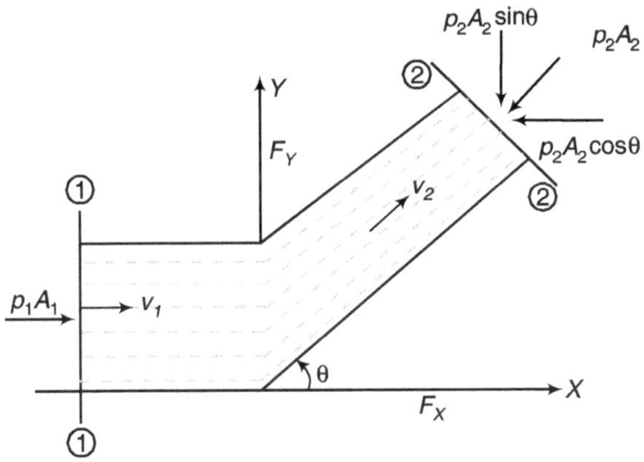

Fluid flow through bend pipe

Discharge through Large Rectangular Orifice

With a large rectangular orifice, the velocity of the liquid particles is not constant, because there is a considerable variation of effective pressure head over the height of a orifice.

$$\text{Area of strip} = b \cdot dh$$

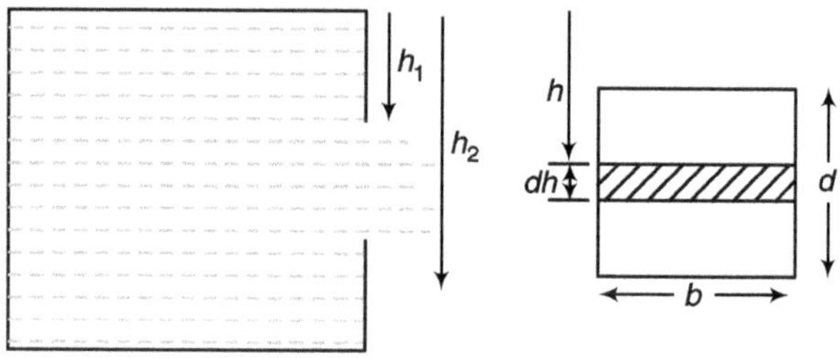

Large rectangular orifice

Discharge, $$Q = \frac{2}{3} b \sqrt{2g} \, (h_2^{3/2} - h_1^{3/2})$$

If C_d is coefficient of discharge then,

$$Q_{\text{actual}} = \frac{2}{3} C_d \cdot b\sqrt{2g} \, (h_2^{3/2} - h_1^{3/2})$$

Notches and Weirs

A notch is a device used for measuring the rate of flow of a liquid through a small channel. A weir is a concrete or masionary structure placed in the open channel over which the flow occurs. It is generally in the form of vertical wall, with a sharp edge at the top running all the way across the open channel. The notch is of small size while the weir is of bigger size.

Discharge over a Rectangular Notch or Weir

Consider a rectangular notch in one side of a tank over which water is flowing as shown in figure.

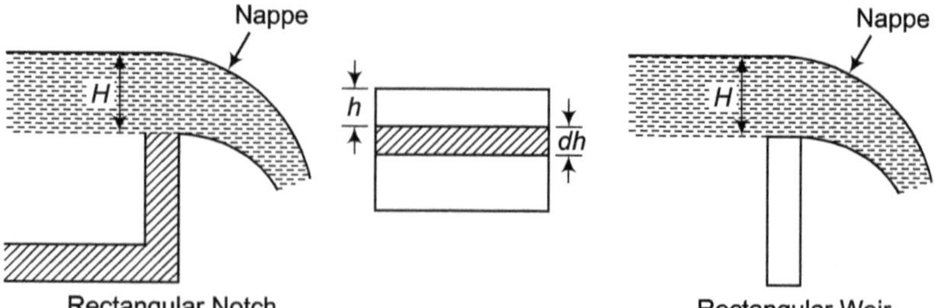

Rectangular Notch Rectangular Weir

Total discharge for the whole notch or Weir

$$Q = \frac{2}{3} C_d \times L \times \sqrt{2g}\, H^{3/2}$$

where, H = Head of water over the crest , L = Length of notch or weir.

Time required to empty a reservoir or a tank with a rectangular weir or notch

$$T = \frac{3A}{C_d L\sqrt{2g}} \left[\frac{1}{\sqrt{H_2}} - \frac{1}{\sqrt{H_1}}\right]$$

where, L = Length of crest of weir or notch

C_d = Coefficient of discharge

H_1 = Initial height of liquid above the crest of notch

H_2 = Final height of liquid above the crest of notch

T = Time required in seconds to lower the height of liquid from H_1 to H_2.

Time required to empty a reservoir or a tank with a triangular weir or notch.

$$T = \frac{5A}{4 C_d \times \tan\frac{\theta}{2} \sqrt{2g}} \left[\frac{1}{H_2^{3/2}} - \frac{1}{H_1^{3/2}}\right]$$

θ = Angle of notch

Discharge over a Triangular Notch or Weir

A triangular notch is also called a ***V*-notch**. Consider a triangular notch, in one side of the tank, over which water is flowing as shown in figure.

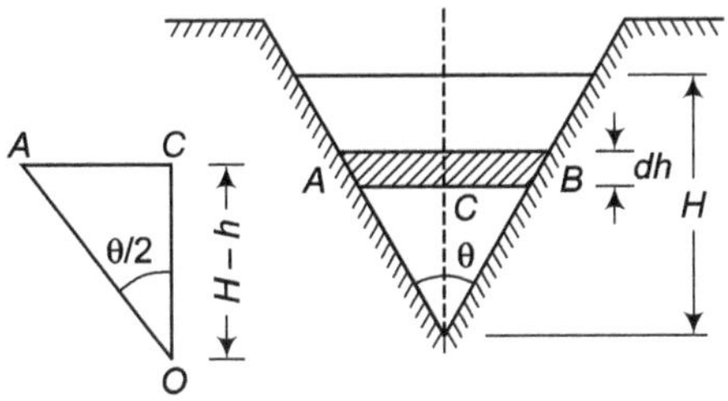

Triangular notch

Let H = Head of water above the V notch, Q = Angle of notch

Discharge over a triangular notch or weir

$$Q = \frac{8}{15} C_d \tan\frac{\theta}{2} \times \sqrt{2g}\, H^{5/2}$$

Hydraulic Jump

The rise of water level which takes place due to the transformation of the unstable shooting flow (super-critical) to the stable streaming flow (subcritical flow) is known as **hydraulic jump.**

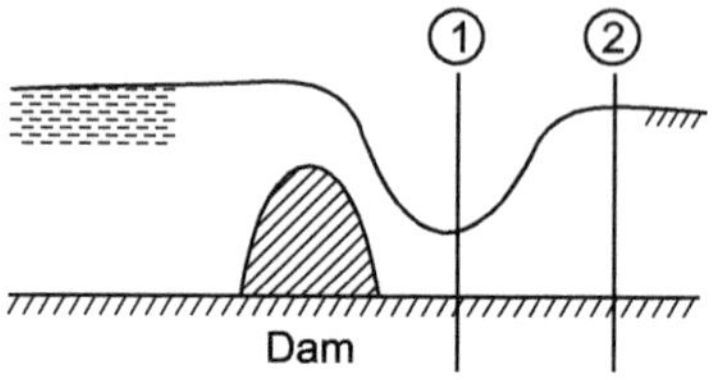

Hydraulic jump

Depth of flow after the jump,

$$d_2 = -\frac{d_1}{2} + \sqrt{\frac{d_1^2}{4} + \frac{2v_1^2 d_1}{g_1}} = \frac{d_1}{d_2}\left(\sqrt{1 + 8\,(\text{Fe})_1^2} - 1\right)$$

Depth of hydraulic jump $= d_2 - d_1$

v_1 = Velocity of flow on the upstream side

d_1 = Depth of flow on upstream side

$$F_e = \frac{v_1}{\sqrt{gd_1}}$$

F_e = Fraud number, d_2 = Depth of flow after the hydraulic jump.

Length of Hydraulic Jump

It is total length between two section taking one before hydraulic jump and two after hydraulic jump. Its value for a rectangular channel is approximately, 5 to 7 times to the height of the jump.

Energy loss due to hydraulic jump

$$h_L = \left(d_1 + \frac{v_1^2}{2g}\right) - \left(d_2 + \frac{v_2^2}{2g}\right) = \frac{(d_2 - d_1)^2}{4d_1 d_2}$$

Flow through Syphon

It is a long bent pipe which is used to transfer liquid from a reservoir at higher level to another reservoir at lower level when these two are separated by a hill or high level ground. At point C in given figure,

$$p_C < p_A(p_a)$$

because $p_A = p_D = p_a$ and C is h is distance above D then obviously at C, pressure will be less than atmospheric pressure Applying Bernoulli's equation to points A and B, we get

$$\frac{p_A}{\rho g} + \frac{v_A^2}{2g} + z_A = \frac{p_B}{\rho g} + \frac{v_B^2}{2g} + z_B + (h_f)$$

e.g., $(h_f)_{\text{between } A \text{ and } B}\ (p_A = p_B \pm v_A = v_B = 0)$

$$z_A - z_B = h_f = \frac{4f \cdot Lv^2}{d \cdot 2g} \quad \ldots(i)$$

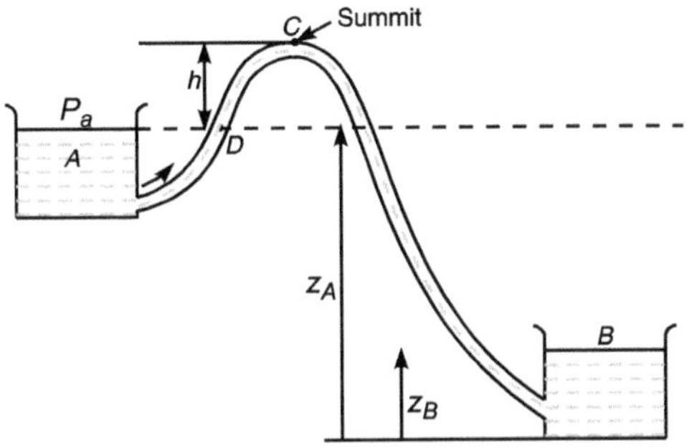

Flow through syphon

Now, applying Bernoulli's equation between A and C,

$$\frac{p_A}{\rho g} + \frac{v_A^2}{2g} + z_A = \frac{p_C}{\rho g} + \frac{v_C^2}{2g} + z_C + (h_f),$$

$$(h_f)_{\text{between } A \text{ and } C} \quad 0 + 0 + z_A = \frac{p_C}{\rho g} + \frac{v^2}{2g} + z_A + h + \frac{4fLv^2}{d \times 2g}$$

$$\frac{p_C}{\rho g} + \frac{v^2}{2g} + h + \frac{4fLv^2}{d \times 2g} = 0 \qquad (\because p_A = 0, v_A = 0, v_C = v)$$

Note As f is given, so we have to consider friction loss in Bernoulli's equation

Momentum Correction Factor (β)

It is defined as the ratio of momentum per second based on actual velocity to the momentum per second based on average velocity.

$$\beta = \frac{1}{Av^2} \int u^2 dA$$

where, v = Average velocity; u = Local velocity at distance r

For laminar flow $\beta = 1.33$; for turbulent flow $\beta = 1.2$.

Kinetic Energy Correction Factor (α)

It is defined as the ratio of kinetic energy per second based on actual velocity to the kinetic energy/second based on average velocity.

$$\alpha = \frac{1}{A \cdot v^3} \int u^3 dA$$

For laminar flow $\alpha = 2$, for turbulent flow $\alpha = 1.33$

Boundary Layer Theory

When a real fluid flows over a solid body, the velocity of fluid at the boundary will be zero. If boundary is stationary. As we move away from boundary in perpendicular direction velocity increases to the free stream velocity. It means velocity gradient $\left(\frac{du}{dy}\right)$ will exist.

Boundary layer is a thin layer where the variation of velocity from $u = 0$ at solid boundary to $u = u_\infty$ free stream velocity in perpendicular direction to boundary.

Note Velocity gradient $\frac{du}{dy}$ does not exist outside the boundary layer as outside the boundary layer velocity is constant and equal to free stream velocity.

Development of Boundary Layer

Development of boundary layer can be divided in three regions—laminar, transition, turbulent.

$$\text{Reynolds number} = \frac{\rho V \cdot x}{\mu} \; (\text{Re})_x = \frac{V \cdot x}{\nu}$$

- For laminar boundary layer

$$(\text{Re})_x < 5 \times 10^5 (\text{For flat plate}) \text{ and if } (\text{Re})_x > 5 \times 10^5$$

where R_e = Reynold number.

Then, flow is **turbulent**.

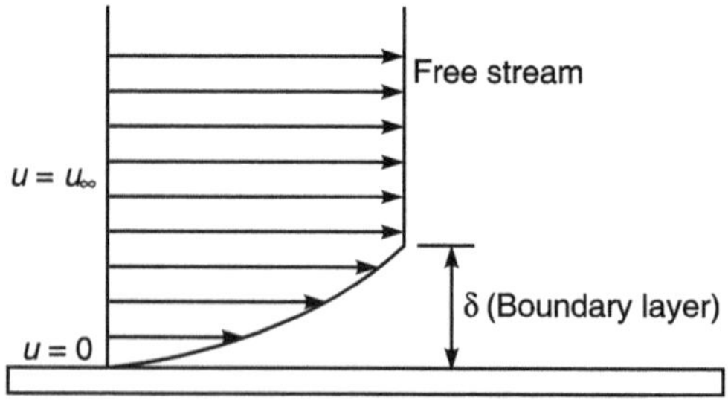

Boundary layer theory

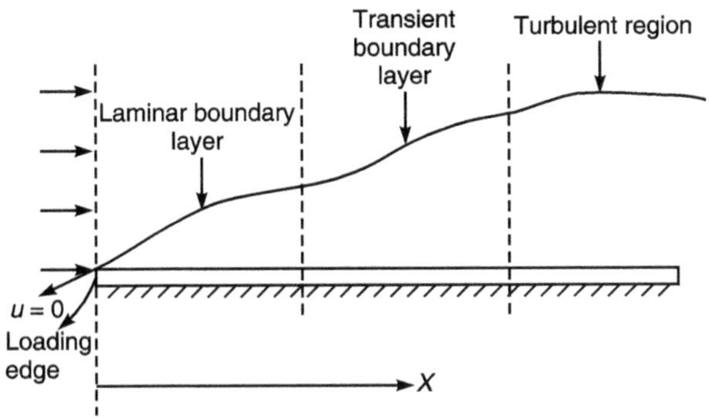

Boundary layer development

Here, x is distance from leading edge in horizontal direction.

Boundary Layer Thickness (δ)

It is the distance from the boundary to the point where velocity of fluid is approximately equal to 99% of free stream velocity. It is represented by δ.

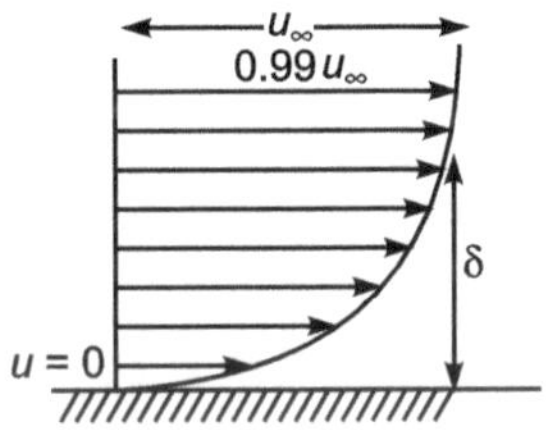

Boundary Layer thickness δ

Displacement Thickness (δ^*)

It is observed that inside the boundary layer velocity of fluid is less than free stream velocity hence, discharge is less in this region. To compensate for reduction in discharge the boundary is displaced outward in perpendicular direction by some distance. This distance is called displacement thickness (δ^*).

$$\delta^* = \int_0^{\delta} \left(1 - \frac{u}{U}\right) dy$$

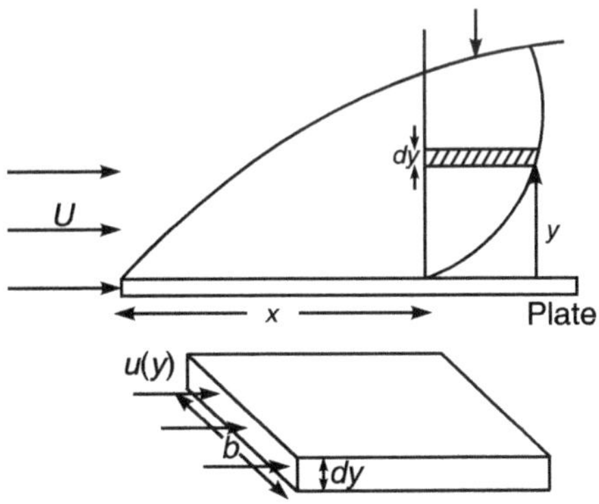

Boundary layer displacement thickness (δ^*)

Momentum Thickness (θ)

As due to boundary layer reduction in velocity occurs so, momentum also decreases. Momentum thickness is defined as the distance measured normal to boundary of solid body by which the boundary should be displaced to compensate for the reduction in momentum of flowing fluid.

$$\theta = \int_0^{\delta} \frac{u}{U}\left(1 - \frac{u}{U}\right) dy$$

Energy Thickness (δ^{**})

It is defined as distance measured perpendicular to the boundary of solid body by which the boundary should be displaced to compensate for reduction in kinetic energy of flowing fluid (KE decreases due to formation of boundary layer)

$$\delta^{**} = \int_0^{\delta} \frac{u}{U}\left(1 - \frac{u^2}{U^2}\right) dy$$

Boundary Conditions for the Velocity Profile

Boundary conditions are as

(a) At $y = 0, u = 0, \frac{du}{dy} \neq 0$ (b) At $y = \delta, u = U, \frac{du}{dy} = 0$

Laminar Flow

A flow in which fluid flows in layer and no intermixing with each other is known as laminar flow. For circular pipe, flow will be laminar.

If $$\text{Re} = \frac{\rho v D}{\mu} < 2000$$

where, ρ = Density of fluid, v = Velocity of fluid
D = Diameter of pipe, μ = Viscosity of fluid,

- For flat plate, flow will be laminar.

If $$\text{Re} = \frac{\rho v L}{\mu} < 5 \times 10^5$$

where L is length of plate.

Turbulent Flow

In this flow, adjacent layer of fluid cross each other (particles of fluid move randomly instead of moving in stream line path), for flow inside pipe. If Re > 4000, the flow is considered turbulent, for flat plate, Re > 5×10^5.

Von Karman Momentum Integral Equation

$$\frac{\tau_0}{\rho U^2} = \frac{d\theta}{dx}$$

where, θ = momentum thickness

- Shear stress $$\tau_0 = \mu\left(\frac{du}{dy}\right)_{y=0}$$

where, U = Free stream velocity; ρ = Density of fluid.

Local Coefficient of Drag (C_D^*)

It is defined as the ratio of the shear stress τ_0 to the quantity $\frac{1}{2}\rho U^2$.

It is denoted by $$C_D^* = \frac{\tau_0}{\frac{1}{2}\rho U^2}$$

Average Coefficient of Drag (C_D)

It is defined as the ratio of the total drag force to the $\frac{1}{2}\rho AU^2$.

$$C_D = \frac{F_D}{\frac{1}{2}\rho AU^2}$$

where, A = Area of surface; U = Free stream velocity
ρ = Mass density of fluid.

Blassius Experiment Results

- For laminar flow, $$\frac{f}{x} = \frac{5.48}{\sqrt{\text{Re}_x}}$$

  Coefficient of drag $$C_{fx} = \frac{0.664}{\sqrt{\text{Re}_x}}$$

  Average coefficient of drag $$C_D = \frac{1.46}{\sqrt{\text{Re}_L}}$$

- For turbulent flow, $\frac{f}{x} = \frac{0.37}{(\text{Re}_x)^{\frac{1}{5}}}$, where x = Distance from leading edge

  Re_x = Reynold number for length x

Re_x = Reynold number at end of plane

$$\text{Coefficient of drag } C_{fx} = \frac{0.059}{(Re_x)^{\frac{1}{5}}}$$

$$\text{Average coefficient of drag } C_D = \frac{0.072}{(Re_L)^{\frac{1}{5}}}$$

For laminar flow, $f \propto \sqrt{x}$

f = Boundary layer thickness, $\tau_0 \propto \frac{1}{\sqrt{x}}$

τ_0 = Shear stress at solid surface
x = Distance from where solid surface starts.

- **Velocity profile** for turbulent boundary layer is

$$\frac{u}{U} = \left(\frac{y}{f}\right)^{\frac{1}{n}} \Rightarrow \frac{1}{n} = \frac{1}{\tau}$$

$$\Rightarrow \quad 5 \times 10^5 < Re < 107$$

Conditions for Boundary Layer Separation

Let us take curve surface *ABCSD* where fluid flow

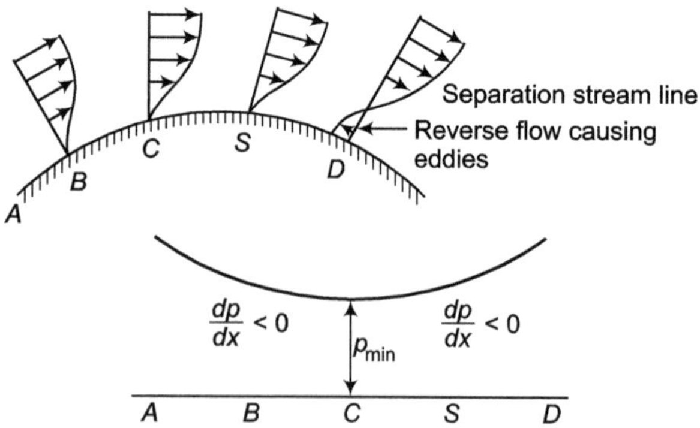

Separation boundary layers ABCSD

Separation print S is determined from the condition $\left(\frac{du}{dy}\right)_y = 0$

- If $\left(\frac{du}{dy}\right)_{y=0}$ = – ve, the flow is separated $\left(\frac{dp}{dx} > 0\right)$

- If $\left(\frac{du}{dy}\right)_{y=0}$ = 0, the flow is on the average of separation $\left(\frac{\partial p}{\partial x} = 0\right)$
- If $\left(\frac{\partial u}{\partial y}\right)_{y=0}$ = +ve, the flow will not separate or flow will remain attained $\left(\frac{dp}{dx} < 0\right)$.

Methods of Preventing Separation of Boundary Layer

- Suction of slow moving fluid by a suction slot
- Supplying additional energy from a blower.
- Providing a bypass in the slotted?.
- Rotating boundary in the direction of flow.
- Providing small divergence in a diffuser.
- Providing guide blades in a bend.
- Providing a trip wire ring in the laminar region for the flow over a sphere.

Flow Through Pipes

It is clear them when fluid flows through a pipe, the fluid experience some resistance due to which some energy of fluid is lost. *Energy loss is categorised as*

Major loss	**Minor loss**
↓	↓
Due to friction	Due to
	(a) sudden expansion of pipe
	(b) sudden contraction of pipe
	(c) bend in pipe
	(d) any obstruction in pipe

Major Loss

It is calculated by Darcy Weisbach formulas

Loss of head due to friction $h_f = \frac{4fLv^2}{d \cdot 2g}$

where, L = Length of pipe, v = Mean velocity of flow
d = Diameter of pipe, f = Coefficient of friction

or $$h_f = \frac{f' L v^2}{d 2g} \quad f' = 4f = \text{friction factor}$$

$$\frac{\text{for laminar flow}}{\text{frictional factor }(4f)} f' = \frac{64}{\text{Re}}$$

$$\text{Coefficient of friction } f = \frac{16}{\text{Re}}$$

$$\text{For turbulent flow, coefficient of friction } f = \frac{0.079}{\text{Re}^{\frac{1}{4}}}$$

Chezy's Formula

In fluid dynamics, Chezy's formula describes the mean flow velocity of steady, turbulant open channel flow.

- **Mean velocity of flow,**

$$v = c\sqrt{mi}\ , c = \text{Chezy constant } = \frac{\sqrt{\rho g}}{f'}$$

i = Loss of head per unit length of pipe

$$= \frac{h_f}{L} \text{ (hydraulic slope } \tan\theta)$$

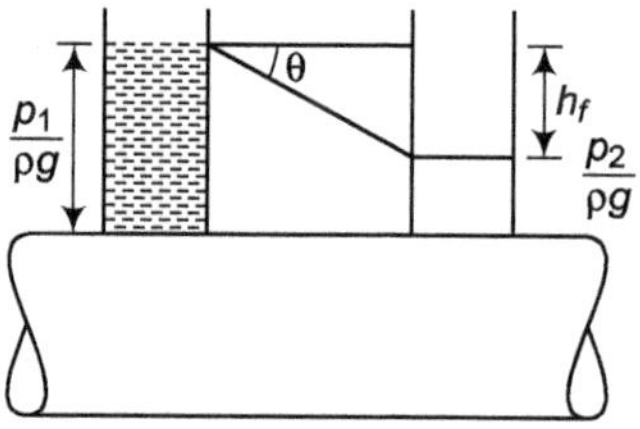

Chezy's formula of steady flow

$$m = \text{Hydraulic mean depth} = \frac{\text{Area (A)}}{\text{Wetted perimetr }(p)}$$

Relation Between Coefficient of Friction and Shear Stress

We get, $f = \dfrac{2\tau_0}{\rho v^2}$

where, f = Coefficient of friction

τ_0 = Shear stress

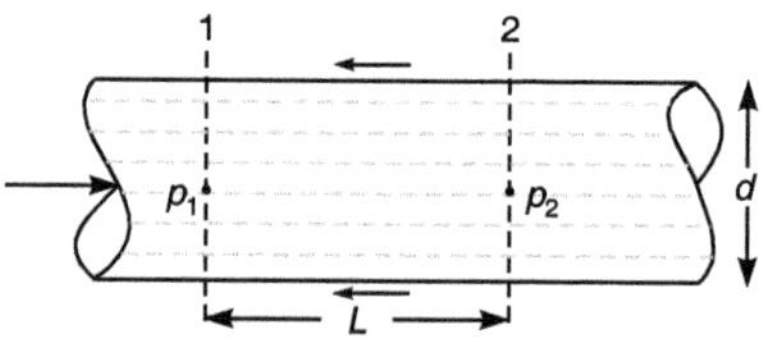

Coefficient of friction and shear stress

Minor Loss

The another type of head loss in minor loss is induced due to following reasons

Loss due to Sudden Enlargement

$$\text{Head loss, } h_L = \frac{(v_1 - v_2)^2}{2g}$$

Loss due to Sudden Contraction

$$\text{Head loss, } h_L = \frac{0.5v_1^2}{2g}$$

Remember v_1 is velocity at point which lies in contracted section.

Loss of Head at Entrance to Pipe

$$\text{Head loss, } h_L = 0.5\frac{v_1^2}{2g}$$

Loss at Exit from Pipe

$$\text{Head loss, } h_L = \frac{v_1^2}{2g}$$

Note In case 1 and 2, flow occurs between pipe to pipe, while in case 3 and 4, flow occurs between tank and pipe. We are taking entry or exit w.r.t pipe. So, be careful.

Combination of Pipes

Pipes may be connected in series, parallel or in both. Let see their combinations.

Pipes in Series

As pipes are in series, the discharge through each pipe will be same.

$$Q = A_1v_1 = A_2v_2 = A_3v_3$$

Total loss of head = Major loss + Minor loss

$$H = h_{L_1} + h_{L_2}$$

Major loss = Head loss due to friction in each pipe

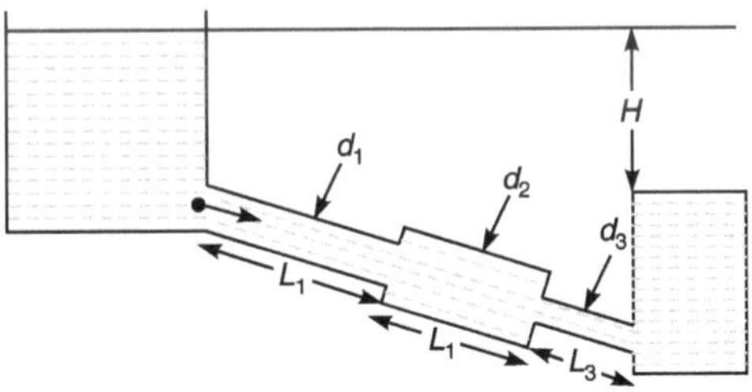

Pipe in series

$$h_{L_1} = h_{f_1} + h_{f_2} + h_{f_3}$$

$$= \frac{f_1L_1v_1^2}{d_1\cdot 2g} + \frac{f_2L_2v_2^2}{d_2\cdot 2g} + \frac{f_3L_3v_3^2}{d_3\cdot 2g}$$

While, minor loss = Entrance loss + Expansion loss + Contraction loss + Exit loss

$$h_{L_2} = \frac{0.5v_1^2}{2g} + \frac{(v_2 - v_1)^2}{2g} + \frac{0.5v_3^2}{2g} + \frac{v_3^2}{2g}$$

If minor loss are neglected then,

$$H = \frac{f_1 L_1 v_1^2}{d_1 \cdot 2g} + \frac{f_2 L_2 v_2^2}{d_2 \cdot 2g} + \frac{f_3 L_3 v_3^2}{d_3 \cdot 2g}$$

$$H = \frac{f_1 L_1 Q^2}{12 \cdot 1 d_1^5} + \frac{f_2 L_2 Q^2}{12 \cdot 1 d_2^5} + \frac{f_3 L_3 Q^2}{12 \cdot 1 d_3^5}$$

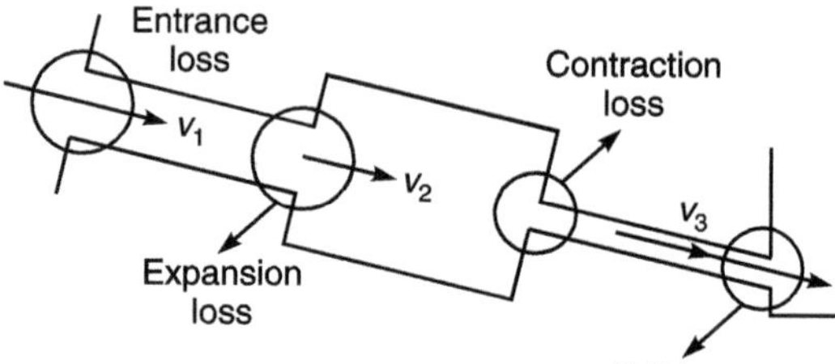

Head loss when minor loss are neglected

Pipes in Parallel

In this discharge in main pipe is equal to sum of discharge in each of parallel pipes.

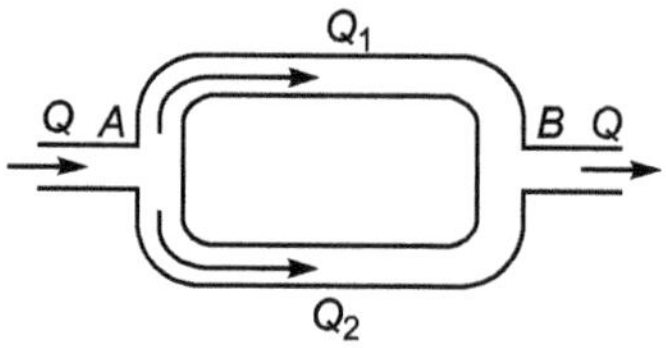

Pipes in parallel

Hence, $Q = Q_1 + Q_2$

Loss of head in each parallel pipe is same

$$h_{f_1} = h_{f_2}$$

$$\frac{f_1 L_1 v_1^2}{d_1 \cdot 2g} = \frac{f_2 L_2 v_2^2}{d_2 \cdot 2g} \text{ or } \frac{f_1 L_1 Q_1^2}{12 \cdot 1 d_1^5} = \frac{f_2 L_2 Q_2^2}{12 \cdot 1 d_2^5}$$

where, h_{f_1} and h_{f_2} are head loss at 1 and 2 respectively.

Equivalent Pipe

A compound pipe which consists of several pipes of different lengths and diameters to be replaced by a pipe having uniform diameter and the same length as that of compound pipe is called as equivalent pipe.

$$h_{Le} = h_{f_1} + h_{f_2} + h_{f_3}$$

$$\frac{fLQ^2}{12.1 d^5} = \frac{f_1 L_1 Q^2}{12.1 d_1^5} + \frac{f_2 L_2 Q^2}{12.1 d_2^5} + \frac{f_3 L_3 Q^2}{12.1 d_3^5} \quad (\text{where, } L = L_1 + L_2 + L_3)$$

If $f = f_1 = f_2 = f_3$

Then, $$\frac{L}{d^5} = \frac{L_1}{d_1^5} + \frac{L_2}{d_2^5} + \frac{L_3}{d_3^5} \Rightarrow \frac{L}{d^5} = \frac{L_1}{d_1^5} + \frac{L_2}{d_2^5} + \frac{L_3}{d_3^5}$$

Hydraulic Gradient Line (HGL) and Total Energy Line (TEL)

- HGL $\rightarrow$ It joins piezometric head $\left(\frac{p}{\rho g} = z\right)$ at various points.
- TEL $\rightarrow$ It joins total energy head at various points.

$$\left(\frac{p}{\rho g} + z + \frac{v^2}{2g}\right)$$

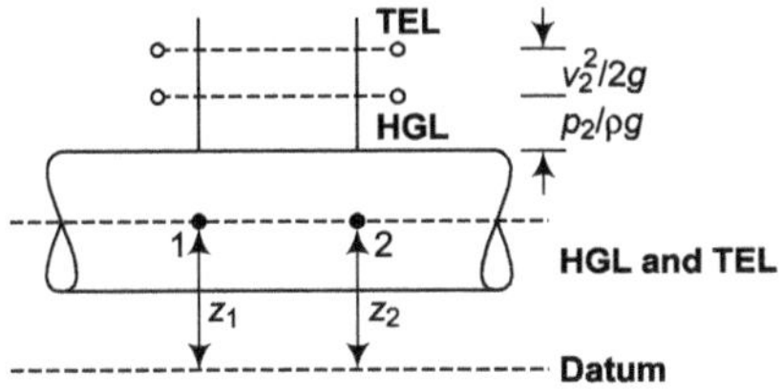

Equivalent pipe diagram

Note HGL is always parallel but lower than TEL.

Power Transmission through Pipe (P)

$$P_{\text{ideal}} = \rho g Q H$$
$$P_{\text{actual}} = \rho g Q(H - h_f)$$
$$h_f = \text{Head loss}$$
$$\text{Efficiency } \eta = \frac{H - h_f}{H}$$

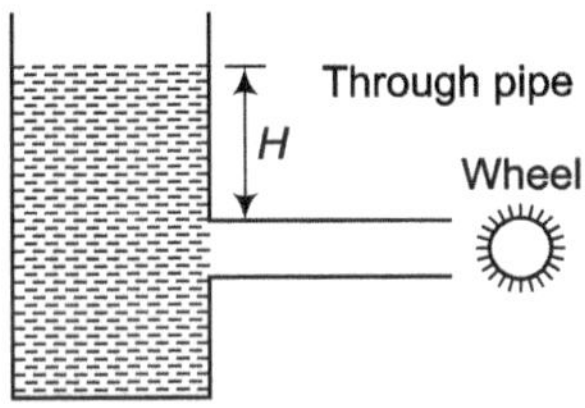

Power transmission through pipe

- Power delivered by a given pipe line is maximum when the flow is such that one third of static head is consumed in pipe friction. Thus, efficiency is limited to only 66.66%
- Maximum efficiency, $\eta_{\max} = \frac{H}{3}$

Water Hammer

When a liquid is flowing through a long pipe fitted with a vale at the end of the pipe and the valve is closed suddenly a pressure wave of high intensity is produced behind the valve. This pressure wave of high intensity is having the effect of hammering action on the walls of the pipe. This phenomenon is known as water hammer.

- Intensity of pressure rise due to water hammer,

$$p = -\frac{\rho L v}{t}$$

- When valve is closed gradually when valve closed suddenly with rigid pipe.

$$p = v\sqrt{K\rho}$$

When valve closed suddenly with plastic pipe

$$p = v \times \sqrt{\frac{p}{\frac{1}{K} + \frac{D}{Et}}}$$

- If the time required to close the valve

$$t > \frac{2L}{C} \Rightarrow \text{Valve closur is said to be gradual.}$$

$$t < \frac{2L}{C} \Rightarrow \text{The valve closur is said to be sudden.}$$

where, L = Length of pipe
D = Diameter of pipe
C = Velocity of pressure wave produced due to water hammer $\sqrt{\frac{K}{\rho}}$
v = Velocity of flow
K = Bulk modulus of water
E = Modulus of elasticity for pipe material
t = Time required to choose the valve.

Turbulent Flow

Turbulent flow is a flow regime characterised by the following points as given below

- **Shear stress** in turbulent flow

$$\tau = \tau_v + \tau_t = \mu \frac{d\bar{u}}{dy} + \eta \frac{d\bar{u}}{dy}$$

where, τ_v and τ_t = shear stress due to viscosity and turbulence.
η = eddy viscosity coefficient.

- **Turbulent shear stress** by Reynold.

$$\tau = \rho u' v'$$

u' and v' fluctuating component of velocity.

- Shear stress in turbulent flow due to Prandtle is

$$\tau = \rho \ell^2 \left(\frac{du}{dy}\right)^2$$

where, ℓ = Mixing length.

- The velocity distribution in the turbulent flow for pipes is given by the expression

$$u = u_{max} + 2.5 u * \log_e \frac{y}{R}$$

$$u_{max} = \text{centre velocity}$$

where, y = Distance from the pipe wall, R = radius of the pipe

u^* = Shear velocity $\left(\sqrt{\frac{\tau_0}{\rho}}\right)$.

- Velocity defect is the difference between the maximum velocity (u_{max}) and local velocity (u) at any point is given by

$$u_{max} - u = 5.75 u * \log_{10} \frac{R}{y}$$

- **Karman-Prandtle velocity distribution equation.**

Hydrodynamically pipe

$$\frac{u}{u*} = 5.75 \log_{10}\left(\frac{u^* y}{\nu}\right) + 5.5 \qquad \text{(for smooth pipe)}$$

$$= 5.75 \log_{10}\left(\frac{y}{k}\right) + 8.5 \qquad \text{(for rough pipe)}$$

where, u = velocity at any point in the turbulent flow

u^* = shear velocity = $\sqrt{\frac{\tau_0}{\rho}}$

ν = Kinematic viscosity of fluid

y = Distance from pipe wall

k = Roughness factor.

- Velocity distribution in terms of average velocity

$$\frac{\bar{u}}{u*} = 5.75 \log_{10} \frac{u * R}{\nu} + 5.5 \qquad \text{(for smooth pipe)}$$

$$= 5.75 \log_{10} \frac{R}{k} + 4.75 \qquad \text{(for rough pipe)}$$

$$\frac{u - \bar{u}}{u*} = 5.75 \log_{10} \frac{y}{k} + 3.75 \text{ (for both rough and smooth pipe)}$$

- Coefficient of friction $f = \frac{16}{\text{Re}}$ (for laminar flow)

$$f = \frac{0.0791}{(\text{Re})^{\frac{1}{4}}},\ 4000 \le \text{Re} < 10^5 \qquad \text{(for smooth pipe)}$$

$$= 0.0008 + \frac{0.05525}{\text{Re}^{0.257}}, \quad 4 \times 10^7 \le \text{Re} \le 10^5 \quad \text{(for smooth pipe)}$$

$$\frac{1}{\sqrt{4f}} = 2 \log_{10}\left(\frac{R}{k}\right) + 1.74 \quad \text{(for rough pipe)}$$

Dimensional *and* **Model Analysis**

One of the most useful applications of dimensional and model analysis is in designing physical and numerical experiment, and in reporting the results of such theorem.

Buckingham π Theorem

It states that if all the *n*-variables are described by *m* fundamental dimensions, they may be grouped into $(n - m)$ dimensionless π term.

$$f(\pi_1, \pi_2 \ldots \pi_{n-m}) = 0$$

$$\pi_1 = X_2^{a_1} X_3^{b_1} X_4^{c_1} X_1$$

$$\pi_2 = X_2^{a_2} X_3^{b_2} X_4^{c_2} X_5$$

$$\vdots \quad \vdots \quad \vdots \quad \vdots$$

$$\pi_{n-m} = X_2^{a_{n-m}} X_3^{b_{n-m}} X_4^{c_m} X_n$$

where, X_2, X_3 and $X_4 \rightarrow$ repeating variables.

Different Numbers and their Equations in Fluid Mechanics

Number	Equation	Signification
Reynold number	$\frac{F_i}{F_v} = \frac{\rho v l}{\mu}$	Flow in closed conduit pipe
Fraude number	$\sqrt{\frac{F_i}{F_g}} = \frac{v}{\sqrt{gL}}$	For free surface where gravity predominant *e.g.*, weir spillway, channel
Euler number	$\sqrt{\frac{F_i}{F_g}} = \frac{v}{\sqrt{\frac{p}{\rho}}}$	In cavitation studies
Mach number	$\sqrt{\frac{F_i}{F_e}} = \frac{v}{C}$	For compressible fluid
Weber number	$\sqrt{\frac{F_i}{F_6}} = \frac{v}{\sqrt{\frac{6}{\rho^{\ell}}}}$	In capillary studies

Model Laws or Similarity Laws

The laws on which the models are designed for dynamic similarity are called model laws or laws of similarity.

Reynold's Model Law

Use where viscous force predominant

$$(\text{Re})_m = (\text{Re})_p$$

where, v_m = Velocity of fluid in model, ρ_m = Density of fluid in model

L_m = Length in model, μ_m = Viscosity of fluid in model

v_p, ρ_p, L_p and μ_p are corresponding values of velocity, density, linear dimension and viscosity of fluid in prototype.

$$\frac{\rho_m v_m L_m}{\mu_m} = \frac{\rho_p v_p L_p}{\mu_p}, \frac{\rho_r v_r L_r}{\mu_r} = 1$$

Fraud's Model Law

Use where gravity force is predominant

$$\frac{v_m}{\sqrt{g_m L_m}} = \frac{v_p}{\sqrt{g_p L_p}}$$

Scale ratio for velocity $v_r = \sqrt{L_r}$ (L_r is the scale ratio for length)

Time $T_r = \sqrt{L_r}$

- Acceleration $a_r = 1$
- Discharge $Q_r = L_r^{2.5}$
- Force $F_r = L_r^3$
- Scale ratio for power $p_r = L_r^{3.5}$
- Scale ratio for energy, work torque moment

Ship partially submerged in rough and turbulent sea.

Euler's Model Law

Use where the pressure force are dominant force

$$\frac{v_m}{\sqrt{\frac{p_m}{\rho_m}}} = \frac{v_p}{\sqrt{\frac{p_p}{\rho_p}}}$$

p_m and p_p are pressure of fluid in model and prototype.
If same fluid for model and prototype,

$$\rho_m = \rho_p$$
$$\frac{V_m}{\sqrt{p_m}} = \frac{V_p}{\sqrt{p_p}}$$

Weber Model Law

Use where the surface tension is predominant.

$$\frac{V_m}{\sqrt{\frac{\sigma_m}{\rho_m L_m}}} = \frac{V_p}{\sqrt{\frac{\sigma_p}{\rho_p L_p}}}$$

where, σ_m = Surface tensile force in model.

Mach Model Law

Use where the elastic force is predominant.

$$\frac{V_m}{\sqrt{\frac{K_m}{\rho_m}}} = \frac{V_p}{\sqrt{\frac{K_m}{\rho_p}}}$$

where, K_m = Elastic stress for model.

Hydraulic Turbine *and* Pumps

Hydraulic machines (turbines) are defined as those machines which convert either hydraulic energy into mechanical energy or mechanical energy into hydraulic energy.

Head

Head is defined as energy content of fluid per unit weight of fluid.

Head H = specific energy of liquid = $\dfrac{\text{Total energy}}{\text{Weight of liquid}}$

Its unit is metre or centimetre.

Dynamic Head

It is also known as kinetic head. It is the ratio of kinetic energy to weight of fluid.

$$\text{Dynamic head} = \frac{\frac{1}{2}mv^2}{mg} = \frac{v^2}{2g}$$

Potential or Static Head

It is the ratio of potential energy of fluid to its weight.

$$\text{Static head} = \frac{mgh}{mg} = h$$

Pressure Head

It is the ratio of pressure energy to its weight.

$$\text{Pressure head} = \frac{\text{Pressure energy}}{\text{Weight}}$$

$$= \frac{p \cdot V}{w} = \frac{p \cdot V}{\rho \cdot V \cdot g} = \frac{p}{\rho g}$$

$$\therefore \quad \text{Pressure head} = \frac{p}{\rho g}$$

Key Points

- Hydraulic turbines extract energy from water which has a high head.
- *These turbines are basically of two types*
  (a) Reaction turbines (b) Impulse turbines
- Reaction turbines are smaller than impulse turbines
- Reaction turbines are low head, high-flow devices.

Piezometric Head

It is the sum of pressure head and potential head.

$$\text{Piezometric head} = \frac{p}{w} + h$$

Centrifugal Head

When liquid particles rotate, a centrifugal force acts on these particles which increases pressure forces on the liquid particles. In figure, an annular ring is shown which is rotated with angular velocity ω. Consider a thin ring of thickness dr at r distance from centre.

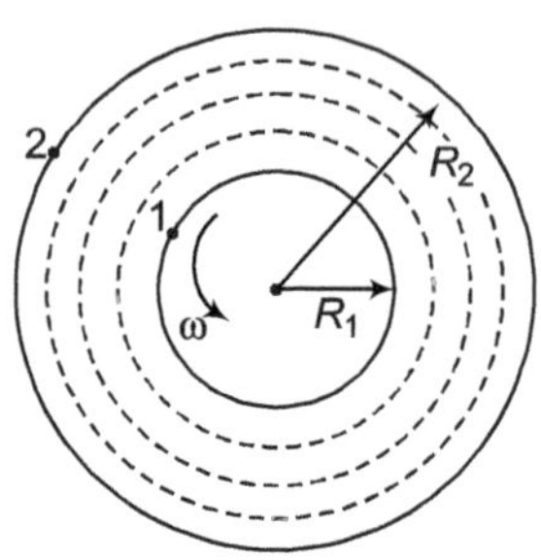

Angular ring

Then, net force due to pressure difference = change in centrifugal force

$$\Rightarrow \quad (p+dp)\cdot dA - p\cdot dA = (dm)\cdot r\cdot \omega^2$$

$$\Rightarrow \quad \frac{1}{\rho}\int_1^2 dp = \omega^2 \int_{R_1}^{R_2} r\cdot dr$$

$$\frac{p_2 - p_1}{\rho g} = \frac{\omega^2}{g}\left(\frac{R_2^2}{2} - \frac{R_1^2}{2}\right)$$

$$\Rightarrow \quad h_c = \frac{u_2^2 - u_1^2}{2g} \qquad (\because u = r\omega)$$

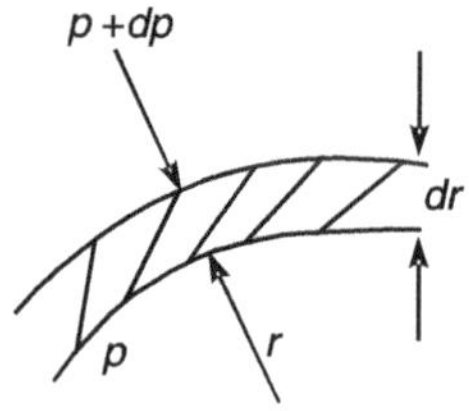

Thin ring

It can be said that centrifugal head represents the energy per unit weight of fluid which is lost or gained by the fluid due to displacement of fluid caused by centrifugal force.

Impact of Jet

When a jet strikes to plate, its momentum is changed. Hence, it exerts force on the plate (Newton's second law) *i.e.*,

As m = constant, $\quad F = m\cdot\frac{dv}{dt} \quad$ or $\quad F = \frac{m}{t}(v_2 - v_1) = \rho\cdot Q\cdot(v_2 - v_1)$

Q is volume flow rate of fluid.

Force due to Impact of Jet on a Flat Fixed Plate

We assume that plate is smooth (friction between jet and plate is neglected), there is no loss of energy during impact of jet and velocity of jet is uniform throughout.

Let, v = Velocity of jet

ρ = Density of fluid

As plate is fixed, its velocity $v = 0$

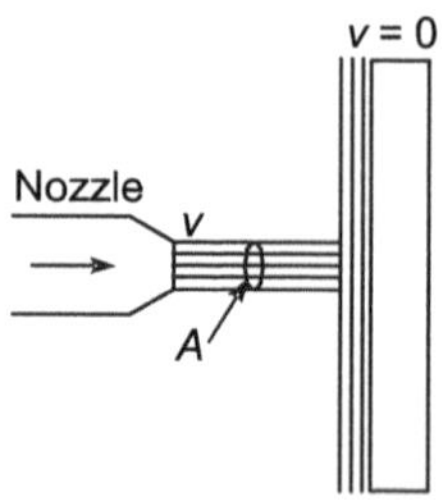

Jet on a flat fixed plate

Force = Rate of change of momentum

$$\Rightarrow \qquad F = m(v_1 - v_2)\left[\dot{m} = \frac{m}{t}, \dot{m} = \rho\,\frac{A \times \ell}{t} = \rho \cdot A \cdot v = \rho \cdot Q\right]$$

$$v_1 = v, v_2 = 0$$

$$\Rightarrow \qquad F = \rho\, Av^2$$

Work done = 0, as plate is stationary

Force on Moving Flat Plate in Direction of Jet

Here, v = Velocity of jet

u = Velocity of plate

$$F = m\,(v_{\text{initial}} - v_{\text{final}})$$

$$m = \rho \cdot A(v - u)$$

$$F = \rho A\,(V - u)[(V - u) - 0]$$

$$\therefore \qquad F = \rho\, A\,(v - u)^2$$

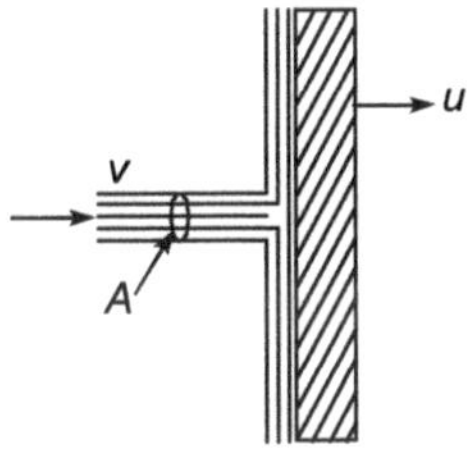

Flat plate in direction of jet

$(v - u)$ is the relative velocity with which water strikes to plate

$$v_{\text{initial}} = v - u$$

$$v_{\text{final}} = 0$$

As we are taking jet velocity in X-direction after striking to the plate

Rate of work done = F_n × Distance moved by plate per second

$$= \rho\, A \cdot (v - u)^2 \cdot u$$

If Jet Strikes at Centre of a Stationary Symmetrical Curved Plate

Symmetrical curved plate means equal blade angles. Blade angles are achieved by drawing tangent at end of blade.

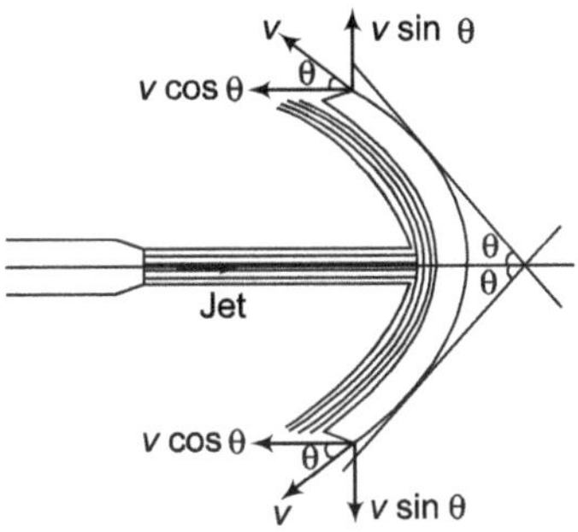

Symmetrical curved plate

If Plate is Moving with Some Velocity (u)

$$F_n = m\,(v_{\text{initial}} - v_{\text{final}}) \qquad \ldots\text{(i)}$$

$$m = \rho\, A(v - u)$$

As plate is single and moving with velocity u,

$$v_{\text{initial}} = v - u,\ v_{\text{final}} = -(v-u)\cos\theta$$

By Eq. (i) $F_n = \rho\, A\,(v-u)\,[(v-u) - \{-(v-u)\cos\theta\}]$

$$= \rho A\,(v-u)^2\,(1+\cos\theta)$$

Rate of work done $= F_n \cdot u$

$$= \rho A\,(v-u)^2\,(1+\cos\theta)\cdot u$$

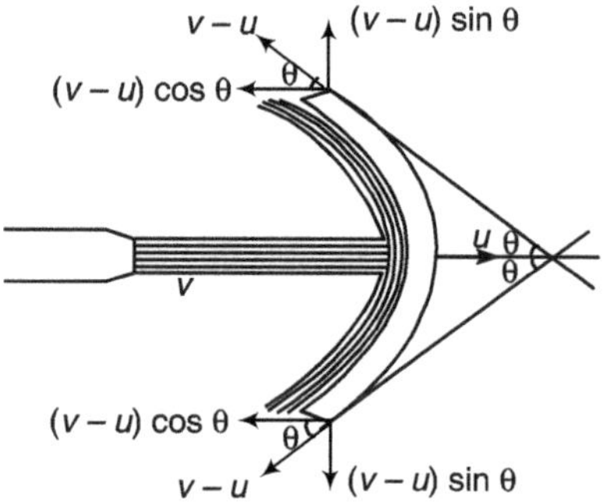

Moving plate

Series of Moving Vanes

This is the case of impulse turbine as shown in figure, $F_n = m = (v_{\text{initial}} - v_{\text{final}})$

$$m = \rho A v \qquad \text{(as series of plate)}$$

$$v_{\text{initial}} = v - u$$

$$v_{\text{final}} = -(v-u)\cos\theta$$

$$= \rho\, Av\,(v-u)\,(1+\cos\theta)$$

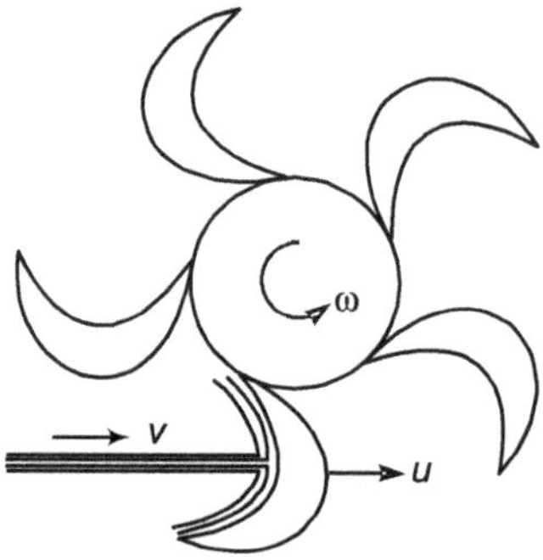

Moving vanes

- Work done per second
  $$W = F_n \cdot u = \rho Av\,(v-u)\,(1+\cos\theta)\,u$$
- KE supplied $= \dfrac{1}{2}\,(\rho Av)\,v^2 = \dfrac{1}{2}\rho A v^3$
- Efficiency, $\eta = \dfrac{\text{Work done per second}}{\text{KE per second}}$

$$\eta = \frac{2\,(v-u)\,(1+\cos\theta)\cdot u}{v^2}$$

For maximum efficiency, $\dfrac{d\eta}{du} = 0$

We get $v = 2u$

$$\eta_{\max} = \frac{1+\cos\theta}{2}$$

(for $\theta = 0$) $\eta_{\max} = 100\%$

Impact of Jet on a Single Unsymmetrical Moving Curved Vane Tangentially

Here, θ = Inlet angle of vane ,ϕ = Outlet angle of vane

These are obtained by drawing tangent at the tip.

where, α = Angle at which jet strikes the vane

β = Angle at which jet leaves the vane

v_1 = Absolute velocity of jet at inlet

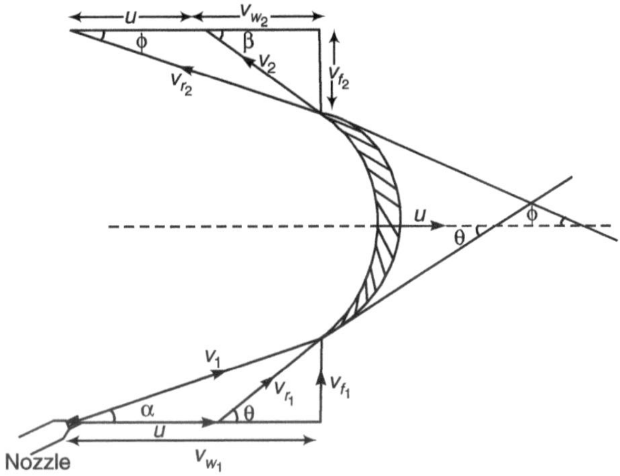

Unsymmetrical moving vane

v_2 = Absolute velocity of jet at outlet, v_{r_1} = Relative velocity at inlet

v_{r_2} = Relative velocity at outlet

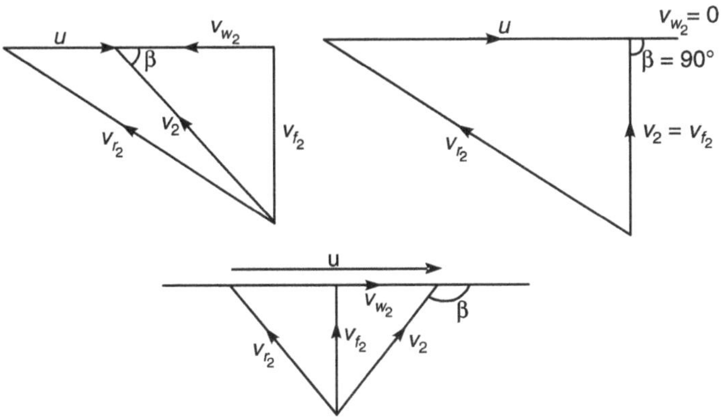

Graphs at different angle

$$v_{\text{initial}} = v_{w_1}, v_{\text{final}} = -v_{w2}$$
$$F_x = \rho A v_{r_1} (v_{w_1} + v_{w_2})$$

where, v_{w_2} is in opposite direction if $\beta < 90°$. v_{w_2} is in direction of v_{w_1}, if $\beta > 90°$.

v_{w_1}, is zero, if $\beta = 90°$.

- Work done per second $= F \cdot u = \rho \cdot A \cdot v_{r_1} (v_{w_1} + v_{w_2}) \cdot u$

$$\eta = \frac{\text{Work output per second}}{\text{Input KE per second}} = \frac{\rho A v_{r_1} (v_{w_1} + v_{w_2}) \cdot U}{\frac{1}{2} \rho A \cdot (v_1) \cdot v_1^2} = \frac{2 v_{r_1}(v_{w_1} + v_{w_2}) \cdot u}{v_1^3}$$

Force Exerted on a Series of Moving Radial Curved Vanes

This is the case of reaction turbine. In this case, velocity of tip of inlet and outlet of vane is different.

$$u_1 = R_1\omega \quad \text{and} \quad u_2 = R_2\omega$$

where, $\omega = \dfrac{2\pi N}{60}$ rad/s

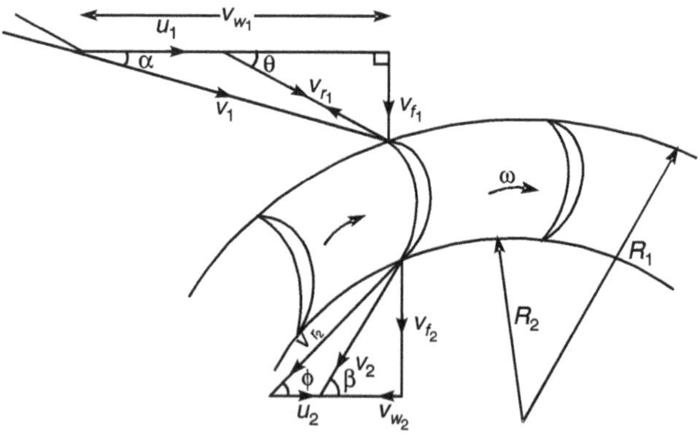

Moving radial vanes

$$m = \rho A v_1 \qquad \text{[Don't take } v_{r_1} \text{ as it is series of moving vanes]}$$

Rate of angular moment of jet at inlet $= m v_{w_1} \cdot R_1 = A v_1 \cdot v_{w_1} \cdot R_1$

Rate of angular moment of jet at outlet $= m v_{w_2} - R_2 = \rho A v_1 \cdot v_{w_2} \cdot R_2$

We know, $\quad \text{torque} = \dfrac{dJ}{dt}$

$$T = \rho A v_1 \cdot v_{w_1} R_1 - \rho A v_1 (-v_{w_2}) \cdot R_2$$
$$= \rho A v_1 (v_{w_1} \cdot R_1 + v_{w_2} \cdot R_2)$$

$$W = F \cdot u = T \times \omega$$
$$= \rho A v_1 (v_{w_1} \cdot u_1 + v_{w_2} \cdot u_2)$$
$$\eta = \frac{\text{Rate of work done}}{\text{KE per second}} = \frac{2(v_{w_1} \cdot u_1 + v_{w_2} \cdot u_2)}{v_1^2}$$

Impulse Turbine

It works on the principle of impulse. In this turbine, the whole pressure energy of water is converted into kinetic energy. Fluid is brought out by a penstock from dam.

At end it is ejected through a nozzle. The fluid coming out of the nozzle in the form of a free jet strikes on a series of buckets mounted on the periphery of the wheel. Pelton wheel is the example of impulse turbine.

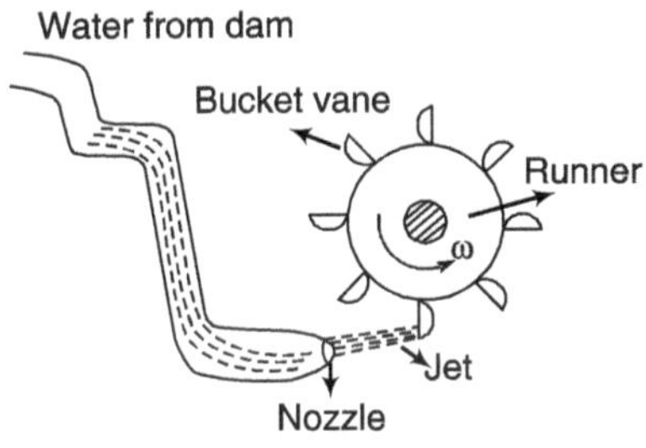

Pelton wheel

Velocity Diagram of Pelton Wheel

By inlet velocity diagram,

- Hydraulic efficiency

$$\eta_h = \frac{\text{Work done per second}}{\text{KE supplied per second}}$$
$$\eta_h = \frac{2(v_{w_1} + v_{w_2}) \cdot u}{v_1^2}$$

- Maximum efficiency,

$$\therefore \quad \eta_{max} = \frac{1 + \cos\phi}{2}$$

- Mechanical efficiency

$$\eta_{mech} = \frac{\text{Power available at turbine shaft}}{\text{Power developed by runner}}$$

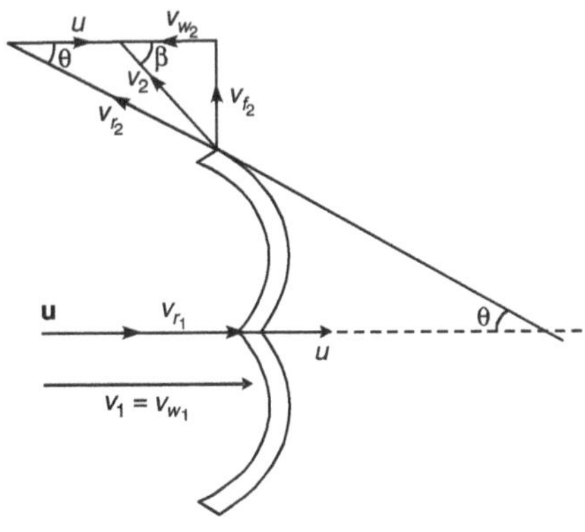

Intel velocity diagram

- Power developed by runner = power available at turbine shaft + Friction loss (mechanical)
- Volumetric efficiency $\eta_{vol} = \dfrac{\text{Actual volume of water striking the bucket}}{\text{Volume of water issued by jet}}$
- Overall efficiency $\eta_{overall} = \dfrac{\text{Shaft power } (P_s)}{\text{Power input by water jet}}$
- Power input by jet $= \rho Q \cdot gH$, $P_s = \dfrac{2\pi NT}{60 \times 1000}$

where, N is rpm of wheel $\eta_{overall} = \eta_{vol} \cdot \eta_{hydraulic} \, \eta_{mech}$

Some Important Terms belonging to Pelton Wheel

- **Velocity of jet** $v_1 = C_v \sqrt{2gH}$
  where, C_v is coefficient of discharge.
  H is net head on turbine.
- **Velocity of wheel** $u = \dfrac{D}{2} \cdot \omega = \dfrac{D}{2} \times \dfrac{2\pi N}{60}$,
  $u = \dfrac{\pi DN}{60}$ (Here, N is in rpm)

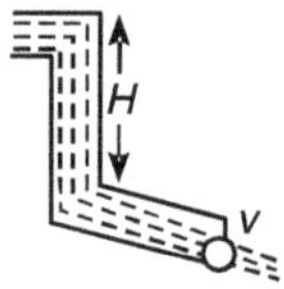

Turbine

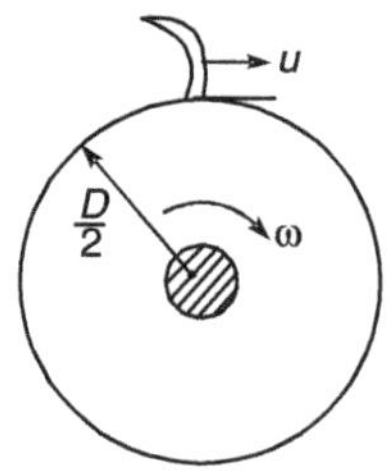

Runner wheel

- **Speed ratio** $K_u = \dfrac{\text{Ratio of runner speed}}{\text{Velocity of jet}} = \dfrac{u}{v_1}$
- **Jet ratio** $m = \dfrac{\text{Diameter of wheel}}{\text{Jet diameter}} = \dfrac{D}{d}$
- **Number of jets** Total flow rate of water = (Number of jet) × (Discharge through a single nozzle)
  Number of jet $= \dfrac{Q}{\dfrac{\pi}{4} d^2 \times v_1}$

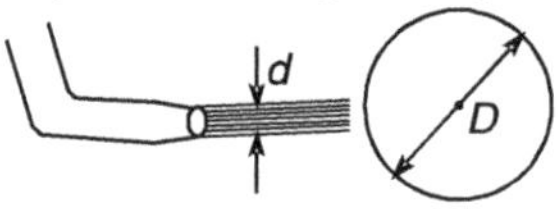

Jet with wheel

- **Specific speed** It is the speed of a geometrically similar turbine which would produce unit power under a unit head $N_s = \dfrac{N\sqrt{P}}{H^{5/4}}$

Reaction Turbine

It is based on impulse reaction. In this turbine, a part of pressure energy is converted into kinetic energy before entry to runner. Remaining part of pressure energy is gradually converted into KE until pressure is reduced to atmospheric pressure. This pressure difference and impulse action causes reaction which rotates the runner and produces mechanical work.

e.g., Francis Turbine, Kaplan Turbine.

Francis Reaction Turbine

It is radially inward flow turbine. In this turbine, water enters at the outer periphery of runner and flows radially inwards towards the centre of runner.

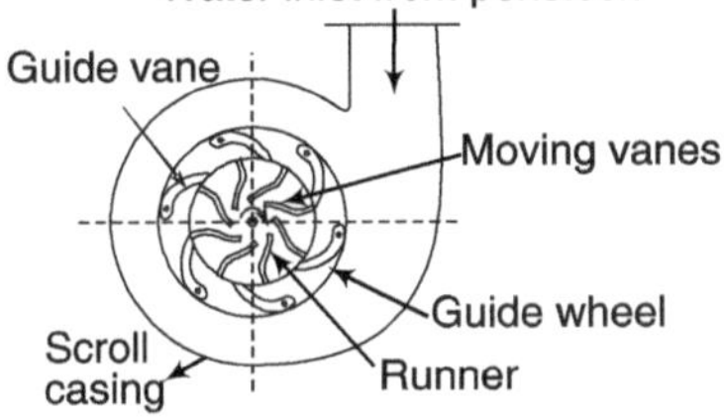

Francis reaction turbine

Its main components are as follow

- **Scroll casing** It surrounds the guide vanes, the runner, the shaft etc. It is necessary in reaction turbine because in reaction turbine water flows under pressure over the runner.
- **Guide vanes** It consists of a stationary ring in the form of wheel. It's function is to make the shockless entry of water from casing to moving vanes. It also regulates the discharge according to load on turbine.
- **Runner** It consists of blades of airfoil section so that water enters and leaves the blades without shock.
- **Draft tube** After passing over the runner blades water is discharged to tail race through a tube which has gradually increasing area. This tube is known as draft tube. Its function is as

  It converts the kinetic energy of water into useful pressure energy so water discharged into tail race or at exit of draft tube at very low velocity. So, draft tube increases the pressure head.

Key Points

- Reaction turbines are acted on by water.
- They must be encased to contain the water pressure.
- They must be fully submerged in the water flow.
- Newton's third law describes the transfer of energy for reaction turbines.

Velocity Diagram of Francis Reaction Turbine

Water enters at outer periphery and exists to centre of runner. We know

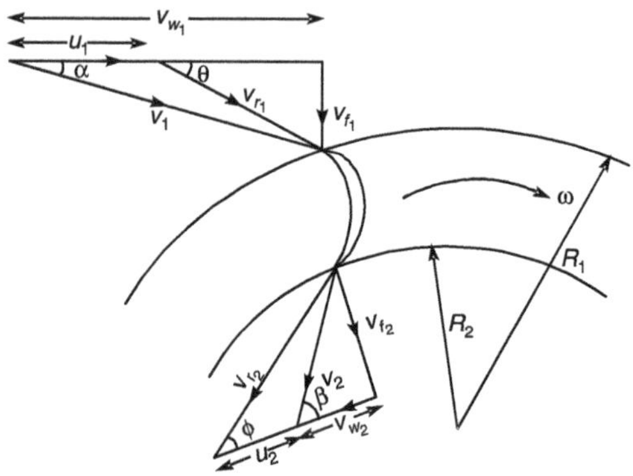

Velocity diagram of Francis

If $\beta > 90°$, then $\quad W = m\,(v_{w_1} \cdot u_1 - v_{w_2} \cdot u_2)$

If $\beta = 90°$, then absolute velocity at exit is radial ($v_{w_2} = 0$).

- Hydraulic efficiency $= \dfrac{\text{Power developed by runner}}{\text{Hydraulic power input}}$

$$\eta_H = \frac{v_{w_1}u_1 \pm v_{w_2}u_2}{gH}$$

- Mechanical efficiency $= \dfrac{\text{Shaft power}}{\text{Power developed by runner}}$

$$\eta_m = \frac{2\pi NT}{60 \times 1000} \bigg/ \rho Q(v_{w_1}u_1 \pm v_{w_1}u_2)$$

- Overall efficiency $= \dfrac{\text{Output}}{\text{Input}}$

$$\eta_0 = \frac{\dfrac{2\pi NT}{60} \times 1000}{\rho Q\, gH}$$

Important Terms belonging to Reaction Turbine

- **Speed ratio** $k_u = \frac{u_1}{v_1} = \frac{u_1}{\sqrt{2gH}}$
- **Flow ratio** $k_f = \frac{v_{f_1}}{v_1} = \frac{v_{f_1}}{\sqrt{2gH}}$
- **Blade thickness and width** Let widths of blade at inlet and outlet are B_1 and B_2 respectively and thickness is t.
- **Discharge** Q = Area of flow × flow velocity
  or $Q = \pi D_1 B_1 v_{f_1} = \pi D_2 B_2 v_{f_2}$
- If thickness of blade is considered then number of blades affect the discharge. If number of blades is n_1, then $Q = (\pi D_1 - n_1 t) \cdot B_1 \cdot v_{f_1}$

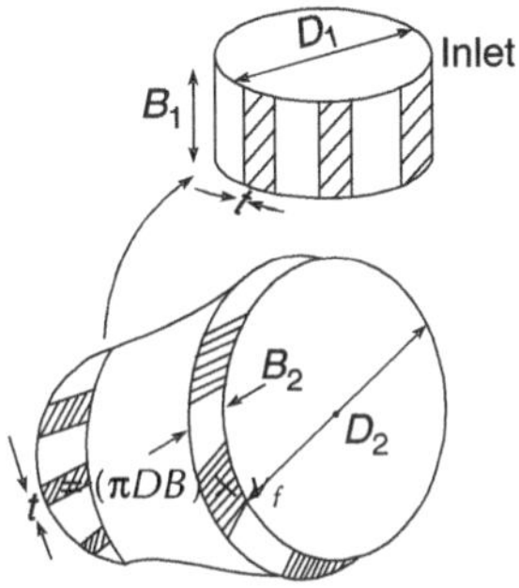

Blade at inlet and outlet

Specific Speed of Turbine

Specific speed of a turbine is defined as the speed at which a turbine runs when it is working under a unit head and develops unit power.

$$N_S = \frac{N\sqrt{P}}{H^{5/4}}$$

where, P = Shaft power in kW, H = Net head on the turbine.

Unit Quantities of Turbine

Unit quantities are the quantities which are obtained when the head on turbine is unity. They are unit speed (N_u), unit power (P_u) and unit discharge (Q_u)

$$N_u = \frac{N}{\sqrt{H}}, \quad Q_u = \frac{Q}{\sqrt{H}}, \quad P_u = \frac{P}{H^{3/2}}$$

Centrifugal Pump

A pump is a device which transfers the mechanical energy given by motor or engine into pressure energy or KE or both of fluid. In centrifugal pump, the rotating impeller of pump impresses a centrifugal head or pressure on the liquid which leaves the impeller at high velocity.

In centrifugal pump, a certain mass of liquid is made to rotate along the impeller from central axis of rotation. Centrifugal head causes the water to move radially outwards at higher velocity and causes the water to rise to a higher level.

Velocity Diagram of Centrifugal Pump

For best efficiency, liquid enters the impeller vane radially so,

$$\alpha = 90^\circ$$

$$\therefore \qquad v_1 = v_{f_1} \quad \text{and} \quad v_{w_1} = 0$$

- Outlet diagram

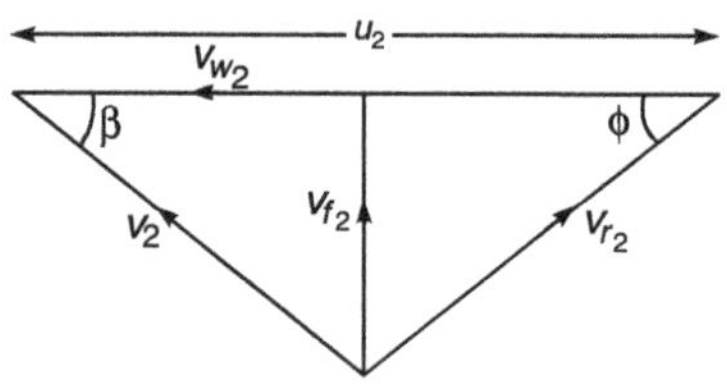

Outlet diagram

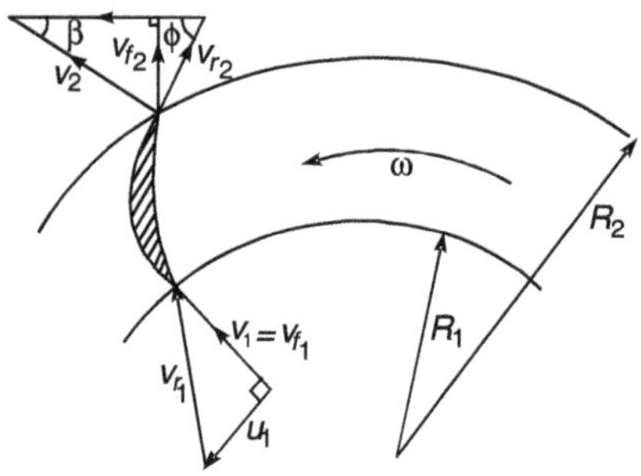

Outlet diagram

- Inlet diagram,

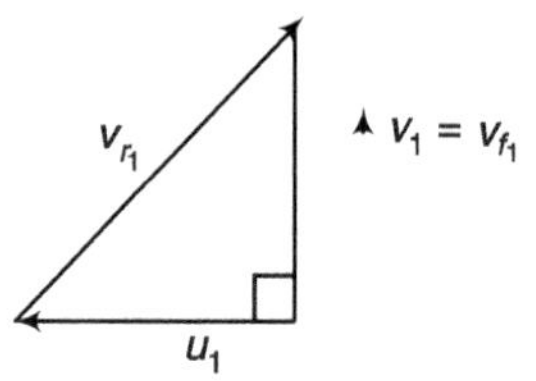

Inlet diagram

- Work done by impeller on water $= \rho Q\,(v_{w_2} \cdot u_2 - v_{w_1} u_1)$

  As entry is radial $\qquad v_{w_1} = 0$

  $\therefore \qquad W = (v_{w_2} u_2) \cdot \rho Q$

- Discharge rate = Area × Velocity of flow

  $$Q = (\pi\, D_1 B_1)\, v_{f_1} = \pi D_2 B_2 \cdot v_{f_2}$$

  B_1 and B_2 are widths of impeller.

Key Points

- There are no drive seals, therefore the risk of leaks is completely eradicated.
- No heat transfer from the motor.
- Magnetic coupling can be broken, if the load of the pump is too great.
- Liquids containing ferrous particles are problematic when a magnetic drive pump is used.

Schematic Diagram of Centrifugal Pump

Let, h_s = Suction lift, h_d = Delivery lift

H_s = Static head = $h_s + h_d$

$$HG = \text{Gross head} = H_s + (h_{fs} + h_{fd}) + \left(\frac{V_s^2}{2g} + \frac{V_d^2}{2g}\right)$$

Manometric head H_m is defined as the minimum amount of head against which the pump has to work to deliver the required discharge

$$H_m = \frac{V_d^2}{2g} + H_s + h_{fs} + h_{fd}$$

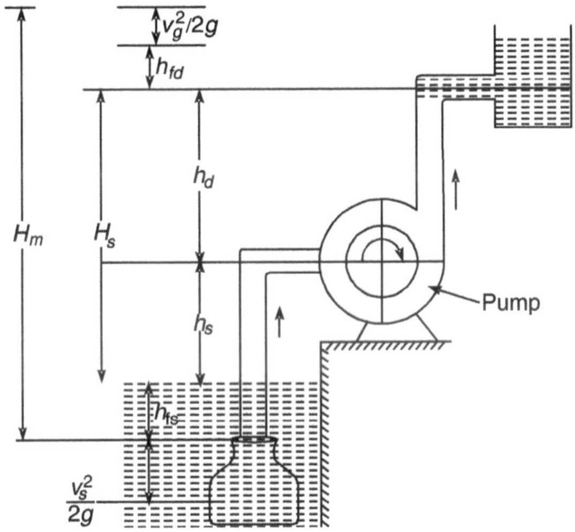

Centrifugal pump

- Mechanical efficiency $\eta_m = \dfrac{\text{Power available at impeller}}{\text{Shaft power}}$

$$\eta_m = \frac{P}{P_{shaft}}$$

- Manometric efficiency $= \dfrac{\text{Manometric head } (H_m)}{\text{Head imparted by impeller}}$

$$\eta_{mano} = \frac{H_m}{\left(\frac{V_{w_2} \cdot u_2}{g}\right)} = \frac{gH_m}{V_{w_2}u_2}$$

- Overall efficiency $\eta_o = \dfrac{\text{Power output}}{\text{Shaft power}} = \dfrac{\rho QgH_m}{P_s}$

Cavitation

It is defined as the phenomenon of formation of vapour bubbles in the region of flowing liquid where its pressure falls below the vapour pressure of liquid. In this phenomenon, liquid vaporises and flow does not remain continuous. A schematic view of centrifugal pump is shown.

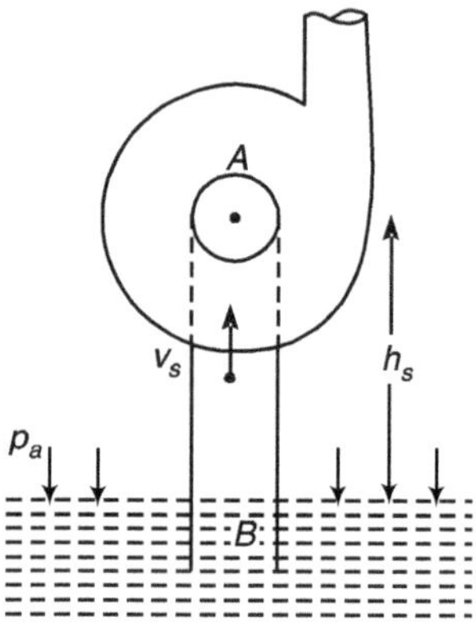

Cavitation

Assume, p_a = Atmospheric pressure

p_s = Pressure at eye of impeller

h_s = Suction lift

h_{fs} = Friction loss in suction pipe

V_s = Velocity in suction pipe

By Pascal's law, $p_B = p_{atm}$

By Bernoulli's theorem between A and B,

$$\frac{p_a}{\rho g} = \frac{p_s}{\rho g} + h_s + \frac{V_s^2}{2g} + h_{fs}$$

So,

$$\frac{p_s}{\rho g} = \frac{p_a}{\rho g} - \left(h_s + h_{fs} + \frac{V_s 2}{2g}\right)$$

NPSH (Net Positive Section Head)

It is the absolute pressure head at the inlet to the pump to force the liquid into the pump.

$$\frac{p_a}{w} = \frac{p_v}{w} + h_s + \frac{V_s^2}{2g} + h_{fs} + \text{NPSH} \qquad (\because w = \rho g)$$

(minimum value of p_s is p_{vapour})

$$\text{NPSH} = h_a - \left(h_v + h_s + \frac{V_s^2}{2g} + h_{fs}\right) \qquad \left(i.e., h = \frac{p}{\rho g}\right)$$

Reciprocating Pump

In reciprocating pump, mechanical energy is converted into hydraulic energy by sucking the liquid into a cylinder in which a piston is reciprocating which exerts the thrust on the liquid and increases its hydraulic energy.

Discharge Through a Pump Per Second

$$Q = \frac{ALN}{60} \quad \text{(for single acting)}$$

$$Q = \frac{2ALN}{60} \quad \text{(for double acting)}$$

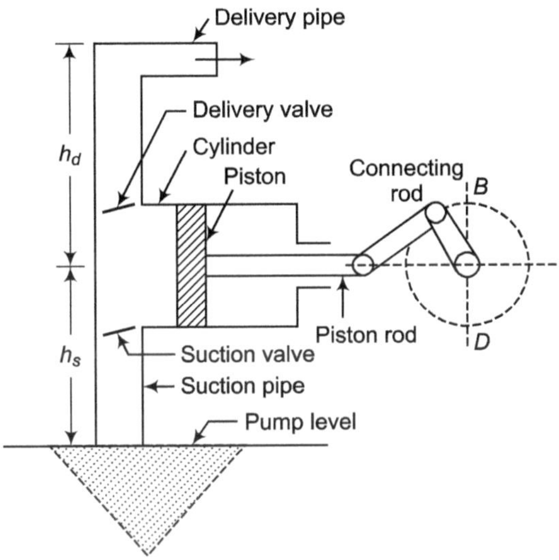

Reciprocating pump

A = Cross-sectional area of the piston or cylinder $= \frac{\pi}{4}D^2$

L = Length of stroke = 2 × radius of crank (2r)

N = Rpm of crank.

h_s = Height of the axis of cylinder from water surface in pump

h_d = Height of delivery outlet above the cylinder axis.

- Work done by Reciprocating Pump

$$W = \frac{\rho g\ ALN}{60}(h_s + h_d) \quad \text{[for single acting]}$$

$$W = \frac{2\rho g\ ALN}{60}(h_s + h_d) \quad \text{[for double acting]}$$

Slip

Slip of a pump is defined as the difference between the theoretical discharge and actual discharge of the pump.

Comparison between Centrifugal Pump and Reciprocating Pumps

S. No.	Centrifugal Pump	Reciprocating Pump
1.	Continuous and smooth and can handle large quantity of liquid	Fluctuating and pulsating and can handle small quantity of liquid only.
2.	Used for large discharge and small head.	Used for small discharge and large head.
3.	It runs at high speed.	It runs at low speed.
4.	Efficiency of centrifugal pump is high.	Low efficiency.
5.	It run smoothly without much noise.	It run with much noise.

Fluid System

Fluid system is defined as the device in which power is transmitted with the help of fluid which may be liquid or a gas under pressure.

The Hydraulic Press

It is a device used for lifting heavy weights by the application of a much smaller force.

$$\frac{F}{a} = \frac{w}{A}$$

where,

w = Weight to be lifted
F = Force applied on the plunger
a = Area of plunger
p = Pressure intensity produced by force $= \frac{F}{a}$.

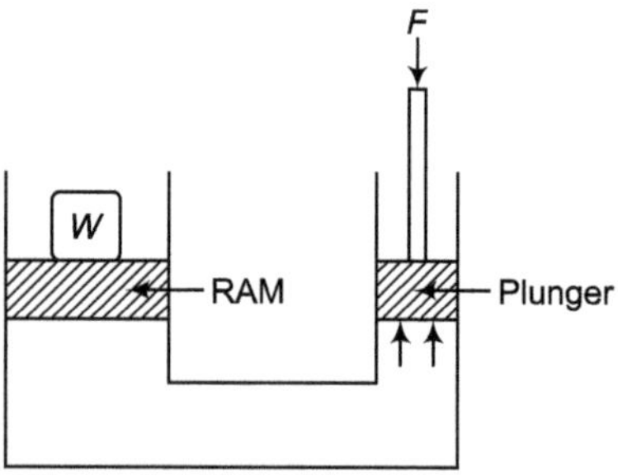

Hydraulic press

Mechanical Advantage (MA)

The ratio of weight lifted to force applied on the plunger is defined as mechanical advantage.

$$\text{MA} = \frac{w}{F}$$

The Hydraulic Ram

The hydraulic ram is a pump which raises water without any external power for its operation. It works on the principle of water hammer.

- Efficiency of hydraulic ram $\eta = \dfrac{\text{Energy delivered by ram}}{\text{Energy supplied to the ram}}$

$$= \frac{w \times H}{W \times h}$$

where, W = Weight of water flowing per second into chamber
w = Weight of water raised per second
h = Height of water in supply tank above the chamber
H = Height of water raised from the chamber.

The Hydraulic Lift

The hydraulic lift is a device used for carrying passenger of goods from one floor to another in multi-storied building.

The Hydraulic Crane

Hydraulic crane is a device used for raising or transferring heavy loads. It is widely used in work shop, warehousing and clock siding.

The Fluid or Hydraulic Coupling

The fluid or hydraulic coupling is a device used for transmitting power from driving shaft to driven shaft with the help of connection between the two shafts.

The Hydraulic Torque Converter

It is a device used for transmitting increased torque at the driven shaft. The torque transmitted at the drive shaft may be more or less than the torque available at the driving shaft.

6

Heat *and* Mass Transfer

Steady *and* Unsteady State Conductions

Steady state conduction is the form of conduction that happens when the temperature difference driving the conduction are constant, so that, the spatial distribution of temperatures in the conducting object does not change any further.

In general, during any period in which temperatures are changing in time at any place within an object, the mode of thermal energy flow is termed transient conduction or non-steady state conduction.

Fourier's Law of Heat Conduction

The rate of heat flow through homogeneous solid is directly proportional to the area measured normal to the direction of heat flow and the temperature gradient in the direction of heat flow.

$$Q = -kA\frac{dT}{dx}$$

where, Q = Heat transfer in given direction

A = Cross-sectional area perpendicular to heat flow direction

dT = Temperature difference between two ends of a block of thickness dx

dx = Thickness of solid body

$\frac{dT}{dx}$ = Temperature gradient in direction of heat flow.

General Heat Conduction Equation

(i) **Cartesian Coordinates** (side parallel to x, y and z-directions)

Let q_g = Internal heat generation per unit volume per unit time

t = Temperature at left face of differential control volume

k_x, k_y, k_z = Thermal conductivities of the material in x, y and z-directions respectively

c = Specific heat of the material

ρ = Density of the material

α = Thermal diffusivity $\left(\frac{k}{\rho c}\right)$

$d\tau$ = Instantaneous time.

$$\frac{\partial}{\partial x}\left(k_x \frac{\partial t}{\partial x}\right) + \frac{\partial}{\partial y}\left(k_y \frac{\partial t}{\partial y}\right) + \frac{\partial}{\partial z}\left(k_z \frac{\partial t}{\partial z}\right) + q_g = \rho c \frac{\partial t}{\partial \tau}$$

- For homogeneous and isotropic material,

$$k_x = k_y = k_z = k, \alpha = \frac{k}{\rho c}$$

$$\frac{\partial^2 t}{\partial x^2} + \frac{\partial^2 t}{\partial y^2} + \frac{\partial^2 t}{\partial z^2} + \frac{q_g}{k} = \frac{1}{\alpha}\frac{\partial t}{\partial \tau}$$

- For steady state condition (Poisson's equation),

$$\frac{\partial^2 t}{\partial x^2} + \frac{\partial^2 t}{\partial y^2} + \frac{\partial^2 t}{\partial z^2} + \frac{q_g}{k} = 0$$

- For steady state and absence of internal heat generation (Laplace equation),

$$\frac{\partial^2 t}{\partial x^2} + \frac{\partial^2 t}{\partial y^2} + \frac{\partial^2 t}{\partial z^2} = 0$$

- For unsteady heat flow with no internal heat generation,

$$\frac{\partial^2 t}{\partial x^2} + \frac{\partial^2 t}{\partial y^2} + \frac{\partial^2 t}{\partial z^2} = \frac{1}{\alpha}\frac{\partial t}{\partial \tau}$$

(ii) **Cylindrical Coordinates**

- For homogeneous and isotropic material,

$$\left[\frac{\partial^2 t}{\partial r^2} + \frac{1}{r}\frac{\partial t}{\partial r} + \frac{1}{r^2}\frac{\partial^2 t}{\partial \phi^2} + \frac{\partial^2 t}{\partial \tau^2}\right] + \frac{q_g}{k} = \frac{1}{\alpha}\frac{\partial t}{\partial \tau}$$

- For steady state unidirectional heat flow in radial direction with no internal heat generation,

$$\frac{\partial^2 t}{\partial r^2} + \frac{1}{r}\frac{\partial t}{\partial r} = 0$$

$$\frac{1}{r}\frac{d}{dr}\left(r\frac{dt}{dr}\right) = 0 \Rightarrow r\frac{dt}{dr} = \text{Constant}$$

(iii) Spherical Coordinates

- For homogeneous and isotropic material,

$$\frac{1}{r^2 \sin^2\theta}\frac{\partial^2 t}{\partial \phi^2} + \frac{1}{r^2 \sin\theta}\frac{\partial}{\partial \theta}\left(\sin\theta \frac{\partial t}{\partial \theta}\right) + \frac{1}{r^2}\frac{\partial}{\partial r}\left(r^2\frac{\partial t}{\partial r}\right) + \frac{q_g}{k} = \frac{1}{\alpha}\frac{\partial t}{\partial \tau}$$

- For steady state unidirection heat flow in radial direction with no internal heat generation,

$$\frac{1}{r^2}\frac{d}{dr}\left(r^2\frac{dt}{dr}\right) = 0$$

Thermal Resistance of a Hollow Cylinders

Let, r_1 = Internal radius of cylinder

r_2 = Outer radius of cylinder

L = Length of cylinder.

Thermal resistance of a hollow cylinders is given as

$$R = \frac{\log_e \frac{r_2}{r_1}}{2\pi k.L}$$

Heat transfer $\quad Q = \dfrac{T_i - T_0}{R} = \dfrac{\Delta T}{R}$

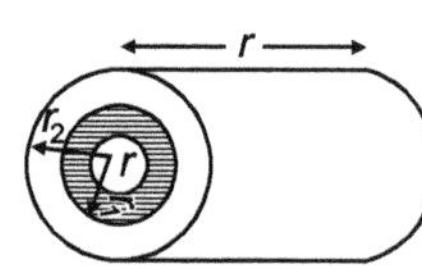

Hollow cylinder

Thermal Resistance of a Hollow Sphere

Consider a hollow cylinder of internal radius r_1 and external radius r_2 with respective internal and external temperatures of T_i and T_0 as shown in figure

$$R = \frac{r_2 - r_1}{4\pi k r_1 r_2}$$

where, k = Thermal conductivity

$$Q = \frac{\Delta T}{R}$$

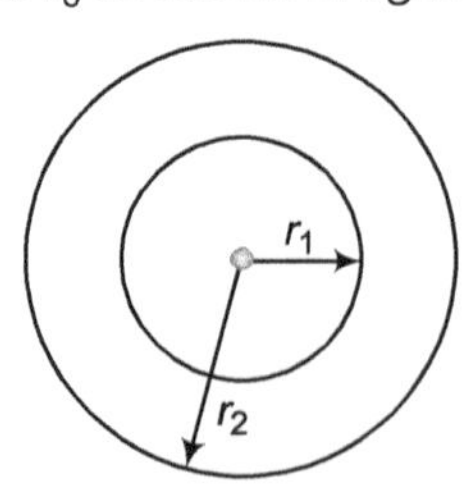

A hollow sphere

Heat Transfer through a Composite Cylinder

Assume heat transfer coefficient for liquid and inner surface of cylinder is h_i and for gas and outer surface of cylinder is h_o then heat transfer will be as

Convection → conduction → conduction → convection

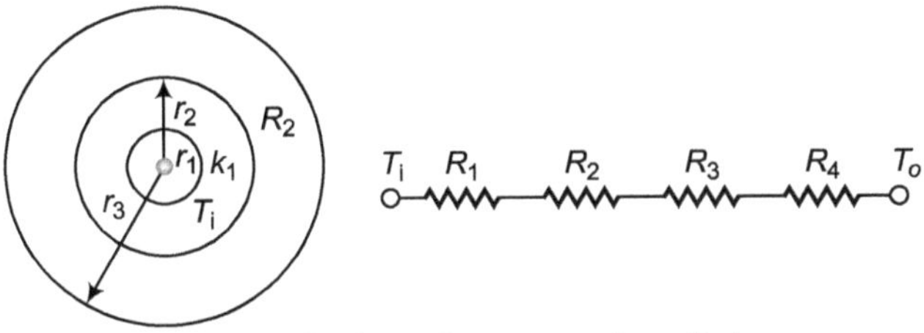

Heat transfer through a composite cylinder

$$Q = \frac{T_i - T_0}{R_{eq}}$$

$$R_{eq} = R_1 + R_2 + R_3 + R_4$$

$$R_1 = \frac{1}{h_i 2\pi r_1 l}, \quad R_2 = \frac{\ln r_2/r_1}{2\pi k_1 l}$$

$$R_3 = \frac{\ln r_3/r_2}{2\pi k_2 l}, \quad R_4 = \frac{1}{h_o\, 2\pi r_2 l}$$

Heat Transfer through a Composite Sphere

All the arrangements are same as given in composite cylinder only resistance will change.

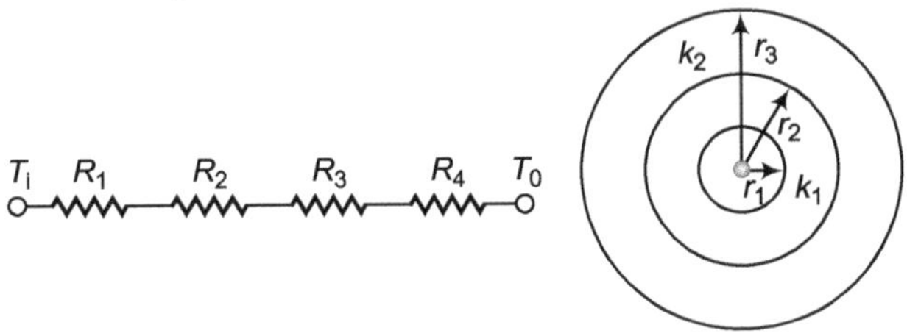

Heat transfer through a composite sphere

$$R_1 = \frac{1}{h_i A_i} = \frac{1}{h_i\, 4\pi\, r_1^2}$$

$$R_2 = \frac{r_2 - r_1}{4\pi\, k\, r_1\, r_2}, \quad R_3 = \frac{r_3 - r_2}{4\pi\, k\, r_2\, r_3}$$

$$R_4 = \frac{1}{h_o\, 4\pi\, r_2^2}, \quad Q = \frac{T_i - T_0}{R_1 + R_2 + R_3 + R_4}$$

Critical Thickness of Insulation

Critical thickness of insulation graph is shown in adjoining figure.

In case of cylinder, $$r_c = \frac{k_0}{h}$$

where, k_0 = Thermal conductivity
h = Heat transfer coefficient

In case of sphere, $$r_c = \frac{2k_0}{h}$$

Key Points

- In critical radius of insulation, total resistance first decreases and attain a minimum value and then increases.
- Critical radius is independent of pipe radius.
- For purpose of insulation to the radius, the heat transfer rate for tube, Outside radius > critical radius (r_c)
- For purpose of insulation to the increase, the heat transfer rate for tube. Outside radius < critical radius (r_c)
- Heat transfer will always reduce on application of insulation of wall as convective resistance will remain unchanged while conductive resistance will increase with thickness of insulation.

Heat Transfer from Extended Surface (Fin)

A fin is a surface that extends from an object to increase the rate of heat transfer to or from the environment by increase convection. Adding a fin to an object, increases the surface area and can sometimes be an economical solution to heat transfer problems.

The rate of heat transfer from a solid surface to atmosphere is given by $Q = hA\,\Delta T$ where, h and ΔT are not controllable.

So, to increase the value of Q surface area should be increased. The extended surface which increases the rate of heat transfer is known as **fin**.

Steady Flow of Heat along a Rod

Circular fin

$$p = \pi d$$

$$A_c = \frac{\pi}{4} d^2$$

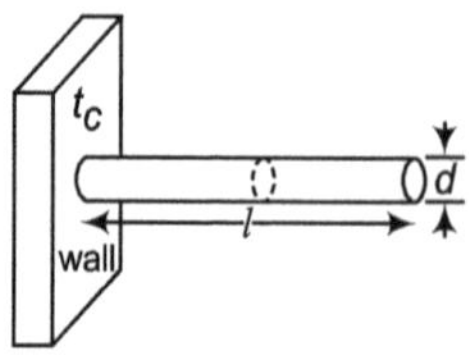

Circular fin diagram

Generalised Equation for Fin

Rectangular fin

$$\frac{d^2\theta}{dx^2} + \frac{1}{A_c}\frac{dA_c}{dx}\frac{d\theta}{dx} - \frac{h}{kA_c}\frac{dA_s}{dx}\theta = 0$$

where A_c and A_s are cross-sectional and surface area.

And $\theta(x) = t(x) - t_a$

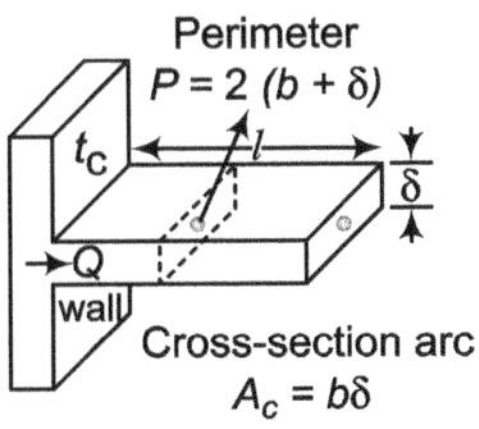

Rectangular fin diagram

Heat balance equation if A_c constant and $A_s \propto P(x)$ linear

$$\frac{d^2t}{dx^2} - \frac{hp}{kA_c}(t - t_a) = 0$$

$$\frac{d^2\theta}{dx^2} - m^2\theta = 0$$

$$m = \sqrt{\frac{hp}{kA_c}}$$

General equation of 2nd order $\theta = c_1 e^{mx} + c_2 e^{-mx}$

Heat dissipation can be takes place on the basis of three cases.

Case 1 Heat Dissipation from an Infinitely Long Fin ($l \to \infty$)

In such a case, the temperature at the end of Fin approaches to surrounding fluid temperature t_a as shown in figure. The boundary conditions are given below

At $x = 0, t = t_0$

$\theta = t_0 - t_a = \theta_0$

at $x = l \to \infty$

$t = t_a, \theta = 0$

$$\frac{t - t_a}{t_0 - t_a} = e^{-mx}$$

$$\theta = \theta_0 e^{-mx}$$

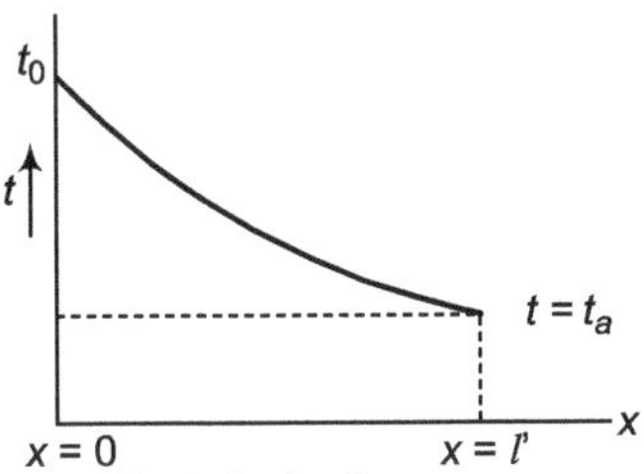

Heat dissipation curve

Heat transfer by conduction at base

$$Q_{fin} = \sqrt{kA_cPh}\,(t_0 - t_a)$$

Case 2 Heat Dissipation from a Fin Insulated at the End Tip

Practically, the heat loss from the long and thin film tip is negligible, thus the end of the tip can be. considered as insulated.

$$\text{At } x = 0, t = t_0 \quad \text{and} \quad \theta = t_0 - t_a = \theta_0$$

$$\text{at } x = l, Q = 0 \quad i.e., \quad \frac{d\theta}{dx} = 0$$

$$\frac{\theta}{\theta_0} = \frac{t - t_a}{t_0 - t_a} = \frac{\cosh m (l - x)}{\cosh m\, l}$$

$$Q_{\text{fin}} = \sqrt{Phk\, A_c}\, (t_0 - t_a) \tanh ml$$

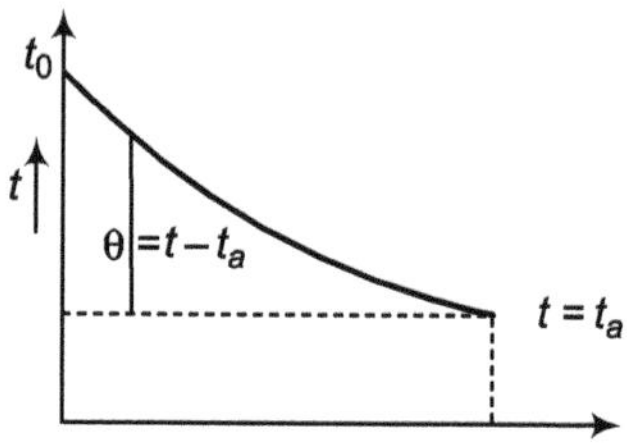

Case 3 Heat Dissipation from a Fin loosing Heat at the End Tip

The boundary conditions are given below.

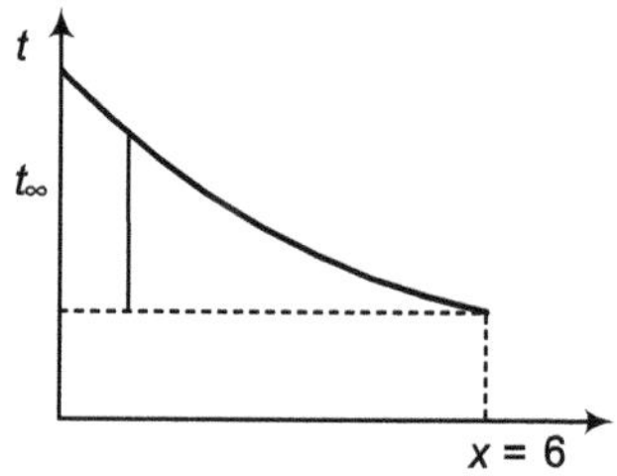

$$\text{At } x = 0, t = t_0 \text{ and } \quad \theta = \theta_0$$

$$\text{at } \; x = l, Q_{\text{conduction}} = Q_{\text{convection}}$$

$$-kA\left(\frac{dt}{dx}\right)_{x=l} = h\, A_s\, (t - t_a)$$

$$\theta = \theta_0 \left(\frac{\cosh m\,(l - x) + \frac{h}{mk} \sinh m\,(l - x)}{\cosh ml + \frac{h}{mk} \sinh ml} \right)$$

$$Q_{\text{fin}} = \sqrt{hPkA_c}\, \theta_0 \frac{\tanh ml + \frac{h}{mk}}{1 + \frac{h}{mk} \tanh ml}$$

Fin Efficiency

Fin efficiency is given by $\eta = \dfrac{\text{Actual heat rate from fin } Q}{\text{Maximum heat transfer rate } Q_{\text{max}}}$

- If $l \to \infty$ (infinite length of fin), $\eta = \dfrac{\sqrt{hPkA_c}\, \theta_0}{h\,(Pl + b\delta)\, \theta_0} = \dfrac{1}{l}\sqrt{\dfrac{kA_c}{hP}}$
- If fin is with insulated tip, $\eta = \dfrac{\theta_0 \sqrt{hPkA_c}\, \tanh ml}{hPl\theta_0}$
- If finite length of fin, $\eta = \dfrac{\theta_0 \sqrt{h\, PkA_c} \left[\dfrac{\tanh ml + \frac{h}{mk}}{1 + \frac{h}{mk} \tanh ml} \right]}{h\,(Pl + b\delta)\, \theta_0}$

Fin Effectiveness

Fin effectiveness is dented by ε.

$$\varepsilon = \frac{\text{Actual heat transfer from fin surface } (Q)}{\text{Rate of heat transfer without fin}}$$

$$\varepsilon = \frac{Q}{hA\theta_0} = \frac{\theta_0 \sqrt{PhkA_c}\,\tanh ml}{h\,A\,\theta_0} = \frac{\tanh ml}{\sqrt{\frac{h\,A_c}{Pk}}} \qquad (\text{if } l \to \infty)$$

If $\sqrt{\frac{Pk}{hA_c}} \le 1$, then $\varepsilon \le 1 \Rightarrow$ It means there will be reduction in rate of heat transfer due to its very high convective coefficient of heat transfer. (in case of boiling condensation and high velocity fluid).

Key Points

- Heat conduction is along the x-axis under steady state.
- There is no internal heat generation within the fin.
- Base temperature is maintained constant at T_0.
- Convective heat transfer coefficient is uniform over the entire surface of fin.

Unsteady State

If temperature of a body does not vary with time, it is said to be in steady state. But if there is an abrupt change in its surface temperature, it attains an equilibrium temperature or a steady state after some period. During this period, the temperature varies with time and body is said to be in unsteady or transient state. This phenomenon is known as transient heat conduction.

Heat conduction (Transient heat conduction) $\left(\frac{dt}{d\tau}\right) \neq 0$

Let V = Volume of body
c = Specific heat of body
ρ = Density of body
R = Thermal conductivity of material
h = Convective heat transfer coefficient
t_i = Initial temperature of body
t_a = Surrounding temperature.

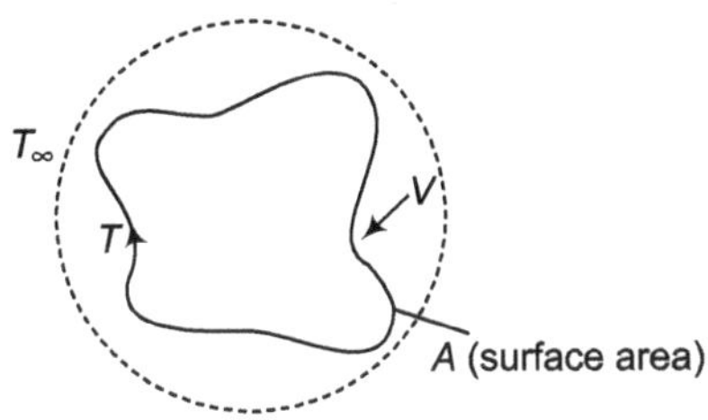

Heat conduction body

$$-\rho Vc\frac{dt}{d\tau}=hA(t-t_a),\ \frac{t-t_a}{t_i-t_a}=e^{\frac{-hA\tau}{\rho Vc}}$$

Total heat flow rate,

$$Q=-\rho V_C\,(t_i-t_a)\left[e^{\left(\frac{-hA\tau}{\delta V_C}\right)}\right]=\rho V_C\,(t_i-t_a)\left[e^{\left(\frac{-hA\tau}{\rho V_C}\right)}-1\right]$$

$$\text{Biot number Bi}=\frac{\text{Internal conductive resistance } \frac{l_c}{kA}}{\text{Surface or convective thermal resistance}\left(\frac{1}{hA}\right)}$$

$$\text{Bi}=\frac{hl_c}{k}$$

where, l_c = Characteristic length

$$\text{Fourier number}\quad \text{Fo}=\frac{\alpha\tau}{l_c^2}\left(\alpha=\frac{k}{\rho c}\right),\ \frac{t-t_a}{t_i-t_a}=e^{(-\text{Bi Fo})}$$

$$Q=-hA\,(t-t_a)\,e^{(-\text{Bi Fo})}\ \text{or}\ Q=\rho Vc\,(t_i-t_a)\,[e^{-\text{Bi Fo}}-1]$$

- Temperature of a body in unsteady state can be calculated at any time only when Biot number < 0.1.

Characteristic Length

Characteristic length is denoted by l_c.

$$l_c=\frac{\text{Volume}(V)}{\text{Surface area exposed to surrounding }(A)}$$

Characteristic Length for different Section

Characteristic length for sphere	$l_c=\dfrac{\frac{4}{3}\pi R^3}{4\pi R^2}=\dfrac{R}{3}$
Characteristic length for solid cylinder	$l_c=\dfrac{\pi R^2L}{2\pi R\,(L+R)}$ if $l>>R,\ l_c=\dfrac{R}{2}$
Characteristic length for cube	$l_c=\dfrac{L^3}{6L^2}=\dfrac{L}{6}$
Characteristic length for rectangular plate	$l_c=\dfrac{lbt}{2lb}=\dfrac{t}{2}$
Characteristic length for hollow cylinder	$l_c=\dfrac{\pi\,(r_0^2-r_i^2)l}{2\pi r_0 l+2\pi r_i l+2\pi\,(r_0^2-r_i^2)}$

Convection

Convection is heat transfer by mass motion of a fluid such as air or water when the heated fluid is caused to move away from the source of heat, carrying energy with it. Convection can be qualified in terms of being natural, forced, gravitational, granular or thermomagnetic. Generally, it occurs between a solid surface and fluid in motion.

Forced Convection

If fluid motion is imparted by external means like pump, fan, compressor etc., the convection is called **forced convection**.

Natural or Free Convection

When the fluid moves due to density difference caused by the heat transfer between solid surface and fluid. It is said to be **natural** or free convection.

Mean Bulk Temperature (t_b)

It denotes the equilibrium temperature that would result if the fluid at a cross-section was thoroughly mixed in an adiabatic container.

Mean Film Temperature (t_f)

It is the arithmetic mean of the surface temperature t_s of a solid and the undisturbed temperature t_∞ of the fluid which flows.

$$t_f = \frac{t_s + t_\infty}{2}$$

Average Heat Transfer Convective Coefficient

It is given by

$$h = \frac{1}{l}\int_0^l h_x \, dx$$

Dimensionless Number

There are following types of dimensionless number

Nusselt Number (Nu) It is defined as

$$\text{Nu} = \frac{\text{Rate of heat transfer by convection}}{\text{Rate of heat transfer by conduction}} = \frac{h \cdot A \cdot \Delta T}{kA \dfrac{\Delta T}{l}}$$

$$\text{Nu} = \frac{hl}{k}$$

Nu measure energy transfer by convection occurring at the surface. Larger the value of Nu, larger will be the rate of heat transfer by convection.

Reynold Number (Re)

Reynold number is defined as, $\text{Re} = \dfrac{\text{Inertia force}}{\text{Viscous force}}$

$$\text{Re} = \frac{\rho v l}{\mu}$$

$$\text{Re} = \frac{v l}{\nu}$$

where, ρ = Density of fluid

v = Velocity of fluid passing through length (l)

μ = Kinematic viscosity

ν = Dynamic viscosity $\left(\dfrac{\mu}{\rho}\right)$

Critical Reynold Number

It represents the number where the boundary layer changes from laminar to turbine flow.

For flat plate, $\text{Re} < 5 \times 10^5$ (laminar)

$\text{Re} > 5 \times 10^5$ (turbulent)

For circular pipes, $\text{Re} < 2300$ (laminar flow)

$2300 < \text{Re} < 4000$ (transition to turbulent flow)

$\text{Re} > 4000$ (turbulent flow)

Prandtl Number (Pr)

It provides a measure of relative effective of momentum and energy transport by diffusion in velocity and thermal boundary layers respectively. Higher Pr means higher Nu and it shows higher heat transfer ($\text{Nu} \propto \bar{h}$)

$$\text{Pr} = \frac{\text{Momentum diffusivity through the fluid}}{\text{Thermal diffusivity through the fluid}}$$

$$\text{Pr} = \frac{\nu}{\alpha} = \frac{\mu/\rho}{k/\rho c_p} \Rightarrow \text{Pr} = \frac{\mu c_p}{k}$$

For liquid metal, $\text{Pr} < 0.01$

For air and gases, $\text{Pr} \approx 1$

For water, $\text{Pr} \approx 10$

For heavy oil and grease, $\text{Pr} > 10^5$

Stanton Number (St)

It is defined as
$$St = \frac{\text{Heat transfer coefficient}}{\text{Heat flow per unit temperature rise}}$$
$$St = \frac{Nu}{Re \times Pr}$$

Grashof Number (Gr)

It is used for free convection.
$$Gr = \frac{\text{Inertia force} \times \text{buoyancy force}}{\text{viscous force}}$$
$$Gr = \frac{\rho^2 \beta g \, \Delta t \, l^3}{\nu^2}$$

where, β = Coefficient of volumetric expansion $= \frac{1}{T}$

Rayleigh Number (Ra)

It is product of Grashof number and Prandtl number. It is used for free convection.
$$Ra = Gr \cdot Pr \; ; \; Ra = \frac{g \, \beta \, l^3 \, \Delta t}{\nu \cdot \alpha}$$

where, g = Acceleration due to gravity
β = Thermal expansion coefficient
ν = Kinematic viscosity
α = Thermal diffusivity
Pr = Prandtl number
Gr = Grashof number

If free or natural convection

$10^4 < Ra < 10^9$ (laminar flow)

$Ra > 10^9$ (turbulent flow)

Empirical Correlation for Free Convection

(a) Heated surface up or cooled surface down

Laminar flow $2 \times 10^5 < Gr.Pr < 2 \times 10^7$

$Nu = 0.54 \, (Gr\,Pr)^{0.25}$

Turbulent flow $2 \times 10^7 < Gr\,Pr < 3 \times 10^{10}$

$Nu = 0.14 \, (Gr\,Pr)^{0.33}$

(b) **Heated surface down or cooled surface up**

Laminar flow $3 \times 10^5 < \text{Gr Pr} < 7\times 10^8$

$\text{Nu} = 0.27\,(\text{Gr Pr})^{0.25}$

Turbulent flow $7 \times 10^8 < \text{Gr Pr} < 11\times 10^{10}$

$\text{Nu} = 0.107\,(\text{Gr Pr})^{0.33}$

Long cylinder $l/d > 60$

Laminar flow $10^4 < \text{Gr Pr} < 10^9$

$\text{Nu} = 0.53\,(\text{Gr Pr})^{0.25}$

Turbulent flow $10^9 < \text{Gr Pr} < 10^2$

$\text{Nu} = 0.13\,(\text{Gr Pr})^{0.33}$

(c) **For horizontal wire,**

$D < 0.005$ cm

$\text{Nu} = 0.4$

Vertical plates and Large cylinder

Laminar flow $10^4 < \text{Gr Pr} < 10^9$

$\text{Nu} = 0.59\,(\text{Gr Pr})^{0.25}$

Turbulent flow $10^9 < \text{Gr Pr} < 10^{12}$

$\text{Nu} = 0.13\,(\text{Gr Pr})$

Empirical Correlation for Forced Convection

Laminar Flow over Flat Plate

$$\text{Nu}_x = \frac{hx}{k} = 0.332\,(\text{Re}_x)^{\frac{1}{2}}\,\text{Pr}^{\frac{1}{3}}$$

(average of Nu)

$$\overline{\text{Nu}} = \frac{\bar{h}l}{k} = 0.64\,(\text{Re})^{\frac{1}{2}}\,\text{Pr}^{\frac{1}{3}}$$

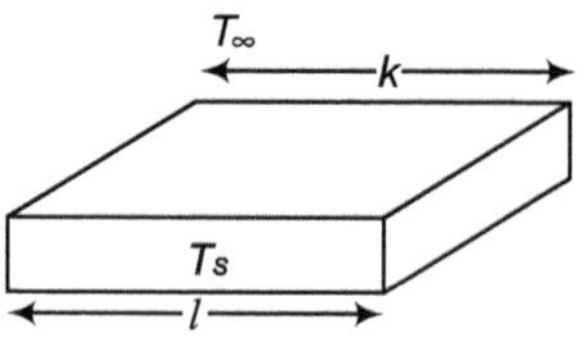

Rectangular plate

Hydrodynamic boundary layer thickness

$$\rho = \frac{5x}{\sqrt{\text{Re}_x}}$$

Laminar Flow over Inside Tube

$$\text{Nu} = \frac{hd}{k}, \ T_\infty = \frac{T_o + T_i}{2}$$

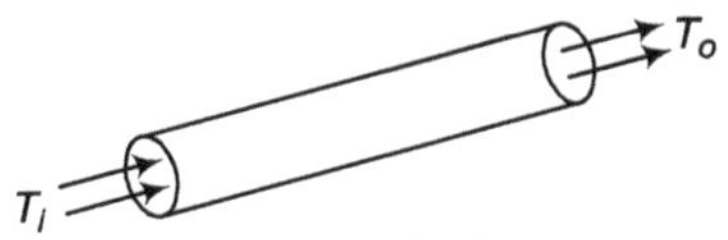

Cylindrical tube

For constant heat flux, Nu = 4.36

For constant wall temperature,

Fully developed flow Nu = 3.66

Slug flow Nu = 5.18

- **Turbulent flow over flat plate**

$$\text{Nu}_x = 0.0292\,(\text{Re})^{0.8}\,(\text{Pr})^{0.33}$$

$$\overline{\text{Nu}} = 0.036\,(\text{Re})^{0.8}\,(\text{Pr})^{0.33}$$

$$\delta = \frac{0.37\,x}{(\text{Re}_x)^{\frac{1}{5}}}$$

- **Turbulent flow in tubes**

$$\text{Nu} = 0.023\,(\text{Re})^{0.8}\,(\text{Pr})^n$$

where, $n = 0.4$ if fluid is being heated, $= 0.3$ if fluid is being cooled.

$$\overline{\text{Nu}} = 0.0036\,(\text{Re})^{0.8}\,(\text{Pr})^n$$

Hydrodynamic Boundary Layer

Over flat plate $\frac{\delta}{x} = \frac{1}{\sqrt{\text{Re}_x}}$, $\text{Re}_x = \frac{u_\infty x}{v}$

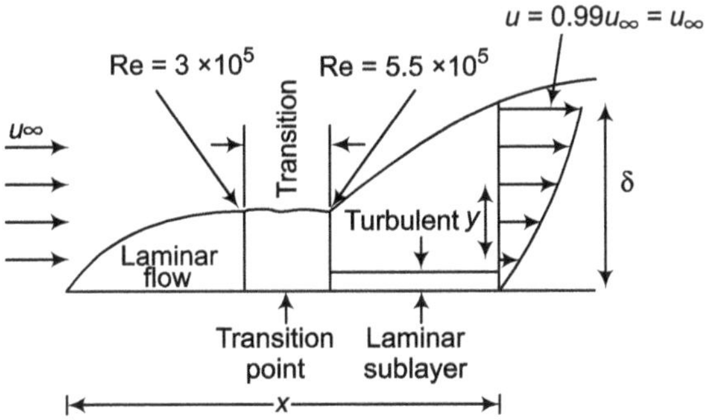

Hydrodynamic Boundary Layer

Blassius solution gives the ratio

$$\frac{\delta}{x} = \frac{5}{\sqrt{\text{Re}_x}} \qquad \text{(for laminar flow)}$$

local skin friction coefficient $C_{fx} = \dfrac{\tau_w}{\delta_{Kt}\dfrac{1}{2}Pu_\infty^2}$, where $\tau_w = \mu\left(\dfrac{\partial u}{\partial y}\right)_{y=0}$

$$\delta_{kt} = \frac{0.664}{\sqrt{\text{Re}_x}} \quad \text{(for laminar flow)}$$

Average skin coefficient upon total length l of the plate (for laminar flow)

$$\overline{C_f} = \frac{1.328}{\sqrt{\text{Re}_x}}$$

For turbulent flow

$$\frac{\delta}{x} = \frac{0.576}{(\text{Re}_x)^{\frac{1}{5}}}$$

$$C_{fx} = \frac{0.059}{(\text{Re}_x)^{\frac{1}{5}}}$$

$$\overline{C_f} = \frac{0.074}{(\text{Re}_x)^{\frac{1}{5}}}$$

Diagrammatic Representation of Thermal Boundary Layer over a Plate

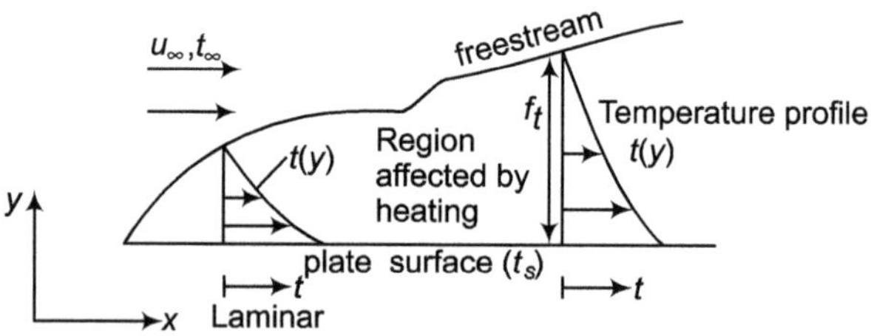

Thermal boundary layer during flow of cool fluid over a warm plate

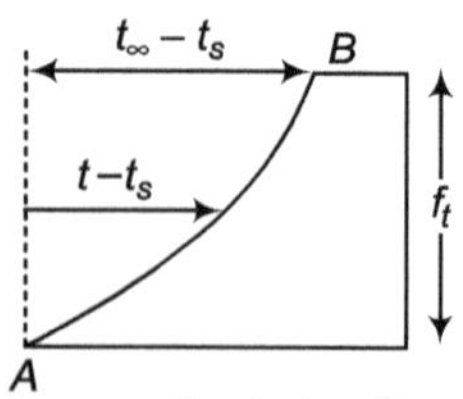

Temperature profile during flow of warm (Thermal boundary layer) fluid over a cool plate.

Relation between δ and δ_t

δ and δ_t graph can be shown in the figure.

Thermal boundary layer thickness

$$\delta_t = \frac{\delta}{(\text{Pr})^{1/3}}$$

$\delta \rightarrow$ Hydrodynamic boundary layer thickness

(i) When Pr = 1, $\delta_t = \delta$

(ii) When Pr > 1, $\delta_t < \delta$

(iii) When Pr < 1, $\delta_t > \delta$

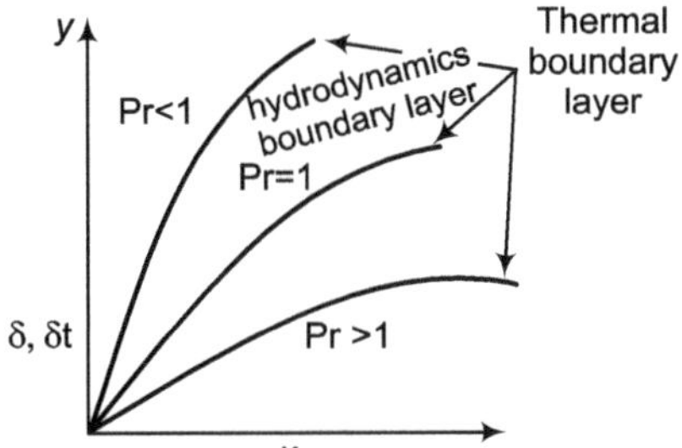

Various Prandtl curve

Key Points

- The heat is conducted by conduction.
- The thermal gradient is set up in the fluid layers, thereby a temperature profile is developed on the fluid flow.
- In the boundary layer region, the temperature of fluid varies from surface up to the thermal boundary layer thickness.

Some Important Equations

At the surface of the plate, there is no fluid motion and the heat transfer can occur only through conduction, then heat flux is

$$\frac{Q}{A} = h_x (t_s - t_a) = -k\left(\frac{dt}{dy}\right)_{y=0}$$

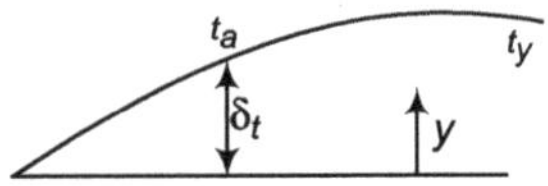

(a) at $y = 0, t = t_s$ (b) at $y = 0, \dfrac{\partial^2 t}{\partial y^2} = 0$

(c) at $y = \delta, t = t_a$ (d) at $y = \delta, \dfrac{dt}{dy} = 0$

t = Temperature of fluid at y distance measured in normal direction from the plate.

Reynolds Analogy

The following equations are used for Reynold analogy

$$\tau_w = \mu \frac{du}{dy} \qquad \ldots(i)$$

$$Q = -k A \frac{dt}{dy} \qquad \ldots(ii)$$

$$\frac{\mu C_p}{k} = 1 \qquad \ldots(iii)$$

$$C_{fx} = \frac{\tau_w}{\frac{1}{2} \rho u_\infty^2} \qquad \ldots(iv)$$

From Eqs. (i), (ii), (iii) and (iv), we get

$$\frac{h_x}{\rho c_p u_\infty} = \frac{C_{fx}}{2}$$

Since, $$S_t = \frac{h_x \Delta t}{\rho C_p u_\infty \Delta t}, \; St_x (\Pr)^{\frac{2}{3}} = \frac{C_{fx}}{2}$$

Heat Exchangers

A heat exchanger is a piece of equipment built for efficient heat transfer from one medium to another. Biolers and condensers in thermal power plants are example of large industrial heat exchangers. Heat exchangers are also abundant in chemical and process industries.

Fouling Factor

The phenomenon of rust formation and deposition of fluid impurities on surface which responsible to increase thermal resistance with a corresponding drop in the performance of the heat exchange equipment.

Fouling factor

$$R_f = \frac{1}{v_{\text{dirty}}} - \frac{1}{v_{\text{clear}}}$$

where, v = Overall heat transfer coefficient.

Logarithmic Mean Temperature Difference (LMTD)

LMTD is used to determine the temperature driving force for heat transfer in flow systems, most notably in heat exchangers. LMTD is a logarithmic average of the temperature difference between the hot and cold streams at each end of the exchanger.

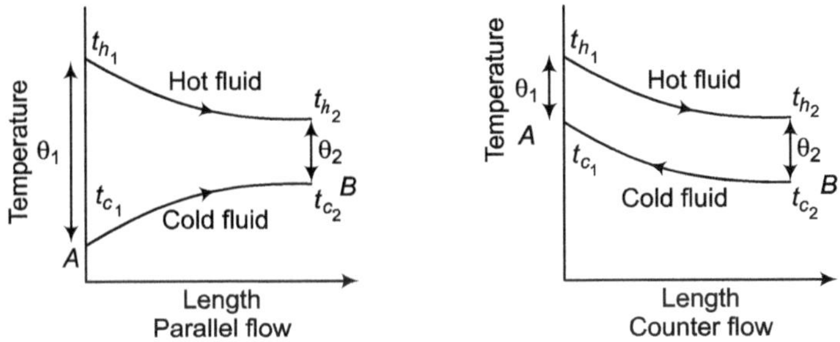

Temperature distribution for parallel and counter flow heat exchanger

Heat transfer,
$$Q = UA\frac{\theta_1 - \theta_2}{\log_e\left(\frac{\theta_1}{\theta_2}\right)} = UA\theta_m$$

$$\text{LMTD } \theta_m = \frac{\theta_1 - \theta_2}{\log_e \frac{\theta_1}{\theta_2}}$$

$$\theta_1 = t_{h_1} - t_{c_1}$$
$$\theta_2 = t_{h_2} - t_{c_2}$$

If the heat capacity of two fluid equal

$$\theta_m = \theta_1 - \theta_2$$

where, θ_1 = Temperature difference between two streams at end A

θ_2 = Temperature difference between two streams at end B.

θ_m = Logarithmic mean temperature difference

Capacity Ratio

The capacity ratio is defined as the ratio of the minimum to maximum capacity rate.

Capacity ratio $c = mc$, where c = Specific heat

If $m_h c_h > m_c c_c$, $R = \dfrac{m_c c_c}{m_h c_h}$

If $m_h c_h < m_c c_c$, $R = \dfrac{m_h c_h}{m_c c_c}$

Effectiveness of Heat Exchanger

The effectiveness of a heat exchanger is defined as the ratio of the energy actually transferred to the maximum theoretical energy transfer.

$$\varepsilon = \frac{(Q_{act})\text{ actual heat transfer}}{(Q_{max})\text{ maximum possible heat transfer}}$$

$$Q_{act} = m_h c_h (t_{h_1} - t_{h_2})$$
$$= m_c c_c (t_{c_2} - t_{c_1})$$
$$Q_{max} = c_{min} (t_{h_1} - t_{c_1})$$

If $m_c c_c < m_h c_h \Rightarrow c_{min} = m_c c_c$

$\Rightarrow$ $Q_{max} = m_c c_c (t_{h_1} - t_{c_1})$

$$\varepsilon = \frac{m_c c_c (t_{c_2} - t_{c_1})}{m_c c_c (t_{h_1} - t_{c_1})} = \frac{t_{c_2} - t_{c_1}}{t_{h_1} - t_{c_1}}$$

If $m_c c_c > m_h c_h \Rightarrow c_{min} = m_h c_h$

$\Rightarrow$ $Q_{max} = m_h c_h (t_{h_1} - t_{c_1})$

$$\varepsilon = \frac{m_h c_h (t_{h_1} - t_{h_2})}{m_h c_h (t_{h_1} - t_{c_1})} = \frac{t_{h_1} - t_{h_2}}{t_{h_1} - t_{c_1}}$$

where, C_h = Heat capacity rates for hot fluids

C_c = Heat capacity rates for cold fluids $t_{h_1}, t_{h_2}, t_{c_1}$ and t_{c_2} are the temperature streams for hot and cold fluids

m_h = Mass of hot fluid

m_c = Mass of cold fluid

Number of Transfer Units (NTU)

NTU is a measure of the size of heat exchanger, it provides some indication of the size of the heat exchanger.

$$\text{NTU} = \frac{UA}{c_{min}}$$

where, U = Overall heat transfer coefficient

A = Surface area

C_{min} = Minimum capacity rate

If $m_h c_h > m_c c_c \Rightarrow c_{min} = m_c c_c$

$\Rightarrow$ $$\text{NTU} = \frac{UA}{m_h c_c}$$

If $\quad m_h c_h < m_c c_c$

$\Rightarrow \quad C_{min} = m_h c_h$

$$\text{NTU} = \frac{UA}{m_h c_h}$$

Effectiveness for Parallel Flow Heat Exchanger

It is given by
$$\varepsilon = \frac{1 - \exp[-\text{NTU}(1+R)]}{1+R}$$

where, $R = \dfrac{C_{min}}{C_{max}}$

$$\text{NTU} = \frac{UA}{C_{min}}$$

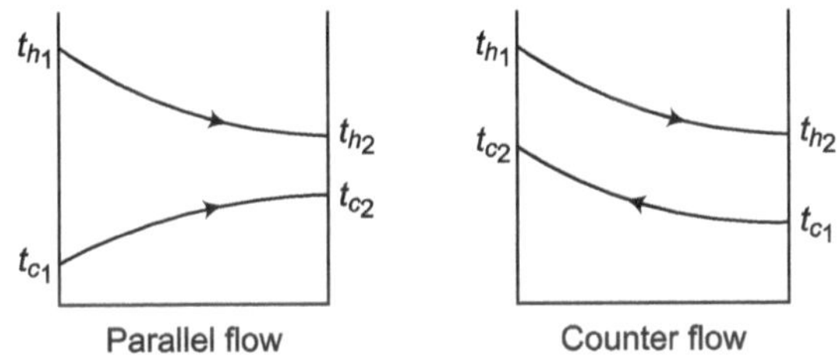

Heat exchanger for parallel and counter flow

Effectiveness for the Counter Flow Heat Capacity

It is given by
$$\varepsilon = \frac{1 - \exp[-\text{NTU}(1-R)]}{1 - R\exp[-\text{NTU}(1-R)]}$$

Radiation

Radiation is the process in which energetic particles or energetic waves travel through a medium or space.

Surface Emission Properties

Emission of radiation by a body depends upon the following factors

(a) Temperature of the surface

(b) The nature of the surface

(c) The wavelength or frequency of radiation

The parameters which deals with the surface emission properties are as given below

Total Emissive Power (E)

It is defined as the total amount of radiation emitted by a body per unit time and area.

or $$E = \sigma T^4 \text{ W/m}^2$$

σ = Stefan Boltzmann constant $= 5.67 \times 10^{-8}$ W / m^2 K^4

$$E_b = \sigma A T^4$$

Monochromatic (Spectral) Emissive Power (E_λ)

It is defined as the rate of energy radiated per unit area of the surface per unit wavelength.

$$E = \int_0^\infty E_\lambda d\lambda \text{ W/m}^2$$

Emission from Real Surface

The emissive power from a real surface is given by

$$E = \varepsilon\sigma A T^4 \text{ W}$$

where, ε = Emissivity of the surface

T = Surface temperature

Emissivity (ε)

It is defined as the ratio of the emissive power of any body to the emissive power of a black body of same temperature.

$$\varepsilon = \frac{E}{E_b}$$

- For black body, $\varepsilon = 1$
- For white body, $\varepsilon = 0$
- For gray body, $0 < \varepsilon < 1$

Reflectivity (ρ)

It is defined as the fraction of total incident radiation that are reflected by material.

$$\text{Reflectivity } (\rho) = \frac{\text{Energy reflected } (Q_r)}{\text{Total incident radiation } (Q)}$$

Absorptivity (α)

It is defined as the fraction of total incident radiation that are absorbed by material.

$$\text{Absorptivity } (\alpha) = \frac{\text{Energy absorbed } (Q_a)}{\text{Total incident radiation } (Q)}$$

Transmissivity (τ)

It is defined as the fraction of total incident radiation that are transmitted through the material.

$$\text{Transmissivity } (\tau) = \frac{\text{Energy transmitted } (Q_t)}{\text{Total incident radiation } (Q_w)}$$

$$Q_0 = Q_a + Q_r + Q_t$$

$$\frac{Q_0}{Q_0} = \frac{Q_a}{Q_0} + \frac{Q_r}{Q_0} + \frac{Q_t}{Q_0}$$

$$\alpha + \rho + \tau = 1$$

- For black body $\alpha = 1, \rho = 0, \tau = 0$
- For opaque body $\tau = 0, \alpha + \rho = 1$
- For white body $\rho = 1, \alpha = 0$ and $\tau = 0$

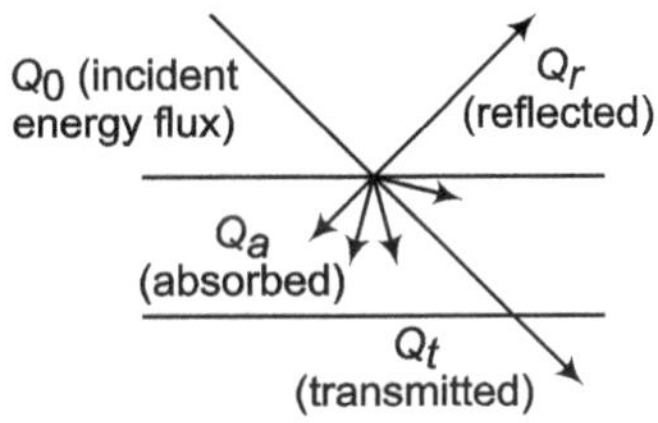

Diagram for Q_o, Q_r, Q_a, Q_t

Kirchoff's Law

The emissivity ε and absorptivity α of a real surface are equal for radiation with identical temperature and wavelength.

$$\alpha = \varepsilon = \frac{E}{E_b}$$

Stefan Boltzmann Law

Emissive power of a black body is directly proportional to the fourth power of its absolute temperature. $E_b = \sigma T^4$

E_b = Emissive power of a black body

σ = Stefan Boltzmann constant

Wien's Displacement Law

Wien's displacement law state that the product of λ_{max} and T is constant.

$$\lambda_{max} T = \text{constant}$$

where, λ_{max} = Wavelength at which the maximum value of monochromatic emissive power occurs.

Planck's Law

Planck suggested following formula, monochromatic emissive power of a black body.

$$(E_\lambda)_b = \frac{2\pi c^2 h \lambda^{-5}}{\exp\left(\frac{ch}{\lambda kT}\right) - 1}$$

$$\text{Total missive power } E_b = \int_0^\infty (E_\lambda)_b \, d\lambda$$

where, c = Velocity of light in vacuum = 2.998×10^8 m/s
h = Planck's constant = 6.625×10^{-34} J-s
λ = Wavelength, $k = 1.3805 \times 10^{-23}$ J / K, T = absolute temperature.
$(E_{\lambda b})$ = monochromatic emissive power of a black body.

Electrical Network Approach for Radiation Heat Exchange

Consider the radiant heat exchange between two non-black surfaces. An electric network representing space and surface resistance to radiation are following

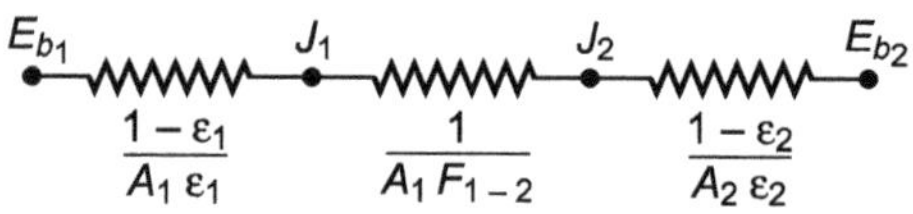

An electrical network between two non black surfaces

$$(Q_{1-2})_{net} = \frac{E_{b_1} - E_{b_2}}{\frac{1-\varepsilon_1}{A_1 \varepsilon_1} + \frac{1}{A_1 F_{1-2}} + \frac{1-\varepsilon_2}{A_2 \varepsilon_2}}$$

or $$(Q_{1-2})_{net} = (F_g)_{1-2} A_1 \sigma_b (T_1^4 - T_2^4)$$

Here, New Gray Body Factor

$$(F_g)_{1-2} = \frac{1}{\frac{1-\varepsilon_1}{A_1 \varepsilon_1} + \frac{1}{A_1 F_{12}} + \frac{1-\varepsilon_2}{A_2 \varepsilon_2}}$$

where, E_1 = Emissivity for body 1
E_2 = Emissivity for body 2

Cases

- In case of black surfaces, $\varepsilon_1 = \varepsilon_2 = 1$, $(F_g)_{12} = F_{1-2}$

$$Q_{net} = F_{1-2}\, A_1\, \sigma\, (T_1^4 - T_2^4)$$

- In case of parallel planes, $A_1 = A_2$ and $F_{1-2} = 1$

$$(F_g)_{1-2} = \frac{1}{\dfrac{1}{\varepsilon_1} + \dfrac{1}{\varepsilon_2} - 1}$$

- In case of concentric cylinder or sphere, $F_{1-2} = 1$

$$(F_g)_{1-2} = \frac{1}{\dfrac{1-\varepsilon_1}{\varepsilon_1} + 1 + \dfrac{1-\varepsilon_2}{\varepsilon_2}\dfrac{A_1}{A_2}}$$

where, $\dfrac{A_1}{A_2} = \dfrac{r_1}{r_2}$ (for concentric cylinder)

$\dfrac{A_1}{A_2} = \dfrac{r_1^2}{r_2^2}$ (for concentric sphere)

- When a small body lies inside a large enclosure

$$F_{1-2} = 1,\ A_1 << A_2 \ \Rightarrow\ \frac{A_1}{A_2} = 0$$

$$(F_g)_{1-2} = \frac{1}{\dfrac{1-\varepsilon_1}{\varepsilon_1} + 1}$$

Radiation Shield

Radiation shields reduce the radiation heat transfer by effectively increasing the surface resistance without actually removing of heat from overall system.

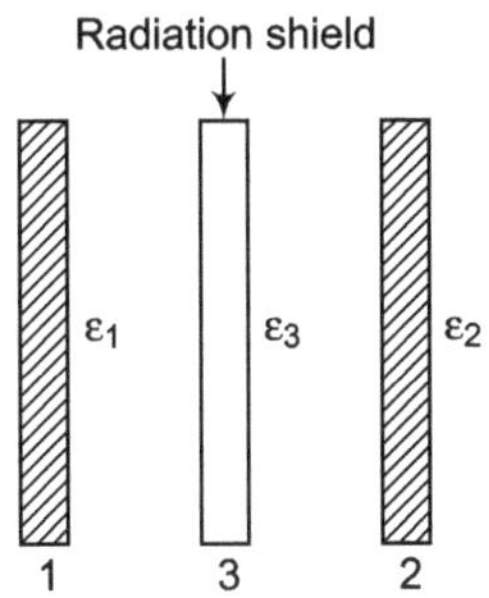

Radiation shield diagram

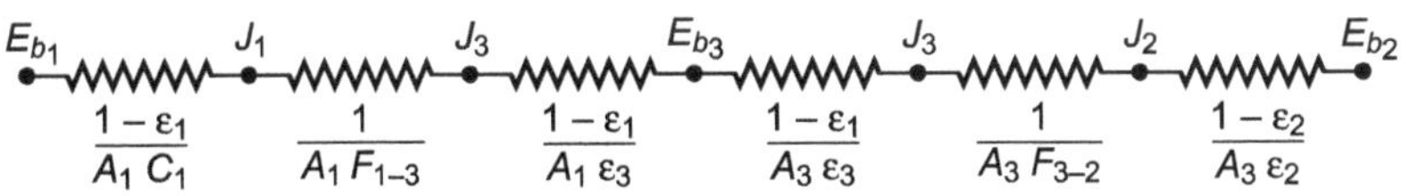

Radiation network for 2 parallel infinite places separated by one shield

$$(Q_{1-3})_{net} = (Q_{3-2})_{net} \qquad (A_1 = A_3 = A_2)$$

$$\frac{A\,\sigma\,(T_1^4 - T_3^4)}{\dfrac{1}{\varepsilon_1} + \dfrac{1}{\varepsilon_3 - 1}} = \frac{A\,\sigma\,(T_3^4 - T_2^4)}{\dfrac{1}{\varepsilon_3} + \dfrac{1}{\varepsilon_2} - 1}$$

$$(Q_{1-2})_{net} = \frac{A\,\sigma\,(T_1^4 - T_2^4)}{\left(\dfrac{1}{\varepsilon_1} + \dfrac{1}{\varepsilon_3} - 1\right) + \left(\dfrac{1}{\varepsilon_3} + \dfrac{1}{\varepsilon_2} - 1\right)}$$

$$\frac{[(Q_{1-2})_{net}] \text{ with shield}}{[(Q_{1-2})_{net}] \text{ without shield}} = \frac{\left(\dfrac{1}{\varepsilon_1} + \dfrac{1}{\varepsilon_2} - 1\right)}{\left(\dfrac{1}{\varepsilon_1} + \dfrac{1}{\varepsilon_3} - 1\right) + \left(\dfrac{1}{\varepsilon_3} + \dfrac{1}{\varepsilon_2} - 1\right)}$$

If $\qquad \varepsilon_1 = \varepsilon_2 = \varepsilon_3$

Then, $\qquad (Q_{1-3})_{net} = (Q_{3-2})_{net} = \dfrac{1}{2}(Q_{1-2})_{net}$

and $\qquad T_3^4 = \dfrac{1}{2}(T_1^4 + T_2^4)$

n Shields having Equal Emissivities

In general case when there are n shields with equal emissivities ($\varepsilon_1 = \varepsilon_2 \ldots = \varepsilon_n$). There will be $(n + 1)$ spaces resistance and shape factor ($F_{1-2} = 1$) is unity for each infinite parallel plans.

$$Q_n \text{ shield} = \frac{1}{(n+1)\left(\dfrac{2}{\varepsilon} - 1\right)} A\,\sigma\,(T_1^4 - T_2^4)$$

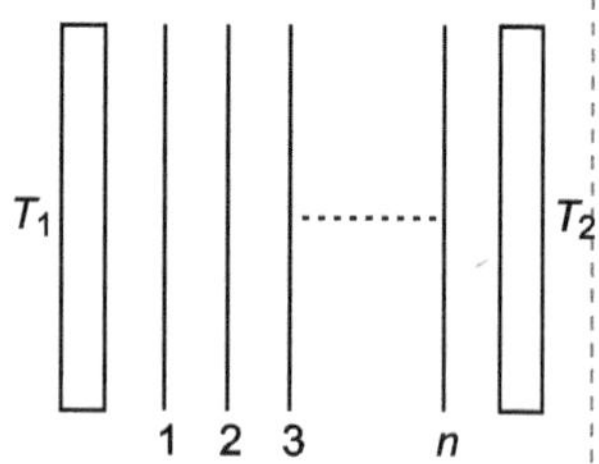

$$Q \text{ without shield} = \frac{1}{\dfrac{2}{\varepsilon} - 1} A\,\sigma\,(T_1^4 - T_2^4)$$

$$\frac{Q_n \text{ shield}}{Q_n \text{ without shield}} = \frac{1}{n+1}$$

$\therefore$ It provide that reduction in heat flow due to insertion of shield does not affect the wall temperature.

7

Thermodynamics

Basic Concepts of Thermodynamics

Thermodynamics is a physical science that studies the effects on material bodies and on radiation in regions of space of transfer of heat and of work done by the bodies. Thermodynamics is built on the study of energy transfers that can be strictly resolved into two distinct components, heat and work.

Thermodynamics System

It is defined as the quantity of matter or a region in space chosen for study. The mass and region outside the system is called surrounding. Real or imaginary surface (mathematically thickness zero) that separates the system from the surrounding is called the boundary.

Closed System (Control Mass)

It consists of a fixed amount of mass and no mass can cross its boundary or leave or enter a closed system. However, energy in the form of heat or work can cross the boundary and the volume of closed system does not have to be fixed. *e.g.*, piston cylinder device.

Open System (Control Volume)

It is a properly selected region in the space and both mass and energy region can cross the boundary. The boundary of a control volume is called a **control surface** and they can be real or imaginary. *e.g.*, compressor, turbine, nozzle.

Note A control volume can be fixed in size and shape as in case of nozzle or it may involve moving boundaries as shown in Fig. (b). Most control volume, however have fixed boundaries and thus do not involve any moving boundaries.

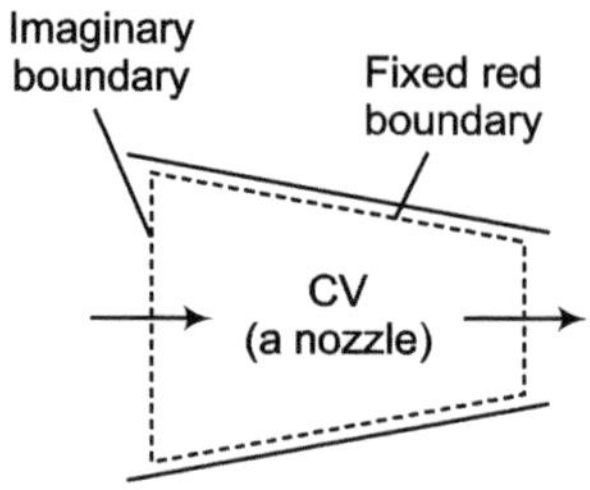

(a) Nozzle

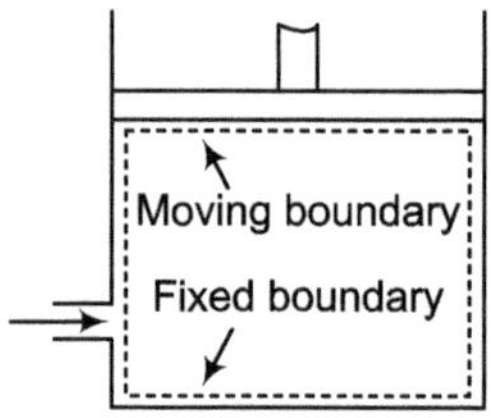

(b) Pistion

Key Points

- In closed system, energy transfers but mass does not transfer across the boundary.
- In open system, both mass and energy transfer across the boundary.
- In isolated system, neither mass nor energy transfer across the boundary.

Properties of a System

Any characteristic of a system is called a property. It can be independent or dependent.

Intensive and Extensive Properties of a System

Intensive properties are those that are independent of the size of system such as temperature, pressure and density.

Extensive properties are those whose values depend on the size or extent of the system such as mass m, volume V and total energy E.

An easy way to determine whether the property is intensive or extensive is to divide the system into two equal parts with a partition.

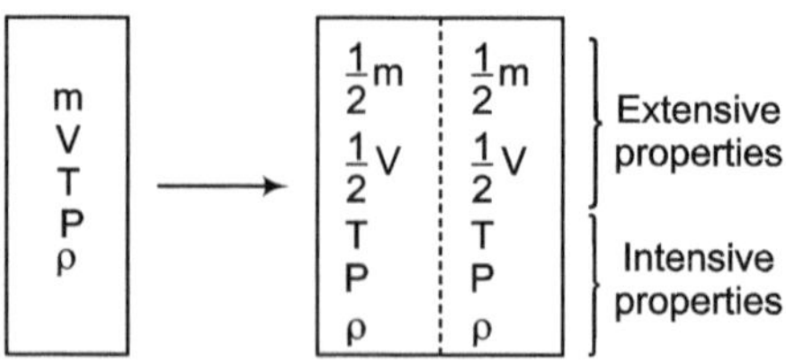

Distribution of extensive and intensive properties

Generally, upper case letters are used to denote extensive properties (exception m) and lower case letters are used for intensive properties (exception P, T).

Specific Properties of a System

Extensive properties per unit mass are called **specific properties**.*e.g.*,

$$\text{Specific volume } v = \frac{V}{m}$$

$$\text{Specific energy } e = \frac{E}{m}$$

Term **specific gravity** (or relative density) is used to define the density of a substance with respect to density of some standard substance at a specified temperature.

(Usually, water at 4°C, $\rho_{H_2O} = 1000 \text{ kg/m}^3$)

Key Points

- Such properties which can be measured by simple instrument and sensed by human eye are known as **macroscopic** properties.
- Such properties which cannot be sensed by human eye and are observed at molecular level are known as **microscopic** properties.

Continuum

Continuum is defined as a continuous, homogeneous matter with no holes. The continuum idealisation allows us to treat properties as point function and to assume properties to vary continually in space with no jump discontinuities.

State

State of system is described by its properties. At a given state, all the properties of a system have fixed values. However, there is no need to specify all the properties in order to fix a state. The number of properties required to fix the state of a system is given by the **state postulates**. It means that properties should be independent to each other *i.e.*, when one property varies, other should be held constant. In other words, the condition of a system at any instant of time is called its state.

Equilibrium

The word equilibrium implies a state of balance. In an equilibrium state, there are no unbalanced potentials (or driving forces) within the system.

A system in equilibrium experience has no changes when it is isolated from its surrounding. Thermodynamics deals with equilibrium states.

- A system is in thermal equilibrium if the temperature is same throughout the entire system.
- In mechanical equilibrium, there is no change in pressure at any point of the system with time.
- In chemical equilibrium, no chemical reactions occur.

Processes

Any change that a system goes from one equilibrium state to another equilibrium state is called a process and series of states through which a system passes during a process is called the path of the process.

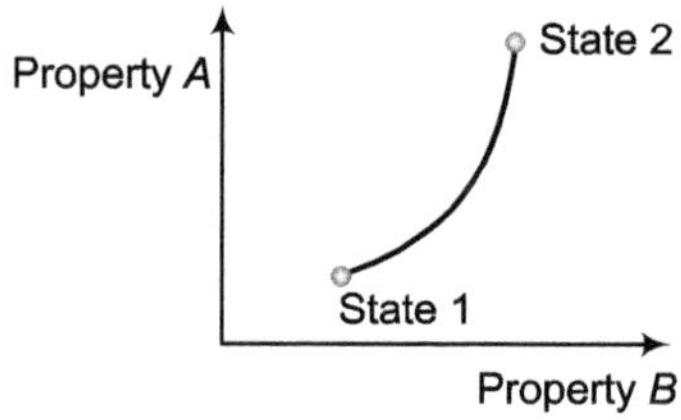

Changes in states

Quasi Equilibrium Process

When a process proceeds in such a manner that the system remains infinitesimally close to an equilibrium state at all times, it is called a quasi static or quasi equilibrium process.

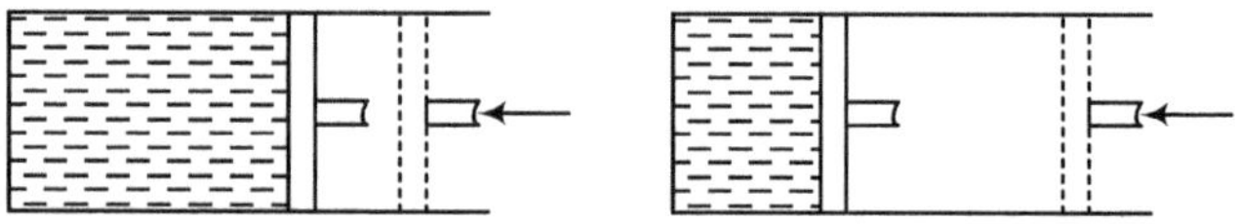

Slow compression (quasi equilibrium)

Very fast compression (Non-quasi equilibrium)

Key Points

- Quasi equilibrium is an idealised process and is not a true representation of an actual process.
- Quasi equilibrium process are easy to analyse.
- Quasi equilibrium process are work producing devices deliver the maximum work when they operate on quasi equilibrium processes.
- Quasi equilibrium processes as standards to which actual processes can be compared.

Cycle

A system is said to have undergone a cycle if it returns to its initial state at the end of process *i.e.*, for a cycle the initial and final states are identical.

Steady Flow Process

Steady flow process is defined as a process during which a fluid flows through a control volume steadily. It means at any fixed point in the control volume, fluid properties remain the same during the entire process. However, they can change from point to point.

Temperature

The temperature is a measure (degree) of **hotness** or **coldness**. Generally, we express the level of temperature qualitatively with words like freezing cold, warm hot.

e.g., a metal chair will feel much colder than a wooden one, even both are at the same temperature.

Temperature Scales

A scale of temperature independent of the thermometric substance is called a thermodynamic temperature scale. The temperature scales used in the **SI** and **English system** today are the celsius scale and the fahrenheit scale, respectively. They are related to absolute temperature scales.

(i) Kelvin scale is related to celsius scale by

$$T(\text{K}) = T(\text{C}) + 273.16$$

(ii) Rankine scale is related to fahrenheit scale by

$$T(\text{R}) = T(\text{F}) + 459.67$$

The temperature scales in the two unit systems are related by

$$T(\text{R}) = 1.8\,T(\text{K})$$

$$T(°\text{F}) = 1.8\,T\,(°\text{C}) + 32$$

$$\Rightarrow \qquad \Delta T°\,(\text{F}) = 1.8\,\Delta T(\text{K}) = \Delta T(\text{R})$$

Pressure

Pressure is defined as the force exerted by a fluid per unit area. We speak of pressure only when we deal with gas or liquid. The counterpart of pressure is stress *i.e.*, force per unit area.

$$1 \text{ pascal} = 1\,\text{N/m}^2$$

$$1 \text{ bar} = 10^5 \text{ Pa} = 0.1 \text{ MPa} = 100 \text{ kPa}$$

$$\text{atm} = 101325 \text{ Pa} = 101.325 \text{ kPa} = 1.01325 \text{ bar}$$

$$1 \text{ kg f/cm}^2 = 9.807 \text{ N/cm}^2$$

$$= 9.807 \times 10^4 \text{ Pa} = 0.9807 \text{ bar}$$

$$= 0.96788 \text{ atm}$$

The actual pressure at a given position is called **absolute pressure** and it is measured relative to absolute vacuum (*i.e.*, absolute zero pressure)

$$p_{gauge} = p_{abs} - p_{atm}$$

(for pressures above p_{atm})

$$p_{vac} = p_{atm} - p_{abs}$$

(for pressures below p_{atm})

where, p_{gauge} = gauge pressure

p_{abs} = absolute pressure

p_{atm} = atomic pressure

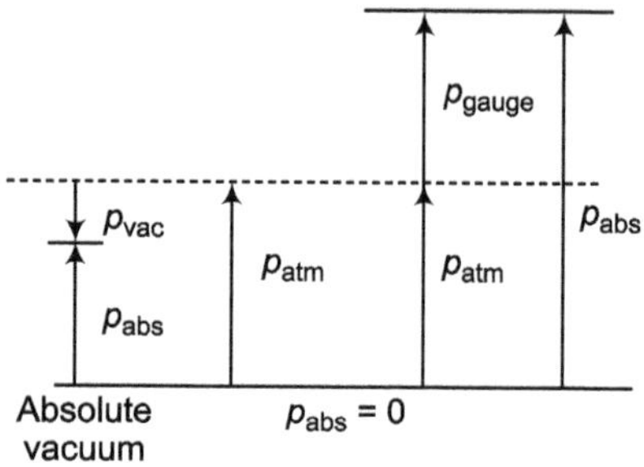

Key Points

- Isolated system is special case of closed system in which mass and energy both are not allowed to cross the system.
- Concepts of continuum is valid only from the macroscopic point of view (not macroscopic point of view). It is not valid for similar dimensions of system and molecules and valid only when size of system is large relative to space between the molecules.
- Thermal equilibrium between two or more bodies exists when they are brought together, there is no change of temperature, density pressure.

Laws of Thermodynamics

The four laws of thermodynamics define fundamental physical quantities (temperature, energy and entropy) that characterise thermodynamic systems. The laws described how these quantities behave under various circumstances and forbid certain phenomena.

Zeroth Law of Thermodynamics

The zeroth law of thermodynamics states that if two bodies are in thermal equilibrium with a third body, they are also in thermal equilibrium with each other. It may seem silly that such an obvious fact is called one of the basic laws of thermodynamics. By replacing the third body with a thermometer, the zeroth law can be restated as "two bodies are in thermal

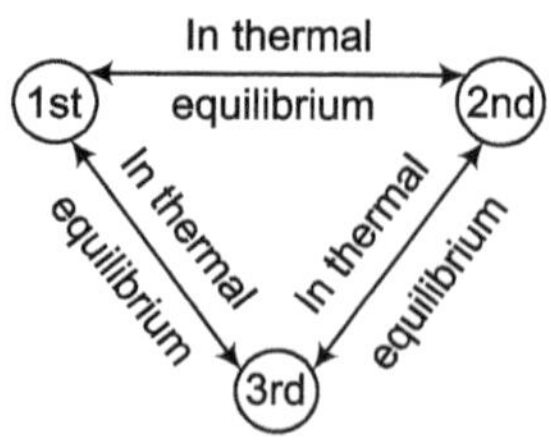

Representation of zeroth law of thermodynamics

equilibrium if both have the same temperature reading even if they are not in contact".

Heat and Work

Heat is defined as the form of energy that is transferred between two systems (or system and its surrounding) by virtue of a temperature difference.

And if energy interaction between two systems (or system and its surrounding) is not caused by a temperature difference then it is called **work**.

First Law of Thermodynamics

It is based on conservation of energy according to which total amount of energy in any thermodynamics system remains constant. Energy cannot be destroyed or produced, it can only be converted from one form to another form.

It states that whenever a closed system executes a cyclic process, the algebraic sum of work transfer is proportional to algebraic sum of heat transfers.

Mathematically, $$\oint dQ = \oint dW$$

Heat is given to a closed system is equal to sum of work done by a system and change in internal energy of the system.

$$\Delta Q = \Delta U + \Delta W$$

where ΔQ and ΔW are infinitesimal amounts of heat supplied to the system by its surroundings and work done by the system on its surroundings, respectively and ΔU is an infinitesimal change in internal energy.

Key Points

- If heat is given to the system then in equation we put ΔQ as positive. In case of heat is given by the system we put ΔQ as negative.
- If work is done by system then we take ΔW as positive. However, if work is done on the system then we put ΔW as negative.
- ΔU is known as internal energy. It is path independent and it depends on the final stage (temperature) and initial stage (temperature).

$$\Delta U = mC_V\ (\Delta T)$$

Work Done in Moving the Boundaries of a Closed System

Suppose a gas contained in piston cylinder expands from position 1 to 2, then work done by the gas

$$\Delta W = \int_{V_1}^{V_2} p\,dV$$

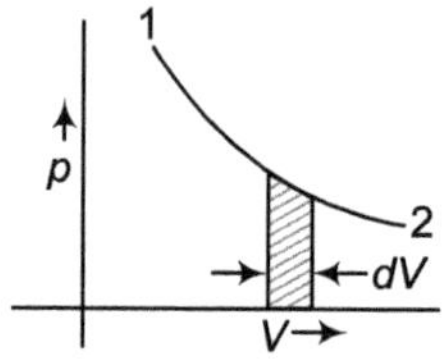

Piston cylinder expands from 1 to 2

Constant Volume Process (Isochoric Process)

In isochoric process, volume is constant during process.

Work done $dW = p\,dV = 0$

Heat supplied to system $dQ = mC_V\,dT$

$$\Delta Q = mC_V(T_2 - T_1)$$

where, C_v = Specific heat for constant volume process

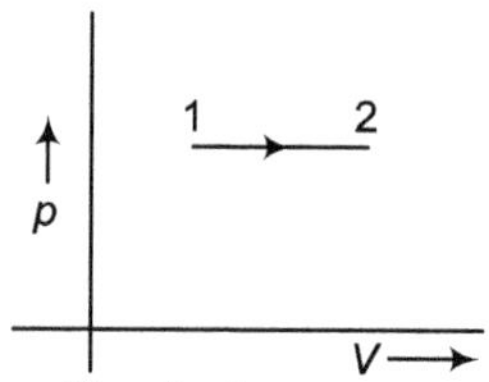

Constant pressure process

Change in internal energy of system

$$dU = mC_V\,dT$$

$$\Delta U = mC_V\,(T_2 - T_1)$$

Constant Pressure Process (Isobaric Process)

In isobaric process, pressure is constant throughout process

$$\Delta W = p_2V_2 - p_1V_1$$

$$= mR\,(T_2 - T_1)$$

$$\Delta Q = mC_p(T_2 - T_1)$$

$$\Delta U = mC_V(T_2 - T_1)$$

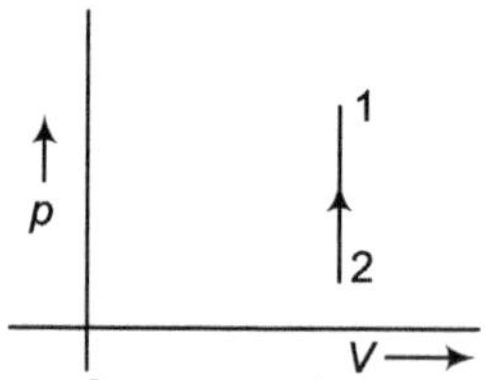

Constant volume heating process

Constant Temperature (Isothermal Process)

In isothermal process, temperature is constant during process. If an ideal gas follows a constant temperature process, then Boyle's law is

$$pV = \text{constant}$$

$$\Delta W = p_2V_2 \ln\frac{V_2}{V_1}$$

or $$\Delta W = mRT \ln\frac{V_2}{V_1}$$

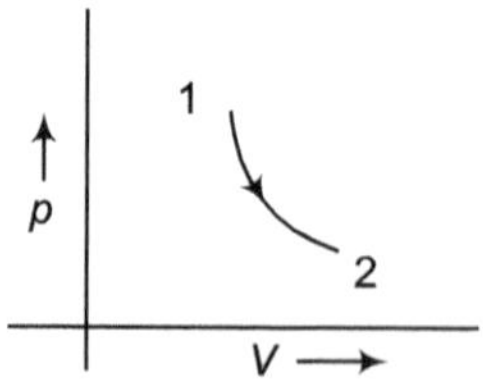

Isothermal process

$$\Delta U = 0 \quad (\text{as } T_2 = T_1)$$

$$\Delta Q = \Delta W = p_2V_2 \ln \frac{V_2}{V_1} = p_2 V_2 \ln \frac{V_2}{V_1} = mRT \ln \frac{V_2}{V_1}$$

Reversible Adiabatic Process

In reversible adiabatic process there is no heat transfer between the system and the surrounding *i.e.*, $Q = 0$

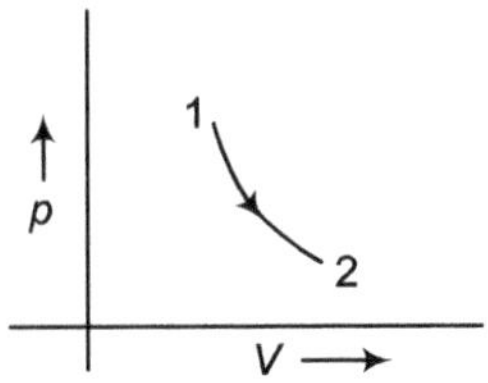

Reversed adiabatic process

$$pV^{\gamma} = \text{constant}, \qquad \gamma = \frac{C_p}{C_V}$$

where, r = adiabatic index

C_p = specific heat for constant pressure

$$\Delta W = \frac{p_1V_1 - p_2V_2}{\gamma - 1} = \frac{mR\,(T_1 - T_2)}{\gamma - 1}$$

$$\Delta Q = 0 \quad \text{(No heat transfer from surrounding)}$$

$$\Delta U = -\,\Delta W, \qquad (0 = \Delta U + \Delta W)$$

Reversible Polytropic Process

In polytropic process, both heat and work transfers take place. The process equation is represented by $p.V^n = C$ where n is called the index of process or polytropic index.

$$\Delta W = \frac{p_1V_1 - p_2\,V_2}{n - 1} = \frac{mR\,(T_1 - T_2)}{n - 1}$$

$$\Delta U = mC_V\,(T_2 - T_1)$$

or
$$\Delta Q = \frac{\gamma - n}{\gamma - 1} \times \Delta W$$

where, n = polytropic index

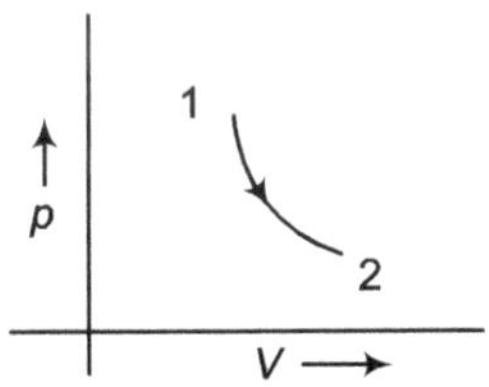

Reversible polytropic process

Second law of Thermodynamics

Second law has been mentioned by two statements "Clausius and Kelvin-Planck statement". Each statement is based on irreversible process.

Clausius Statement

It is impossible to construct a device operating on a cycle whose sole effect is the transfer of heat from a low temperature body to high temperature body.

Kelvin-Planck Statement

It is impossible to construct a device operating on a cycle whose sole effect is transfer of heat energy from a single heat reservoir and deliver equivalent amount of energy by work.

Perpetual Motion Machines (PMM)

Any device that violates either first law or second law of thermodynamics is known as perpetual motion machine.

These machines are of two types given in table

Perpetual Motion Machines

S. No.	PMM 1	PMM 2
1.	PMM 1 is that machine which violates first law of thermodynamics.	PMM 2 is that machine which violates second law of thermodynamics.
2.	It produces continuous work without receiving energy from other systems $Q_{=0} \rightarrow \text{PMM1} \rightarrow \text{W}$	This machine exchange heat from a single reservoir and do equal amount of work $Q \rightarrow \text{PMM1} \rightarrow \text{W}$

Note Violation of Kelvin-Planck statement leads to violation of Clausius statement and *vice-versa.*

Third Law of Thermodynamics

It states that entropy of all perfect crystalline solids is zero at absolute zero temperature. It means zero entropy datum is chosen at 0°C or 273 K.

Free Expansion Process

The expansion of a gas against vacuum is called free expansion.

$$\Delta W = 0, \Delta Q = \Delta U$$

In case of reversible adiabatic free expansion process,

$$\Delta W = 0\,;\ \Delta Q = 0\,;\ \Delta U = 0$$

Reversible Polytropic Process

Processes in which pV^n = constant

These material should be given before second law of thermodynamics (after reversible polytropic process)

$$n = \pm \infty \ (V = C)$$

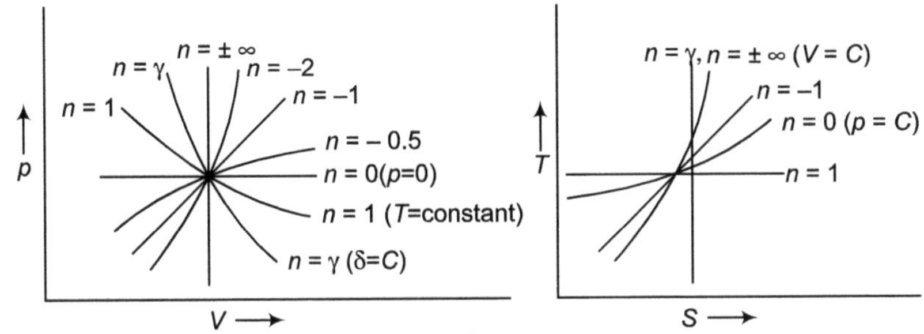

p-V and T-S diagrams

$$pV^n = C, \frac{dp}{dV} = -n\frac{p}{V}$$

The slope of curve increases in the negative direction with increase of n for

Isobaric process $(p = C)$ $n = 0$

Isothermal process $(p = C)$ $n = 1$

Isentropic process $(S = C)$ $n = \gamma$

Isometric or isochoric process $(V = C)$ $n = \infty$

Ideal Gas Equation (IGE)

$$pV = nRT, \; pV = \frac{m}{M}RT$$

where, m = Weight of gas, M = Molecular weight

(R = 8.3143 kJ/kg mol-K)

or

$$pV = m\left(\frac{R}{M}\right)T$$

$$pV = m\bar{R}T \Rightarrow \bar{R} = \frac{R}{M}$$

$$R_{air} = \frac{8.3143}{28.96} = 0.287 \text{ kJ/kg-K} \Rightarrow R_{O_2} = \frac{8.3143}{32} = 0.262 \text{ kJ/kg-K}$$

Key Points

- For calculating heat transfer in reversible polytropic process, if change in potential energy is given use, formula $\Delta Q = \frac{n-\gamma}{n-1}(\Delta U)$

  If work done is given, use formula $\Delta Q = \frac{\gamma - n}{\gamma - 1}(\Delta W)$
- In case of reversible adiabatic free expansion process $\Delta W = 0, \Delta Q = 0, \Delta U = 0$
- For an isolate system $\Delta Q = 0, \Delta W = 0$

  $\Delta U = 0$

  $\Rightarrow$ U = constant
- Internal energy is function of temperature only $U = f(T)$

Steady Flow Open System

If mass and energy both transfer across the boundary of system. Then, the system is categorised as open system.

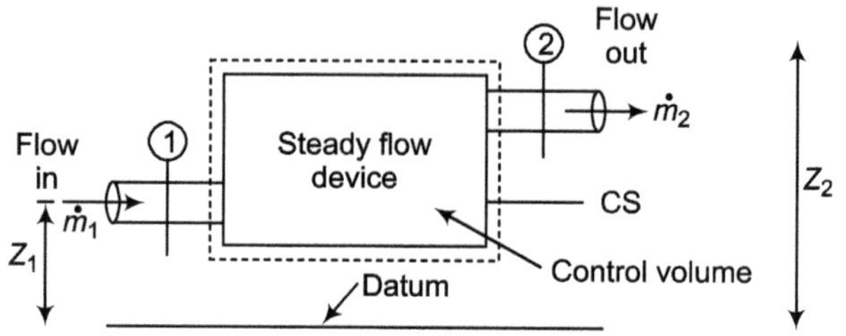

Steady flow open system

In case of steady flow, there is no accusation of mass or energy within the control volume and property at any location within the control volume are steady with time.

So, $$m_{in} = m_{out}$$

where, m_{in} = Mass flow rate within the system

m_{out} = Mass flow rate outside the system

Many flow processes, such as filling up and evacuating gas cylinders, are not steady. Such processes can be analysed by the control volume technique.

Let $A_1, A_2 \rightarrow$ Cross-sections of stream (m^2)

$\dot{m}_1, \dot{m}_2 \rightarrow$ Mass flow rate (kg/s)

$p_1, p_2 \rightarrow$ Absolute pressure (N/m^2)

$V_1, V_2 \rightarrow$ Specific volume (m^2/kg)

$u_1, u_2 \rightarrow$ Specific internal energy (J/kg)

$V_1, V_2 \rightarrow$ Velocities in m/s

$Z_1, Z_2 \rightarrow$ Elevation above an arbitrary datum, (m).

$\dfrac{dQ}{d\tau} \rightarrow$ Net rate of heat transfer through the control surface (J/s)

$\dfrac{dW_x}{d\tau} \rightarrow$ Net rate of work transfer through the control surface (J/s)

For steady flow process,

Mass balance $$\dot{m}_1 = \dot{m}_2 \Rightarrow \frac{A_1V_1}{v_1} = \frac{A_2V_2}{v_2}$$

Energy balance

$$\dot{m}_1\left(h_1+\frac{v_1^2}{2}+Z_1 g\right)+\frac{dQ}{d\tau}=\dot{m}_2\left(h_2+\frac{v_2^2}{2}+Z_2 g\right)+\frac{dW}{d\tau}$$

Steady Four Energy Equation (SFEE)

$$h_1+\frac{v_1^2}{2}+Z_1 g+\frac{dQ}{dm}=h_2+\frac{v_2^2}{2}+Z_2 g+\frac{dW}{dm}$$

Key Points

- The first law of thermodynamics may be applied to flow processes.
- A control surface is the boundary of a control volume which is a fixed region in space upon which attention is concentrated in the analysis of a problem.
- A steady flow process is a process in which all conditions within the control volume remain constant with time.

Variable Flow Process

Consider a device through which a fluid is flowing under non-steady state conditions. The rate at which the mass of fluid within the control volume is accumulated is equal to the net rate of mass flow across the control surface as given below.

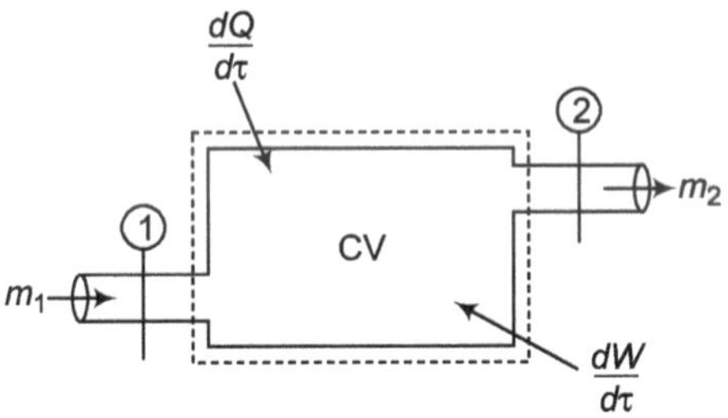

Variable flow process

$$\frac{dm_V}{d\tau}=m_1-m_2$$

$$\frac{dm_V}{d\tau}=\frac{dm_1}{d\tau}-\frac{dm_2}{d\tau}$$

where m_V is the mass of fluid within the control volume at any instant over any finite period of time.]

$$\Delta m_V=\Delta m_1-\Delta m_2$$

where m_V = Mass of fluid within the control volume

The rate of accumulation of energy within the control volume is equal to the net rate of energy flow across the control surface. If E_V is the energy of fluid within the control volume at any instant.

Rate of energy increases

= Rate of energy inflow – Rate of energy outflow

$$\frac{dE_V}{d\tau}=m_1\left(h_1+\frac{v_1^2}{2}+Z_1 g\right)+\frac{dQ}{d\tau}-m_2\left(h_2+\frac{v_2^2}{2}+Z_2 g\right)-\frac{dW}{d\tau}$$

$$E_V = \left(U + \frac{mv^2}{2} + mgZ\right)_V$$

where, E_V = Energy of fluid within the control volume at any instant.

$$\frac{d}{d\tau}\left(U + \frac{m\,v^2}{2} + mg\,Z\right)_V = \left(h_1 + \frac{v_1^2}{2} + Z_1 g\right)\frac{dm_1}{d\tau} + \frac{dQ}{d\tau} - \left(h_2 + \frac{v_2^2}{2} + Z_2 g\right)\frac{dm_2}{d\tau} - \frac{dW}{d\tau}$$

or $$\Delta E_V = Q - W_x + \int\left(h_1 + \frac{v_1^2}{2} + Z_1 g\right)dm_1 - \int\left(h_2 + \frac{v_2^2}{2} + Z_2 g\right)dm_2$$

The above equation is general equation for steady flow

$$\frac{dE_V}{d\tau} = 0$$

For closed system, $m_1 = 0, m_2 = 0$

$$\frac{dE_V}{d\tau} = \frac{dQ}{d\tau} - \frac{dW}{d\tau} \Rightarrow dE_V = dQ - dW$$

Carnot Cycle

The carnot cycle can be described that it is the most efficient cycle for converting a given amount of thermal energy into work, or conversely, creating a temperature difference by doing a given amount of work.

The cycle consists of two isothermal and two irreversible adiabatic processes.

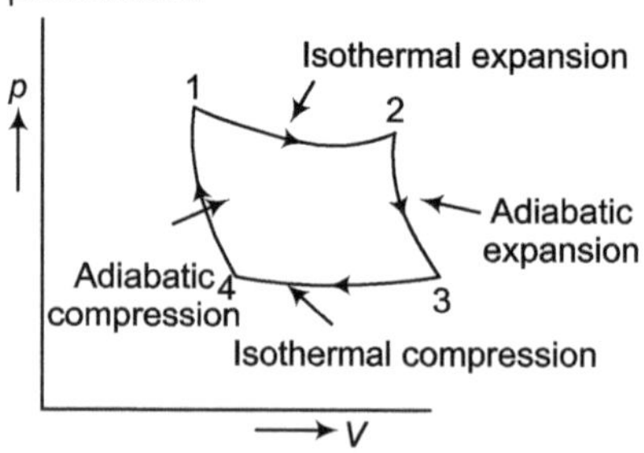

Carnot engine

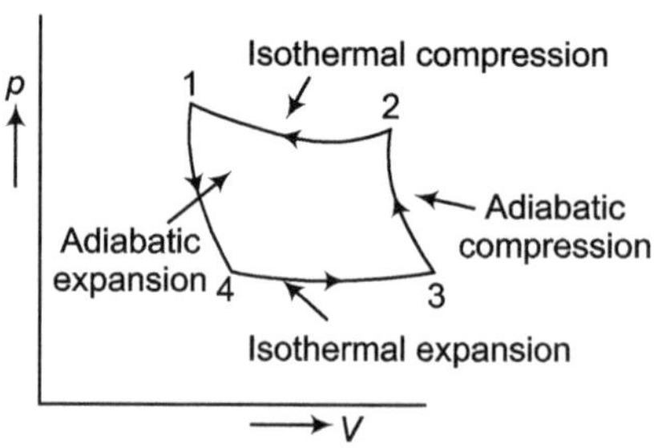

Carnot heat pump

In process (1-2), heat Q, is gained by fluid at constant temperature. In process (3-4), heat Q_2 is rejected by fluid at constant temperature.

For Carnot engine,

Thermal efficiency,

$$\eta_{\text{thermal}} = \frac{Q_1 - Q_2}{Q_1}\left(\frac{\text{Work done}}{\text{Heat supplied}}\right)$$

$$\eta_{\text{thermal}} = 1 - \frac{T_2}{T_1}$$ is called the efficiency of Carnot cycle.

Carnot's Theorem

It states that any heat engine operating on a cycle between two heat reservoirs at different fixed temperature limits cannot be more efficient than a reversible engine.

Corollary of Carnot's Theorem

Efficiency of all reversible heat engines operating between the same temperature level is the same.

The Inequality of Clausius

Inequality of Clausius provide criterion of the reversibility of cycle.

$$\oint \frac{dQ}{T} \le 0$$

where dQ = Heat transfer at a part of system boundard

T = Absolute temperature

If $\oint \frac{dQ}{T} = 0 \Rightarrow$ the cycle is reversible

$\oint \frac{dQ}{T} < 0 \Rightarrow$ the cycle is irreversible and possible

$\oint \frac{dQ}{T} > 0 \Rightarrow$ the cycle is impossible since it violates the second law.

Note For an irreversible process, change in entropy

$$dS \ge \left(\frac{dQ}{T}\right)$$

$$\Rightarrow \qquad S_2 - S_1 > \left(\int_1^2 \frac{dQ}{T}\right)_{\text{irreversible path}}$$

$$\Rightarrow \qquad S_2 - S_1 = \left(\int_1^2 \frac{dQ}{T}\right)_{\text{reversible path}}$$

where, S = Strength of the inequality

Entropy

Entropy is a function of a quantity of heat which shows the possibility of conversion of that heat into work. In other words, entropy shows the randomness of molecules. So, entropy of solid is almost equal to zero while entropy of gases is large. Entropy is property of system and it is an extensive property.

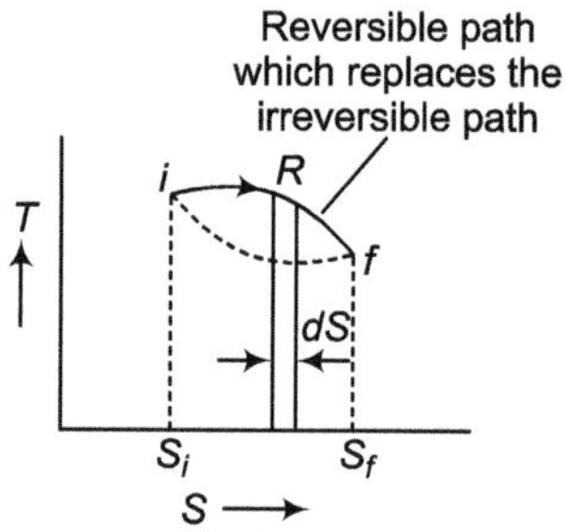

T-S diagram

Mathematically,

$$dS = \left(\frac{dQ}{T}\right)_R \quad (R \rightarrow \text{Reversible process})$$

dS is an exact differential because S is point function.

Integration can be done only on a reversible path

$$S_f - S_i = \int_i^f \frac{dQ_{\text{rev}}}{T} = (\Delta S)_{\text{rev path}}$$

In the case of process reversible and adiabatic,

$$dS = \frac{dQ_{\text{rev}}}{T} \Rightarrow dS = 0 \quad (dQ_{\text{rev}} = 0)$$

So, a reversible adiabatic process is an isentropic process.

- Entropy shows the randomness of molecules.
- Entropy of solid is considered almost zero and gases have greater value of entropy.
- The increment in entropy is greater when heat is added at low temperature.

Principle of Increase of Entropy

Entropy change of system and surrounding during a process between two equilibrium states is always equal to or greater than zero.

$$(\Delta S)_{\text{system}} + (\Delta S)_{\text{surrounding}} \geq 0$$

$$\Delta S_{\text{universe}} \geq 0, - \Delta S_{\text{universe}} = 0$$

If process is reversible, $\Delta S_{\text{universe}} = 0$

If process is irreversible, $\Delta S_{\text{universe}} > 0$

Entropy Changes for a Closed System

Let m kg of gas at pressure p_1, volume V_1, temperature T_1 and entropy S_1 be heated and these state variable quantities become final pressure p_2, volume V_2, temperature T_2 and entropy S_2, respectively.

$$S_2 - S_1 = mC_V \ln\left(\frac{T_2}{T_1}\right) + mR \ln\left(\frac{V_2}{V_1}\right)$$

or $$\Delta S = mC_V \ln \frac{p_2}{p_1} + mC_p \ln \frac{V_2}{V_1}$$

or $$\Delta S = mC_p \ln \frac{T_2}{T_1} - mR \ln \frac{p_2}{p_1}$$

Important Formulae

- In constant volume process, $\Delta S = mC_V \ln \frac{T_2}{T_1}$
- In constant pressure process, $\Delta S = mC_p \ln \frac{T_2}{T_1}$
- In isothermal process, $\Delta S = mR \ln \frac{V_2}{V_1}$
- In adiabatic process (reversible), $\Delta S = 0$
- In polytropic process (reversible), $\Delta S = mC_V \left(\frac{n-\gamma}{n-1}\right) \ln \frac{T_2}{T_1}$

Applications of Entropy Principle

The principle of increase of entropy is one of the most important laws of physical science. The higher the entropy increase of the universe, the higher will be irreversibility of the process.

The *few applications* of *the entropy principle are illustrated in the following*

Transfer of Heat through a Finite Temperature Difference

Let Q be the rate of heat transfer from reservoir A at T_1 to reservoir B at T_2 $(T_1 > T_2)$.

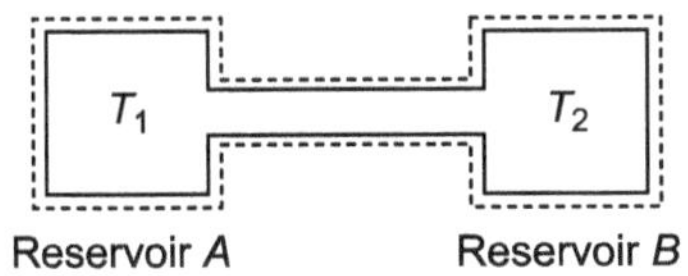

Transfer of heat in reservoirs *A* and *B*

$$\Delta S_{\text{universal}} = \Delta S_A + \Delta S_B$$

$$\Delta S_{\text{universal}} = -\frac{Q}{T_1} + \frac{Q}{T_2}$$

Mixing of Two Fluids

Subsystem 1 having a fluid of mass m_1, specific heat C_1 and temperature t_1 and subsystem 2 consisting of mass m_2, specific heat C_2 and temperature t_2 comprise a composite system in an adiabatic enclosure.

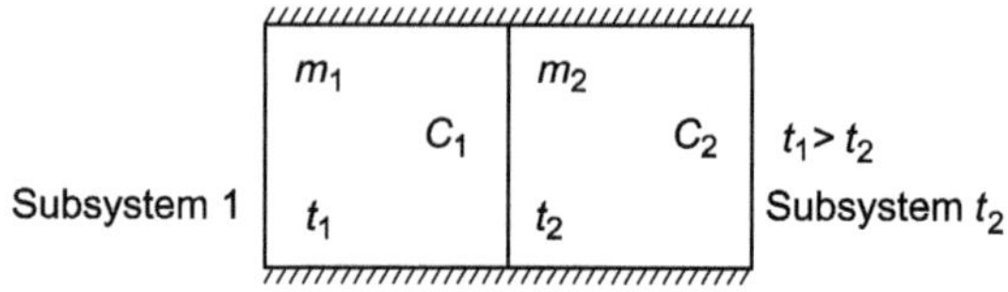

Mixing of two fluids of mass m_1 and m_2

$$t_f = \frac{m_1 C_1 t_1 + m_2 C_2 t_2}{m_1 C_1 + m_2 C_2}$$

$$\Delta S_{\text{universal}} = m_1 C_1 \ln \frac{T_f}{T_1} + m_2 C_2 \ln \frac{T_f}{T_2}$$

$$\Delta S_{\text{universal}} = m_1 C_1 \ln \frac{T_f}{T_1} + m_2 C_2 \ln \frac{T_f}{T_2}$$

$\Delta S_{\text{universal}}$ will be positive definite and mixing process is irreversible.

Two Finite Bodies at Temperatures T_1 and T_2

Maximum work obtainable from two finite bodies at temperature T_1 and T_2.

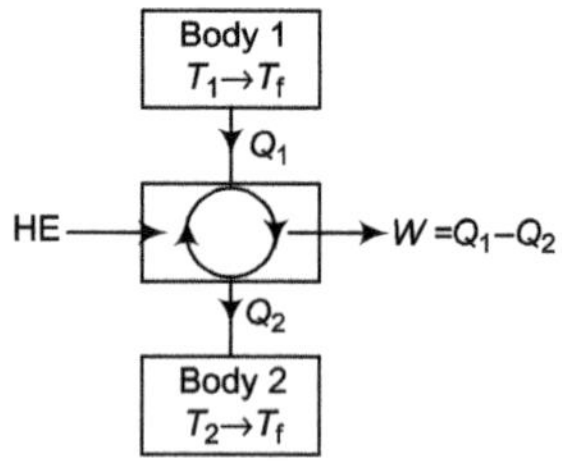

Two bodies at temperatures T_1 and T_2

$$W = Q_1 - Q_2$$

$$W = C_p(T_1 + T_2 - 2T_f)$$

$$\Delta S_{\text{universal}} = C_p \ln \frac{T_f}{T_1} + C_p \ln \frac{T_f}{T_2} \geq 0$$

For W_{max} T_f,

$$T_f = \sqrt{T_1 T_2}$$

$$W_{max} = C_p\,(\sqrt{T_1} - \sqrt{T_2})^2$$

Maximum Work Obtainable from a Finite Body and a TER

$$\Delta S_{\text{universal}} = C_p \ln \frac{T_0}{T} + \frac{Q - W}{T_0}$$

$$W_{\max} = C_p \left[(T - T_0) - T_0 \ln \frac{T}{T_0} \right]$$

Here, TER stands for Thermal Energy Reservior.

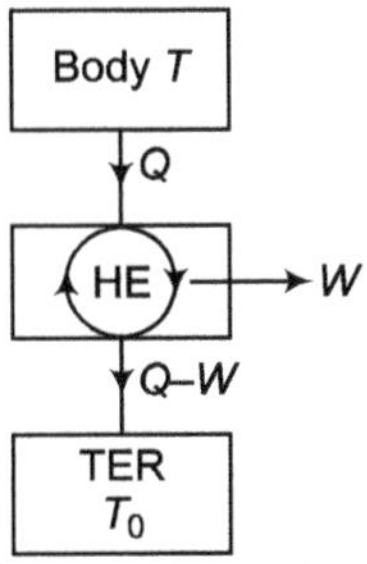

Work W obtained from a finite body and TER

Entropy Generation in a Closed System

The entropy of any closed system can increase in two ways

- By heat interaction in which there is entropy transfer.
- By internal irreversibilities or dissipative effects in which work is dissipated into internal energy increase.

Entropy production

$$S_{\text{gen}} = \underbrace{S_2 - S_1}_{\text{Entropy change}} - \underbrace{\int_1^2 \frac{dQ}{T}}_{\text{Entropy transfer}}$$

$$dS_{\text{gen}} = dS - \frac{dQ}{T} \quad \text{or} \quad dS = dS_{\text{gen}} + \frac{dQ}{T}$$

On a time rate basis, the entropy balance can be written as

$$\frac{dS}{d\tau} = \Sigma \frac{Q_j}{T_j} + S_{\text{gen}}$$

Entropy Generation in an Open System

From the second law inequality

$$\underbrace{\frac{dS}{d\tau}}_{\text{Rate of increase of entropy}} \geq \underbrace{\Sigma m_i S_i - \Sigma m_e S_e + \frac{Q}{T}}_{\text{Net rate of entropy transfer}}$$

$$\frac{dS}{d\tau} = \Sigma m_i S_i - \Sigma m_e S_e + \frac{Q}{T} + S_{\text{gen}}$$

At the steady state,

$$\sum_i m_i = \sum_e m_e = m$$

$$0 = \frac{Q}{T} + \Sigma\, m_i S_i - \sum_e m_e S_e + S_{gen}$$

$$S_2 - S_1 = \frac{1}{m}\left(\frac{Q}{T}\right) + \frac{S_{gen}}{m}$$

First and Second Law Combination

From first and second law equation we can relate the properties of the system.

From the second law,

$$dQ_{rev} = T \cdot dS$$

From first law ,

For a closed non-flow system,

$$dQ = dU + pdV \;;\; TdS = dU + pdV$$

The enthalpy $H = U + p \cdot V$ (for reversible process)

$$dH = dU + pdV + V\,dp$$

$$dH = TdS + V\,dp \;\{dU + pdV = dQ_{rev}\}$$

$$TdS = dH - V\,dp$$

or $Tds = dh - V\,dp$

Reversible Adiabatic Work in a Steady Flow System

Reversible adiabatic work can be find out by using equation given as,

$$W_x = h_1 - h_2 = -\int_1^2 V\,dp = \text{Area } 12ab$$

where, h_1 and h_2 are the enthalpy at 1 and 2, respectively.

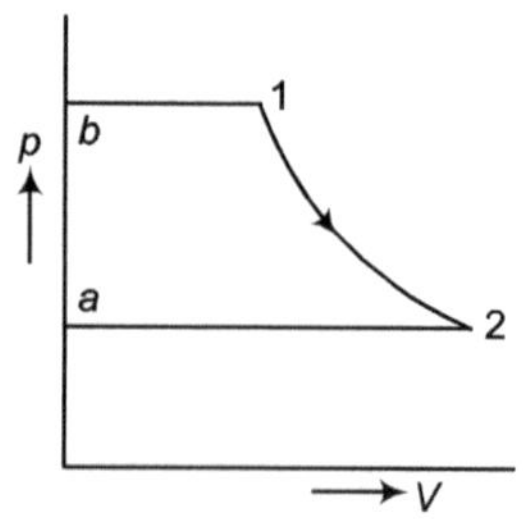

Reversible adiabatic work

Availability *and* Irreversibility

The sources of energy can be divided into two groups *i.e.*, high grade energy (mechanical work, electrical energy, water power, wind power) and low grade energy (heat or thermal energy, heat derived from nuclear fission or combustion of fossil fuels). That part of the low grade energy which is available for, conversion is referred to as available energy, while the part which is not available is known as **unavailable energy**.

Availability

When a system is subjected to a process from its original state to dead state the maximum amount of useful work that can be achieved under ideal conditions is known as **available energy** or **availability of the system**.

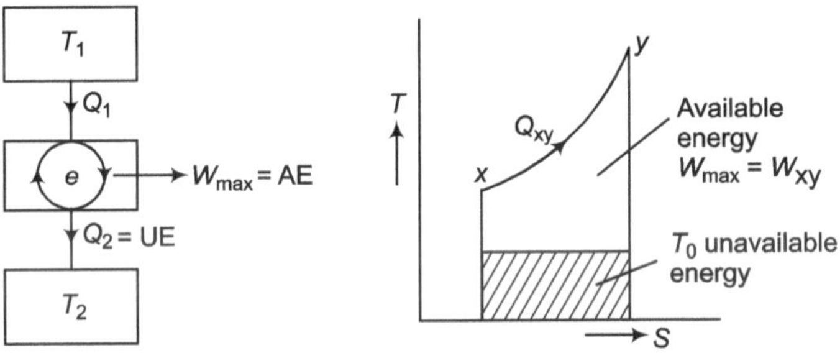

Availability and unavailability of energy

$$W_{max} = AE = Q_{xy} - T_0(S_y - S_x)$$

Unavailable Energy $\quad UE = T_0\,(S_y - S_x)$

where, S_x and S_y are the entropy at x and y, respectively.

The Available Energy (AE) is also known as exergy and the Unavailable Energy (UE) as anergy.

Available Energy from a Finite Energy Source

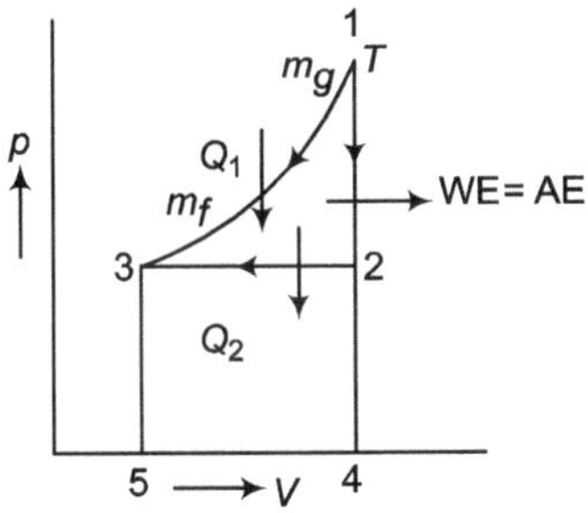

Availability of an energy

Let us consider a hot gas of mass m_g at temperature T when the environmental temperature is T_0.

Let us gas be cooled from state 1 to state 3 and heat given by the gas Q_1 be utilised in heating up reversibly a working fluid of mass m_{wf} from state 3 to state 1. The working fluid expands reversibly and adiabatically in an engine or turbine from state 1 to state 2 and then return to state 3 to complete heat engine cycle.

Available energy of a finite energy source

$$\text{Available energy} = W_{max} = Q_1 - Q_2$$

$$= m_g\, C_{pg}(T - T_0) - T_0\, m_g C_{pg} \ln \frac{T}{T_0}$$

$$= m_g C_{pg}\left[(T - T_0) - T_0 \ln \frac{T}{T_0}\right]$$

Law of Degradation of Energy

First law states that energy is always conserved quantity wise while second law emphasises that energy always degrades quality wise. Second law is called the law of **degradation of energy**. Energy is always **conserved** but its quality is always **degraded**.

$$\begin{bmatrix}\text{all the energy}\\ \text{that enters}\\ \text{the system}\end{bmatrix} = \begin{bmatrix}\text{all the energy}\\ \text{that leaves}\\ \text{the system}\end{bmatrix} + \begin{bmatrix}\text{the gain in}\\ \text{reversible}\\ \text{energy}\end{bmatrix} + \begin{bmatrix}\text{the gain in}\\ \text{irreversible}\\ \text{energy}\end{bmatrix}$$

Reversible work by an open system exchanging heat only with the surroundings.

$$dW_{max} = dm_1\left(h_1 - T_0 S_1 + \frac{v_1^2}{2} + gZ_1\right) - dm_2\left(h_2 - T_0 S_2 + \frac{v_2^2}{2} + gZ_2\right) - d\left(U - T_0 S + \frac{m v^2}{2} + mgZ\right)$$

where, Z_1 and Z_2 are the elevation at 1 and 2, respectively.

For steady flow process, $\Rightarrow$ $dm_1 = dm_2 = dm$

$$dE_V = 0$$

The maximum reversible work,

$$dW_{max} = dm\left[\left(h_1 - T_0 S_1 + \frac{v_1^2}{2} + gZ_1\right) - \left(h_2 - T_0 S_2 + \frac{v_2^2}{2} + gZ_2\right)\right]$$

The expression $(H - T_0 S)$ is called Kennan function B. In term of per unit mass $(h - T_0\ \&)$ is called Kennan function B.

$$W_{max} = \left(b_1 + \frac{mv_1^2}{2} + mg\,Z_1\right) - \left(b_2 + \frac{mv_2^2}{2} + mgZ_2\right)$$

$$= \psi_1 - \psi_2$$

where, ψ is called the availability function of a steady flow process given by

$$\psi = b + \frac{mv^2}{2} + mgZ$$

If $\Delta KE = 0$ and $\Delta PE = 0$

$$W_{max} = b_1 - b_2 = (h_1 - h_2) - T_0 (S_1 - S_2)$$

Maximum reversible work for a closed system,

$$dm_1 = dm_2 = 0$$

$$dW_{max} = -d\left[U - T_0 S + \frac{mv^2}{2} + mgZ\right]$$

$$= -d\,[E - T_0 S]$$

where, $E = u + \frac{mv^2}{2} + mgZ$; $\quad W_{max} = (E_1 - T_0 S_1) - (E_2 - T_0 S_2)$

$\Delta KE = 0$, $\Delta PE = 0$; $\quad W_{max} = (U_1 - T_0 S_1) - (U_2 - T_0 S_2)$

where, ΔKE = Change in kinetic energy

ΔPE = Change in potential energy

Maximum reversible work for unit mass of fluid,

$$W_{max} = (U_1 - T_0 S_1) - (U_1 - T_0 S_2)$$

Useful Work

Maximum useful work $\quad (W_u)_{max} = W_{max} - p_0(V_2 - V_1)$

V_1 and V_2 are the initial and final volumes of the systems and p_0 is the atmospheric pressure.

Useful Work for Unsteady Open System

$$(dW_u) = dm_1\left(h_1 - T_0 S_1 + \frac{v_1^2}{2} + gZ_1\right) - dm_2\left(h_2 - T_0 S_2 + \frac{v_2^2}{2} + gZ_2\right) - d\left[U + p_0 V - T_0 S + \frac{mv^2}{2} + mgZ\right]$$

Useful Work for Unsteady Closed System

$$(dW_u)_{max} = -d\left[U + p_0 V - T_0 S + \frac{mv^2}{2} + mgZ\right]$$

$$= -d\,[E + p_0 V - T_0 S]$$

$$(W_u)_{max} = E_1 - E_2 + p_0(V_1 - V_2) - T_0 (S_1 - S_2)$$

$$\Delta KE = 0, \Delta PE = 0$$
$$(W_u)_{max} = U_1 - U_2 + p_0 (V_1 - V_2) - T_0 (S_1 - S_2)$$
$$(W_u)_{max} = (U_1 + p_0 V - T_0 S_1) - (U_2 + p_0 V_2 - T_0 S_2)$$
$$(W_u)_{max} = \phi_1 - \phi_2$$

ϕ is called the availability function for a closed system,

$$\phi = U + p_0 V - T_0 S$$

or $$\phi = u + p_0 v - T_0 s \quad \text{(per unit mass basis)}$$

Maximum useful work obtainable when the system exchanges heat with a thermal reservoirs in addition to the atmosphere

$$(W_u)_{max} = \left(h_1 - T_0 S_1 + \frac{mv_1^2}{2} + mgZ_1\right) - \left(h_2 - T_0 S_2 + \frac{mv_2^2}{2} + mgZ_2\right) - \left(U - T_0 S + \frac{mv^2}{2} + mgZ\right) + Q_R\left(1 - \frac{T_0}{T_R}\right)$$

$\left(\begin{array}{l} T_R \rightarrow \text{Temperature of reservoir} \\ Q_R \rightarrow \text{Heat received by system} \end{array}\right)$

Dead State

When the system is in equilibrium with surrounding with temperature T_0 and pressure p_0 and the system also in chemical equilibrium with zero velocity and minimum potential energy. Then, it is called the **dead state**. All spontaneous processes terminates at the dead state.

Irreversibility

The actual work done by a system is always less than idealised reversible work and the difference between the two is called the irreversibility of the process.

$$I = W_{max} - W$$
$$I = T_0 (\Delta S_{system} + \Delta S_{surrounding})$$
$$I = T_0 (\Delta S)_{universal}$$

$T_0(\Delta S)_{universal}$ represent an increase in unavailable energy.

Goby-Stodola Theorem

It states that the rate of loss of available energy or exergy in a process is proportional to the rate of entropy generation S_{gen}

$$I = W_{lost} = T_0 \, \Delta S_{universal}$$
$$I = T_0 S_{gen}$$

where, I = irreversibility

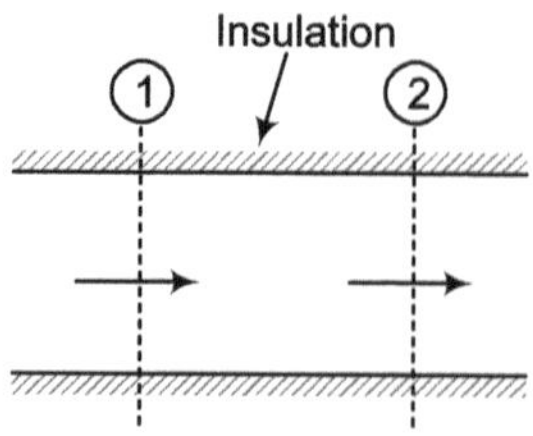

Goby-Stodola theorem

Irreversibility for adiabatic flow of an ideal gas through the segment of pipe with friction decreases in availability and is proportional to pressure drop and mass flow rate.

$$I = W_{lost} = T_0 S_{gen} \Rightarrow I = mRT_0 \frac{\Delta p}{p_1}$$

Key Points

- The term Kennan function B $(= H - T_o S)$ is used in steady flow process and availability function is given by
$$\Psi = B + \frac{mv^2}{2} + mgZ$$
- For closed system availability function is given by
$\oint = U + p_0 V - T_0 S$ and $a = \oint - \oint_0$ where $\oint_0 = U_0 + p_0 V_0 - T_0 S_0$
- Property Helmholty function F–be defined by the relation $F = U - TS$
- Given function G is defined as $G = H - TS = U + pV - TS$

Properties of Pure Substance

A pure substance is a material with homogeneous and invariable composition. Pure substances can have multiple phases, an ice water mixture is still a pure substance but an air-steam is not a pure substance.

Pure Substance

A substance that has a fixed chemical composition throughout is called a **pure substance** *i.e.,* water, nitrogen, helium, and CO_2. Substances which are mixture of various elements or compounds also qualifies as a pure substances as long as mixture is homogeneous.

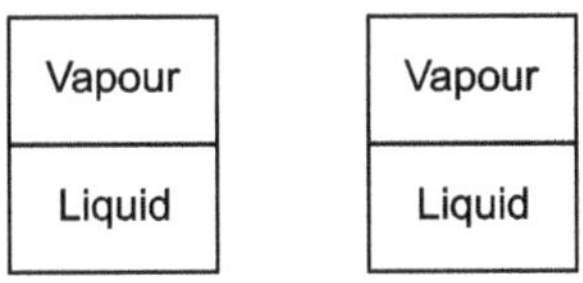

(a) H_2O (b) Air

Saturation Temperature and Saturation Pressure

At given pressure, the temperature at which a pure substance changes phase is called the **saturation temperature** T_{sat}. Likewise at a given temperature, the pressure at which a pure substance changes phase is called the **saturation pressure** p_{sat}. *e.g.*, for water at a pressure of 101.325 kPa, T_{sat} is 100°C, conversely at a temperature of 100°C, p_{sat} is 101.325 kPa.

Latent Heat

The amount of energy absorbed or released during a phase change process is called the **latent heat.** More specifically, the amount of energy absorbed during melting is called the latent heat of vaporisation. Similarly, the amount of energy absorbed during vaporisation is called latent heat of vaporisation and is equivalent to the energy released during condensation.

Liquid-Vapour Saturation Curve

From the figure, it is clear that T_{sat} increases with p_{sat}. Thus, a substance at higher pressure will boil at higher temperatures.

$$T_{sat} = f(p_{sat})$$

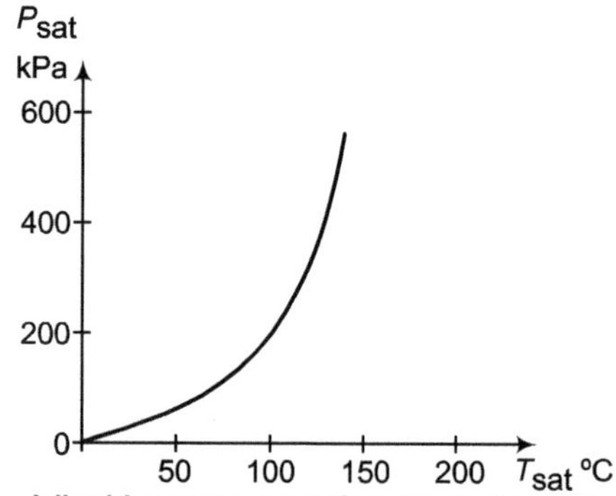

A liquid-vapour saturation curve of a pure substance

Key Points

- In the kitchen, higher boiling temperature means shorter cooking time and energy saving.
- The atmospheric pressure, and thus the boiling temperature of water, decreases with elevation. Therefore, it takes longer time to cook at higher altitudes than it does at sea level.

Property Diagrams for Phase-change Process

The T-V Diagram

- Consider a piston cylinder device containing liquid water at 20°C and 1 atm.
- Water will start boiling at a much higher temperature (179.9°C) at inside pressure of the cylinder reaches at 1 MP.
- The specific volume of the saturated liquid is larger and the specific volume of the saturated vapour is smaller than the corresponding values at 1 atm pressure. That is, the horizontal line that connects the saturated liquid and saturated vopour states is much shorter.
- As the pressure is increased further, this saturation line will continue to get shorter as shown in figure and it will become a point when the pressure reaches 22.09 MPa for the case of water. This point is called the **critical point** and it is defined as the point at which the saturated liquid and saturated vapour state are identical.

  For water, critical pressure $p_{cr} = 22.09$ MPa

  critical temperature $T_{cr} = 374.14°$ C

  critical volume $V_{cr} = 0.003155 \text{ m}^3/\text{ kg}$
- At pressure above the critical pressure, there will not be a distinct phase change. Instead, the specific volume of the substance will continually increase and at all times there will be only one phase present. It is

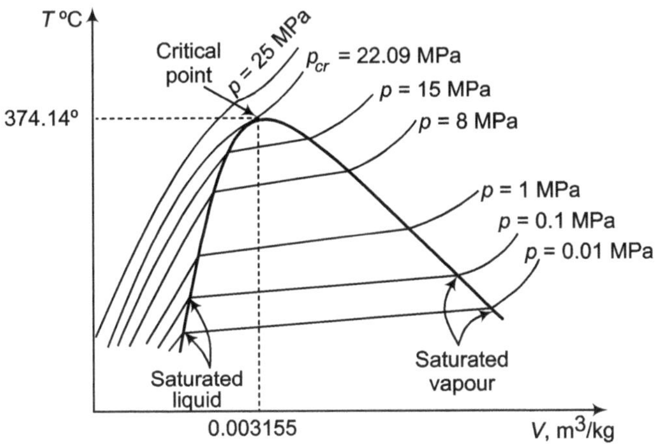

T-V diagram of constant pressure phase-change process of a pure substances at various pressures.

customary to refer to the substance as superheated vapour at temperature above the critical temperature and as compressed liquid at temperatures below the critical temperature.

The *p*-*V* Diagram

The general shape of the p-V diagram of a pure substance is very much like the T-V diagram but the T = constant lines on this diagram have a downward trend.

Consider again a piston cylinder device that contains liquid water at 1 MPa and 150°C, water at this state exists as a compressed liquid. Now, the weights on top of the piston are removed one by one so that the pressure inside the cylinder decreases gradually. The water is allowed to exchange heat with the surroundings so its temperature remains constant. As the pressure decreases, the volume of the water will increase slightly, when the pressure reaches the saturation pressure volume at the specific temperature, the water will start to boil.

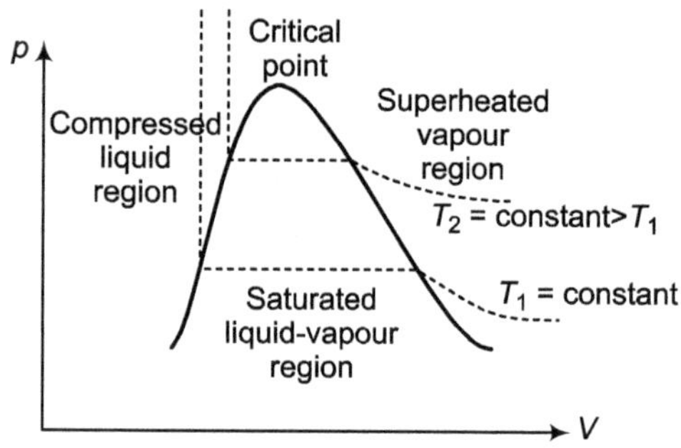

p-V diagram of pure substance

During this vaporisation process, both the temperature and the pressure remain constant but the specific volume increases. Once the last drop of liquid is vaporised further reduction in pressure results in a further increase in specific volume.

If the process is repeated for other temperatures similar paths will be obtained for the phase change processes.

Triple Phase

When all three phases of a pure substance co-exist in equilibrium. It is called triple phase.

Triple phase states form a line called the triple line. The triple line appears as a point on the p-T diagram and therefore is often called the **triple point**.

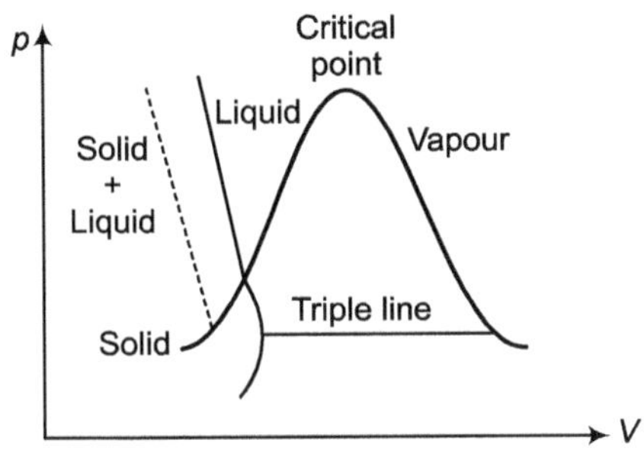

For triple phase

Key Points

- No substance can exist in the liquid phase in stable equilibrium at pressure below the triple point pressure.
- The same can be said for temperature for substance that contract on freezing.
- Substances at high pressure can exist in the liquid phase at temperatures below the triple point temperature.

The *p-T* Diagram

Solid – Liquid = Fusion

Liquid – Vapour = Vaporisation

Solid – Vapour = Sublimation

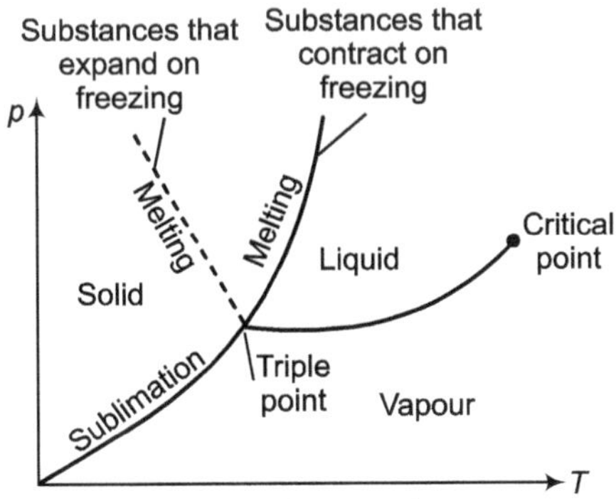

p-T diagram of pure substances

Enthalpy

Enthalpy is a measure of the total energy of a thermodynamic system. It includes energy required to create a system and the amount of energy required to make room for it by displacing its environment and establishing its volume and pressure.

For certain type of processes particularly in power generation and refrigeration. We define the new combination property enthalpy.

Enthalpy, $H = U + pV$ (kJ)

or per unit mass $h = u + pV$ (kJ /kg)

Professer Richard Mollier recognised the importance of the group $(u + pV)$ in the analysis of steam turbine and in the representation of properties of steam in tabular and graphical form.

Saturated Liquid and Saturated Vapour States

Subscript $f \rightarrow$ properties of saturated liquid

Subscript $g \rightarrow$ properties of saturated vapour

$V_f \rightarrow$ specific volume of saturated liquid

$V_g \rightarrow$ specific volume of saturated vapour

$V_{fg} \rightarrow$ difference between V_g and V_f

$$V_{fg} = V_g - V_f$$

$h_{fg} \rightarrow$ enthalpy of vaporisation or latent heat of vaporisation (amount of energy need to vaporise a unit mass of saturated liquid at a given temperature of pressure).

Key Points

- The magnitude of latent heat depends on temperature and pressure at which phase change occurs.
  for e.g., at 1 atm pressure, the latent heat of fusion of water is 333.7 kJ/kg and latent heat of vaporisation is 2257.1 kJ/kg.
- At below triple point pressure, substance begins to change directly gas.
- Enthalpy of vaporisation decreases as the temperature or pressure increases and become zero at the critical point.

Saturated Liquid Vapour Mixture

During vaporisation process, a substance exists as part liquid and part vapour. A new property quality x is the ratio of mass of vapour to the total mass of the mixture

Quality or dryness fraction

$$x = \frac{m_{\text{vapour}}}{m_{\text{total}}}$$

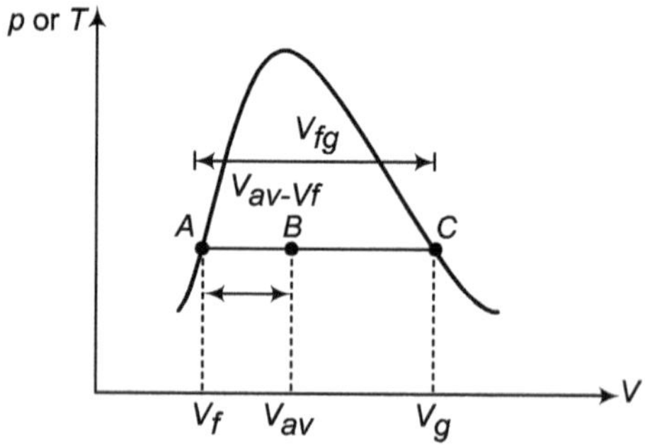

p-V diagram

where, $m_{total} = m_{liquid} + m_{vapour}$
$= m_f + m_g$

where, m_f = mass of the saturated liquid

m_g = mass of the saturated vapour

Quality has significance for saturated mixtures only. It has no meaning in the compressed liquid or superheated region. Its value lie between 0 (saturated liquid) and 1 (saturated vapour).

> **Note** The properties of the saturated liquid are the same whether it exists alone or in a mixture with saturated vapour. During vaporization process, only the amount of saturated liquid changes not its properties. The same can be said about a saturated vapour.

Quality $$x = \frac{\overline{AB}}{\overline{\overline{AC}}} \Rightarrow x = \frac{V_{av} - V_f}{V_{fg}}$$

$$V_{av} = (1 - x) V_f + x V_g \quad \text{or} \quad V_{av} = V_f + x V_{fg}$$

Similarly, $U_{av} = U_f + x U_{fg}$

$\Rightarrow$ $h_{av} = h_f + x h_{fg}$

Superheated Vapour

Since, the superheated region is a single phase region (vapour phase only) temperature and pressure are no longer dependent properties and they can conveniently be used as the two independent properties.

Superheated vapour is characterised by

Lower pressures ($p < p_{sat}$ at a given T)
Higher temperatures ($T > T_{sat}$ at a given p)
Higher specific volumes ($U > V_g$ at a given p or T)
Higher internal energies ($U > U_g$ at a given p or T)
Higher enthalpies ($h > h_g$ at a given p or T)

Compressed Liquid

A compressed liquid may be approximated as a saturated liquid at the given temperature. This is because the compressed liquid properties depend on temperature much more strongly than they do on the pressure.

So, given p and T $V \cong V_f$

$u \cong u_f$

$h \cong h_f$

In general, *a compressed liquid is characterised by*

Higher pressures	($p > p_{sat}$ at a given T)
Lower temperatures	($T < T_{sat}$ at a given p)
Lower specific volumes	($V < V_f$ at a given p or T)
Lower internal energies	($U < U_f$ at a given p or T)
Lower enthalpies	($h < h_f$ at a given p or T)

Specific Heat

Specific heat is defined as the energy required to raise the temperature of a unit mass of a substance by one degree. When this energy is required at constant volume then it is denoted by C_V. Similarly, when energy is required at constant pressure then it is denoted by C_p.

Note $C_p > C_V$

Because at constant pressure, the system is allowed to expand and the energy for this expansion work must also be supplied to the system.

Key Points

- $C_V = \left(\frac{\partial U}{\partial T}\right)_V$ = the change in internal energy with temperature at constant volume.
- $C_p = \left(\frac{\partial h}{\partial T}\right)_p$ = the change in enthalpy with temperature at constant pressure.
- These two equations are property relations and as such are independent of the type of processes. They are valid for any substance undergoing any process.

Internal Energy

For an ideal gas internal energy is a function of the temperature only.

$$U = U(T)$$

However, for gases that deviate significantly from ideal gas behaviour, the internal energy is not a function of temperature only.

Using definition of enthalpy and the equation of state of an ideal gas

$$h = U + pV \quad h = U + RT$$

$$pV = RT$$

So, $$h = h(T)$$

So, we can express the change in internal energy

$$dU = C_V(T)\,dT$$

$$\Delta U = U_2 - U_1 = \int_1^2 C_V(T)\, dT$$

and $$\Delta h = h_2 - h_1 = \int_1^2 C_p\,(T)\, dT$$

Since, $h = U + RT\,;\, dh = dU + R\,dT$

Replacing $dh = C_p\, dT, dU = C_V\, dT$

$$C_p\, dT = C_V\, dT + R\, dT$$

$$C_p = C_V + R \text{ (kJ/kg K)}$$

Specific heat ratio $k = \dfrac{C_p}{C_V}$

Air at 300 K $C_V = 0.718 \text{ kJ/kg-K}$

$R = 0.287 \text{ kJ/kg-K}$

$C_p = 1.005 \text{ kJ/kg-K}$

or on molar basis $C_v = 20.80 \text{ kJ/kmol-K}$

$R_u = 8.314 \text{ kJ/kmol-K}$

$C_p = 29.114 \text{ kJ/kmol-K}$

The Ideal Gas Equation of State

Any equation that relates the pressure, temperature and specific volume of a substance is called an **equation of state.** Property relations that involve other properties of a substance at equilibrium states are also referred to as equation of state.

Ideal gas equation of state

$$pV = RT,$$

where, p = Absolute pressure, T = Absolute temperature

V = Specific volume

Key Points

- Gas and vapour are often used as synonymous words.
- The vapour phase of a substance is customarily called a gas when it is above the critical temperature.
- Vapour usually implies a gas that is not far from a state of condensation.

Gas Constant

It has been experimentally observed that the ideal gas relation given closely approximately the p-V-T behaviour of real gases at low densities. At low pressure and high temperature, the density of a gas decreases and the gas behaves as an ideal gas under these conditions.

$$R = \frac{R_u}{M}$$

where, R_u = Universal gas constant, M = Molar mass, R = Gas constant.

Compressibility Factor

Compressibility factor (correction factor) is measurement of deviation of gases from ideal gas behaviour.

Compressibility factor, $z = \frac{pV}{RT}$ or $pV = zRT$

It can also be expressed as

$$z = \frac{V_{\text{actual}}}{V_{\text{ideal}}}, \text{ where } V_{\text{ideal}} = \frac{RT}{p}$$

For ideal gases $\Rightarrow z = 1$

For real gases $\Rightarrow z$ is away from unity (> 1or < 1)

Reduce Pressure and Temperature

Gases behave differently at a given temperature and pressure, but they behave very much the same at temperature and pressures normalised with respect to their critical temperatures and pressures.

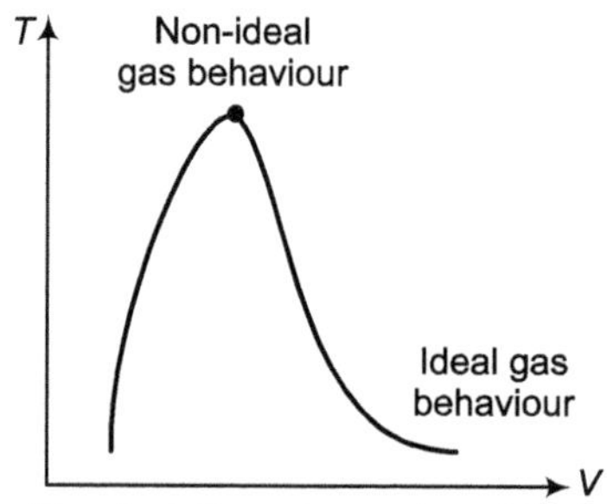

The deviation of a gas from ideal gas behaviour

The normalisation is done as introducing new terms,

Reduce pressure $p_R = \frac{p}{p_{cr}}$

and reduce temperature $T_R = \frac{T}{T_{cr}}$

where, T_{cr} = critical temperature

p_{cr} = critical pressure

z factor for all gases is approximately the same at the same reduced pressure and temperature is greatest in the vicinity of the critical point.

- Gases deviate from the ideal gas behaviour most in the neighbourhood of the critical point. So, we can say that at critical point. Compressibility factor is constant for all substances.

- As mention above *z* factor for all gases is approximately the same at the same reduced pressure and temperature.
  pseudo reduced specific volume V_R

$$V_R = \frac{V_{\text{actual}}}{RT_{Cr} / p_{Cr}}$$

V_R is defined differently from p_R and T_R. It is related to T_{Cr} and p_{Cr} instead of V_{Cr}.

Key Points

- All substances have same critical compressibility factor $Z_c = \frac{p_c V_c}{RT_c}$

  Experimental value of Z_c for most substances fall within a narrow range of 0.20 – 0.33.
- Z the compressibility factor is the same function of P_r and T_r for all gases.
- Specific heat at constant pressure and volume are the properties of the substances and are always properties since $C_p > C_V$, $\frac{C_p}{C_V} > 0$

Other Equations of State

Some other equations of state are given below

Van der Waal's Equation of State

$$\left(p + \frac{a}{V^2}\right)(V - b) = RT$$

Two effects

(i) $\frac{a}{V^2}$, Inter molecular attraction forces

(ii) b accounts for volume occupied by the gas molecules

Above equation can also be written as $p = \frac{RT}{V - b} - \frac{a}{V^2}$

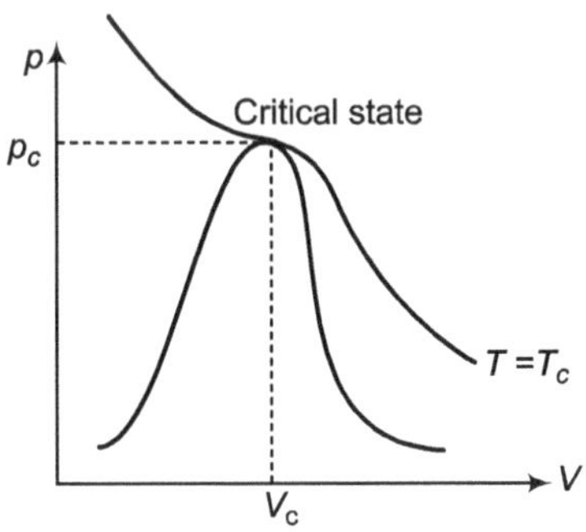

Critical properties on p-V diagram

$$pV^3 - (pb + RT)V^2 + aV - ab = 0$$

$p_r = 0$; $p_r V_r = 0, 9$

$$p_r = \frac{9(p_r V_r) - (p_r V_r)^2}{6}$$

Differentiating p_r w.r.t. p_rV_r

$$\frac{dp_r}{d(p_rV_r)} = 9 - 2(p_rV_r) = 0\ ;\ p_rV_r = 4.5$$

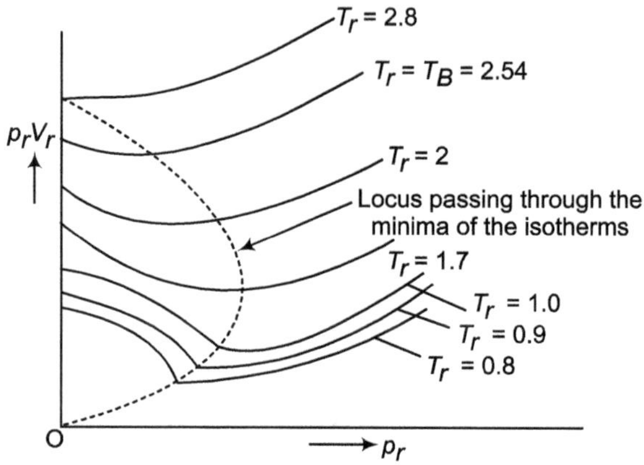

p_r *versus* p_rV_r graph

$$p_r = \frac{9 \times 4.5 - (4.5)^2}{6} = 3.375$$

The parabola has the vertex at $p_rV_r = 4.5$ and $p_r = 3.375$ and it intersects the ordinate at 0 and 9.

Boyle Temperature

Boyle temperature is the temperature at which a non-ideal gas behaves most like an ideal gas.

Boyle temperature T_B is determined by making

$$\left[\frac{\partial(p_rV_r)}{\partial p_r}\right]_{T_r = T_B} = 0, \qquad \text{when } p_r = 0$$

Above the Boyle temperature, the isotherms slope upward and show no minima.

Virial Equation of State

van der Waal's equation of state can be expressed in the virial form as given below

$$\left(p + \frac{a}{V^2}\right)(V - b) = RT$$

$$\left(pV + \frac{a'}{V}\right)\left(1 - \frac{b}{V}\right) = RT$$

$$pV + \frac{a}{V} = RT\left(1 - \frac{b}{V}\right)^{-1}$$

$$= RT\left(1 + \frac{b}{V} + \frac{b^2}{V^2} + \frac{b^3}{V^3} + \ldots\right) \quad \left(\text{where } \frac{b}{V} < 1\right)$$

$$pV = RT\left[1 + \left(b - \frac{a}{RT}\right)\frac{1}{V} + \frac{b^2}{V^2} + \frac{b^3}{V^3} + \ldots\right]$$

This is called the **virial equation of state.**

The second virial coefficient $B = b - \dfrac{a}{RT}$

The third virial coefficient $\quad C = b^2$

To determine Boyle temperature T_B

$$\left[\frac{\partial(pV)}{\partial p}\right]_{\substack{T=c \\ p=0}} = 0 = \frac{B}{RT}$$

$$B = 0$$

or $\qquad T_B = \dfrac{a}{bR}$

Because $\qquad B = b - \dfrac{a}{RT}$

The point at which B is equal to zero gives the Boyle temperature.

Beattie-Bridgeman Equation of State

$$p = \frac{R_u T}{\bar{V}^2}\left(1 - \frac{C}{\bar{V}T^3}\right)(\bar{V} + B) - \frac{A}{\bar{V}^2}$$

where, $A = A_0\left(1 - \dfrac{a}{\bar{V}}\right)$ and

$$B = B_0\left(1 - \frac{b}{\bar{V}}\right)$$

The above equation is known to be reasonably accurate for densities up to about $0.8\,\rho_{cr}$.

where, ρ_{cr} is the density of the substance at the critical point.

Benedict Webb -Rubin Equation of State

$$p = \frac{R_u T}{\overline{V}} + \left(B_0 R_u T - A_0 - \frac{C_0}{T^2}\right)\frac{1}{\overline{V}^2} + \frac{b R_u T - a}{\overline{V}^3}$$

$$- + \frac{a\alpha}{\overline{V}^6} + \frac{C}{V^3 T^2}\left(1 + \frac{\gamma}{\overline{V}^2}\right) e^{-\frac{\gamma}{\overline{V}^2}}$$

- Valid upto about 2.5 ρ_{cr}.

Thermodynamics Relations *and* Joule Thomson Coefficient

Thermodynamic relations is based on the condition of exact differentia.

If a relation exists among the variable x, y and z.

$$z = f(x, y)$$

$$dz = \left(\frac{\partial z}{\partial x}\right)_y dx + \left(\frac{\partial z}{\partial y}\right)_x dy$$

Let $\left(\frac{\partial z}{\partial x}\right)_y = M$ and $\left(\frac{\partial z}{\partial y}\right)_x = N$

$$\left(\frac{\partial M}{\partial y}\right)_x = \left(\frac{\partial N}{\partial x}\right)_y$$

$$\left(\frac{\partial x}{\partial y}\right)_z \left(\frac{\partial z}{\partial x}\right)_y \left(\frac{\partial y}{\partial z}\right)_x = -1$$

For thermodynamic variables p, V & T

$$\left(\frac{\partial p}{\partial V}\right)_T \left(\frac{\partial V}{\partial T}\right)_p \left(\frac{\partial T}{\partial p}\right)_V = -1$$

Maxwell's Equation

Using thermodynamics relations and applying to each of those relations results in four equations that are known as Maxwell's relation.

For a pure substance undergoing an infinitesimal reversible process,

$$dU = T dS - p dV$$

$$dH = dU + pdV + V\,dp$$

or $$dH = T\,dS + V\,dp$$

$$dF = dU - T\,dS - S\,dT$$

or $$dF = -\,pdV - S\,dT$$

$$dG = dH - T\,dS - S\,dT$$

or $$dG = V\,dp - S\,dT$$

$$\left(\frac{dT}{dV}\right)_S = -\left(\frac{\partial p}{\partial S}\right)_V \qquad \ldots\text{(i)}$$

$$\left(\frac{\partial T}{\partial p}\right)_S = \left(\frac{\partial V}{\partial S}\right)_p \qquad \ldots\text{(ii)}$$

$$\left(\frac{\partial p}{\partial T}\right)_V = \left(\frac{\partial S}{\partial V}\right)_T \qquad \ldots\text{(iii)}$$

$$\left(\frac{\partial V}{\partial T}\right)_p = -\left(\frac{dS}{\partial p}\right)_T \qquad \ldots\text{(iv)}$$

of the four Maxwell's relations, the last two Eqs. (iii) and (iv) are more valuable since they relate entropy derivatives $\left(\frac{\partial S}{\partial V}\right)_T$ and $\left(\frac{\partial S}{\partial p}\right)_T$ to derivatives of pressure and volume $\left(\frac{\partial p}{\partial T}\right)_V$ and $\left(\frac{\partial V}{\partial T}\right)_V$.

TdS Equation

First *TdS* equation, $TdS = C_V\,dT + T\left(\frac{\partial p}{\partial T}\right)_V dV$

Second *TdS* equation, $TdS = C_p dT - T\left(\frac{\partial V}{\partial T}\right)_p dp$

Difference in Heat Capacity

$$\underbrace{C_p - C_V}_{+\text{ve}} = -T\underbrace{\left(\frac{\partial V}{\partial T}\right)_p^2}_{+\text{ ve}}\ \underbrace{\left(\frac{\partial p}{\partial V}\right)_T}_{-\text{ ve}}$$

as $T \to 0 \Rightarrow C_p \to C_V$ or absolute zero expansivity

$$\beta = \frac{1}{V}\left(\frac{\partial V}{\partial T}\right)_p$$

Isothermal compressibility k_T is defined as

$$k_T = -\frac{1}{V}\left(\frac{\partial V}{\partial p}\right)_T$$

Adiabatic compressibility is defined as

$$k_s = -\frac{1}{V}\left(\frac{\partial V}{\partial p}\right)_S$$

$$C_p - C_V = \frac{TV\beta^2}{k_T}$$

Joule Thomson Coefficient

If a real gas undergoes throttling process then its temperature changes. Slope of curve drawn between pressure and temperature is known as **Joule Thomson coefficient** generally denoted as μ_J.

Sentropic curve and inversion curve $-\mu_J = \left(\frac{\partial T}{\partial p}\right)_h$

Here, $\mu_j < 0 \Rightarrow$ temperature increases
$\mu_j = 0 \Rightarrow$ temperature constant
$\mu_j > 0 \Rightarrow$ temperature decreases

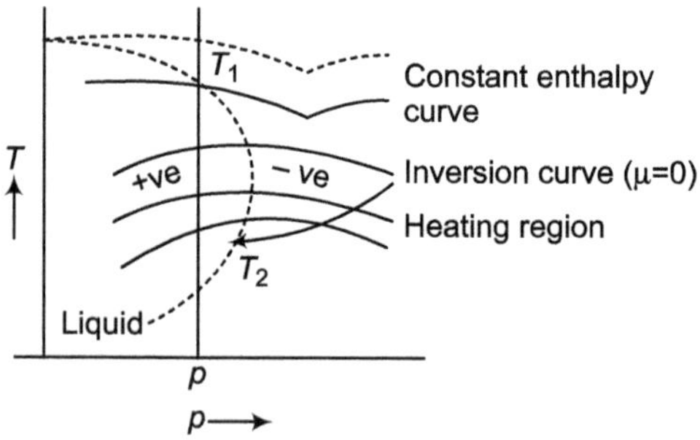

A typical phase diagram

$$\mu_j = \left(\frac{\partial T}{\partial p}\right)_h = \frac{1}{C_p}\left[T\left(\frac{\partial V}{\partial T}\right)_p - V\right]$$

or $$\mu_j = \frac{V}{C_p}(\beta T - 1)$$

For an ideal gas, $\mu_j = 0$

Clausius-Clapeyron Equation

Clausius-Clapeyron equation is a way of characterising a discontinuous phase transition between two phase of matter of a single constituent. On a *p*-*T* diagram, the line separating two phases is known as the coexistence curve.

$$\frac{dp}{dT} = \frac{S_f - S_i}{V_f - V_i} = \frac{l}{T(V_f - V_i)} = \frac{l}{T\Delta V}$$

where, dp / dT is the slope of tangent to the co-existence curve at any point, l is the specific latent heat, T is the temperature and ΔV is the specific volume change and S stands for specific entropy.

where, S_f = entropy of the final phase

S_1 = entropy of the initial phase

V_f = volume of the final phase

V_i = volume of the initial phase

8

Power Plant Engineering

Analysis of Steam Cycles

In steam cycles, the working fluid is water which is sometimes in liquid phase and sometimes in vapour phase during its cycle of operation.

Rankine Cycle (Vapour Power Cycle)

In Rankine cycles, each process is assumed to ideal process. Rankine cycle is an idealised thermodynamic cycle of a heat engine that converts heat into mechanical work.

Rankine cycle is a reversible cycle with following components.

Boiler

In steam boiler, steam is generated by reversible constant pressure & heating process of water.

Turbine

In steam turbine, there would be reversible adiabatic expansion of steam.

Condenser

In the steam condenser, there would be reversible constant pressure heat rejection as the steam condenses till it become saturated liquid.

Pump

In pump, the saturated liquid is compressed, and reversible adiabatically ending at initial pressure.

- Steady Flow Energy Equation (SFEE) for the boiler (as control volume)

$$h_4 + Q_1 = h_1 \quad \Rightarrow \quad Q_1 = h_1 - h_4$$

- SFEE for turbine (as control volume)

$$W_T = h_1 - h_2$$

where, W_T = work done by turbine.

- SFEE for the condenser

$$Q_2 = h_2 - h_3$$

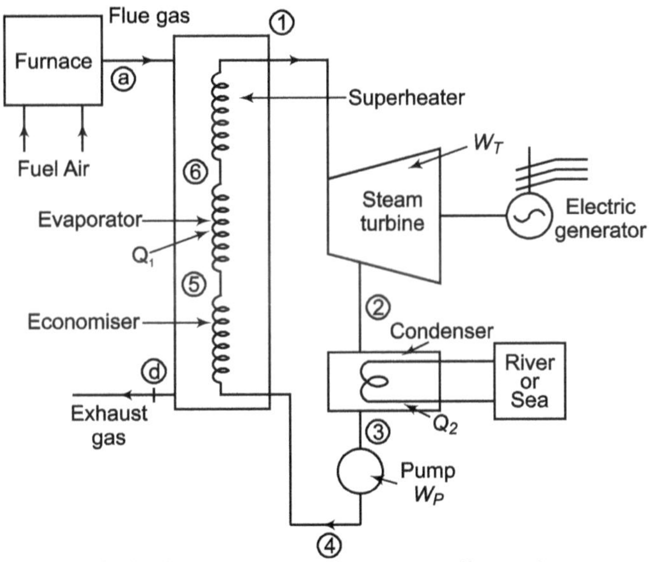

A simple steam power plant representing cycle

- SFEE for the pump ($ds = 0$)

$$W_P = h_4 - h_3 \ \left(\int_3^4 dh = \int_3^4 V dp\right)$$

$$W_{net} = W_T - W_P$$

- Efficiency of the Rankine cycle,

$$\eta = \frac{W_{net}}{Q_1} = \frac{W_T - W_P}{Q_1} = \frac{(h_1 - h_2) - (h_4 - h_3)}{(h_1 - h_4)}$$

- Steam rate $= \dfrac{1}{W_{net}} \dfrac{\text{kg}}{\text{k}_{\text{ws}}}$

- Heat rate $= \dfrac{Q_1}{W_T - W_P} = \dfrac{1}{\eta} = \dfrac{\text{kJ}}{\text{kWS}}$

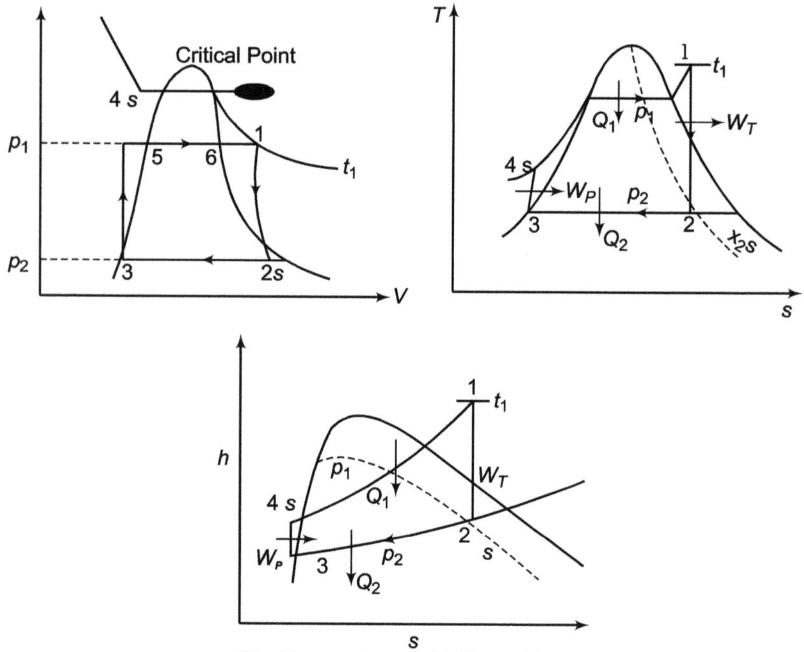

Rankine cycle on *p-V*, *T-s* and *h-s* coordinates

$$Q_{\text{Eco}} = h_5 - h_4 \quad \text{(Economiser)}$$
$$Q_{\text{Eva}} = h_6 - h_5 = h_{fg} \quad \text{(Evaporator)}$$
$$Q_{\text{SH}} = h_1 - h_6 \quad \text{(Superheater)}$$

$$\eta = \frac{\text{Heat supplied} - \text{Heat rejected}}{\text{Heat supplied}}$$

$$= \frac{T_3\,(S_3 - S_2) - T_1)\,(S_4 - S_1)}{T_2(S_3 - S_2)}$$

$$= 1 - \frac{T_1}{T_2} \;(S_3 - S_2 = S_4 - S_1)$$

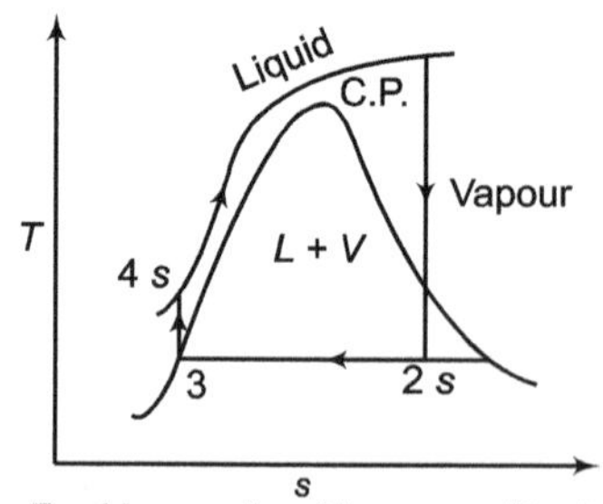

Rankine cycle with supercritical boiler pressure

Reheating of Steam

In the reheat cycle, the expansion of steam from the initial state 1 to the condenser pressure is carried out in two or more steps depending upon the number of reheats used.

Cycle efficiency improves with reheat, however the cycle efficiency in a single reheat plant is influenced by pressure at which steam is reheated. The efficiency increases as the reheat pressure is lowered and reaches a peak at a **pressure ratio** p_{rh}/p_1 between 0.20 and 0.25

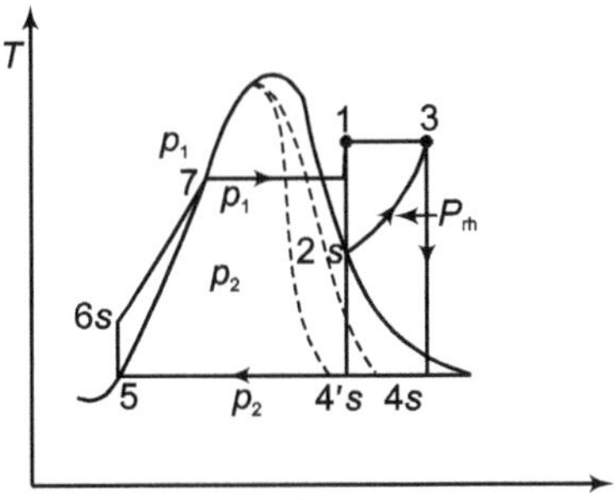

Reheating cycle

$$\eta = \frac{W_T - W_P}{Q_1} = \frac{(h_1 - h_{2s} + h_3 - h_{4s}) - (h_{6s} - h_5)}{h_1 - h_{6s} + h_3 - h_{2s}}$$

Reheating steam also increase the net work output of turbine.

Key Points

- Internal irreversibility of Rankine cycle (Real cycle) is caused by fluid frictions throttling and mixing.
- Externally, irreversibility of the Rankine cycle is caused due to the temperature difference between the combustion gases and the working fluid on the same side and the temperature difference between the condensing working fluid and the condenser cooling water on the sink side.

Regeneration

The mean temperature of heat addition (and so efficiency) can also be increased by reducing the amount of heat added at low temperatures in the economiser section of **steam generator**. In the regeneration process energy is exchanged internally between the expanding fluid in the turbine and the compressed fluid before heat addition.

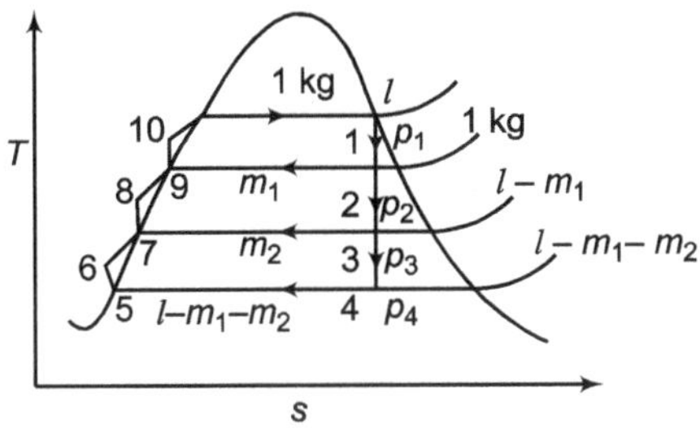

Regeneration cycle with two direct contact feed water heaters

Ideal regenerative cycle done not affect work output from turbine, it is more efficient with high **steam rate.**

Efficiency of Steam Power Plant

Overall efficiency of the steam per plant is given by

$$\eta_{overall} = \eta_{boiler} \times \eta_{cycle} \times \eta_{turbine} \times \eta_{generator} \times \eta_{aux}$$

$$\eta_{overall} = \frac{\text{Power available at the generator terminals}}{\text{Rate of energy released by the combustion of fuel}}$$

(a) $h_{boiler} = \frac{\text{Rate of energy absorption by water from steam}}{\text{Rate of energy released by combustion of fuel}}$

(b) $h_{turbine\ (mechanical)} = \frac{\text{Brake output of the turbine}}{\text{Internal output of the turbine}}$

(c) $h_{auxiliary} = \frac{\text{Net power transmitted by the generator}}{\text{Gross power produced by plant}}$

(d) $h_{cycle} = \frac{\text{Net work output}}{\text{Heat supplied}}$

(e) $h_{generator} = \frac{\text{Electrical output at generator terminal}}{\text{Brake output of turbine}}$

- Reheating of steam improved the thermal efficiency of the plant, net work output of turbine, reduction in blade erosion (or quality of steam improve)
- By regeneration thermal efficiency of the plant can be increased but it does not affect work output from turbine.

Combined Cycle Power Generation

It is an assembly of heat engines that works with the same source of heat. The principle of combined cycle power generation is that the exhaust of one heat engine is used as the source of heat for another heat engine. This process increases the overall efficiency of a **heat engine.**

Characteristics of Ideal Working Fluid

- The fluid should have high critical temperature and the saturation pressure at the temperature of heat rejection should be above the atmospheric pressure.
- Specific heat of liquid should be small.
- The saturated vapour line of *T-s* diagram should be steep, very close to the turbine expansion process.
- The freezing point of the fluid should be below the room temperature.
- The fluid should be chemically stable, non-toxic, non-corrosive, not excessively viscous and low in cost.

Binary Vapour Cycle

In the binary cycle, two cycles with different working fluids are coupled in series and the heat rejected by one being utilised in the other.

$$W_T = m(h_a - h_b) + (h_1 - h_2)$$
$$W_P = m(h_d - h_c) + (h_4 - h_3)$$
$$Q_1 = m(h_a - h_d) + (h_1 - h_6) + (h_5 - h_4)$$

m = flow rate of binary in the necessary cycle per kg of steam cycle.

$$\eta_{cycle} = \frac{W_T - W_P}{Q_1}$$

$$m(h_b - h_c) = h_6 - h_5$$

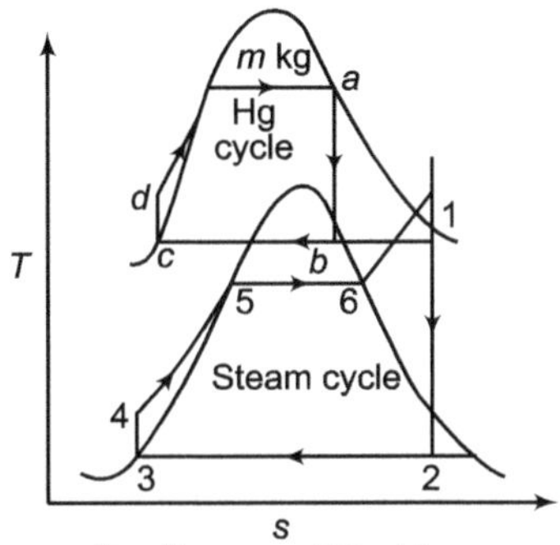

T-s diagram of Rankine binary cycle

Brayton Rankine Combine Cycle Plant

In this cycle, the Brayton and Rankine cycles are combined to increase the efficiency.

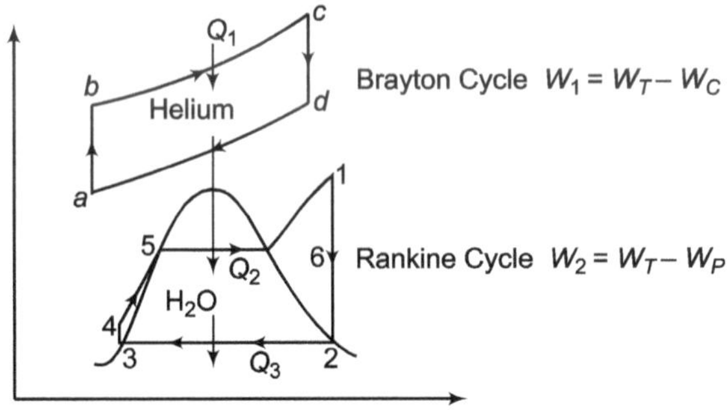

Combine cycle graph

When Plant in Series

Let us consider two cyclic pair plant coupled in series. The toppling plant operating on Brayton cycle and boiling are operating on Rankine cycle.

$$\eta = \eta_1 + \eta_2 - \eta_1\eta_2 - \eta_2 x_L$$
$$\eta_1 = \frac{W_1}{Q_1}, \eta_2 = \frac{W_2}{Q_3}$$

$$x_L = \text{fraction of heat supplied which is loss} \left(\frac{Q_L}{Q_1}\right)$$

When Plants Operate in Parallel

$$\eta = \frac{W_1 + W_2}{Q_1}, \eta_1 = \frac{W_1}{Q_2} \text{ and } \eta_2 = \frac{W_2}{Q_4}$$

$$\eta = \eta_1 - x_1(\eta_1 - \eta_2)$$

where, $x_L = \dfrac{Q_4}{Q_2 + Q_4}$

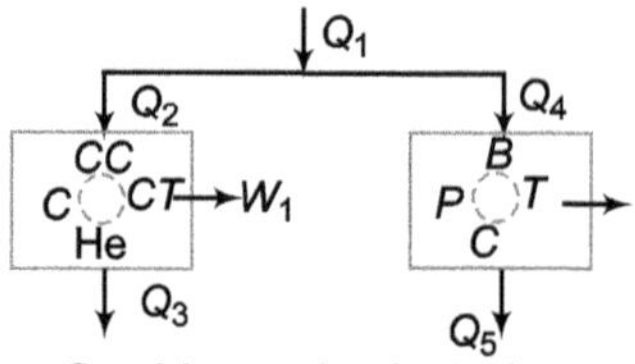

Combine cycle plant when plants operate in parallel

Advantages of Combined Cycle

- High overall plant efficiency
- Great operating flexibility
- Simplicity of operation
- Low investment costs
- Water required in small amount
- Low environmental impact.

Proximate Analysis of coal

It indicates the behaviour of coal when it is heated.

$$FC + VM + M + A = 100\% \text{ by mass}$$

FC = Fixed Carbon

where, VM = Volatile Matter
M = Moisture
A = Ash

Ultimate Analysis of coal

It gives the chemical elements that comprises the coal substances, together with ash and moisture.

$$C + H + O + N + S + M + A = 100\% \text{ by mass}$$

where, C = Carbon, H = Hydrogen, O = Oxygen
N = Nitrogen, S = Sulphur, M = Moisture, A = Ash.

Heating Value by Using Bomb Calorimeter

Heating value is the heat generated when the products of complete combustion of a sample of fuel are cooled to initial temperature of fuel air.

$$\text{LHV} = \text{HHV} - m_w h_{fg}$$

h_{fg} = Latent heat of vaporisation
m_w = Mass of water vapour

$$m_w = M + gH + \gamma_A W_A$$

where, M and H are the mass fractions of moisture and hydrogen in the coal. γ_A is the specific humidity of atmospheric air and W_A is the actual amount of air supplied per kg of coal.

Dulong and Petit Formula

If the ultimate analysis is known, the HHV of anthracite and bituminous coals can be determined approximately by using Dulong and Petit formula as

$$HHV = 33.83\ C + 144.45\left(H - \frac{O}{8}\right) + 9.38\ S$$

where, C = Carbon, O = Oxygen, H = Hydrogen, S = Sulphur

Orsat Gas Analysis

Orsat gas analyser measures the volume or mole fraction of CO_2, CO and O_2 in the dry flue gas.

$$CO_2 + CO + O_2 + N_2 = 100\% \text{ by volume}$$

Height of Chimney

Height of chimney is given by following formula,

$$h = 353\,H\left(\frac{1}{T_a} - \frac{m+1}{m}\frac{1}{T_g}\right) \text{mm of water}$$

where, H = Height of chimney above the fire grate in metres
h = Drought pressure in mm of water
m = Mass of air used for per kg of fuel
T_a = Absolute temperature of air outside the chimney (K)
T_g = Absolute temperature of air inside chimney (K)

Condition for maximum discharge through the chimney.
The drought in mm of water column

$$h = \frac{353\,H}{2T_a} = \frac{176.5H}{T_a}$$

Turbine

A turbine is a rotatory mechanical device that extract energy from a fluid flow and converts it into useful work. A turbine is a turbo machine with at least one moving part called a rotor assembly, which is a shaft or drum with blades attached moving fluid acts on the blades so that they move and impart rotational energy to the rotor.

Steam Turbine

A steam turbine is prime mover which converts high pressure energy and high temperature steam supplied by steam generator into **shaft work** with the low temperature steam exhausted to a condenser.

Impulse Turbines

- All pressure drops of steam occurs in the nozzles.
- No pressure drops as steam flow through the passage between two blades.

Before entering nozzle	p_0, V_1
After	p_1, V_1
After deflection	p_2, V_2

Gas turbines

These turbines are sometimes referred to as turbine engines. Such engines usually feature an inlet, fan, compressor, combustor and nozzle in addition to one or more turbines.

Momentum Conservation in Blades

Momentum of steam jet at inlet to the blades-momentum of steam jets at exit from the blades = momentum absorbed by wheel in producing shaft work.

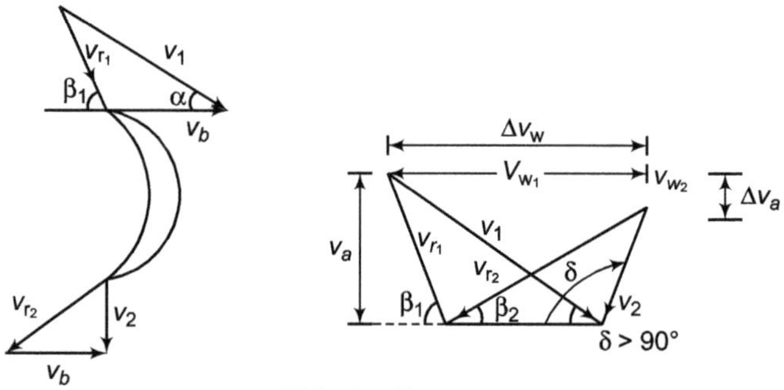

Velocity diagrams

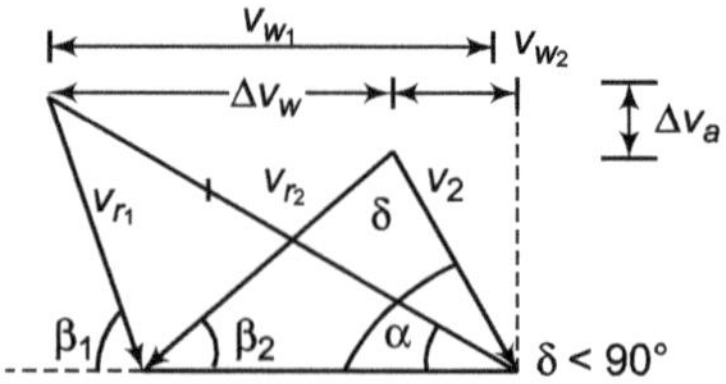

Velocity diagrams for impulse turbine

Δv_w = velocity of whirl at inlet – velocity of whirl at outlet.

α = Nozzle angle subtended by nozzle axis with the direction of rotation of wheel

$$\beta_1 = \text{Inlet blade angle}$$
$$\beta_2 = \text{Exit blade angle}$$
$$\Delta v_w = v_{r_1} \cos \beta_1 (1 + k_b), \text{ blade friction factor}$$
$$k_b = \frac{v_{r_2}}{v_{r_1}} = (v_1 \cos \alpha - v_b)(1 + k_b)$$

- **For symmetrical blades** $\beta_1 = \beta_2$

  v_b = mean peripheral velocity of the blades (mean blade velocity)

  $= \pi \dfrac{D_m N}{60}$

  D_m = mean diameter of the blade
- **Tangential thrust** impressed by jet on the blades $P_t = W_s \cdot \Delta V_w$

  W_s = steam flow rate
- **Axial thrust** $p_a = W_s \cdot \Delta v_a$, where $\Delta v_a = v_{r_1} \sin B_1 - v_{r_2} \sin B_2$
- **Area of the flow** or blade annulus,

  $$A_b = \pi(D_2 + D_1/2) \cdot (D_2 - D_1/2) = \pi D_m h_b$$

  D_1 is the root diameter, D_2 is the tip diameter, h_b is the height of the blades.
- **Blading or diagram efficiency**

  $$\eta_d = \frac{\text{Rate of work done on the blades}}{\text{Rate of energy input to the blades}} = \frac{W_s \Delta v_w v_b}{\frac{1}{2} v_1^2 W_3} = 2 \Delta v_w \frac{v_b}{v_1^2}$$
- **Energy loss due to friction** in the blades is equal to $v_1^2 - \dfrac{v_{r_1}^2}{2}$
- **Optimum velocity ratio** $p \left(= \dfrac{v_b}{v_1} \right)$

  $$p_{opt} = \cos \frac{\alpha}{2} \quad \text{and} \quad \eta_{d\,max} = \cos^2 \alpha \, (1 + k_b)$$
- If the energy loss due to friction is small,

  $$k_b = 1, \text{ then } \eta_{d\,max} = (\cos^2 \alpha)$$
- Nozzle angle has to be maintained within a certain range, which varies from 16° to 22° in impulse turbine.

Compounding of Steam Turbines

Compounding of stream turbines is the method in which energy from the steam is extracted in a number of stages rather than a single stage in a turbine.

- One row of nozzle followed by one row of blades is called a **stage** (compounding) of a **turbine.**
- With a single stage, the velocity of steam at exit is sufficiently high and there is considerable loss of kinetic energy with existing steam.

There are two methods of compounding which are given below

Pressure Compounding (Rateau staging)

The pressure compounding is used to solve the problem of high blade velocity in the single stage impulse turbine.

- Series of nozzle and blades.
- Pressure drop only at nozzles.
- Kinetic energy of steam increases in the nozzle and it is absorbed by the blades in each stage, in producing torque.
- The total enthalpy drop is divided equally in each stages.
- If n stages, isentropic enthalpy drop in per stages

$$(\Delta h)_{\text{stage}} = \frac{(\Delta h)_{\text{total}}}{n}$$

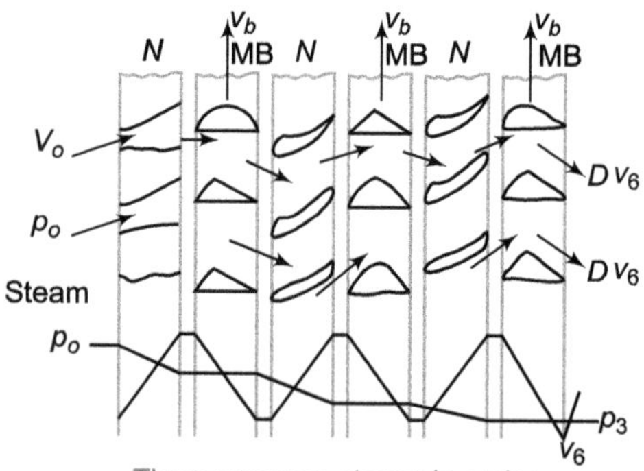

Three pressure stages in series

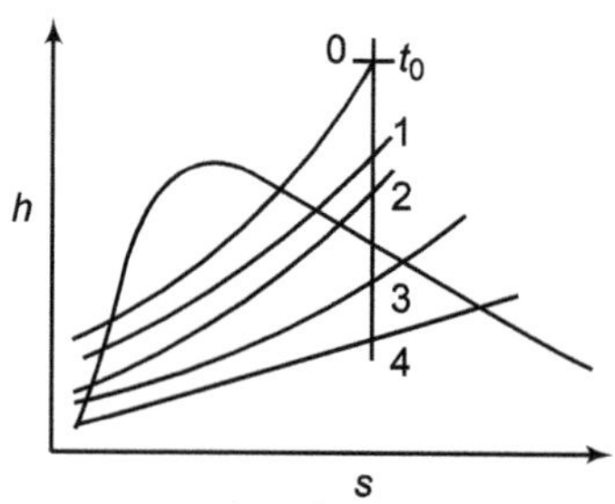

h-s diagram

The velocity of steam at exit from the first row of nozzle

$$V_1 = 44.72\ (h_0 - h_1)^{1/2}$$

$$= 44.72\left(h_0 - \frac{h_4}{4}\right)^{1/2}$$

$$= \left(\frac{1}{2}\right) 44.72\ (h_0 - h_4)^{1/2}$$

where, h_0 and h_1 in kj/kg and velocity in m/s.

- If the kinetic energy of steam is neglected at inlet to each subsequent row of nozzles, the velocity of steam leaving the nozzles at each row will be the same.
- For 4 stages velocity leaving nozzle in each stage = 1/2 of single stage
- For 9 stages velocity leaving nozzle in each stage = 1/3 of single stage
- For each impulse stage operating at its maximum blade efficiency, the blade velocity is given by $v_b = (\cos\alpha/2)\, v_1$

$$(\Delta h)_{stage} = 4\left[\frac{v_b}{(44.72\cos\alpha)}\right]^2$$

Velocity Compounding (Curtis staging)

Velocity compounding is used to solve the problem of single stage impulse turbine for use of high pressure and temperature steam.

- Pattern : nozzle moving blade $\rightarrow$ stationary blades $\rightarrow$ moving blades $\rightarrow$ stationary blade.
- All the pressure drops (enthalpy drop) in a single row of nozzles.
- Resultant kinetic energy of steam is absorbed by wheel in a number of rows of moving blades with guided blades in between two such rows.
- **Blade friction factor** $k_b = \frac{V_{r_2}}{V_{r_1}} = \frac{V_{r_3}}{V_{r_2}} = \frac{V_{r_4}}{V_{r_3}}$

Let,

α_1 = exit angle guide blade

β_1, β_2 = inlet and exit angle of the first row of moving blade

β_3, β_4 = inlet and exit angle of the second row of moving blades

$\Delta v_{w_1}, \Delta v_{w_2}$ = changes in velocity of whirl in the first row and second row of moving blades

$\Delta v_{a_1}, \Delta v_{a_2}$ = changes in axial component of velocity in the first row and second row of moving blades

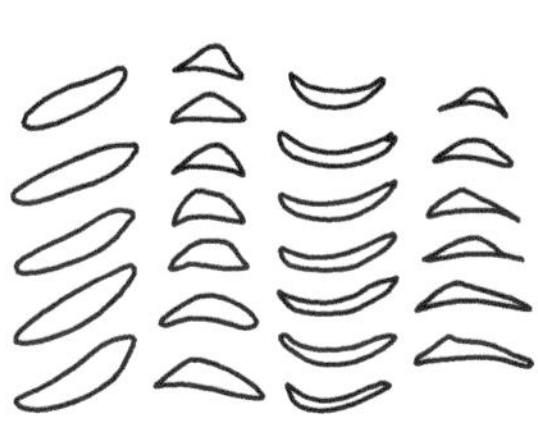

Curtis staging

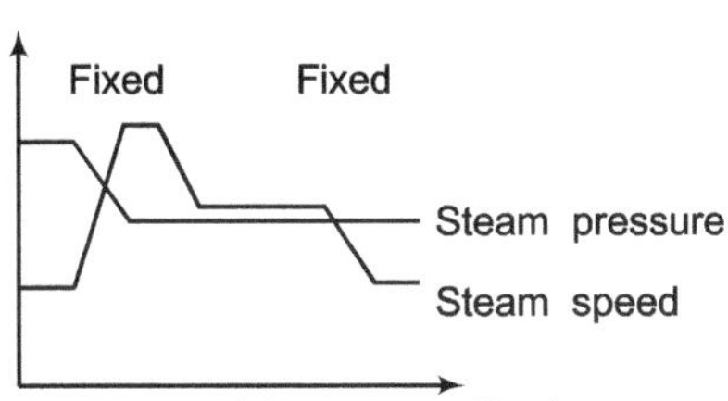

A two row curtis stage

- **The tangential thrust,** $P_t = W_S\ (\Delta v_{w_1} + \Delta v_{w_2})$
- **The axial thrust,** $P_a = W_s\ (\Delta v_{a_1} + \Delta v_{a_2})$
- **Blading or diagram work** $W_D = P_t \cdot v_b$
- **Blading or diagram efficiency,** $\eta_d = 2\ \Sigma\ (\Delta v_w) \cdot \frac{v_b}{v_1^2}$
- Ratio of work done by the steam jets on the moving blades in row.

  In 2-row Curtis stage, $W_{D_1} : W_{D_2} = 3:1$

  It means 3/4 of the work done by the steam jets on the first row of moving blades and 1/4 of the total work done on the second row of moving blades.

  In 3-row Curtis stage, $W_{D_1} : W_{D_2} : W_{D_3} = 3:2:1$

Note As the number of rows of moving blade in a Curtis stage increases, the effectiveness of the later rows decreases.

Optimum Velocity Ratio for a Curtis Stage

For a Curtis stage having z-rows of moving blade

$$\rho_{opt} = \frac{\cos\ \alpha}{2z}$$

where, z = number of rows

For a 2-row Curtis stage, $n_d = (\cos\ \alpha)^2$

$$v_b = (\cos\ \alpha/4)\ v_1$$

$$(\Delta h)_{stage} = 16\left[\frac{v_b}{(44.72\ \cos\ \alpha)}\right]^2$$

where, Δh = Change in drought pressure

Reaction Turbines

Pressure drops occur both in the nozzles or the fixed row of blades, as well as in the moving row of blades. Moving blades also have nozzle shape. Expansion of steam through the blades, increase in kinetic energy give rise to reaction in the opposite direction. Rotation of blades due to both the impulse effect of jets (change in momentum) and reaction force of the existing jets impressed on the blades in the opposite direction. Also called **impulse reaction turbine**.

The degree of reaction $(R) = \Delta h_{MB}/(\Delta h_{MB} + \Delta h_{FB})$

$\Delta h_{MB} = 0$, $R = 0$ pure impulse turbine (no enthalpy drop in the moving blades, all in nozzle)

$\Delta h_{FB} = 0$, $R = 1$(100% pure reaction turbine Hero's turbine)

$\Delta h_{MB} = \Delta h_{stage}/2,$

$R = \frac{1}{2}$ – (50% reaction turbine)

For manufacturing advantage, both fixed blades and moving blades are made similar in shapes so that they can be extruded from the same set of dies.

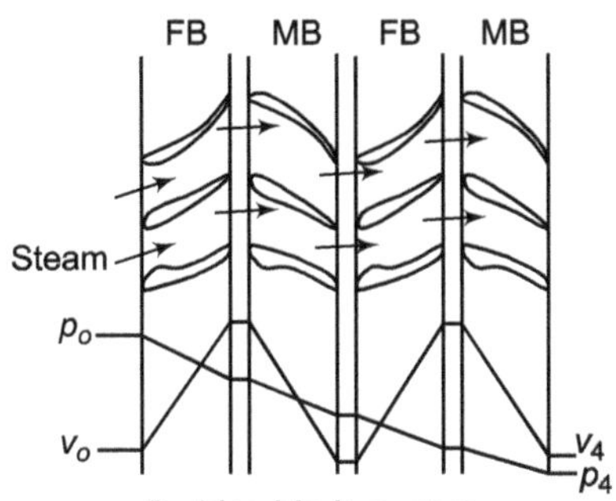

Turbine blade system

Since, $\Delta h_{FB} = \Delta h_{MB}, v_1 = v_{r_2}$

For similar geometry,

$$\alpha = \beta_2 = 180° - \gamma$$
$$v_{r_1} = v_2$$
$$\beta_1 = 180° - \delta$$

β_1 is not equal to β_2 (the blades are unsymmetrical).

$\Delta v_a = 0$ in 50% reaction turbine, there is no axial thrust.

However, there will be considerable axial thrust produced due to the **pressure difference** across the blades in each rotor disc since there is pressure drop of steam across the moving blades.

Key Points

- In reaction turbine, since both fixed and moving blades act as nozzle. So, we refer fixed blades as stator and moving blades as rotor.
- Optimum efficiency for simple impulse, Curtis and reaction turbine blading are all equal. However, when friction is taken into account, the reaction stage found to be most efficient due to low flow velocity followed by Rateau and Curtis in that order.
- Comparison of enthalpy drops in various stages

$$(\Delta h_s)_{50\%R} : (\Delta h_s)_{\text{simple impulse stage}} : (\Delta h_s)_{\text{2- row Curtis stage}} = 1 : 2 : 8$$

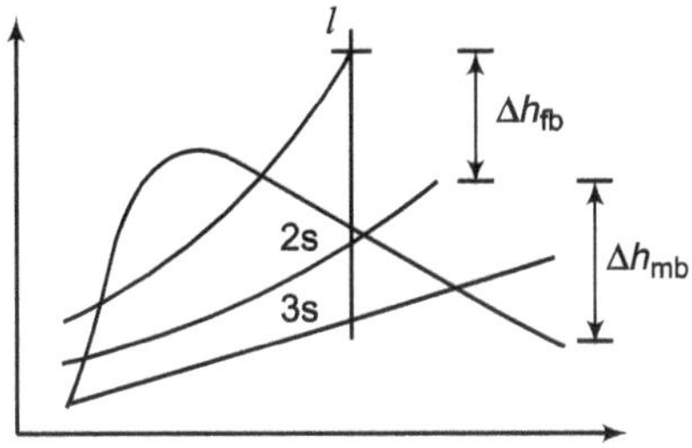

Enthalpy on various stages

The diagram work per kg of steam, $W_D = \Delta v_w \cdot v_b = (2\, v_1 \cos\alpha - v_b)\cdot v_b$

The energy input the blades per kg steam $= v_1^2/2 + \dfrac{(v_{r_1}^2 - v_{r_2}^2)}{2}$

$$= \frac{v_1^2 - v_b^2 + 2v_1 v_b \cos\alpha}{2}$$

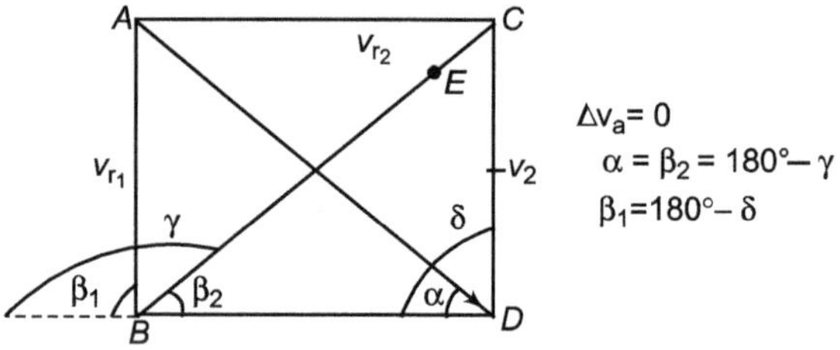

Velocity diagram of a 50% reaction turbine

$$\rho_{opt} = \cos\alpha \text{ or } v_b = v_1 \cos\alpha \text{ and } \eta_{d_{max}} = \frac{2\cos^2\alpha}{H\cos^2\alpha}$$

Specific blading work corresponding maximum blade efficiency

$$W_b = \Delta v_w v_b = v_b^2$$

Gas Turbine

A gas turbine, also called a **combustion turbine**, is a type of internal combustion engine. It has an upstream rotating compressor coupled to a downstream turbine, and a **combustion chamber** in between. The basic operation of the gas turbine is similar to the steam power plant except that air is used instead of water.

Gas Turbine Power Plant

- For power generation, it is attractive due to its low capital cost and its high reliability and flexibility in operation.
- Because of its low weight per unit power, gas turbine is exclusively used to drive aviation systems of all kind of aircraft.

Closed Cycle and Open Cycle Plants

- In open cycle plant which is more common, where the combustion products after doing work in the turbine are exhausted to atmosphere.
- In a closed cycle plant, the working fluid (air, helium, argon, carbon dioxide etc.) is externally heated (by burning fuel or by nuclear reactor) and cooled and it operates in a closed cycle.

Analysis of Gas Turbine Power Plant

Heat supplied $Q_1 = m_a c_p \ (T_3 - T_2)$

Heat rejected $Q_2 = m_a c_p \ (T_4 - T_1)$

Let m_a = mass of air

r_p = pressure ratio $\left(\dfrac{p_2}{p_1}\right)$

$$\eta_{cycle} = 1 - \frac{1}{r_p^{[(\gamma - 1)/\gamma]}} = 1 - (T_1/T_{4s}) \ r_p^{[(\gamma - 1)/\gamma]}$$

$$(r_p)_{max} = (T_3/T_1)^{(y/y - 1)} = (T_{max}/T_{min})^{[\gamma(\gamma - 1)]}$$

$$(r_p)_{opt} = (r_p)_{max}^{1/2}$$

$$W_{net} = m_a c_p \ [(T_{max})^{1/2} - (T_{min})^{1/2}]^2$$

If compressor and turbine efficiencies are considered,

$$(r_p)_{opt} = n_c n_T \ (T_3/T_1)^{[\gamma(\gamma - 1)]}$$

The work ratio r_w is defined as the ratio of net work to work done in turbine.

Variation of cycle efficiency with r_p

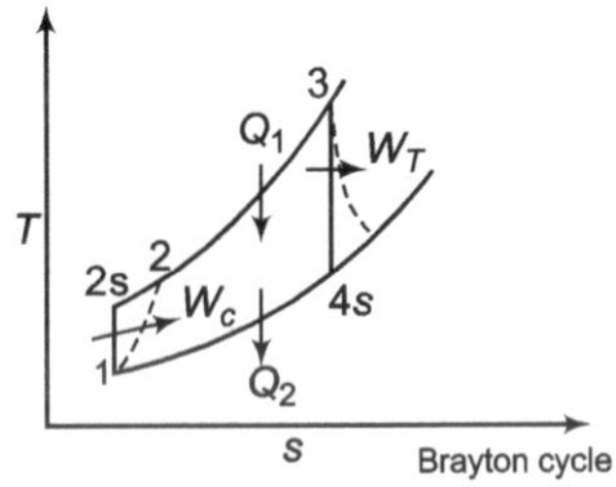

T-s diagram

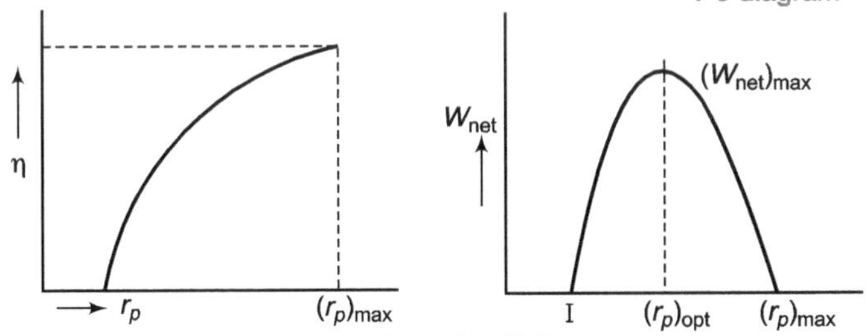

Variation in cycle efficiency

Key Points

- The ideal cycle that the working fluid undergoes in this closed loop is the **Brayton cycle**. (Isentropic compression, constant pressure heat addition, isentropic expansion, constant pressure heat rejection).
- The advantages of the open cycle are the quick and easy starting.
- The advantages of the closed cycle plant are constant efficiency at all loads and higher unit rating permitting the use of higher back pressure.

Effect of Regeneration

Regenerator effectiveness e = (actual temperature rise of air)/(maximum temperature of rise possible)

$$= T_5 - T_2/T_4 - T_2$$

- Q_1 and Q_2 both of which decreases whereas W_{net} remains unchanged due to regeneration. Therefore, efficiency of the cycle W_{net}/Q_1 increases.
- In practice, the regenerator is costly, heavy and bulky and causes pressure losses which may bring about a decreases in cycle efficiency. So, there may be balanced.

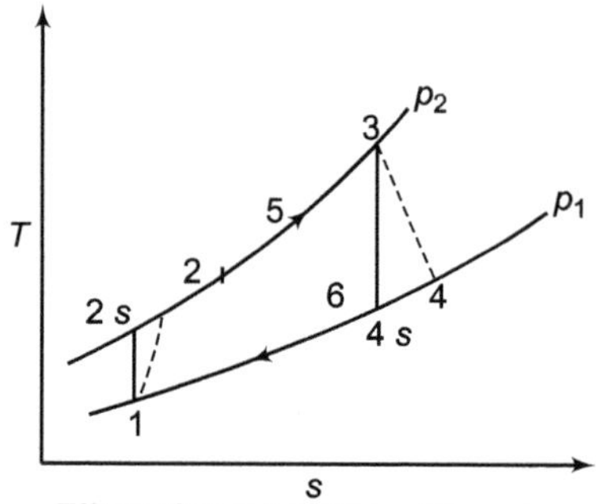

Effect of regeneration with cycle efficiency

Effect of Intercooling

- By staging the compression process with perfect intercooling, the cycle efficiency decreases as shown in figure, where the small cycle 1-2-3-4-4′-1 is added to basic cycle 1–4′-5-6-1, without intercooling.
- For minimum work of compression, the intercooler pressure.

$$p_i = [p_1 p_2]^{1/2}$$

where, p_1 and p_2 are suction and discharge pressures respectively.

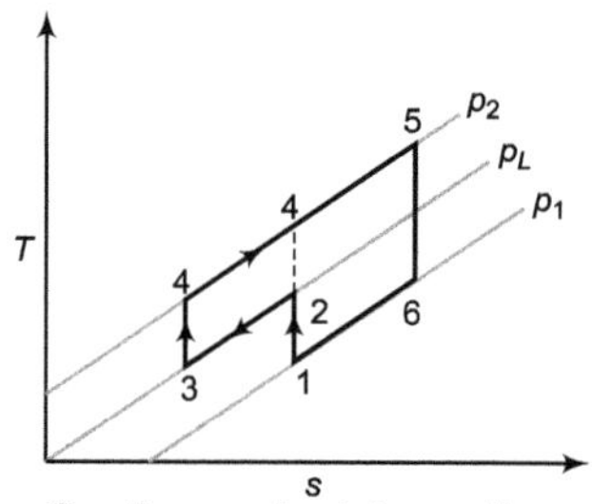

T-s diagram for intercooling

Effect of Reheating

- By staging the heat supply process with a combustor and a reheater, the cycle efficiency decreases.

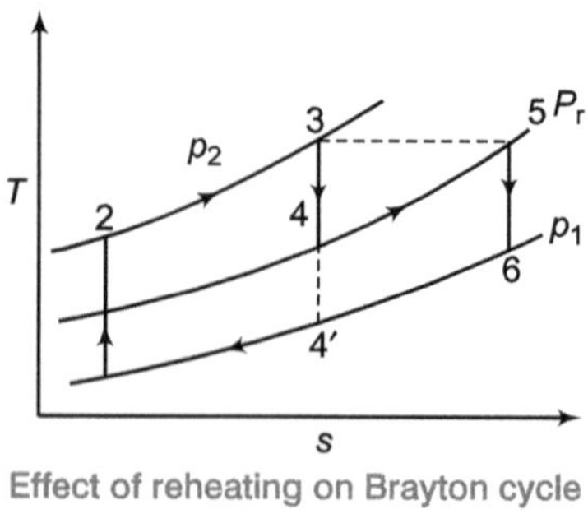

Effect of reheating on Brayton cycle

- But it permits more heat recovery from the turbine exhaust gases, with the result that reheating along with regeneration may bring about an improvement in cycle efficiency.
- It can be shown that the optimum reheat pressure for maximum work is $p_r = (p_1 p_2)^{1/2}$

Components of Gas Turbine Power Plant

Gas turbines have a high power to weight ratio and are lighter and smaller than internal combustion engines of the some power. Their characteristics of high speed and high temperature operation require high precition components and exotic materials.

Some of the various components are as follows

Compressor

The high flow rates of air through the turbine and the relatively moderate pressure ratios necessitate the use of rotary compressors.

Losses in Compressor

There are three losses in compressor as given below

- Profile drag (C_{Dp})
- Annulus drag (C_{Da})
- Secondary loss (C_{Ds})
- For typical axial compressor design, the following empirical formulae have been derived

$$C_{Ds} = 0.018 C_L^2 \text{ and } C_{Da} = 0.020\, s/h$$

where C_L is the lift coefficient and s and h are pitch and blade height of blades respectively.

Compressor further can be classified as

Slip

The inertia of air trapped between the impeller blades, however, causes the actual whirl velocity to be less than ideal whirl velocity. It is known as slip.

$$\text{Under ideal conditions, } C_{w_2} = U_2$$

$$\text{Slip} = U_2 - C_{w_2}$$

$$\text{Slip factor } \sigma = C_{w_2}/U_2 = C_{w_2}/(U - \cot\beta_2)$$

$$= 1 - 0.63\,\pi/n$$

Division of Loss between Impeller and Diffuser

The division of loss at various levels between impeller and diffuser is shown below

$$\text{Actual work input} = h_{03} - h_{01} = \psi\sigma U_2^2 \quad \text{or} \quad T_{03} - T_{01} = \psi\, \sigma U_2^2/C_p$$

For a stage, $p_{03}/p_{01} = (T'_{03}/T_{01})^{\gamma/(\gamma-1)}$

$$= [1 + \eta_c\,(T_{03} - T_{01})/T_{01}]^{\gamma/(\gamma-1)}$$

Because $\eta_c = (T'_{03} - T_{01})/(T_{03} - T_{01})$

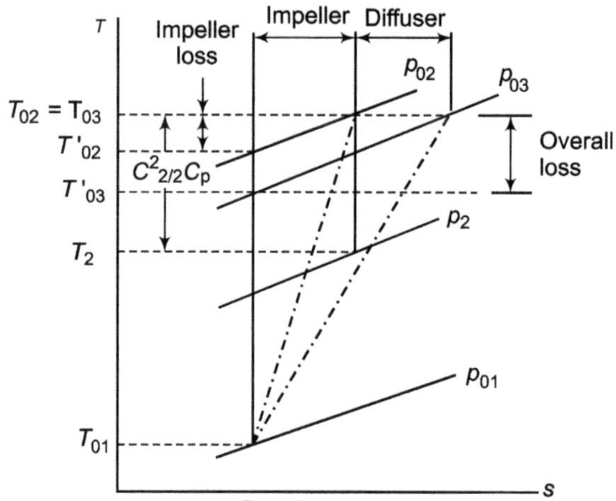

T-s diagram

i.e., $T'_{03} = T_{01} + \eta_c\,(T_{03} - T_{01}) = [1 + \eta_c \psi\sigma U^2 / C_p T_{01}]^{\gamma/(\gamma-1)}$

Stage Loading Factor (ψ)

The stage loading factor ψ is another important design parameter of a compressor stage and is one which strongly affects the off-design performance characteristics. *It is defined by*

$$\psi = W/u^2 = \frac{(h_{03} - h_{01})}{u^2} = (c_{y_2} - c_{y_1})/u = \Delta c_y/u$$

With $c_{y_2} = u - w_{y2}$

Hence, $$\psi = \frac{[u - (c_{y_1} + w_{y_2})]}{u} = 1 - \left(\frac{c_x}{u}\right) \cdot (\tan\alpha_1 + \tan\beta_2)$$

or $$\psi = 1 - \eta\,(\tan\alpha_1 + \tan\beta_2) \qquad \ldots\text{(i)}$$

where, $\eta = \frac{c_x}{u}$ is called **flow coefficient.**

α_1 and β_2 are the fluid inlet and outlet angles.

Simplified Off-design Performance

Horlock (1958) has considered how the stage loading behaves with varying flow coefficient, ϕ and how this off-design performance is influenced by the choice of design conditions. Now cascade data suggests that fluid outlet angles β_2 (for the rotor) and α_1 $(=\alpha_3)$ for the 4 stator do not change appreciably for a range of incidence up to the stall point. In Eq. (i) the quantity $(\tan\beta_2 + \tan\alpha_1)$ can be assumed constant in a wide range of incidence up the stalling limit,

Writing $A = \tan\beta_2 + \tan\alpha_1$ Hence, $\psi = 1 - A\eta$

Work done Factor

For increased axial velocity c_{y_2} is less than and c_{y_1} is more. Work capacity of the stage is reduced with an increase of the axial velocity and *vice-versa*, hence work capacity of the stage is reduced in central region. The expected increase in the hub and tip region is not obtained in actual practice on account of higher losses.

$$W = u\,(c_{y_2} - c_{y_1})$$

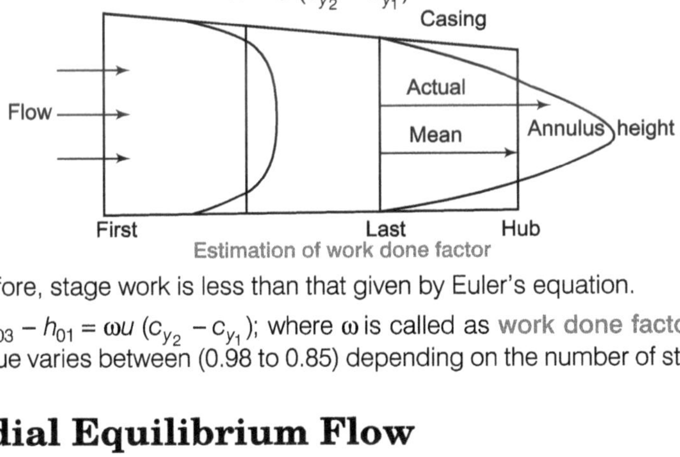

Estimation of work done factor

Therefore, stage work is less than that given by Euler's equation.

$W = h_{03} - h_{01} = \omega u\,(c_{y_2} - c_{y_1})$; where ω is called as work done factor and its value varies between (0.98 to 0.85) depending on the number of stages.

Radial Equilibrium Flow

The flow in an annular passage in which there is no radial component of velocity, whose streamlines lie in circular, cylindrical surfaces and which is axisymmetric, is commonly known as radial equilibrium flow. An analysis called the radial equilibrium method, widely used for three dimensional design calculations in axial compressors and turbines is based upon the assumption that any radial flow which may occur, is completed within a blade row, the flow outside the row then being in radial equilibrium.

Radial Equilibrium Equation

When hub-tip ratio is less than 0.8, radial velocities through a blade row may become appreciable. The basic equation expressing the balance between pressure force and inertial forces can be derived by considering the forces acting on the fluid element as under.

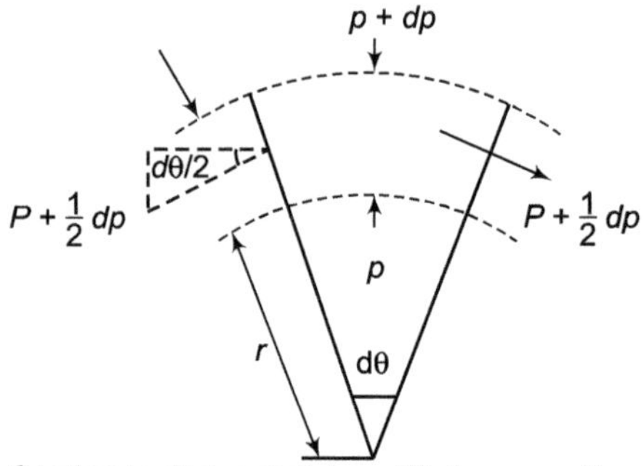

Section to derive radial equilibrium equation

Radial Equilibrium of Fluid Element

Pressure forces balances the centrifugal forces

$$(p+dp)(r+dr)\,d\theta - prd\theta - \left(p+\frac{1}{2}dp\right)dr\,d\theta = dm\,\frac{C_y^2}{r}$$

Ignoring second order terms, we get

$$(1/\rho)\frac{dp}{dr} = C\theta^2/r \qquad \ldots(i)$$

$$C_y \cdot r = \text{constant}$$

- This is known as **free vortex condition.**
- This condition gives air angle and reactions as shown in figure.
- For forced vortex condition, $\frac{C_y}{r}$ = constant

Combustion Chamber

In an open cycle gas turbine plant, combustion may be arranged to take place in one or two large cylindrical can type combustion chambers with ducting to convey the hot gases to the turbine. Can type combustion chambers are as follows

- Can type combustion with swirl flow flame stabiliser
- Can type combustion with bluff- body flame stabiliser

Gas Turbines

- Like steam turbines, gas turbines are also of the axial flow type.
- More stages are always preferred in gas turbine power plants, because it helps to reduce the stresses in the blades and increases the overall life of the turbines.

- The cooling of gas turbine is essential for long life as it continuously subjected to high temperature gases.
- Blade angles of gas turbines follow the axial flow compressor blading, where the degree of reaction is not 50%.

  Degree of Reaction, $R = v_f (\tan \beta_2 - \tan \beta_1)/2v_b$

Key Points

- The duct work consists of ducts between the compressor and the combustion, combustion chamber to the turbine and the exhaust duct.
- The duct must be sized to minimise the pressure losses, as the loss in pressure directly reduces the capacity of the plant.

Steam Condenser

A condenser where the exhaust the steam from the turbine is condensed operates at a pressure lower than atmosphere.

Objective of Steam Condenser

- To reduce turbine exhaust pressure so as increase the specific output of the turbine and efficiency.
- To recover high quality feed water in the form of condensate and feed it back to the steam generator without any further treatment.

Note Use of steam condenser increase the specific work output of turbine from $(h_1 - h_3)$ to $(h_1 - h_2)$

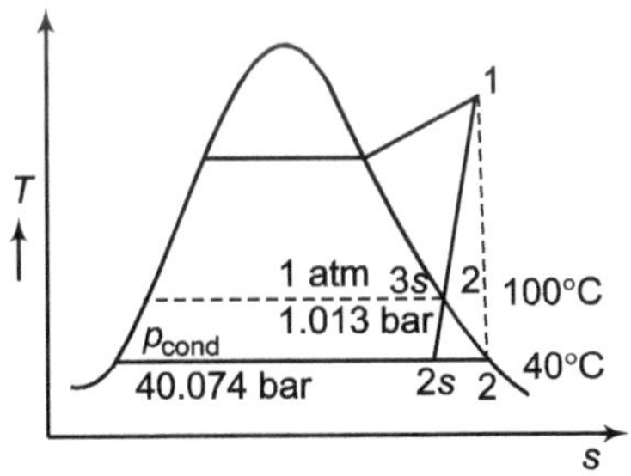

T-s graph of steam condenser

The steam condenser can be classified as

Direct Contact Type

Mixing of condensate and cooling water. *It is of three types*

Spray condenser

Spray of cooling water into the steam.

$$W_2 h_2 + W_5 h_5 = W_3 h_3$$

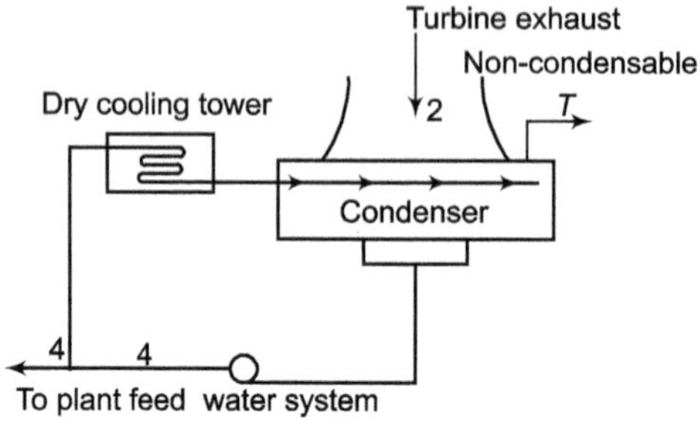

Direct contact type condenser

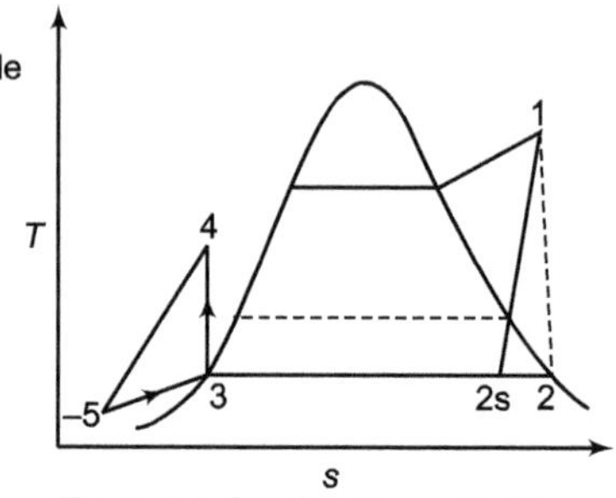

T-*s* curve for direct contact type

Barometric Condenser Falling of cooling water on series of baffles (large surface area) and direct contact with steam fed out from below.

Non-condensable to steam jet air ejector

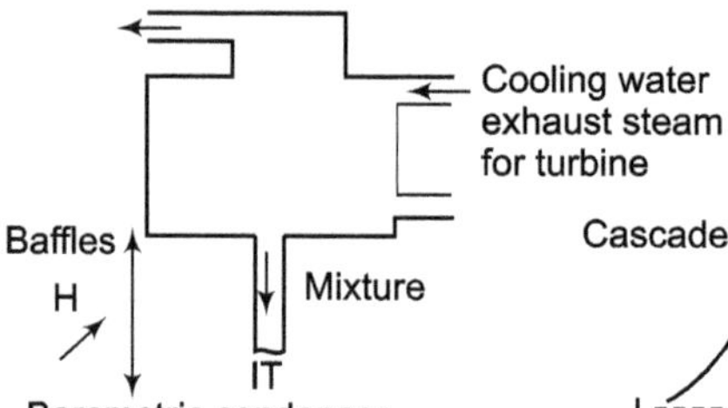

Barometric condenser

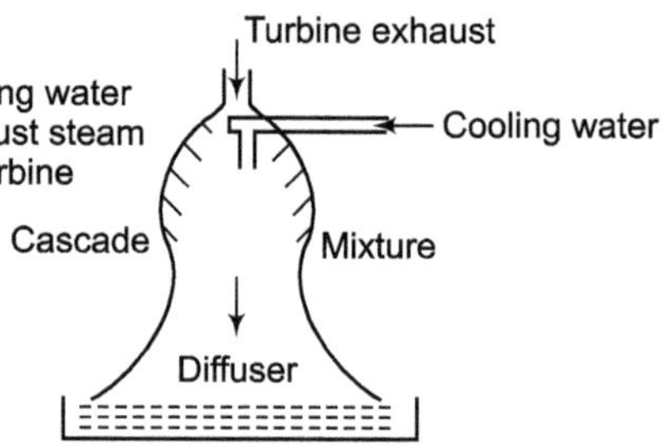

Jet condenser

$$p_{atm} - p_{cond} + \Delta p_f = dgh$$

Jet Condenser Similar as barometric but use diffuser.

Surface Condensers

These are shell and tube heat exchanger.

- For the convenience of cleaning and maintenance, cooling water flows through the tubes steam condenses outside the tubes.
- Single pass condenser is good for overall plant efficiency and reduces thermal pollution but require four times the pumping power and twice the water flow.

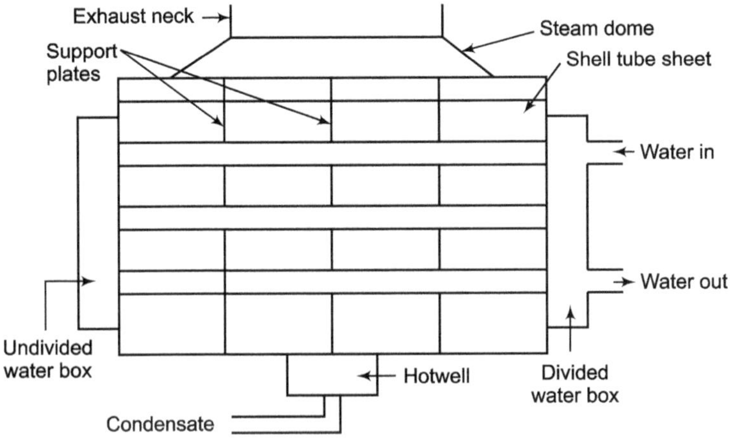

Surface condenser

- The tube materials: cupronickel (70% Cu, 30% Ni),aluminium brass (76% Cu, 22% Zn, 2% Al),muntz metal (60% Cu, 40% Zn),admiralty alloy (71% Cu, 28% Zn, 1% Sn).

Air Removal

Leakage of air from condenser shell, along with steam reduce the condenser vacuum and increases the turbine exhaust pressure, thus reduces the turbine output and also reduces heat transfer (air low k). An air cooler section is provided in the condenser shell, having main bank and smaller bank (air cooler).

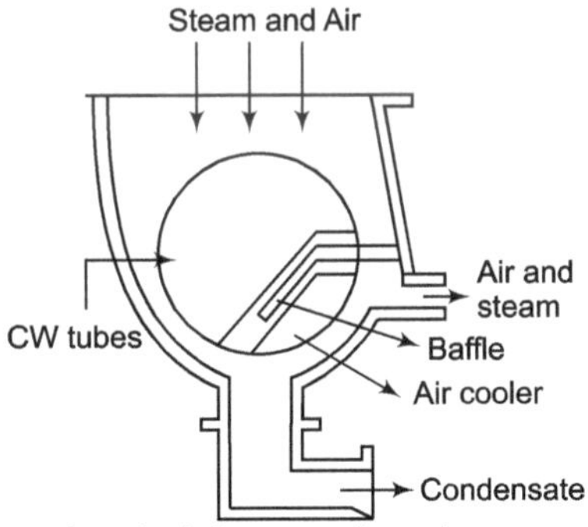

A typical diagram of condenser

Important Formulae

- $$\text{Vacuum efficiency} = \frac{\text{Vacuum produced by steam at condenser inlet}}{\text{Barometric pressure} - \text{saturation pressure at exhaust steam } p}$$

$$\text{Condenser efficiency} = \frac{\text{Actual temperature rise of cooling water}}{\text{Maximum temperature rise of cooling water}}$$

Feed Water Heaters

These are used in steam power plant to improve cycle efficiency. *These heaters can be classified in two ways*

1. **Closed feed water heater** Shell and tube heat exchangers. They are basically small condenser which operate at higher pressures than the main condenser because blade steam is condensed on the shell side, whereas feed water acting like circulating cooling water in the condenser, is heated on tube side.
2. **Open feed water heater** (deaerator) Heating of feed water by direct mixing with the steam blade from turbine.

It is of three types spray type, tray type, combined spray-tray deaerator.

Cooling Towers

These cool the warm water discharged from the condenser and feed the cooled water back to the condenser. *A cooling tower can be specified by*

Wet Cooling Towers

Minimum temperature to which water can be cooled is the adiabatic saturation or wet bulb temperature of the ambient air.

Dry Cooling Towers

These are used where scarcity of cooling water. Warm water from condenser flow through finned tube over which the cooling air is passed. Heat is rejected to the air as water is cooled. Ammonia is also used as the coolant for condenser.

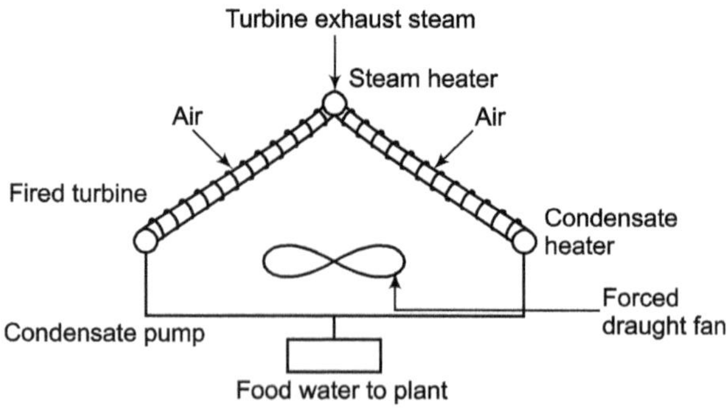

Dry cooling towers

Isentropic Flow *and* Shock Waves

The second law of thermodynamics requires that the adiabatic and frictionless flow of any fluid results in no change in entropy. Constant entropy flow is called **isentropic flow.**

Isentropic condition implies absence of heat transfer and frictional effect. In case of fluid in internal flow, heat transfer is absent while frictional losses are always present. However, it can be neglected when the length of the duct or pipe is small, such as in the case of a nozzle or diffuser.

Nozzle

A nozzle is a duct by flowing through which the velocity of a fluid increases at the expense of pressure drop.

Diffuser

A diffuser is a duct by flowing through which the velocity of a fluid decreases and causes a corresponding increase of pressure.

Stagnation Properties

The isentropic stagnation state is defined as the state a fluid in motion would reach if it were brought to rest isentropically in steady flow, adiabatic, zero work output device.

Stagnation enthalpy $$h_0 = h + \frac{v^2}{2}$$

$$\frac{T_0}{T} = 1 + \frac{\gamma - 1}{2} M^2$$

where, M = Mach number

$$\frac{p_0}{p} = \left(1 + \frac{\gamma - 1}{2} M^2\right) \gamma - 1$$

where p_0 and T_0 are the stagnation pressure and temperature respectively.

Velocity of Sound

Velocity of sound can be defined as

$$c = \sqrt{\gamma RT} = \sqrt{\left(\frac{\partial p}{\partial \rho}\right)_S}$$

$\left(\dfrac{\partial p}{\partial \rho}\right)_S$ = Infinitesimal pressure change of fluid across section with respect to density with reversible and isentropic process.

The bulk modulus or the change of density with pressure in a gas is dependent on the process of compression.

Mach Number

It is defined as the ratio of actual velocity v to the sonic velocity c.

$$M = \frac{v}{c}$$

Subsonic and Supersonic Nozzle

For isentropic flow through nozzle,

$$\frac{dA}{A} = \frac{dp}{\rho v^2}(1 - M^2) \quad \text{or} \quad \frac{dA}{A} = (M^2 - 1)\frac{dv}{v}$$

When $M < 1 \Rightarrow$ Inlet velocity is subsonic as flow area A decreases. The pressure decreases and velocity increases. So, convergent passage is nozzle.

$M > 1 \Rightarrow$ Inlet velocity is supersonic as flow area A increases pressure decreases and velocity increases. So, divergent passage is nozzle.

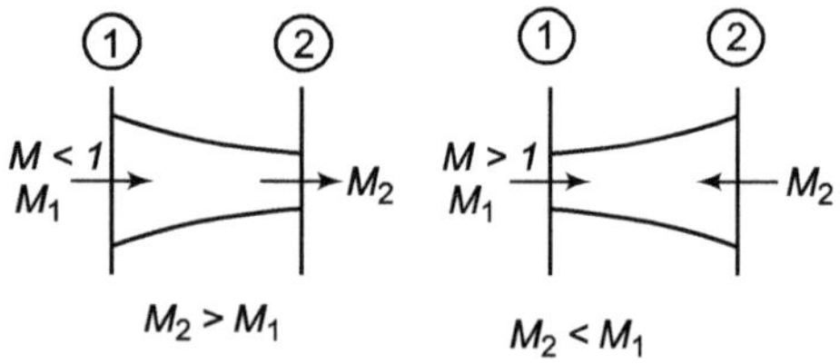

Subsonic nozzle ($M < 1$ and M increase) Supersonic nozzle ($M > 1$ and M increase)

Subsonic and Supersonic Diffuser

When $M < 1$ (subsonic) $\Rightarrow M$ decreases, A increases,

A increases, P increase v decreases

p increase, v decreases. $\Rightarrow$ Divergent passage is diffuser.

A decrease, p increase v decreases

When $M > 1$ (supersonic) $\Rightarrow M$ decreases, A decreases, p increases, v Decreases. $\Rightarrow$ Convergent passage is nozzle.

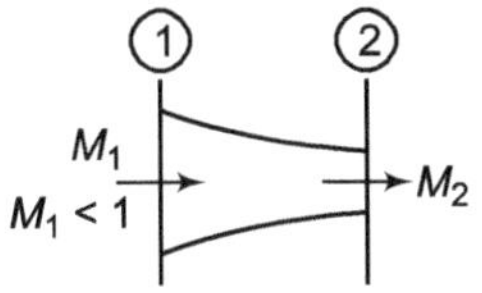

$M_2 < M_1$
Subsonic diffuser
$M < 1$ and M decreases

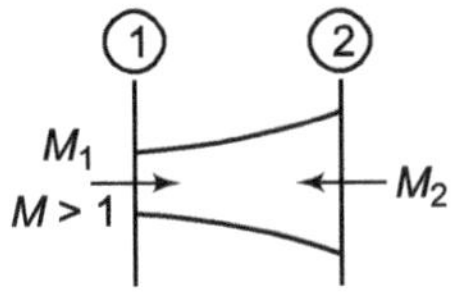

$M_2 < M_1$
Supersonic diffuser
$M > 1$ and M decreases

Note The area dA is zero for sonic condition $M = 1$ indicating a minimum cross-section and so called throat.

Limits of Incompressibility

If we set the limit that a flow is incompressible so long as the density variation does not exceed two per cent.

$M \leq 0.2$ (Range of incompressible)

Isentropic Flow Through a Duct of Varying Cross-Section

Let us consider the isentropic flow of an ideal gas through a convergent-divergent nozzle.

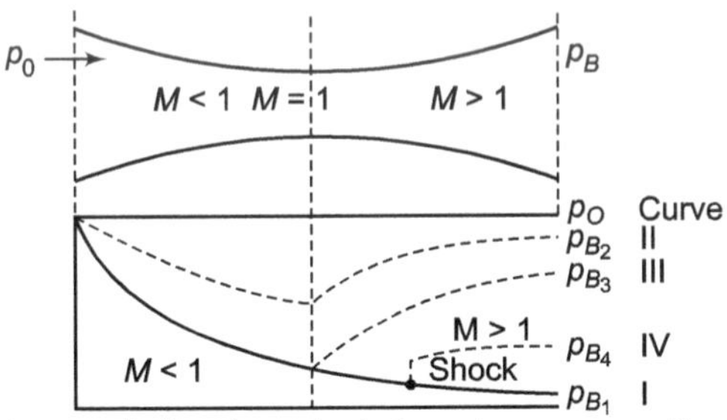

Isentropic flow for a duct of varying cross-section

The following cases will be considered here.

Case 1 $p_{B_2} > p_{B_1}$ $(M < 1)$

When the back pressure p_{B_2} is higher than the value necessary for supersonic exit *i.e.*, p_{B_1} (curve I), the flow will accelerate in the converging portion, the throat Mach number will be less than one and the flow will decelerate in the diverging portion. The flow will remain subsonic all along the nozzle length. The pressure variation is shown by the dashed (curve III).

Case 2 $p_{B_3} > p_{B_1}$ $(M = 1)$

In this case, the flow in the diverging portion can become subsonic if the exit area is not suitable for supersonic discharge (curve III).

Case 3 $p_{B_4} > p_{B_1}$ $(M > 1)$

A strongly irreversible process known as shock may take place in the supersonic portion of the nozzle flow resulting in secure losses and finally subsonic discharge from the converging-diverging nozzle

Mass Flow Rate

Consider isentropic flow through a duct from a reservoir at p_0 and T_0. The mass flow rate $m = \rho AV.20$

The flow rate per unit area

$$\frac{m}{A} = \sqrt{\frac{\gamma}{R}} \frac{p_0 M}{\sqrt{T_0}} \frac{1}{\left(1 + \frac{\gamma - 1}{2} M^2\right)^{\frac{\gamma + 1}{2(\gamma - 1)}}}$$

Flow rate per unit area is maximum, when the sonic condition $M = 1$ is reacted.

$$\left(\frac{m}{A}\right)_{max} = \frac{p_0}{\sqrt{T_0}} \sqrt{\frac{\gamma}{R}} \left(\frac{2}{\gamma + 1}\right)^{\frac{\gamma + 1}{2(\gamma - 1)}}$$

$$\left(\frac{m}{A}\right)_{max} = 0.04044 \frac{p_0}{\sqrt{T_0}} \frac{\text{kg / s}}{\text{m}^2} \qquad (\gamma = 1.4)$$

So, maximum mass flow rate per unit area

$\propto$ stagnation pressure

$$\propto \frac{1}{\sqrt{\text{stagnation temperature}}}$$

Chocking in a Converging Passage

Imaging the discharge from a reservoir at p_0 and T_0 through a converging nozzle in a region where the pressure p_B is variable. The nozzle exit plane pressure is p_ε.

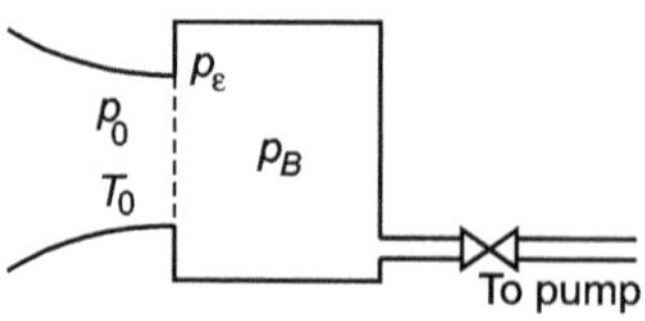

Converging passage

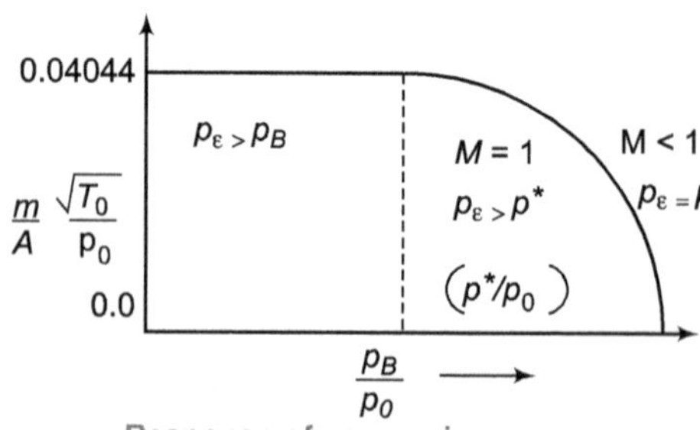

Response of converging passage

- $p_B = p_\varepsilon = p_0$ there shall be no flow through the nozzle.
- As p_B is gradually reduce, the flow rate shall increase. The exit plane pressure p_ε shall remain equal to p_B so long as the maximum discharge condition is not reached.
- By reducing back pressure p_B when $\left(\frac{m}{A}\right)$ has attained its maximum value, the exit plane mach number shall be unity and $p_B = p_\varepsilon = p^*$ (critical pressure).
- Further reduction of back pressure p_B will not accelerate the flow to supersonic condition. As a result the exit pressure p_ε shall continue to remain at p^* even through p_B is lower.

$$p_B < p_\varepsilon \quad (p_\varepsilon = p^*)$$

Frictional Effects on Nozzle

A is inlet stagnation condition at pressure p_0 and throat and exit pressures are respectively p_t and p_e.

Whereas AB is ideal isentropic process to exit pressure p_e, AC is the adiabatic expansion accounting for friction in both converging & diverging portion of nozzle.

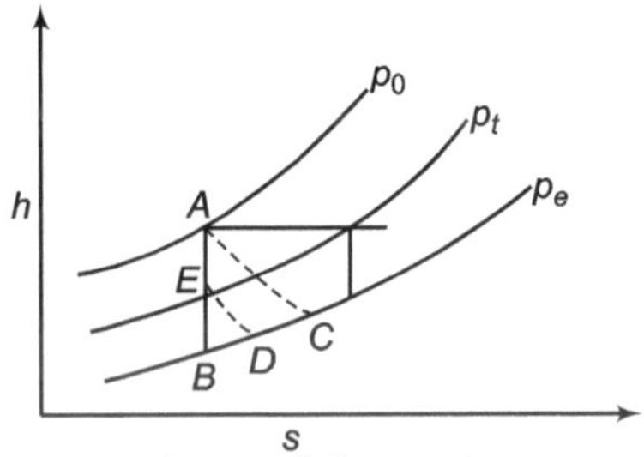

h-s graph for nozzle

A nozzle efficiency is defined as a ratio of the actual kinetic energy in presence of friction *e.g.*, for expansion AC to the ideal one.

$$\eta = \frac{v^2}{2(h_A - h_B)} = \frac{h_A - h_C}{h_A - h_B}$$

and nozzle velocity coefficient $\phi = \sqrt{\eta}$

Nozzle discharge coefficient $C_D = \dfrac{m_{\text{actual}}}{m_{\text{isentropic}}}$

Constant Area Adiabatic Flow

Here, we see the behaviour of the compressible flow of a perfect gas through a constant area duct in the absence of heat transfer, but accounting for frictional effect.

$$m = \rho A v \text{ constant}$$

$$\frac{d\rho}{\rho} = \frac{-dv}{v}$$

$$dh = -v\,dv$$

Mach Number for Different Flow Parameter

Flow parameter	Subsonic $M < 1$	Supersonic
Velocity v	increases (↑)	decreases (↓)
Mach Number M	↑	↓
Static pressure p	↓	↑
Total pressure p_0	↓	↓
Static enthalpy h	↓	↑
Temperature T	↓	↑
Density ρ	↓	↑
Entropy s	↑	↑

Fanno Line

Fanno line show the state of fluid in compressible flow through an adiabatic duct. On enthalpy-entropy diagram.

$$Tds = \frac{v^2 d\rho}{\rho} - \frac{dp}{\rho}$$

$$\left(\frac{\partial p}{\partial \rho}\right)_{ds=0} = v^2 = a^2 \Rightarrow M = 1$$

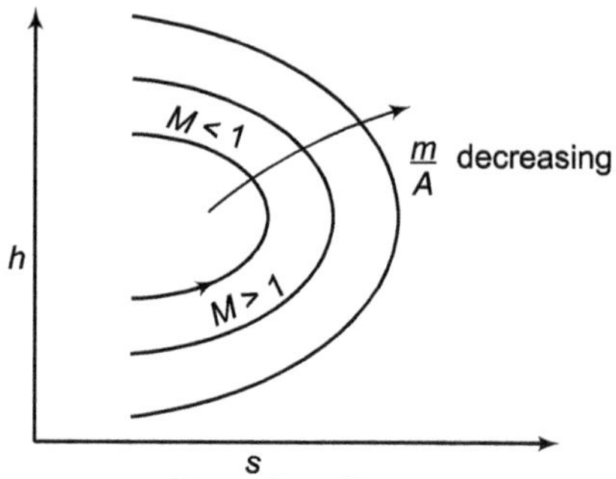

Fanno line diagram

Momentum equation

$$\Rightarrow \int_0^{L_{max}} 4\lambda \frac{dx}{D} = \int_0^1 g(M)\,dM$$

where, λ is frictional factor.

$$4\lambda \frac{L}{D} = \left(\frac{4\lambda L_{max}}{D}\right)_{M_1} - \left(\frac{4\lambda L_{max}}{D}\right)_{M_2}$$

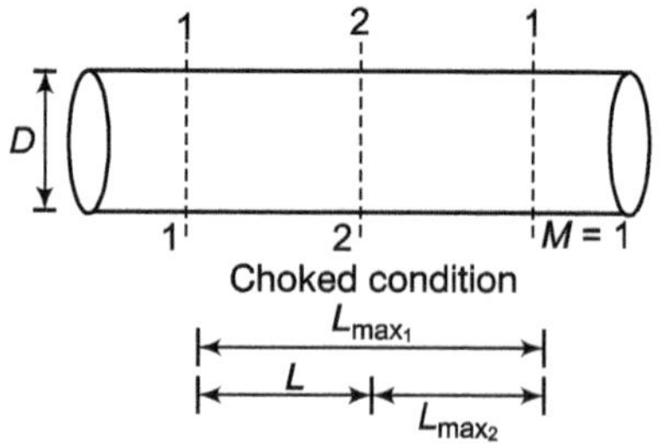

Choked condition

Key Points

- The fanno flow model is often used in the design and analysis of nozzle.
- The fanno flow model is also used extensively with the Rayleigh model.
- These two models intersect at points on the Enthalpy-Entropy and Mach number Entropy diagrams, which is meaningful for many applications.

Constant Area Frictionless Flow with Heat Transfer

A frictionless constant area flow is also known as Rayleigh flow. It is an adiabatic flow where the effect of heat addition or rejection is considered. It can be applied to both compressible and incompressible flow models. *There are some equations in constant area frictionless flow with heat transfer are given below*

Continuity

The continuity equation is given as

$$pV = \text{constant}$$
$$\frac{dp}{p} = -\frac{dV}{V}$$

First Law of Thermodynamics

The first law of thermodynamics states that $dQ = dh$

Second Law of Thermodynamics

The second law of thermodynamics states that

$$dh_0 = dh - Vdp = dh + VdV$$
$$\frac{dV}{V} = \frac{-dp}{pV^2} \quad [\rho V^2 = kpM^2, = \text{specific heat ratio}]$$
$$\frac{dV}{V} = -\frac{1}{kM^2}\frac{dp}{p}$$

Momentum Equation

Impulse function $\quad p + \rho V^2 = \text{constant}$

Equation of State

Equation of state is given as $\quad p = \rho RT$

$$\frac{dp}{p} = \frac{d\rho}{\rho} + \frac{dT}{T}$$

Rayleigh Line

It shows the state of the fluid on Enthalpy-Entropy diagram with constant area frictionless flow with heat transfer. The locus of state point traces a curve known as the Rayleigh line.

$$\frac{dT}{T} = \frac{dp}{p}\left(1 - \frac{1}{kM^2}\right)$$

where, k = specific heat ratio

$$ds = \frac{C_p}{k}\left(\frac{M^2 - 1}{M^2}\right)\frac{dp}{p}$$

$$\frac{dM}{M} = -\frac{1 + kM^2}{2kM^2} \cdot \frac{dp}{p}$$

$$\frac{dp_0}{p_0} = \frac{1 - M^2}{2 + (k-1)M^2} \cdot \frac{dp}{p}$$

Effect of Heat Addition to Constant Area Frictionless Duct Flow

Low parameters	$M < 1$	$M > 1$
Velocity v	increases ↑	decreases ↓
Mach number M	↑	↓
Static pressure p	↓	↑
Static temperature T	(↑) increases for $M < \frac{1}{\sqrt{k}}$ (↑) decreases for $M > \frac{1}{\sqrt{k}}$	
Total pressure p_0	decreases	decreases

Constant Area Isothermal Flow with Friction

Flow parameters	$M < \frac{1}{\sqrt{K}}$ (subsonic)	$M > \frac{1}{\sqrt{K}}$ (subsonic or supersonic)
Velocity v	increases ↑	decreases↓
Mach number M	↑	↓
Static pressure p	↓	↑
Stagnation temperature T_0	↑	↓
Stagnation pressure p_0	↓	(↑) increases for $M < \sqrt{\frac{2}{k+1}}$ (↓) decreases for $M > \sqrt{\frac{2}{k+1}}$

Constant Area Isothermal Flow with Friction

The flow in long pipe lines closely approximates isothermal conditions. The pressure drop in such lines is often large compared to inlet pressure and solution of this problem is outside the limitations of the Darcy equation.

$$\frac{dp}{p} = \frac{-dV}{V} = \frac{-kM^2}{2(1 - kM^2)} 4\lambda \frac{dx}{D} \qquad [\lambda \rightarrow \text{frictional factor}]$$

$$\frac{dT_0}{T_0} = \frac{k(k-1)M^4}{2(1-kM^2)\left(1+\frac{k-1}{2}M^2\right)} 4\lambda \frac{dx}{D}$$

$$\frac{dp_0}{p_0} = \frac{-kM^2}{2(1-kM)^2}\left(\frac{1-\frac{k+1}{2}M^2}{1+\frac{k-1}{2}M^2}\right) 4\lambda \frac{dx}{D}$$

Shocks

Sudden changes occur in the flow properties across a flow discontinuity termed shock.

Strength of a Shock

Strength of a shock wave is defined as the ratio of rise in pressure to upstream pressure.

$$p = \frac{p_2 - p_1}{p_1} = \frac{p_2}{p_1} - 1$$

$$p = \frac{2k}{k+1}(M_1^2 - 1)$$

where, $p_2 - p_1$ = rise in pressure, p_1 = upstream pressure

Normal Shock

The discontinuity is called a normal shock when it is perpendicular to the flow of direction. The Mach number downstream of it is invariably subsonic ($M < 1$).

Oblique Shock

A discontinuity inclined to the flow of direction is called as **oblique shock** downstream of which the Mach number may still be higher than unit.

Governing Equation for Normal Shock

There are following equations for normal shock

$$h_{01} = h_{02} \qquad \text{(for an ideal gas)}$$

$$T_{01} = T_{02}$$

$$\rho_1 V_1 = \rho_2 V_2$$

$$m(V_2 - v_1) = (p_1 - p_2)A$$

Impulse function $\quad p_1 + \rho_1 V_1^2 = p_2 + \rho_2 V_2^2 \; ; ds > 0$

Suffix 1 and 2 refer to conditions before and after shock.

$$\frac{T_1}{T_2} = \frac{1 + \frac{k-1}{2} M_1^2}{1 + \frac{k-1}{2} M_2^2}; \quad \frac{p_1}{p_2} = \frac{\rho_1}{\rho_2} \frac{T_1}{T_2}$$

$$\frac{p_1}{p_2} = \frac{M_2}{M_1} \sqrt{\frac{T_1}{T_2}}; \quad M_2^2 = \frac{M_1^2 + \frac{2}{k-1}}{\frac{2k M_1^2}{k-1} - 1}$$

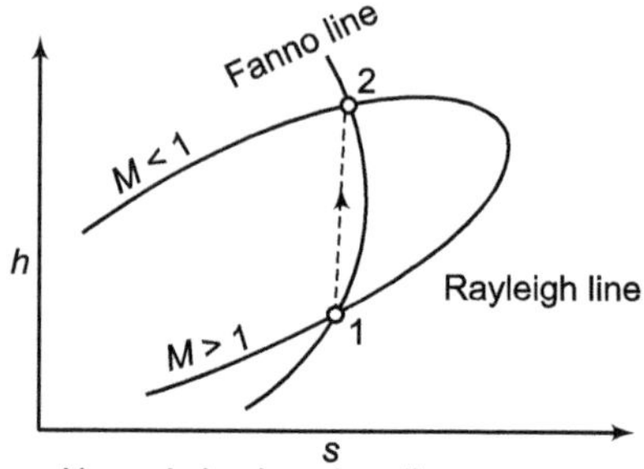

Normal shock on h-s diagram

Flow is always subsonic after a normal shock.

M_2 decreases as M_1 increases for normal shock $M_{2\min}^2 = \frac{k-1}{2k}$

Oblique shocks equations

Consider a supersonic flow at mach M_1 is made to flow along a corner with a not too large deflection angle of v_1 and v_2 are the fluid velocities before and after the shock v_1 makes an angle ε and v_2 an angle β with the shock. ε is known as the **wave angle**.

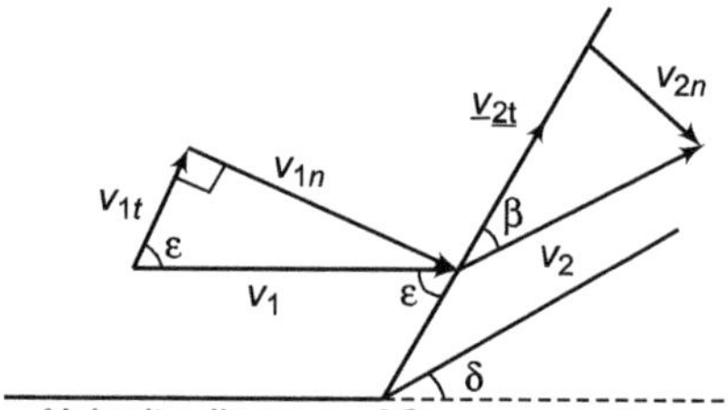

Velocity diagram of Supersonic flow

$$v_{1n} = v_1 \sin \varepsilon$$
$$v_{1t} = v_1 \cos \varepsilon \text{ and } v_{2n} = v_2 \sin \beta$$
$$v_{2t} = v_2 \cos \beta; \ \beta = (\varepsilon - \delta)$$

Mass flow rate, $\rho_1 v_{1n} = \rho_2 v_{2n}; \ \frac{\rho_2}{\rho_1} = \frac{v_1}{v_2} \frac{\sin \varepsilon}{\sin \beta}$

Momentum change, $m v_{1t} = m v_{2t}; \ v_{1t} = v_{2t}$

$$\frac{v_1}{v_2} = \frac{\cos \beta}{\cos \varepsilon}; \ \frac{\rho_2}{\rho_1} = \frac{\tan \varepsilon}{\tan \beta}; \ \frac{p_2}{p_1} = 1 + k M_1^2 \sin^2 \varepsilon \left(1 - \frac{\rho_1}{\rho_2}\right)$$

$$C_p T_1 + \frac{v_{1n}^2}{2} = C_p T_2 + \frac{v_{2n}^2}{2}$$

$$\frac{1}{\tan \delta} = \left(\frac{k+1}{2} \frac{M_1^2}{M_1^2 \sin^2 \varepsilon - 1} - 1\right) \tan \varepsilon$$

where, k = specific heat ratio, ε = wave angle

M_1 = Mach number, δ = angle with respect to ground

9

Refrigeration *and* Air Conditioning

Refrigeration Cycle

Refrigeration is the process in which heat is removed from a body enclosed space so that its temperature is reduced and then maintained at the temperature below the surrounding temperature. The working substance which is used to produce refrigeration is known as **refrigerant**.

Heat Pump

Heat pump is used to remove heat from a body at lower temperature and transfer this heat to a body having high temperature on the expense of external work supplied.

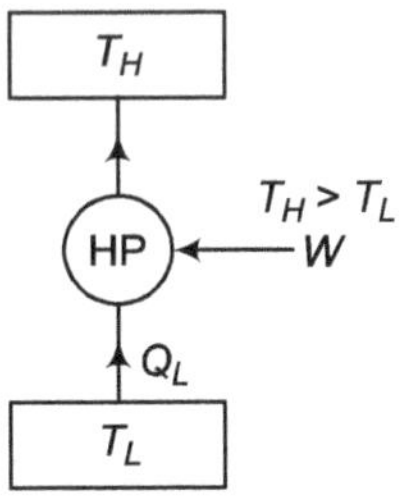

Heat pump process

Coefficient of Performance;

$$\text{COP} = \frac{\text{Desired effect}}{\text{Work input}}$$

Here, the desired effect is to give heat to higher temperature body.

$$\text{COP} = \frac{Q_1}{W} = \frac{Q_H}{Q_H - Q_L}; \text{COP} = \frac{T_H}{T_H - T_L}$$

$$(\text{COP})_{\text{pump}} = \frac{1}{1 - \frac{T_L}{T_H}}$$

where, T_H = high temperature, T_L = lower temperature

Refrigerator

Working principle of refrigerator is same as heat pump, but its aim to keep cool the body whose temperature is below atmospheric temperature.

$$\text{COP} = \frac{\text{Desired effect}}{\text{Work input}}$$

$$\text{COP} = \frac{Q_L}{W}$$

$$\text{COP} = \frac{Q_L}{Q_H - Q_L} = \frac{T_L}{T_H - T_L}$$

$$(\text{COP})_R = \frac{1}{\left(\frac{T_H}{T_L} - 1\right)}$$

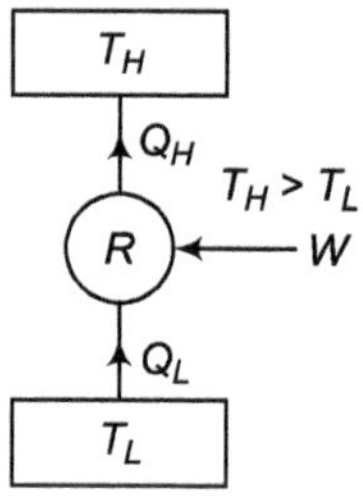

Principle of refrigerator

where T_H and T_L are higher and lower temperature respectively.

$$(\text{COP})_P = (\text{COP})_R + 1$$

where, Q_L = heat at lower temperature
Q_H = heat at higher temperature

Unit of Refrigeration

It has standard unit of TR (Ton of Refrigeration). 1 TR (one ton of refrigeration) means capacity to freeze one ton of water form and at 0°C in 24 h.

$$1\ \text{TR} = 211\ \text{kJ/min}$$
$$= 3.517\ \text{kW}$$
$$= 200\ \text{BTHU/min}$$

Key Points

- A refrigerator is a device which, operating in a cycle, maintains a body at a temperature lower than the temperature of the surroundings.
- There is performance parameter in a refrigerator cycle, called the coefficient of performance.
- Heat pump is a device which operating in a cycle, maintains a body at a temperature higher than the temperature of the surroundings.

Reversed Carnot Cycle

When the processes of Carnot cycle is reversed, then it is called reversed Carnot cycle.

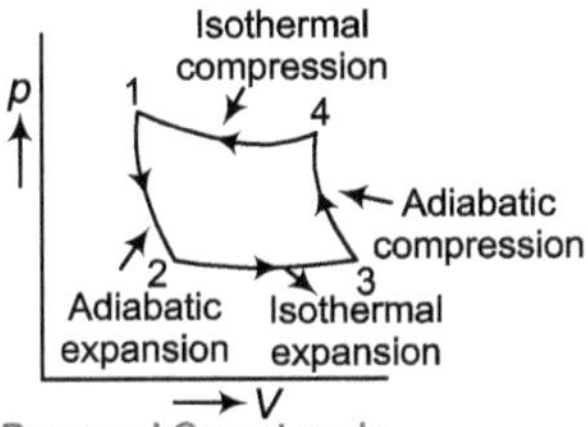

Reversed Carnot cycle on *p-V* coordinate

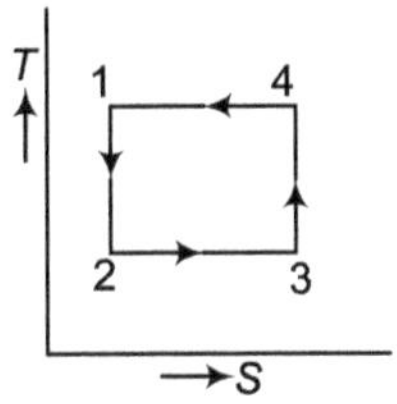

Reversed Carnot cycle on *T-S* coordinate

$$\text{COP} = \frac{\text{Heat absorbed}}{\text{Work input}}$$

$$= \frac{T_2(S_3 - S_2)}{[T_1(S_4 - S_1) - T_2(S_3 - S_2)]} = \frac{T_2(S_3 - S_2)}{(T_1 - T_2)(S_4 - S_1)} \quad \begin{pmatrix} S_4 = S_3 \\ S_1 = S_2 \end{pmatrix}$$

$$\text{COP} = \frac{T_2}{T_1 - T_2}$$

where T_2 and T_1 are the temperature at section 2 and 1 respectively and S_1, S_2, S_3 and S_4 are the entropy at 1, 2, 3 and 4 respectively.

Air Refrigeration System

In air refrigeration, Bell Coleman cycle is used which works on reversed Brayton cycle while Brayton cycle is used for gas turbine.

Brayton cycle (Gas turbine)	Reversed Brayton cycle (Air refrigeration system, Bell Coleman cycle)
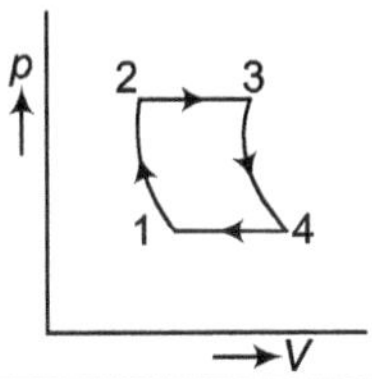	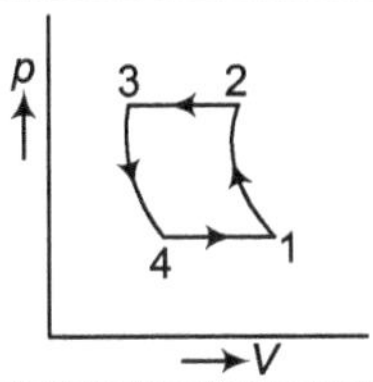
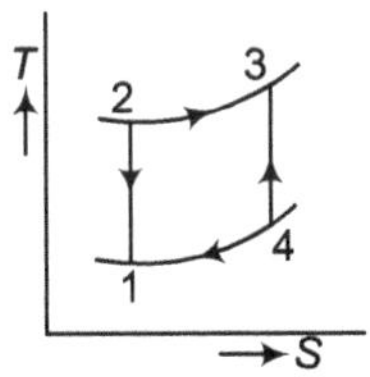	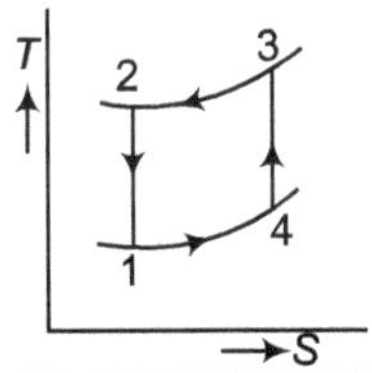

COP of reversed Brayton cycle,

$$\text{COP} = \frac{\text{Refrigeration effect}}{W_{\text{input}}} (R_E)$$

$$= \frac{\text{Heat absorbed by air in cold chamber}}{W_{\text{input}}}$$

$$= \frac{m_a C_p (T_1 - T_4)}{[m_a C_p (T_2 - T_3) - m_a\, C_p\, (T_1 - T_4)]}$$

$$\text{COP} = \frac{(T_1 - T_4)}{(T_2 - T_3) - (T_1 - T_4)}$$

Let pressure ratio is r_p, then

$$r_p = \frac{p_2}{p_1} = \frac{p_3}{p_4}$$

$$\text{COP} = \frac{T_1}{T_2 - T_1} = \frac{T_4}{T_3 - T_1} \qquad \left\{\frac{T_2}{T_1} = \frac{T_3}{T_4}\right\}$$

$$\text{COP} = \frac{1}{(r_p)^{\frac{\gamma - 1}{\gamma}} - 1}$$

where, γ = index, r_p = pressure ratio

Vapour Compression Refrigeration System

In this system, refrigerant (like NH_3, *R*-12, *R*-11etc) changes its phase from liquid to vapour and *vice-versa.* When liquid changes into vapour, it produces refrigerating effect by absorbing latent heat from cold chamber. This process is known as **evaporation**.

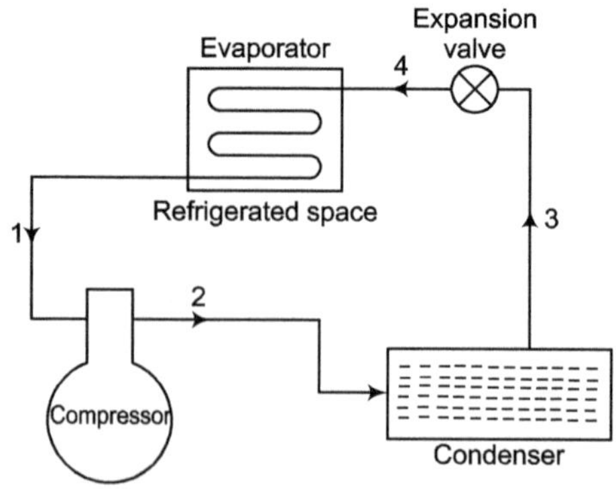

Vapour compression refrigeration system

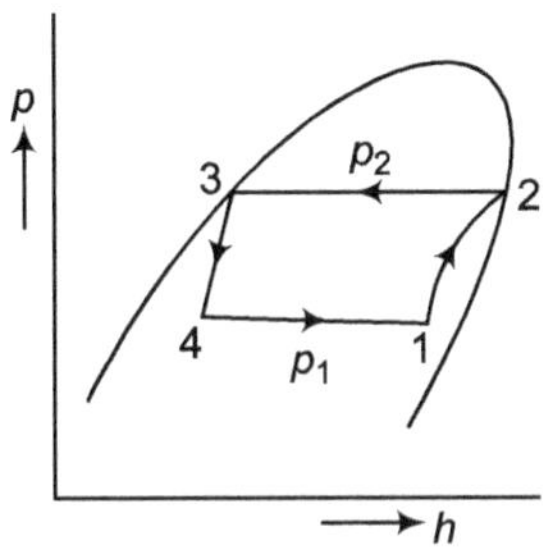

P-h graph

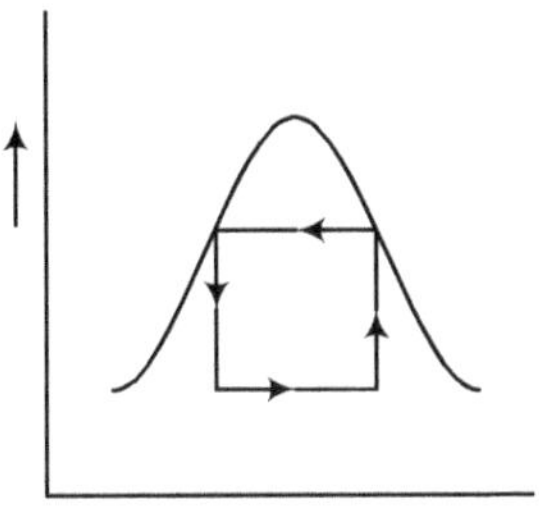

T-S graph

$1 \rightarrow 2 \rightarrow$ compression process	(Isentropic)
$2 \rightarrow 3 \rightarrow$ condensation process	(At constant pressure)
$3 \rightarrow 4 \rightarrow$ expansion process	(Isenthalpic process)
$4 \rightarrow 1 \rightarrow$ evaporation process	(At constant pressure)

$$\text{COP} = \frac{\text{Refrigeration effect}}{\text{Work done by compressor}}$$

$$\text{COP} = \frac{h_1 - h_4}{h_2 - h_1}$$

where h_1, h_2 and h_4 are the enthalpy at section 1, 2 and 4 respectively.

Key Points

- COP of refrigerator working between two fixed temperature T_L and T_H ($T_H > T_L$) will have fixed COP.
- A reversible heat engine can work like a refrigerator or a heat pump.
- Vapour compression refrigeration cycle

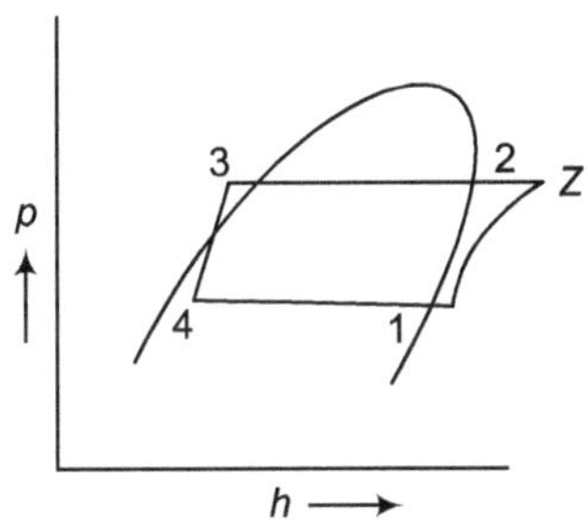

- Entropy $$S_2' = S_2 + C_p \ln \frac{T_2}{T_2'}$$

$$h_2 = h_2' + C_p (T_2 - T_2')$$

For isentropic process, $S_1 = S_2, S_3 = S_4$

Compressor

A compressor is used in refrigerator to increase the pressure of gas.To compress the refrigerant' we use reciprocating compressor generally in the vapour compression refrigeration system.

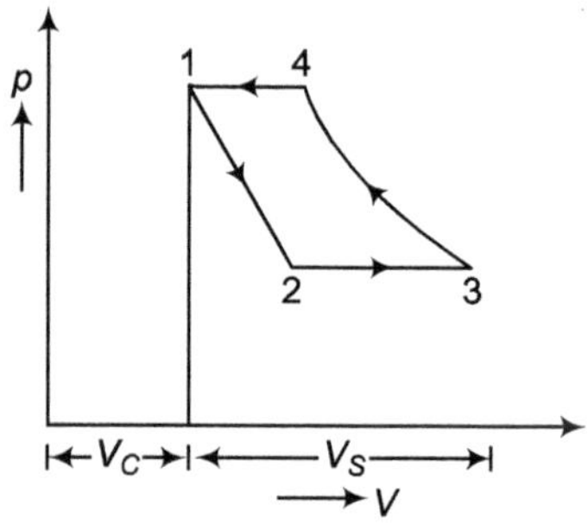

p–V curve

Stroke volume $V_s = \frac{\pi}{4} d^2 L$

where, d = diameter of compressor

L = stroke length

Volume flow rate of refrigerant

$$V = n \cdot \frac{\pi}{4} d^2 L \, \eta_v \cdot N$$

where, N = rotational speed

η_V = volumetric efficiency

n = number of cylinders

$$\eta_V = \frac{\text{Actual volume compressed}}{\text{Stroke volume}}$$

$$\eta_V = \frac{V_c + V_s - V_1}{V_s}$$

where, V_c = clearance volume

$$c = \text{clearance ratio} = \frac{V_c}{V_s} = 1 + \frac{V_c}{V_s} - \frac{V_1}{V_s}$$

$$= 1 + c - c\left(\frac{p_2}{p_1}\right)^{\frac{1}{\gamma}}$$

$$= 1 + c - c\,(r_p)^{\frac{1}{\gamma}}$$

$\left(\text{Compression ratio } r_p = \frac{p_2}{p_1}; \; \gamma = \text{Index of compression}\right)$

Vapour Absorption System

This system utilises low grade form of heat energy to produce cold in compression of vapour compression system. In vapour absorption system, compressor is replaced by absorber pump and generator.

COP of an Ideal Vapour Absorption System

Schematic diagram of energy transfer in vapour absorption system is shown below

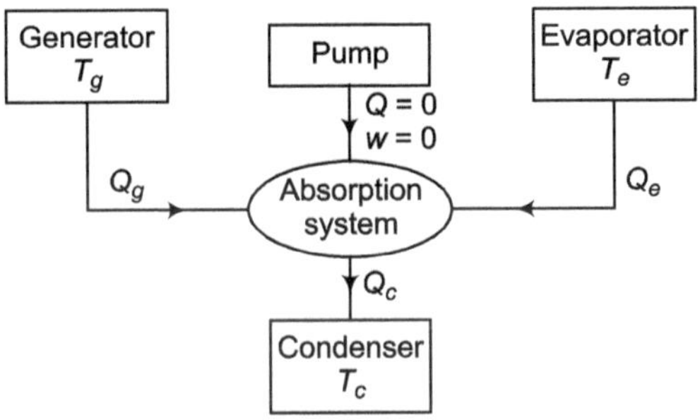

Energy transfer in vapour absorption system

Q_g = heat supplied to refrigerator in generator

Q_e = heat is absorbed by refrigerator in evaporator

Q_c = heat is rejected to condenser or atmosphere

Coefficient of Performance

$$Q_g + Q_e = Q_c$$

For reversible cycle net entropy change is size $\frac{-Q_g}{T_g} \frac{Q_e}{-T_e} + \frac{Q_c}{T_c} = 0$

$$\text{COP} = \left(\frac{T_e}{T_c - T_e}\right)\left(\frac{T_g - T_c}{T_g}\right)$$

$$\text{COP} = (\text{COP})_{\text{Carnot}} \times (\eta_{\text{Carnot}})$$

$$\text{COP of Carnot refrigerator} = \frac{T_e}{T_c - T_e}$$

$$\text{Efficiency of Carnot engine} = \frac{T_g - T_c}{T_g}$$

Here, T_g = generator temperature

T_c = absorber temperature (Generally surrounding temperature)

T_e = evaporator temperature.

Refrigerant

Refrigerant absorbs heat by vaporisation and reject it through condensation in the condenser and so produces refrigerating effect.

Designation of Refrigerant

In International Standards, refrigerants are designated as R followed by some numerals.

For a hydrocarbon Chemical formula is $C_r H_S F_t Cl_y$

- If $S + t + y = 2r + 2$, designation of refrigerant is $R\,(r-1)\,(S+1)t$
- If $S + t + y = 2r$, designation of refrigerant is $R1\,(r-1)\,(S+1)\,t$

For an Inorganic Refrigerant

Designation is R (700 + molecular weight)

e.g., $NH_3 \rightarrow R\,(700 + 17) \Rightarrow R\,717$

Desirable Properties of a Good Refrigerant

- Low boiling point
- Low freezing point
- High latent heat of vaporisation
- High thermal conductivity
- Low viscosity
- Low specific volume in vapour state

Uses and Properties of Some Refrigerant

- In absorption refrigeration system, we use NH_3— H_2O or water lithium bromide or water lithium chloride. In NH_3— H_2O system NH_3 is refrigerant while H_2O is transport medium. In water lithium bromide or water lithium chloride water is refrigerant.
- In ice plant, primary refrigerant is NH_3 or freon while secondary refrigerant is brime solution.
- Freezing temperature of freon-12 is $-157.8°C$, so it is not used for producing temperature below the freezing temperature.
- In dehydrator, silica gel (molecular sieve $\rightarrow SiO_2 + K_2O + NO_2 + Al_2O_3$) is used with calcium chloride. (Not use sodium chloride).
- Hermetically sealed reciprocating compressor is commonly used in room air conditioner.
- For passenger aircraft air is used as refrigerant.
- Refrigerant CO_2 has lowest specific volume.

Psychrometric Charts *and* Its Applications

For comfort air conditioning, properties of air should be known so that we can control the environment in the space. The study of properties of ambient air is known as psychrometric.

Psychrometric terms

There are following terms used in psychrometric study.

Pressure

Pressure is the effect of a force applied to a surface, Pressure is the amount of force acting per unit area. The symbol of pressure is p

Total pressure of air $p = p_v + p_a$

where, p_v = partial pressure of water vapour

p_a = partial pressure of dry air

Partial pressure of water vapour

$$p_v = (p_{vs})_{wb} \frac{-[p-(p_v)_{wb}](t_{db}-t_{wb})}{1527.4 - 1.3\,t_{wb}}$$

or

$$p_v = (p_{vs})_{wb} - \frac{1.8 \times p\,(t_{db}-t_{wb})}{2700}$$

where, $(p_{vs})_{wb}$ = saturation pressure of water vapour corresponding to wet bulb temperature

p = atmospheric pressure of moist air

T_{wb} = wet bulb temperature

t_{db} = dry bulb temperature.

Dry Air

It is the mixture of gases. Generally O_2 and N_2 make up the major part of the combustion.

It consists 21% O_2 and 79% N_2 by volume.

It consists 23% O_2 and 77% N_2 by mass.

Moisture

The water vapour present in the air is called moisture.

Moist Air

It is the mixture of dry air and water vapour.

Absolute Humidity

The weight of water vapour present in unit volume of air is called absolute humidity.

$$\text{Absolute humidity} = \frac{\text{Weight of water vapour}}{\text{Volume of air (mixture)}}$$

Specific Humidity

It is the ratio of mass of water vapour in air (mixture) to mass of dry air in air mixture.

$$\text{Specific humidity} = \frac{\text{Mass of water vapour in air}}{\text{Mass of dry air in air}}$$

$$w = 0.622$$

$$\frac{p_v}{p_s} = 0.622\left(\frac{p_v}{p - p_v}\right)$$

Relative Humidity

It is the ratio of mass of water vapour in a given volume to mass of water vapour in the same volume of saturated air at same temperature.

$$\phi = \frac{m_v}{m_s},\ \phi = \frac{p_v}{p_{vs}}$$

where m_v and m_s are the mass of water vapour and mass of saturated air

p_v and p_{vs} are the partial pressure of water vapour and partial pressure of water vapour in saturated air at same temperature for saturated air relative humidity is 1 or 100%.

Degree of Saturation (μ)

It is the ratio of mass of water vapour in a sample of air to mass of water vapour in saturated air at the same temperature.

$$\mu = \frac{m_v / m_a}{m_{vs} / m_a},\ \mu = \frac{w}{w_s}$$

where w and w_s are specific humidities of air and saturated air.

Dry Bulb Temperature (DBT)

It is the normal temperature of air measured by an ordinary thermometer.

Wet Bulb Temperature

It is the temperature measured by a thermometer when bulb is covered by wet cloth.

Wet Bulb Depression (WBD) = DBT − WBT

Dew Point Temperature (DPT)

It is temperature at which water vapour in air start condensing.

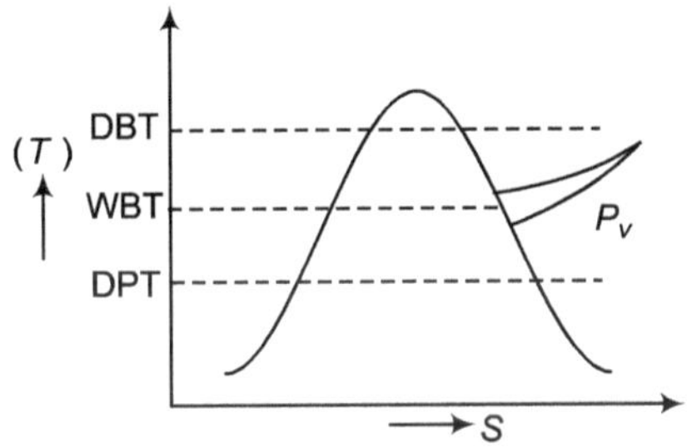

DBT, WBT and DPT graphs

- For unsaturated air, DBT > WBT > DPT
- For saturated air, DBT = WBT = DPT

Dew Point Depression

It is the difference between Dry Bulb Temperature (DBT) and Dew Point Temperature (DPT). *i.e.*, DPD = DBT − DPT

Enthalpy of Air (h)

Enthalpy of air is defined as

$$H = H_a + H_V,\ H = m_a h_a + m_v h_v$$

$$h_a = C_p a(t)$$

$$h_v = 2500 + 1.88\,t$$

t = Dry Bulb Temperature in °C, $h = h_a + \dfrac{m_v}{m_a} h_v$

$$h = C_{pa}\, t_{db} + w\ [2500 + 1.80°\, t_{dht}]\ \text{kJ/kg dry air}$$

or $$h = C_{pm}\, t_{db} + w\,(h_g - C_{pv}\, t_{db})$$

where, $$C_{pm} = C_{pa} + C_{pv} w$$

C_{pa}, C_{pv}, C_{pm} are the specific heat of dry air water vapour and moist air respectively.

$h_g \rightarrow$ enthalpy of saturated water vapour at DPT in kJ/kg.

Psychrometric Chart

- Enthalpy is more or less constant with wet bulb temperature. Enthalpy lines are also downward from left to right.
- Constant specific volume lines are more steeper inclined lines than wet bulb temperature lines or enthalpy lines.
- Adiabatic lines on psychrometric chart are represented along constant enthalpy lines.

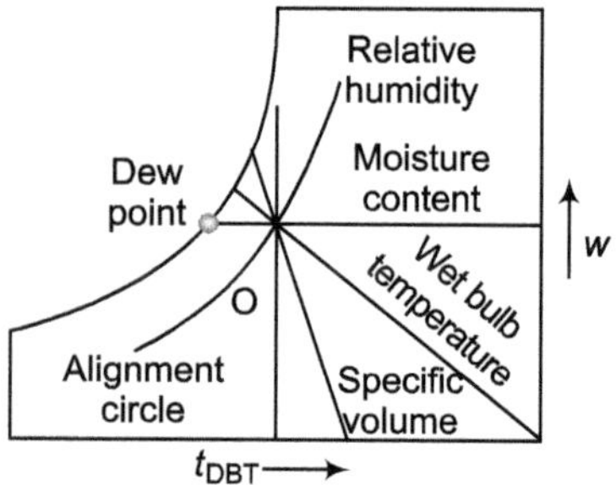

Psychrometric chart

Sensible Heating

It is process of heating air at constant *w*.

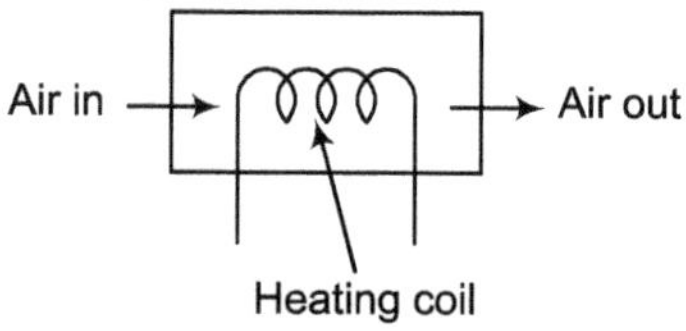

Heating process

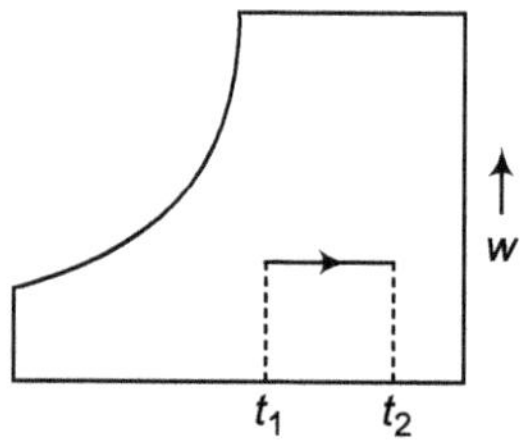

Sensible heating

$$Q = h_2 - h_1 = C_{pm} (t_{dbt_2} - t_{dbt_2})$$

where, c_{pm} = specific heating moist

Sensible Cooling

It is process of cooling air at constant *w*. For sensible cooling, the coil temperature must be greater than dew point temperature and less than dbt.

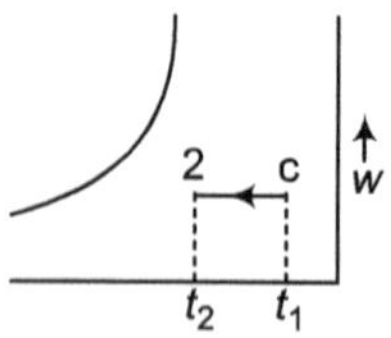

Sensible cooling

Humidification

The addition of moisture to air at a constant Dry Bulb Temperature (DBT) is knows as humidification.

It is process of increasing w at constant DBT.

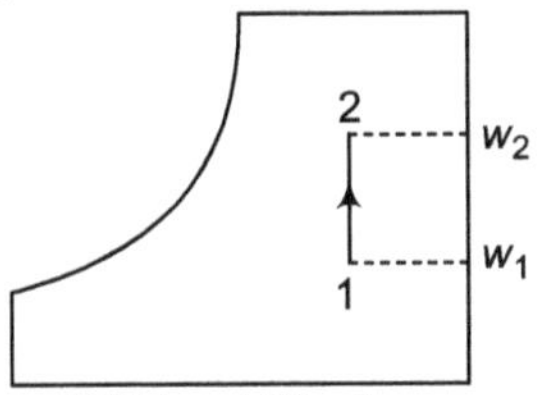

DBT humidification

Dehumidification

The removal of moisture from air at a constant Dry Bulb Temperature is called dehumidification. It is process of decreasing w at constant DBT.

Heating with Humidification

It is process in which air heated with humidification. The heating and humidification process line (1 – 2) as shown in figure below. In this process, Dry Bulb Temperature and specific humidity of incoming air increases.

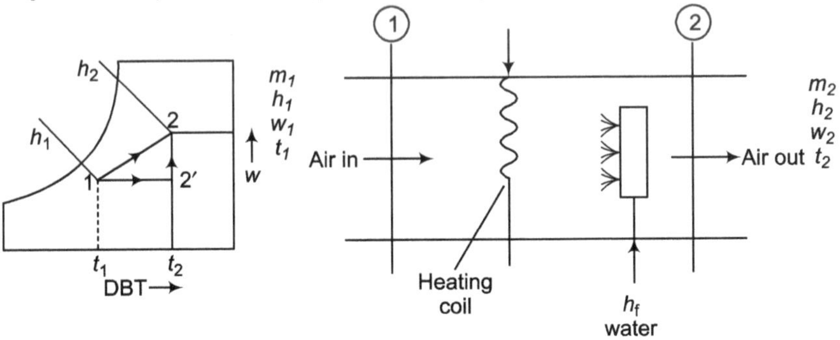

Heating with humidification process

Addition of heat $\quad Q_{1-2} = m_a[(h_1 - h_2) - (w_1 - w_2)\,h_f]$

$$h_2' - h_1 = \text{sensible heat } (S_H)$$

$$h_2 - h_2' = \text{latent heat } (L_H)$$

$$\text{Sensible heat factor} = \frac{\text{Sensible heat added}}{\text{Total heat added}}$$

$$= \frac{S_H}{S_H + L_H} = \frac{h'_2 - h_1}{h_2 - h_1}$$

Heating with Dehumidification

In this process, air is heated with dehumidification.

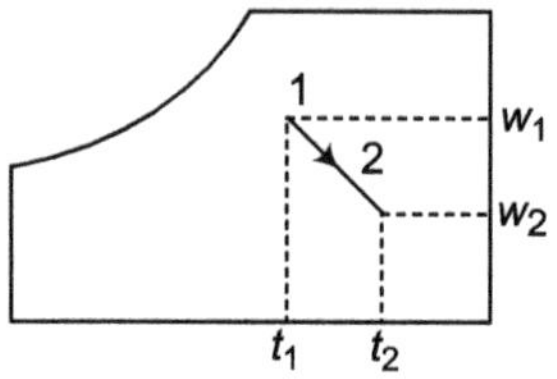

Heating with dehumidification graph

Cooling with Humidification

In this process, an air is cooled with humidification.

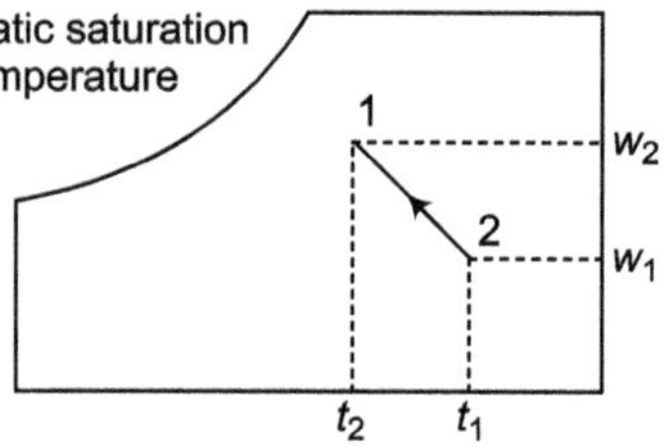

Cooling with humidification graph

Cooling with Dehumidification

In this process air is cooled below the down point with dehumidification. It is necessary to maintain. In a cooling with dehumidification, both the humidity ratio and temperature of most air decrease. Some water vapour is condensed in the form of liquid water, called a condensate. This process is shown by a curve in psychrometric chart.

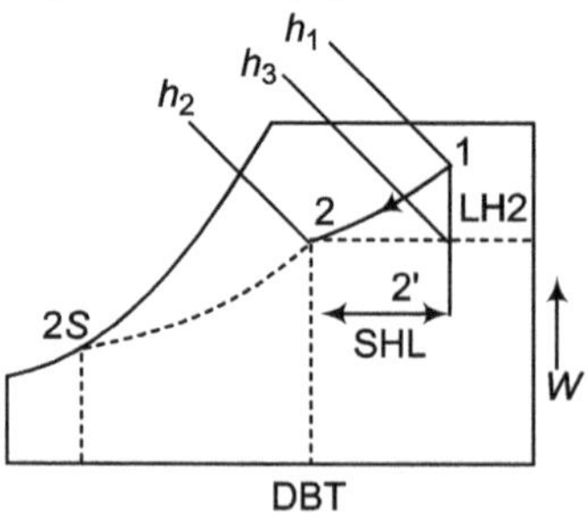

Dehumidification

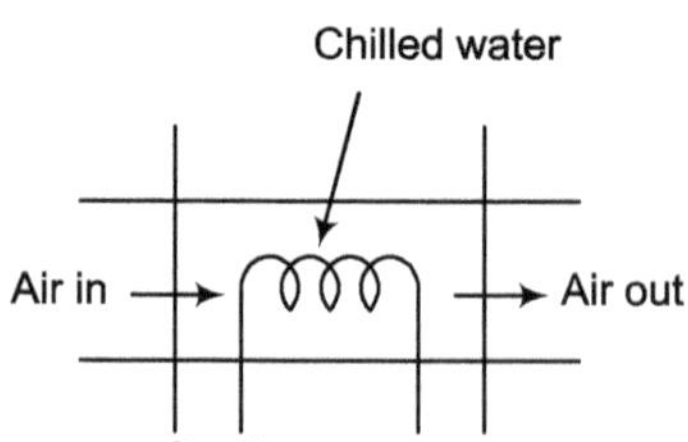

Cooling system

Total heat load on cooling coil,

$$Q_{1-2} = m_a [(h_1 - h_2) - (w_1 - w_2) h_f]$$

Since, $(w_1 - w_2) h_f \approx 0$

So, $$Q_{1-2} = m_a (h_1 - h_2)$$

$$\text{Sensible heat factor} = \frac{h_2' - h_2}{h_1 - h_2}$$

where, h_1 and h_2 = enthalpy of moist air entering and leaving the coil

m_a = mass flow rate of the condensate

h_2' = enthalpy of condensate.

Adiabatic Mixing of Air Streams

Streams of air at different condition are mixed as shown in figure.

By mass balance,

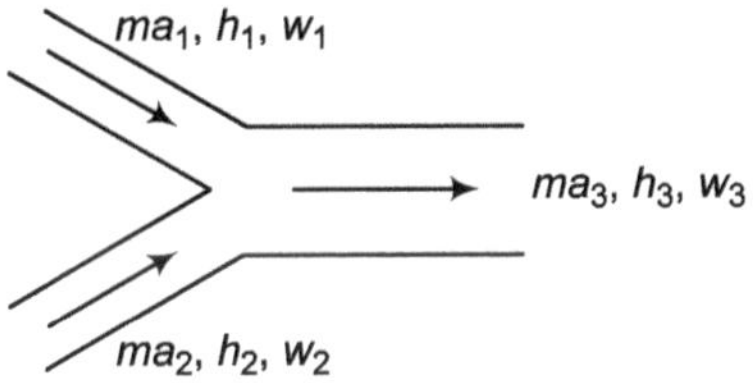

Mixing of air streams

$$m_{a_1} + m_{a_2} = m_{a_3}$$

$$m_{a_1} w_1 + m_{a_2} w_2 = m_{a_3} w_3 \; (m_a w = m_v)$$

$$m_{a_1} h_1 + m_{a_2} h_2 = m_{a_3} h_3$$

$$\frac{m_{a_1}}{m_{a_2}} = \frac{h_3 - h_2}{h_1 - h_2}$$

$$\frac{m_{a1}}{m_{a_2}} = \frac{t_3 - t_2}{t_1 - t_3}$$

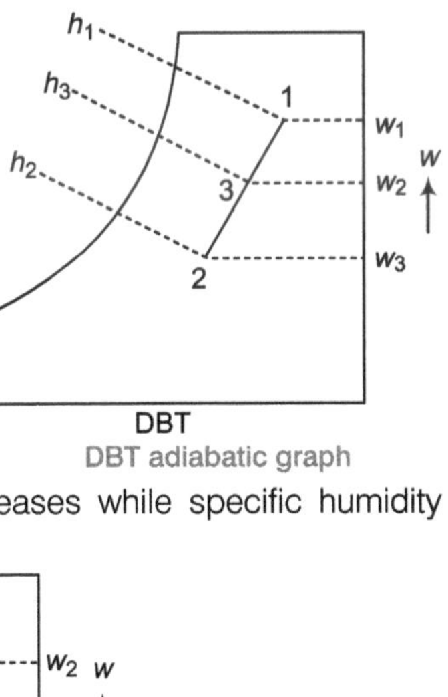

DBT adiabatic graph

Chemical Dehumidification

In chemical humidification DBT of air increases while specific humidity decreases.

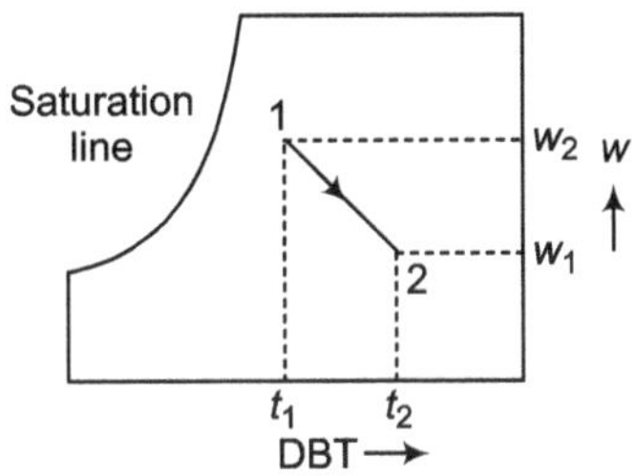

Chemical humidification

Adiabatic Evaporative Cooling

In adiabatic evaporative cooling, the heat removed in lowering the Dry Bulb Temperature of the air is absorbed by the moisture which evaporates and raises the humidity of the air.

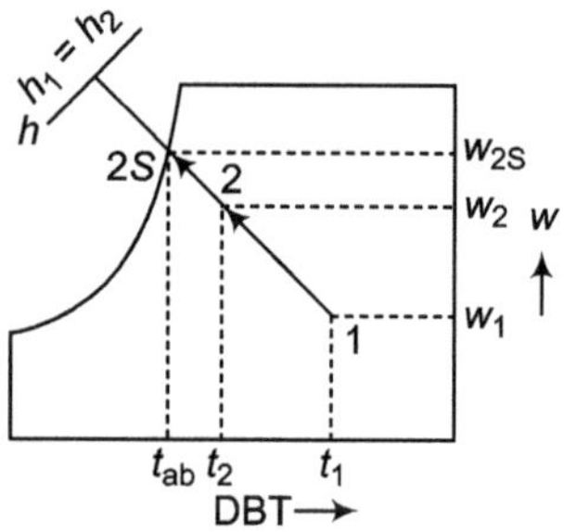

Evaporative cooling graph

By Pass Factor

Let Dry Bulb Temperature of coil surface is t_{coil} and air temperature at entry and exit is t_1 and t_2 respectively.

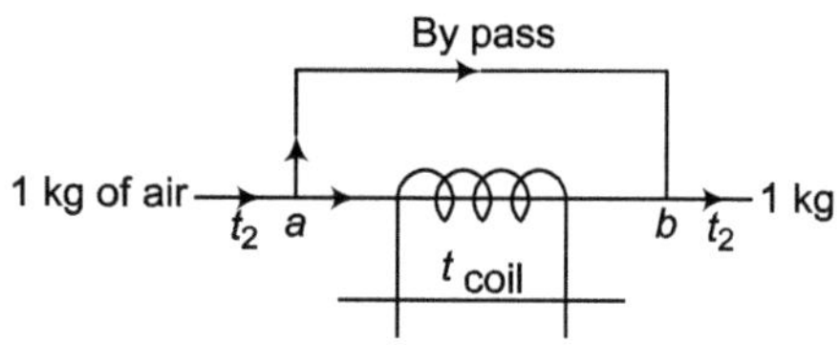

By pass factor system

By pass factor,

$$m = \frac{t_2 - t_{coil}}{t_1 - t_{coil}} \left(\frac{\text{Temperature drop not achieved}}{\text{Temperature drop that could be achieved}} \right)$$

Coil Efficiency

It is determined by the mass coming in perfect contact of coil.

$$\eta = 1 - m = 1 - \text{BPF}$$

From figure,

$$\eta = \frac{t_2 - t_1}{t_3 - t_1}$$

$$\text{BPF} = \frac{t_3 - t_2}{t_3 - t_1}$$

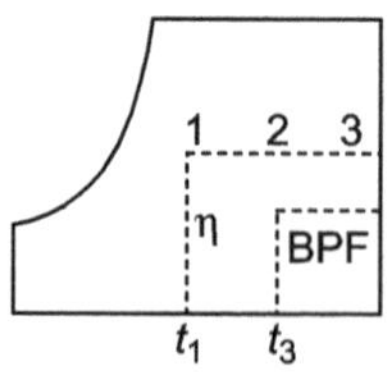

Cooling efficiency diagram

Note If by pass factor is low then coil efficiency will be better.

Summer Air Conditioning with Ventilation Air

- GSHF → the line joining of inlet and exit condition of an A/C equipment
- RSHF → the line joining supply conditions with inside room condition.

The point of interaction of GSHF and RSHF will give supply condition to room.

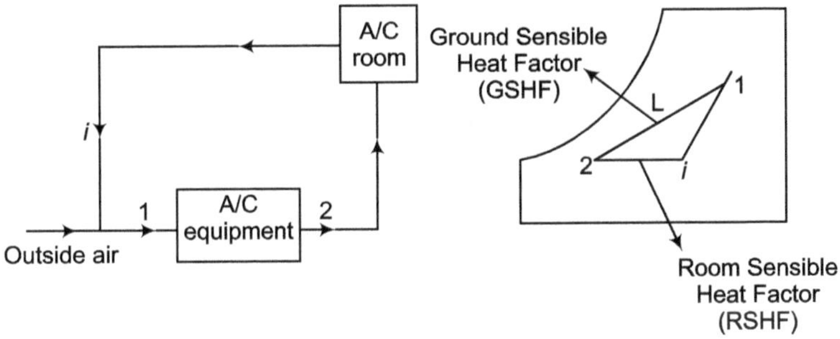

Air conditioning system

Apparated Dew Point (ADP)

The temperature at which cooling and humidification lines meet saturated curve.

$$\text{BPF} = \frac{t_2 - t_{\text{ADP}}}{t_1 - t_{\text{ADP}}}$$

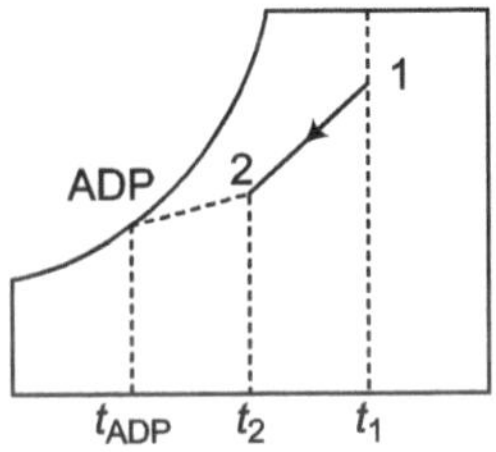

ADP diagram

Comfort Chart

Relative humidity lines are taken from chart. It is found that at an effective temperature of 20°C, 99% of people where comfortable in winter and in summer this temperature is found to be 21.6°C. The comfortable relative humidity conditions are 50% to 60% and hence shaded region shows comfort zone for year round conditioning.

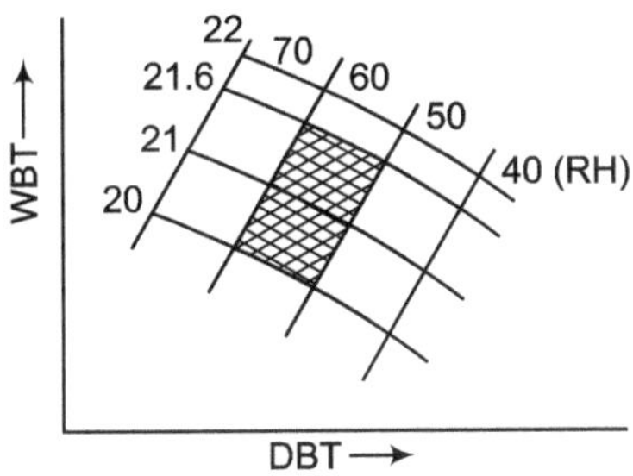

Comfort chart

10

Internal Combustion Engine

Basic Concepts Related to Engine

An engine is a device which transforms one form of energy into another form. **Heat engine** is a device which transforms the chemical energy of a fuel into thermal energy and utilises this thermal energy to perform useful work. *Heat engine can be classified as follows*

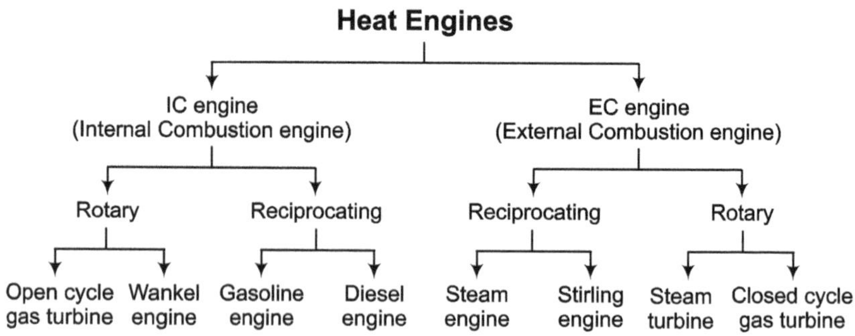

Classification of heat engines

EC and IC Engines

External Combustion (EC) engines are those in which combustion takes place outside the engine whereas in Internal Combustion (IC) engines, combustion takes place within the engine.

Classification of IC Engine

IC engines can be classified as below

Two Strokes Engine

The thermodynamics cycle is completed in two strokes of the piston or in one revolution of the crankshaft. Thus, one power stroke is obtained in each revolution of the crankshaft with uniform turning moment and light flywheel.

Four Strokes Engine

The thermodynamics cycle is completed in four strokes of the piston or in two revolutions of the crankshaft. Thus, one power stroke is obtained in every two revolutions of the crankshaft with non-uniform turning moment and heavier flywheel.

Spark Ignition (SI) Engine

It works on Otto cycle or constant volume heat addition cycle. A gaseous mixture of fuel air introduced during the suction stroke. A carburettor and an ignition system are necessary. Modern engines have gasoline injection.

Compression Ignition (CI) Engine

It works on diesel cycle or constant pressure heat addition cycle. Fuel is injected directly into the combustion chamber at high pressure at the end of the compression stroke. A fuel pump and injector are necessary.

Performance Parameters of Engine

The basic parameters for measuring the performance of engines are given below

Indicated Thermal Efficiency

The total power developed by combustion of fuel in the combustion chamber is called **indicated power.**

This efficiency give an idea of the output generated by the engine with respect to heat supplied in the form of fuel. The indicated thermal efficiency can be defined as indicated power.

$$\eta_{th} = \frac{\text{Indicated power}}{\text{Energy in fuel per second}} = \frac{i_p}{m_f \times C_v}$$

Here, i_p = Indicated power

C_v = Calorific value of fuel

m_f = Mass of fuel/second

Brake Thermal Efficiency

It is defined as ratio of heat equivalent to brake power to heat energy supplied to engine in fuel.

$$\eta_{bth} = \frac{b_p}{m_f \times C_v}, \; b_p \rightarrow \text{Brake power}$$

Mechanical Efficiency

It is the ratio of brake power to indicated power.

$$\eta_m = \frac{b_p}{i_p} = \frac{b_p}{b_p + f_p}, \; f_p \rightarrow \text{frictional power}$$

Volumetric Efficiency

Volumetric efficiency is a measure of the success with which the air supply, and thus the charge, is inducted into the engine. It is a very important parameter, since it indicates the breathing capacity of the engine.

$$\eta_V = \frac{m_a}{\rho_a V_{dis\frac{N}{2}}}, \; \eta_V = \frac{m_a}{\rho_a V_d}$$

Here, ρ_a = Inlet density of air

m_a = Mass of air

V_{dis} = Displaced volume of air

$V_{dis\frac{N}{2}}$ = Volume flow rate for four strokes engine

V_d = Volume flow rate of air.

Relative Efficiency or Efficiency Ratio

It is the ratio of indicated thermal efficiency to thermal efficiency to theoretical cycle at same compression ratio.

$$\eta_{rel} = \frac{\text{Actual thermal efficiency}}{\text{Air standard efficiency}}$$

Key Points

- The indicated thermal efficiency is based on the i_p (input or indicated power).
- The brake thermal efficiency is based on the b_p (brake power).
- Mechanical efficiency takes into account the mechanical losses in an engine.

Mean Effective Pressure (p_m)

The parameter mean effective pressure is measured by the formula which is given below

$$i_p = \frac{p_{im} \, L \, AnK}{60 \times 1000}$$

where, i_p = Indicated power in kW
p_{im} = Indicated mean effective pressure (N/m^2)
L = Length of the stroke (m)
A = Area of the piston (m^2)
N = Speed in revolutions per minute (rpm)
n = Number of power strokes $N/2$ for four strokes and N for two strokes engine
K = Number of cylinders.

Mean Piston Speed

Mean piston speed can be given by

$$\bar{S}_p = 2LN$$

Specific Fuel Consumption

It is the ratio of amount of kilogram of fuel used per hour by engine to power produced by the engine.

$$S_{fc} = \frac{\text{Fuel consumption per unit time}}{\text{Power}}$$

Fuel Air (F/A) or Air Fuel Ratio (A/F)

The ratio of actual fuel air ratio to stoichiometric fuel air ratio is called equivalence ratio.

$$\oint = \frac{\text{Actual fuel air ratio}}{\text{Stoichiometric fuel air ratio}}$$

Accordingly, $\oint = 1$ for stoichiometric mixture

$\oint < 1$ for lean mixture

$\oint > 1$ for rich mixture.

Cycles *and* their Analysis

A cycle is defined as a repeated series of operations occuring in a certain order. In other words, any process or series of processes whose end states are identical is termed as cycle.

Assumptions of Air Standard Cycle

- Assume working medium as perfect gas (Ideal gas).
- No change in mass of working period.
- Reversible processes.
- Supply of heat from a constant high temperature source (not from chemical reactions) and rejection of heat to a constant low temperature source.
- Not heat losses from the system to surroundings.
- Constant specific heats through the cycle.

Cycles

There are different types of cycles which are given below

Carnot Cycle

Carnot cycle is an ideal cycle in which all the processes constituting a reversible cycle. This cycle consists of two isothermal and two reversible adiabatic processes.

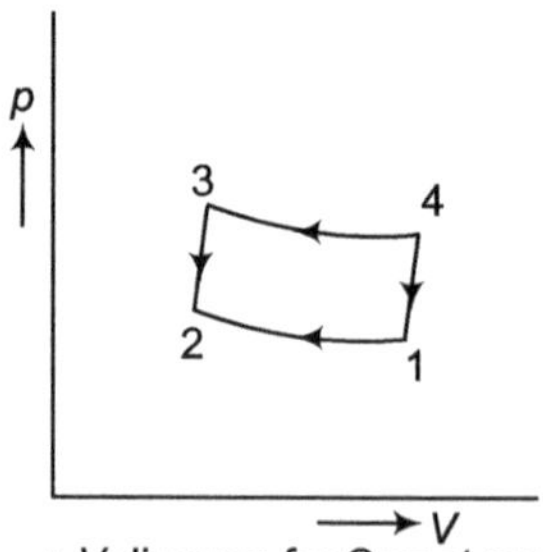

p-V diagram for Carnot cycle

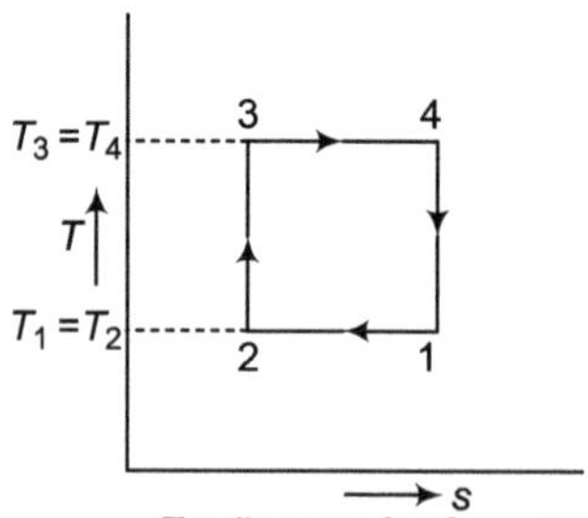

T-s diagram for Carnot cycle

- Process 1 - 2 is isothermal compression process
- Process 2 - 3 is adiabatic process (isentropic process)
- Process 3 - 4 is isothermal expansive process
- Process 4 - 1 is adiabatic process (isentropic process)

All the processes are reversible.

$$\eta_c = 1 - \frac{T_1}{T_3}$$

where, T_1 = Temperature at process 1

T_3 = Temperature at process 3

Carnot cycle has a low mean effective pressure because of very low work output.

The Stirling Cycle

It is modified form of Carnot cycle with high mean effective pressure. It consists of two isothermal and two constant volume processes.

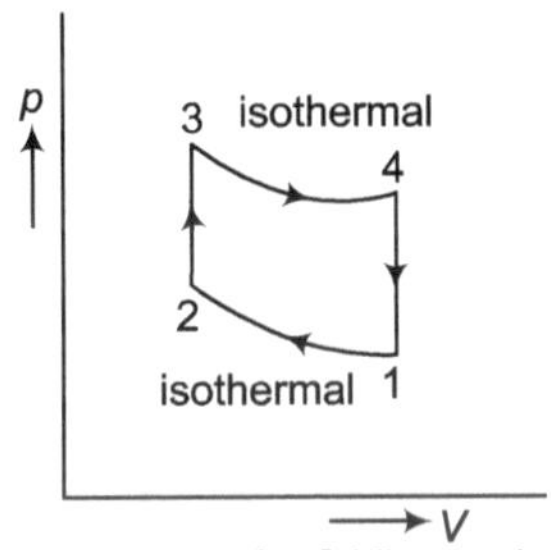

p-V diagram for Stirling cycle

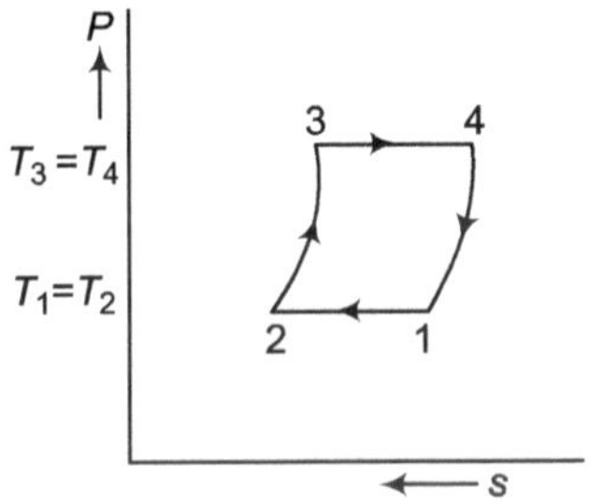

T-s diagram for Stirling cycle

- Isothermal process $\begin{pmatrix} 1 \to 2 \\ 3 \to 4 \end{pmatrix}$
- Constant volume process $\begin{pmatrix} 2 \to 3 \\ 4 \to 1 \end{pmatrix}$

$$\eta_s = 1 - \frac{T_1}{T_3} \qquad \text{(same as Carnot cycle)}$$

- In practical Stirling cycle, efficiency will be less than Carnot efficiency.
- For heat exchange efficiency

$$\eta_s = \frac{R\,(T_3 - T_1)\log_e r}{RT_3 \log_e r + (1 - E)\,C_v\,(T_3 - T_1)}$$

where, E = Heat exchanger effectiveness.

The Ericsson Cycle

It consists of two isothermal and two constant pressure processes. The processes are

- Process 1-2 is reversible isothermal compression
- Process 2-3 is constant pressure heat addition
- Process 3-4 is reversible isothermal expansion
- Process 4-1 is constant pressure heat rejection

The advantage of the Ericsson cycle is its small pressure ratio for given ratio of maximum to minimum specific volume with higher mean effective pressure. Ericsson cycle is applied by a gas turbine employing a large number of stages with heat exchangers, insulators and reheaters.

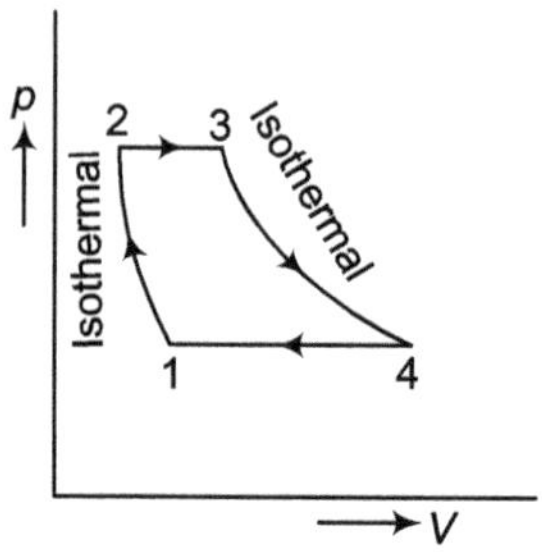

p-*V* diagram for Ericsson cycle

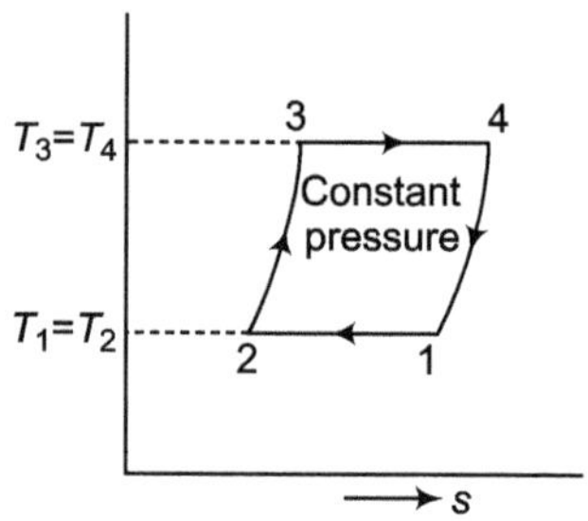

T-*s* diagram for Ericsson cycle

The thermal efficiency of Ericsson cycle is given by,

$$\eta_{th} = \frac{T_H - T_L}{T_H} = \left[1 - \frac{T_L}{T_H}\right]$$

The Otto Cycle

The otto cycle is a set of processes used by spark ignition internal combustion engines (2-stroke or 4-stroke cycles). Petrol engine works on the otto cycle. In petrol engine fuel burnt by spark ignition. It consists two

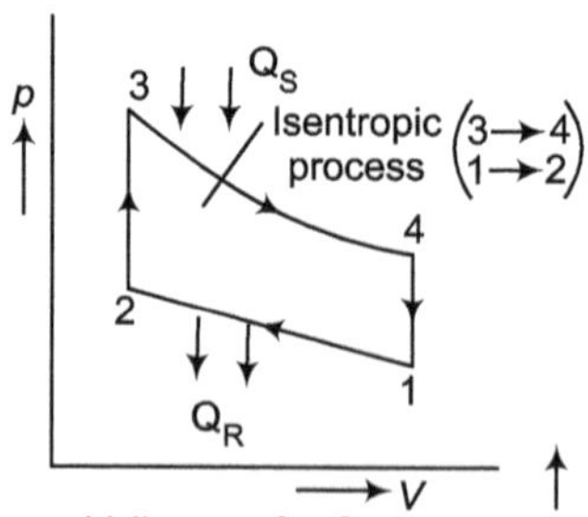

p-*V* diagram for Otto cycle

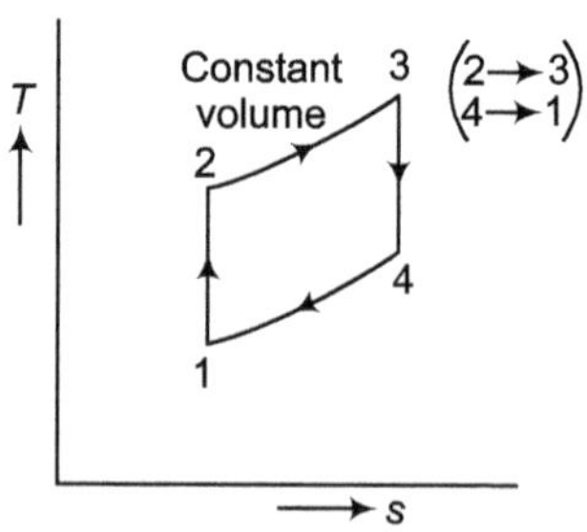

T-*s* diagram for Otto cycle

isentropic processes (reversible adiabatic) and two constant volume proceses.

- Process 1-2 is reversible adiabatic compression of air
- Process 2-3 is heat addition at constant volume.
- Process 3-4 is reversible adiabatic expansion of air
- Process 4-1 is heat rejection at constant volume.

$$\eta_{otto} = \frac{Q_S - Q_R}{Q_S} \Rightarrow \eta_{otto} = 1 - \frac{T_1}{T_2} = 1 - \frac{1}{(r)^{\gamma-1}}$$

$$\text{Compression ratio } r = \frac{V_1}{V_2} = \frac{V_4}{V_3}$$

where, Q_S = Heat supplied during the isothermal process
Q_R = Heat rejected during the isothermal expression process

Work Output (*W*)

It can be measures as $(Q_S - Q_R) = \eta_{otto} Q_S$

where, Q_S = Heat supplied
Q_R = Heat rejected

Heat supplied $Q_S = m C_v (T_3 - T_2)$
Heat rejected $Q_R = mC_v (T_4 - T_1)$

$$W = \frac{p_1 V_1}{\gamma - 1}(r_p - 1)(r^{\gamma-1} - 1)$$

where, γ = Specific heat ratio,
r_p = Pressure ratio

Pressure ratio $r_p = \frac{p_3}{p_2} = \frac{p_4}{p_1}$

Value of γ (Ratio of Specific Heat)

For monoatomic gas $\gamma = 1.67$
For air $\gamma = 1.40$
For exhaust gas $\gamma = 1.30$

Mean Effective Pressure

It is defined as the ratio of the net work done to the displacement volume of the piston.

$$p_m = \frac{\text{Work output}}{\text{Swept volume}}$$

$$\text{Swept volume} = V_1 - V_2 = V_2 (r - 1)$$

$$p_m = \frac{p_1 r (r_p - 1)(r^{(\gamma-1)} - 1)}{(\gamma - 1)(r - 1)}$$

where, p_m = Mean effective pressure
r_p = Pressure ratio
γ = Specific heat ratio.

The Diesel Cycle

This cycle is used in diesel engine. In diesel engine, fuel is burnt by compressing the air upto high pressure. It consists of two isentropic proceses, one constant volume process and one constant pressure process.

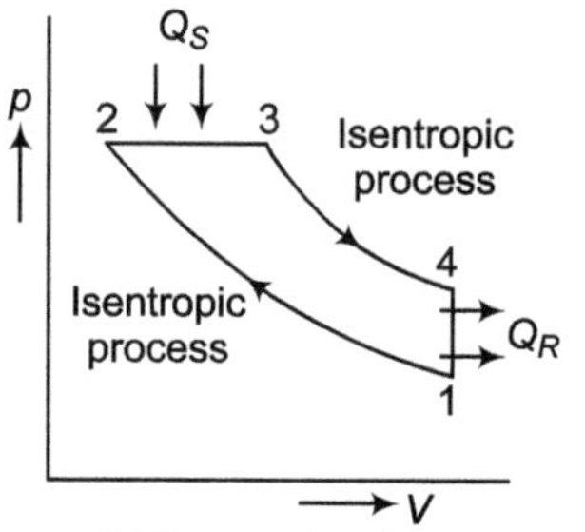

p-V diagram for diesel cycle

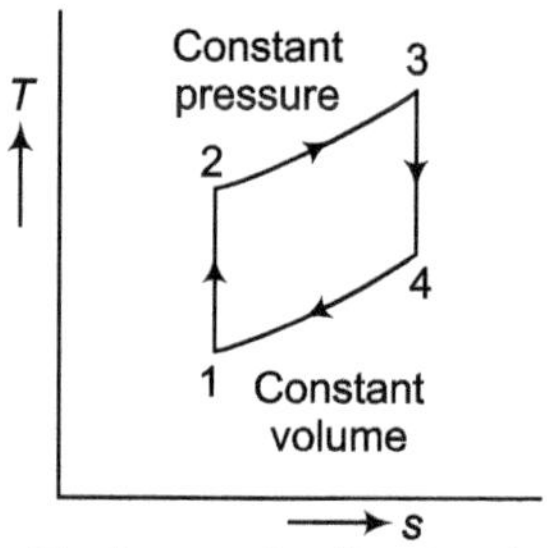

T-s diagram for diesel cycle

- Process 1-2 is reversible adiabatic compression
- Process 2-3 is constant pressure heat addition
- Process 3-4 is reversible adiabatic compression
- Process 4-1 is constant volume heat rejection

$$\eta_{diesel} = 1 - \frac{1}{r^{(\gamma-1)}}\left[\frac{r_c^{\gamma} - 1}{\gamma (r_c - 1)}\right]$$

$$\text{Volume ratio or cut-off ratio } r_c = \frac{V_3}{V_2} = \frac{T_3}{T_2}$$

$$\text{Compression ratio } r = \frac{V_1}{V_2}$$

Work Output

Work output can be calculated by the following relation.

$$W = p_2 (V_3 - V_2) + \frac{p_3 V_3 - p_4 V_4}{\gamma - 1} - \frac{p_2 V_2 - p_1 V_1}{\gamma - 1}$$

$$W = \frac{p_1 V_1 r^{(\gamma-1)} [\gamma (r_c - 1) - r^{1-\gamma} (r_c^{\gamma} - 1)]}{\gamma - 1}$$

Mean effective pressure

$$p_m = \frac{p_1 [\gamma\, r^{\gamma} (r_c - 1) - r\,(r_c^{\gamma} - 1)]}{(\gamma - 1)(r - 1)}$$

where, r = Compression ratio
r_c = Cut-off ratio.

The Dual Cycle

In this cycle, heat is added partly at constant volume and partly at constant pressure. It consists of two isentropic processes, two constant volume processes and one constant pressure process.

- Process 1-2 is reversible adiabatic compression
- Process 2-3 is constant volume heat addition
- Process 3-4 is constant pressure heat addition
- Process 4-5 is reversible adiabatic expansion
- Process 5-1 is constant colume heat rejection

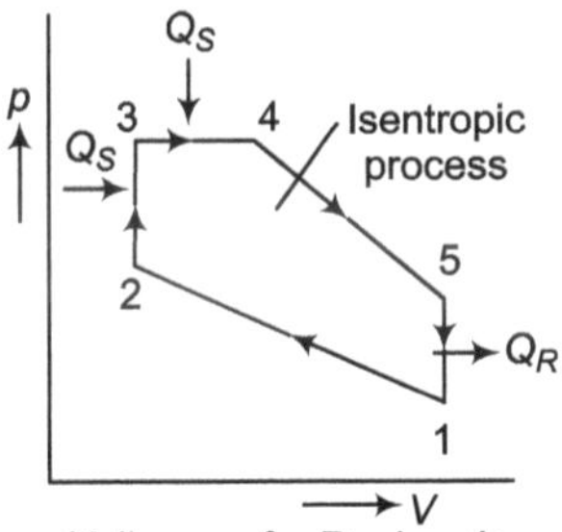

p-*V* diagram for Dual cycle

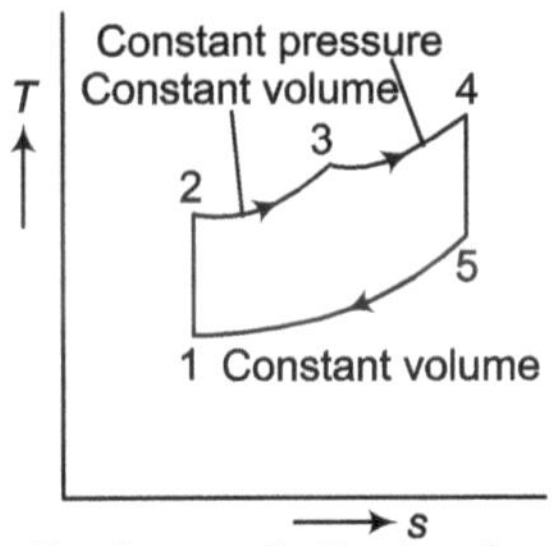

T-*s* diagram for Dual cycle

- Efficiency $\eta = 1 - \frac{1}{r^{(\gamma-1)}}\left[\frac{r_p\, r_c^{\gamma} - 1}{(r_p - 1) + r_p\gamma\,(r_c - 1)}\right]$

- Work output

$$W = \frac{p_1 V_1}{\gamma - 1}[\gamma\, r_p r^{\gamma-1}(r_c - 1) + r^{\gamma-1}(r_p - 1) - (r_p r_c^{\gamma} - 1)]$$

- Mean effective pressure

$$p_m = p_1\left[\frac{\gamma\, r_p r^{\gamma}(r_c - 1) + r^{\gamma}(r_p - 1) - r\,(r_p r_c^{\gamma} - 1)}{(\gamma - 1)(r - 1)}\right]$$

Comparison of Efficiencies in Cycles

- For same compression ratio and heat addition

$$\eta_{otto} > \eta_{dual} > \eta_{diesel}$$

- For same compression ratio and heat rejection

$$\eta_{otto} > \eta_{dual} > \eta_{diesel}$$

- Same peak pressure, peak temperature and heat rejection

$$\eta_{diesel} > \eta_{dual} > \eta_{otto}$$

- Same maximum pressure and heat input

$$\eta_{diesel} > \eta_{dual} > \eta_{otto}$$

- Same maximum pressure and work output

$$\eta_{diesel} > \eta_{dual} > \eta_{otto}$$

The Lenoir Cycle

It consists of one constant volume heat addition, one isentropic expansion and one constant pressure heat rejection.

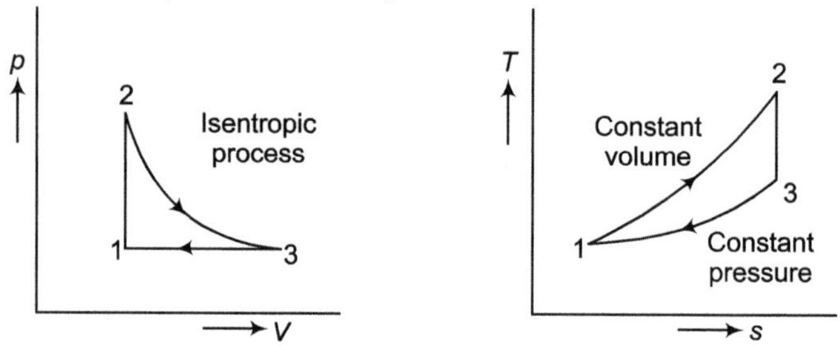

p-V diagram for Lenoir cycle   *T-s* diagram for Lenoir cycle

- Insentropic process (2 → 3)
- Constant volume process (1 → 2)
- Constant pressure process (3 → 1)

The Lenoir cycle is used for pulse jet engines

$$\eta_{Lenoir} = \frac{Q_S - Q_R}{Q_R}$$

Heat supplied, $Q_S = mC_V (T_2 - T_1)$

Heat rejected $Q_R = mC_p (T_3 - T_1)$

- Thermal efficiency $\eta_{th} = 1 - \gamma \left(\frac{(r_p)^{\frac{1}{\gamma}} - 1}{r_p - 1} \right)$

- Pressure ratio, $r_p = \frac{p_2}{p_1} = \frac{T_2}{T_1}$

The Atkinson Cycle

The Atkinson cycle is an ideal cycle for Otto engine exhausting to a gas turbine. In this cycle, the isentropic expansion (3 – 4) of an Otto cycle is further allowed to proceed to the lowest cycle pressure so as to increase the work output.

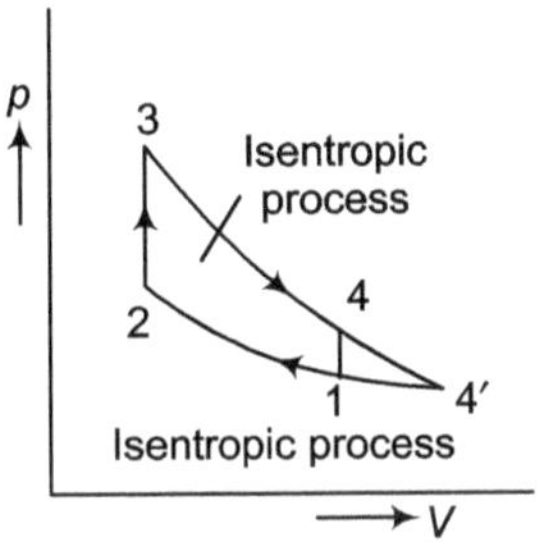

p-V diagram for Atkinson cycle

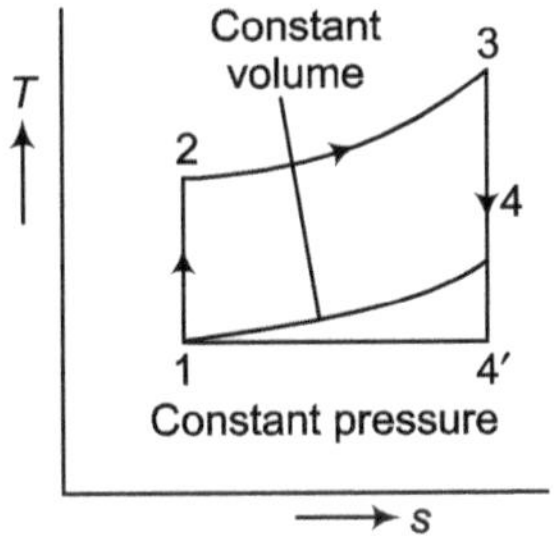

T-s diagram for Atkinson cycle

- Process 1-2 is reversible adiabatic compression
- Process 2-3 is constant volume heat addition
- Process 3-4 is reversible adiabatic expansion (V_3 to V_4)
- Process 4-1 is constant pressure heat rejection

$$\eta_{Atkinson} = 1 - \gamma \left[\frac{e - r}{e^{\gamma} - r^{\gamma}} \right]$$

$$\left(\frac{p_3}{p'_4} \right)^{\frac{1}{\gamma}} = \left(\frac{V'_4}{V_3} \right) = e \qquad \text{(expansion ratio)}$$

$$\left(\frac{p_2}{p_1} \right)^{\frac{1}{\gamma}} = \left(\frac{V_1}{V_2} \right) = r \qquad \text{(compression ratio)}$$

- Heat supplied $Q_3 = C_V (T_3 - T_2)$
- Heat rejected $Q_2 = C_p (T_4 - T_1)$
- Net work done $W = C_V (T_3 - T_2) - C_p (T_4 - T_1)$

The Brayton Cycle

It is a theoretical cycle for a gas turbine. It consists of two reversible adiabatic or isentropic processes and two constant pressure processes.

- Insentropic process ($1 \rightarrow 2$)
- Constant volume process ($4' \rightarrow 1$)
- Constant pressure process ($4 \rightarrow 1$)

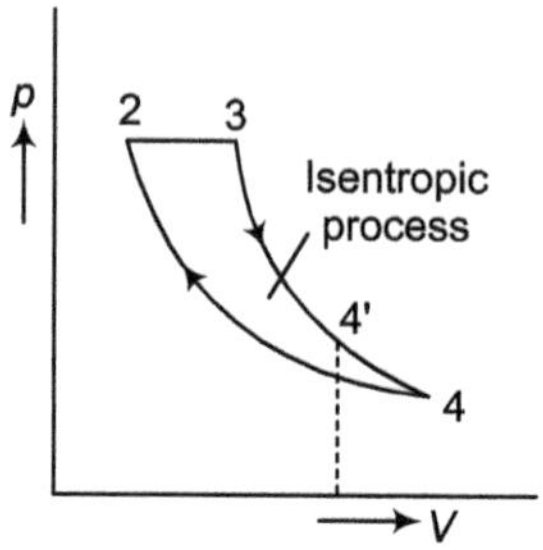

p-V diagram for Brayton cycle

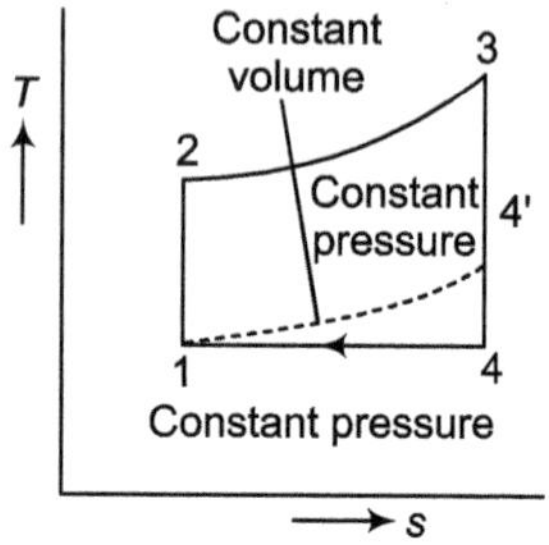

T-s diagram for Brayton cycle

$$\eta_B = 1 - \frac{1}{(r_p)^{\frac{\gamma-1}{\gamma}}};\ r = \frac{V_1}{V_2}$$

$$\frac{T_3}{T_4} = \left(\frac{p_3}{p_4}\right)^{\frac{\gamma-1}{\gamma}} = \left(\frac{p_2}{p_1}\right)^{\frac{\gamma-1}{\gamma}} = \left(\frac{V_1}{V_2}\right)^{\gamma-1} = r^{(\gamma-1)}$$

Actual Cycles

The actual cycle efficiency is much lower than the air standard efficiency due to various losses occuring in the actual engine operation.

The major losses are due to

- Variation of specific heats with temperature.
- Dissociation of the combustion products.
- Progressive combustion.

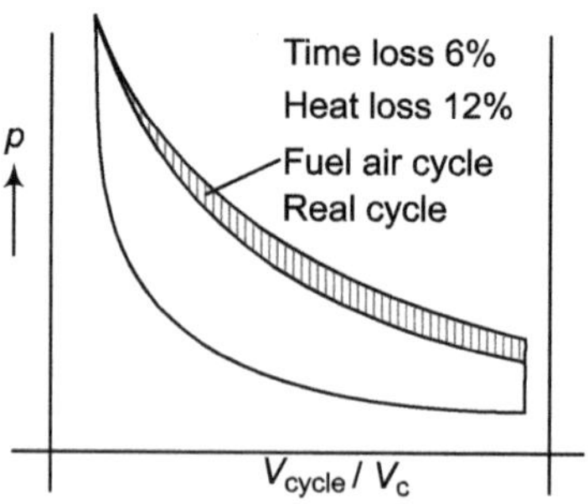

p-V diagram for actual cycle

- Incomplete combustion of fuel.
- Heat transfer into the walls of the combustion chamber.
- Blowdown at the end of exhaust process.
- Gas exchange process.

Fuels

Fuel is any material that stores potential energy in a form that can be practicably released and used as heat energy. Fuel is also a substance that is changed in some way to produce heat, electricity or other form of energy.

Rating of Engine Fuels

The parameters for rating of engine fuels are given below

Octane Number

The octane number is defined as the percentage by volume of iso-octane in a mixture of iso-octane and normal heptane which exactly matches the knocking intensity of the fuel in a standard engine under a set of standard operating conditions. Two reference fuels iso-octane ($C_8 H_{18}$) and normal heptane (C_7H_{16}) have octane number 100 and 0 respectively.

$$\text{Octane number} = 100 + \frac{PN - 100}{3}$$

Here, PN = Performance Number

Cetane Number

Cetane number is defined as the percentage by volume of normal cetane in a mixture of normal cetane and α-methyl napthalene which has the same ignition characteristics (ignition delay) as the test fuel when combustion is carried out in a standard engine under specified operating conditions.

The reference fuels are normal cetane ($C_{16} H_{34}$) and α-methyl naphthalene ($C_{11} H_{10}$) with a cetane number of 100 and 0.

Types of Fuels

Some types of fuels are given below

Solid Fuels

Compare to gaseous and liquid fuels, solid fuels are quite difficult to handle and storage and feeding are quite combustion.

Liquid Fuels

Liquid fuels are preferred for IC engines because they are easy to store and have reasonably good calorific value. The three principal commercial type of fuels are benzyl, alcohol and petroleum products.

Chemical Structure of Petroleum

Family of hydrocarbons	General formula	Molecular structure	Saturated, unsaturated	Stability
Paraffin	C_nH_{2n+2}	Chain	Saturated	Stable
Olefin	C_nH_{2n}	Chain	Unsaturated	Unstable
Nepthalene	C_nH_{2n}	Ring	Saturated	Stable
Aromatic	C_nH_{2n-6}	Ring	Highly Unsaturated	Most unstable

Alcohol

Methanol (methyl alcohol) and ethanol (ethyl alcohol) are two kind of alcohols that seem most promising. It is a high octane fuel with anti-knock index number of over 100. So, it can run more efficiency by using higher compression ratio. Alcohol have low surface content of fuel and overall less emission compared to gasoline.

Main disadvantage of alcohol is having low energy content *i.e.*, the calorific value of fuel is almost half.

Gaseous Fuels

Gaseous fuels are best suited for IC engine since physical delay is almost zero. However, as fuel displaces equal amount of air, the engines may have poor volumetric efficiency.

Alternative Fuels

Some alternative fuels are given below

Hydrogen

Hydrogen fuel is a zero - emission fuel which uses electrochemical cells or combustion in internal engines, to power vehicles and electric devices.

Advantages of Hydrogen Fuel

- No emission of CO or HC in the exhaust. So, exhaust are H_2O, N_2 and NO_x.
- High energy content per volume when stored as a liquid.

Disadvantages of Hydrogen Fuel

- Requirement of heavy, bulky fuel storage both in vehicle and at the service stations.
- High NO_x emission because of high flame temperature.
- High fuel cost.

Compressed Natural Gas (CNG)

Natural gas composed chiefly of methane, can be compressed to a liquid and used as a substitute for other traditional liquid fuels.

Advantages of CNG

- High octane number, high flame speed with high compression ratio.
- Low engine emissions (low CO, HC and NO_x emissions).

Disadvantages of CNG

- Low energy density resulting in low engine performance.
- Low engine volumetric efficiency.

Liquified Petroleum Gas (LPG)

LPG is a mixture of propane and butane, both of which are easily compressible gases under standard atmospheric conditions. It is commonly used for cooking and space heating.

Two types of LPG *are used for automobile engines*

I. Propane II. Butane

Propane and butane are obtained from oil and gas wells.

Advantages of LPG

- LPG contains less carbon than petrol. So, LPG powered vehicle produces 50% less CO per km.
- Due to higher octane rating (110), the combustion of LPG is smoother and knocking is eliminated and the engine runs smoothly.

Disadvantages of LPG

- LPG will produce 10% less horse power for a given engine at full throttle due to fixed volume of the mixture of fuel and air.
- High ignition temperature.

Other Possible Fuels

- Bio-gas
- Producer gas
- Benzol
- Acetone
- Diethyl ether

Carburetion, Fuel Injections *and* Ignition

Fuel injection is a system for admitting fuel into an internal combustion engine. It has become the primary fuel delivery system use in automotive engines, having replaced carburetors during the 1980s and 1990s.

Carburetion

The process of formation of a combustible fuel-air mixture by mixing the proper amount of fuel with air before admission to engine cylinder is called **carburetion** and device which does this job is called a **carburetor**.

Air Fuel Ratio (AFR)

Air fuel ratio is the mass ratio of air to fuel and present in an internal combustion engine.

Air and fuel in ratio can vary quantitatively.

$$AFR = \frac{M_{air}}{M_{fuel}}$$

Air Fuel Ratio for an SI Engine

If exactly enough air is provided to completely burn all of the fuel, the ratio is known as the stoichiometric mixture.

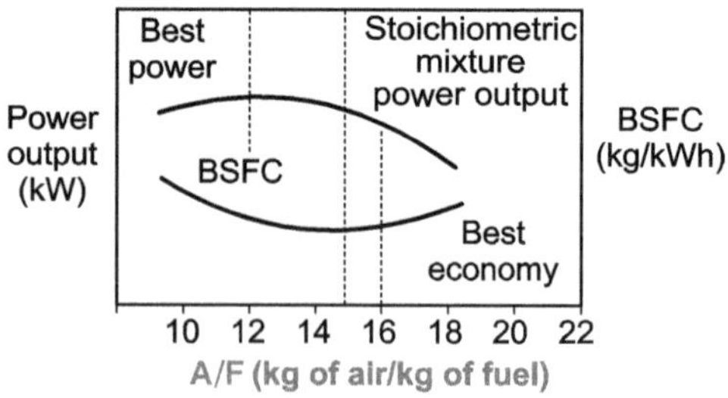

where, BSFC = Brake Specific Fuel Consumption.

Requirement of A/F Ratio Under Different Operations

Operation	A /F ratio
Idling	12-12.5
Cruising/Normal	16-16.5
Maximum power range	12-13
Transient operation	3 - 5

Altitude Compensation

- In the case of different atmospheric conditions, the air fuel ratio changes. The enrichment E due to variation of air density closely follows the relationship

$$E+1=\sqrt{\frac{\rho_0}{\rho}}=\sqrt{\frac{p_0 T}{p T_0}}$$

- At higher altitude, A/F mixture become progressively richer due to decrement of air density.
- At higher altitude, pressure difference $\Delta p\,(p_1-p_2)$ provided by simple carburetor is large as density of air decreases.

The Simple Carburetor

A carburetor is a device that blands air and fuel for an internal combustion engine. The schematic diagram of the simple carburetor is shown in figure below.

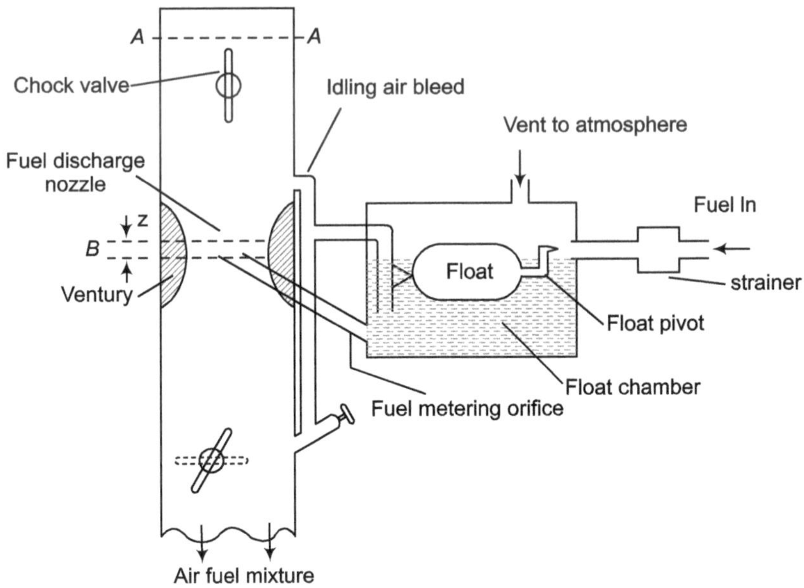

A carburetor structure

Assuming an adiabatic flow of air ($q=0, W=0$)

and velocity of air through section $A-A$ $\quad C_1\approx 0$

Velocity of air across ventury throat (section $B-B$)

$$C_2 = \sqrt{2\, C_p T_1 \left[1 - \left(\frac{p_2}{p_1}\right)^{\frac{\gamma-1}{\gamma}}\right]}$$

- Mass flow rate of air

$$m_a = \rho_1\, A_1 C_1 = \rho_2 A_2\, C_2 \quad \Rightarrow \quad \rho_2 = \left(\frac{p_2}{p_1}\right)^{\frac{1}{\gamma}} \rho_1$$

$$m_a = \frac{A_2 p_1}{R\sqrt{T_1}} \sqrt{2C_p \left[\left(\frac{p_2}{p_1}\right)^{\frac{2}{\gamma}} - \left(\frac{p_2}{p_1}\right)^{\frac{\gamma+1}{\gamma}}\right]}$$

- Actual mass flow rate,

$$m_{a\ \text{actual}} = C_{da} \frac{A_2 p_1}{R\sqrt{T_1}} \sqrt{2C_p \left[\left(\frac{p_2}{p_1}\right)^{\frac{2}{\gamma}} - \left(\frac{p_2}{p_1}\right)^{\frac{\gamma+1}{\gamma}}\right]}$$

- Let C_f is fuel velocity at the nozzle exit and Z is the height of the nozzle exit above the level of fuel in the float bowl and ρ_f is the fuel density

$$C_f = \sqrt{2\left[\frac{p_1 - p_2}{\rho_f} - gZ\right]}$$

- Mass flow rate of fuel

$$m_f = A_f\, C_f\, \rho_f$$

$$m_{f_{\text{actual}}} = C_{df}\, A_f\, C_f\, \rho_f$$

$$m_{f_{\text{actual}}} = C_{df}\, A_f\, \sqrt{2\rho_f\,(p_1 - p_2 - gZ\rho_f)}$$

$C_{df} \rightarrow$ Coefficient of discharge for fuel nozzle

$$\text{A / F ratio} = \frac{m'_{a_{\text{actual}}}}{m'_{f_{\text{actual}}}}$$

where, C_{da} = Coefficient of discharge for air.
T_1 = Temperature of air at section $(A - A)$.
p_1 = Pressure at section $(A - A)$.
p_2 = Pressure at ventury throat section $(B - B)$.
ρ_1 = Density of air at section $(A - A)$.
ρ_2 = Density of air at section $(B - B)$.
$A_1\ A_2$ = Cross-section area at intel and ventury throat

Air-fuel Ratio Neglecting Compressibility of Air

$$m_{a_{\text{actual}}} = C_{da}\, A_2 \sqrt{2\rho_a\, (p_1 - p_2)}$$

$$m'_{f_{\text{actual}}} = C_{df}\, A_2 \sqrt{2\rho_f\, (p_1 - p_2 - gZ\rho_f)}$$

$$\text{A / F ratio} = \frac{C_{da}}{C_{df}} \frac{A_2}{A_f} \sqrt{\frac{\rho_a\,(p_1 - p_2)}{\rho_f\,(p_1 - p_2 - gZ\,\rho_f)}}$$

If $Z = 0$

$$\frac{m'_a}{m'_f} = \frac{C_{da}}{C_{df}} \frac{A_2}{A_f} \sqrt{\frac{\rho_a}{\rho_f}}$$

Here, ρ_a = Density of air
ρ_f = Density of fuel.

Fuel Injections System

Fuel injection is a system for admitting fuel into an internal combustion engine.

The injection system has to perform the important duty of initiating and controlling the combustion process. *Injection systems can be classified as*

1. Air injection system
2. Solid injection system
3. Common rail system
4. Individual pump and nozzle system
5. Unit injector system
6. Distributor system

Injection Pump Governor

Quantity of fuel delivered vary with the speed. Fuel delivered by a pump increases with the speed and so causing excessive carbon deposits and high exhaust temperature. Drastic reduction in load will cause over speeding to dangerous value. Injection pump governor is used to take care of above limitations.

Nozzle

Nozzle is that part of an injector through which the liquid fuel is sprayed into the combustion chamber. Velocity of the fuel through nozzle orifice

$$v_f = C_d \sqrt{2gh},$$

Let, h = Pressure difference between injector and cylinder pressure.

Q = Area of all orifices × fuel jet velocity × time of one injection × number of injections per second for one orifice

$$Q = \left(\frac{\pi}{4} d^2 \times n\right) \times v_f \times \left(\frac{\theta}{360} \times \frac{60}{N}\right) \times \left(\frac{N_i}{60}\right)$$

N_i = Number of injection per minute

$N_i = \frac{N}{2}$ for 4 strokes engine, N for 2 strokes engine

θ = Duration of injection in crank angle degree

d = Diameter of one orifice

n = Number of orifices

$$\text{Velocity of injection } v_{\text{inj}} = C_d \sqrt{\frac{2\,(P_{\text{inj}} - P_{\text{cyl}})}{\rho_f}}$$

where, P_f = Fuel density

p_{inj} = Injection power

P_{cycle} = Power of cycle.

Multi-Point Fuel Injection (MPFI) System

The main purpose of the MPFI system is to supply a proper ratio of gasoline and air to the cylinders. *These systems function under two basic arrangements*

1. Port injection
2. Throttle body injection

Ignition System

Ignition system provide sufficiently large voltage across the spark plug electrodes to effect the spark discharge. It supply the required energy for the spark to ignite the combustible mixture adjacent to the plug electrodes under all operating conditions. *The ignition systems are classified as follows*

1. Battery ignition systems
2. Magneto ignition systems

Dwell Angle

The period during which the contact points in ingnition process remains closed is called **dwell angle** or **cam angle**. Too small dwell angle will result in lower secondary voltage and hence poor sparks or even misfiring. Too large a dwell angle will lead to burning of condenser and contact point due to over saturation of windings.

Firing Order

The firing order is the sequence of power delivery of circular cylinder in a multi-cylinder reciprocating engine.

For four cylinder firing orders

$$1-3-4-2 \quad \text{or} \quad 1-2-4-3$$

For six cylinder firing orders

$$1-5-3-6-2-4$$

or

$$1-2-3-6-5-4$$

or

$$1-5-4-6-2-3$$

or

$$1-2-4-6-5-3$$

Combustion

Combustion is a chemical reaction in which certain elements of the fuel like hydrogen and carbon combine with oxygen liberating heat energy and causing an increase in temperature of gases.

Stages of Combustion in SI (Spark-ignition) Engine

There are 3 stages in SI *engine which are given below*

Stage 1 **Ignition lag** There is a certain time interval between instant of spark there and instant where there is a noticeable rise in pressure due to combustion. This time lag is called **ignition lag.**

Stage 2 **Flame propagation stage** Once the flame is formed at *B*, it should be self sustained and must be able to propagate through the mixture. This is possible when the rate of heat generation by burning is greater than heat lost by flame to surrounding.

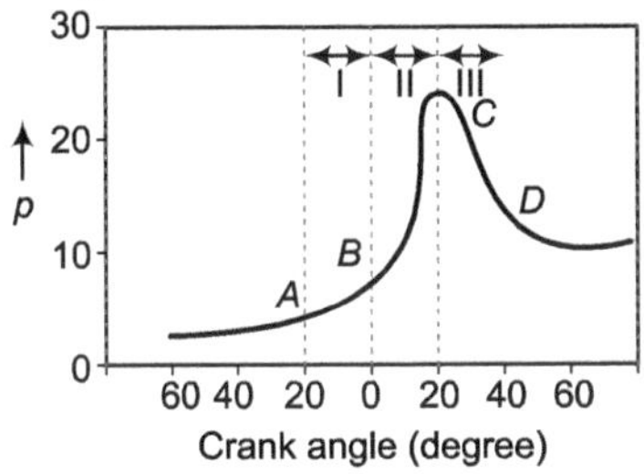

Combustion in SI engine

Stage 3 **After burning** Combustion will not stop at point *C* but continue after attaining peak pressure and this combustion is known as after burning. This generally happens when the rich mixture is supplied to engine.

Knock Phenomenon in SI Engines

If the temperature of the unburnt mixture exceeds the self-ignition temperature of the fuel and remains at or above this temperature during the period of preflame reactions. Spontaneous ignition or autoignition occurs at various pin point locations. This phenomenon is called **knocking**.

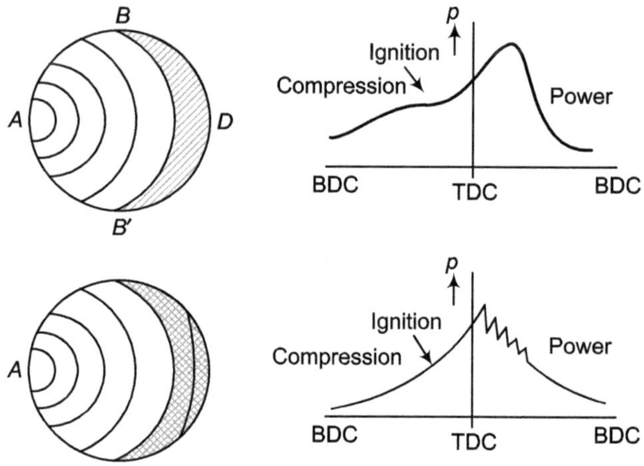

Knock phenomenon in SI engine

Variables Affecting Knock in SI Engine

S.No	Increase in variable	Major effect on unburned reduce charge	Effection on knocking
1.	Compression ratio	Increase *T* and *p*	Decrease knocking
2.	Mass of charge inducted	Increase *p*	Decrease knocking
3.	Inlet temperature	Increase *T*	Decrease knocking
4.	Chamber wall temperature	Increase *T*	Decrease knocking
5.	Spark advance	Increase *T* and *p*	Retard knocking
6.	A/F ratio	Increase *T* and *p*	Reduce knocking
7.	Turbulence	Decrease time factor	Increase knocking
8.	Engine speed	Decrease time factor	Increase knocking
9.	Distance of flame travel	Increase time factor	Reduce knocking

Stages of Combustion in CI (Compression Ignition) Engines

There are four stages in combustion in CI engines

Stage 1 Ignition Delay period /Pre-flame combustion The fuel does not ignite immediately upon injection into the combustion chamber. There is a definite period of inactivity between the time of injection and the actual burning this period is known as the ignition delay period.

Stage 2 Period of Rapid Combustion Period of rapid combustion also called the uncontrolled combustion, is that phase in which the pressure rise is rapid.

Stage 3 Period of Controlled Combustion The temperature and pressure in the second stage are so high that fuel droplets injected burn almost as they enter and find the necessary oxygen and any further pressure rise can be controlled by injection rate. The period of controlled combustion is assumed to end at maximum cycle temperature.

Stage 4 Period of After-Burning Combustion does not stop with the completion of the injection process. The unburnt and fuel partially burnt fuel particles left in the combustion chamber start burning as soon as they come into contact with the oxygen. This process continues for a certain duration called the after-burning period.

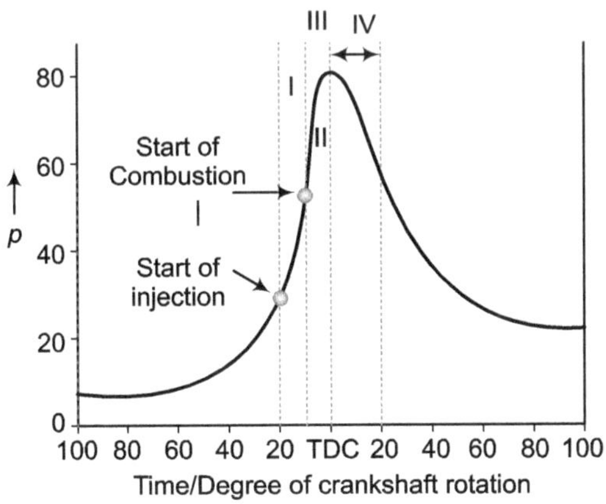

Stages of CI Engines

Knock Phenomenon in CI Engine

If the ignition delay is longer in CI engine, the actual burning of the first few droplets is delayed and a greater quantity of fuel droplets gets accumulated in the chamber. When the actual burning commences, the additional fuel can cause too rapid a rate of pressure rise resulting in a jamming of forces against the piston and rough engine operation. Such a situation produce the extreme pressure differential and violent gas vibrations known as **knocking**. In the CI engine, knocking occur near the beginning of combustion whereas in the SI engine knocking occurs near the end of combustion.

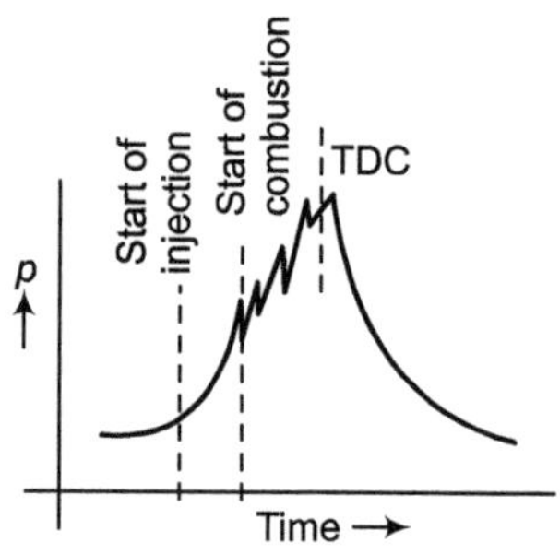

p-T curve for CI engine

Characteristing Tending to Reduce Knock (Detonation)

S.No.	Characteristics	CI Engine	SI Engine	Effect on knock
1.	Ignition temperature of fuel	Low	High	Reduce
2.	Ignition delay	Short	Long	Reduce
3.	Compression ratio	High	Low	Reduce
4.	Inlet temperature	High	Low	Reduce
5.	Inlet pressure	High	Low	Reduce
6.	Combustion wall temperature	Low	High	Reduce
7.	Speed (rpm)	Low	High	Reduce
8.	Cylinder (Size)	Large	Small	Reduce

Engine Emission

Emission standards are requirements that set specific limits to the amount of pollutants that can be released into the environment. Many emissions standard focus on regulating pollutants released by automobiles and other powered vehicles but they can also regulate emissions from industry, power plants, small equipments such as lawn movers and diesel generators.

Engine emission can be classified into two categories

1. Exhaust Emissions
   - Unburnt Hydrocarbon (HC)
   - Oxides of Carbon (CO & CO_2)
   - Oxides of Nitrogen (NO & NO_2)
   - Oxides of Sulphur (SO_2 & SO_3)
   - Particulates
   - Soot and Smoke
2. Non-exhaust Emissions
   1. Fuel tank → The fuel tank emits fuel vapours into the atmosphere.
   2. Carburetor → The carburetor also gives out fuel vapours.
   3. Crank case → It emits flow by gases and fuel vapours into the atmosphere.

Other Emissions

Some other emissions are given below

Aldehydes

When alcohol fuel is used, aldehydes are generated which cause an eye and respiratory irritant.

Sulphur

Many fuels in CI engines contain amount of sulphurs which exhausted in form of SO_2 and SO_3 (SO_x). They contribute to the acid rain problem in the world.

Lead

The additive TEL (Tetra Ethyl Lead) is used to increase gasoline octane number which allowed higher compression ratios and more efficient engines which resulting lead in the engine exhaust. Lead is highly poisonous pollutant.

Emission Control Methods

Emission can be controled by using converter which are defined as below

Thermal Converter

CO and HC emissions can be reduced by thermal converter, but NO_2 emission cannot be reduced using a thermal converter.

Catalyst Converter

Catalyst converters are also called 3-way converters because they are used to reduce the concentration of CO, HC and NO_x in the exhaust.

The catalyst material most commonly used are platinum, polladium, and rhodium. Polladium and Platinum promote, the oxidation of CO and HC with platinum especially active in the hydrocarbon reaction. Rhodium promotes the reaction of NO_x.

Measurements *and* Performance Parameters

With a growing demand for transportation, IC engines have gained lot of importance in automobile industry. It is therefore necessary to produce efficient and economical engine. While developing an IC engine it is required to take in consideration all the parameters affecting the engines design and parameters.

Following methods are used to find the friction power to estimate the performance of the engine

Willan's Line Method

(Fuel Rate Extrapolation Method)

A graph connecting fuel consumption and brake power at constant speed is drawn and is extrapolated on the negative axis of brake power. The intercept of negative axis is taken as friction power of the engine at that speed.

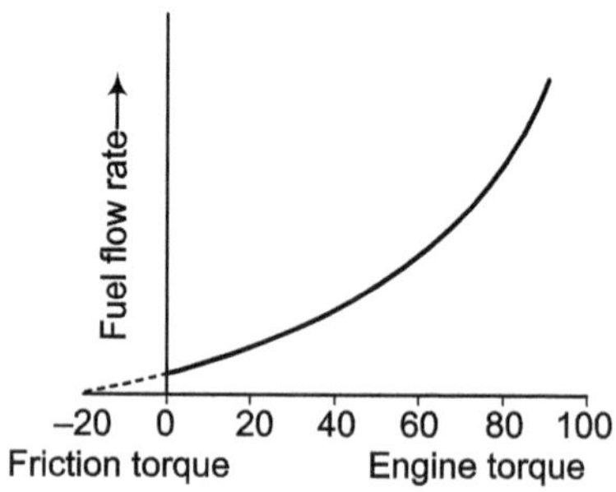

Willan's line method curve

Morse Test

The test consists of making inoperative in turn each cylinder of the engine and note the reduction in brake power developed.

i_p → indicated power
b_p → brake power
f_p → frictional power

If there k cylinder $i_{p_1} + i_{p_2} \ .... + i_{p_k} = \sum_1^k b_{p_k} + f_{p_k}$

First cylinder cut-off $i_{p_2} + ... + i_{p_k} = \sum_2^k b_{p_k} + f_{p_k}$

$$i_{p_1} = \sum_1^k b_{p_k} - \sum_2^k b_{p_k}$$

So, total indicated power $= i_{p_1} + i_{p_2} + \ldots i_{p_k} = \sum_1^k i_{p_k}$

Frictional power of the engine $f_{p_k} = i_{p_k} - b_{p_k}$

Motoring Test

In motoring test, the engine is steadily operated at the rated speed by its own power and allowed to remain under the given speed and load conditions for sufficient time so that the temperature of the engine components, lubricating Oil and cooling water reaches a steady state. This method determine the *fp* at condition very near to the actual operating temperatures at the test speed and load but it does not give the true losses.

Retardation Test

This test involves the method of retarding the engine by cutting the fuel supply.

Let, T_f = Friction torque
T_l = Load torque

Let t_2 and t_3 be time of fall at no load and load conditions respectively. Change in angular speed $d\omega = \frac{T}{mk^2} dt$

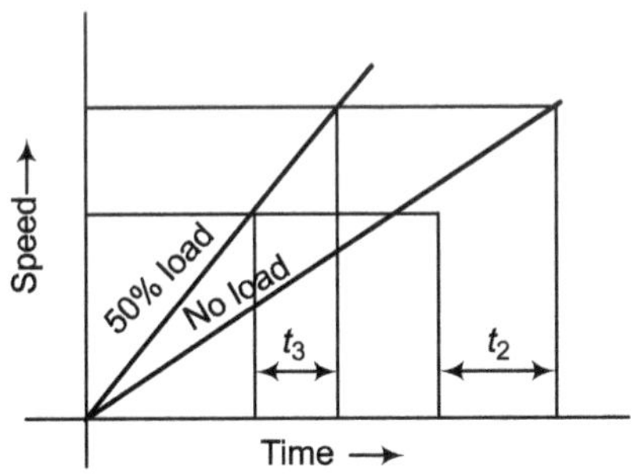

Graph for retardation test

$$\omega_0 - \omega_1 = \frac{T_f}{mk^2}(t_2 - 0)$$

$$\omega_0 - \omega_1 = \frac{T_f + T_l}{mk^2}(t_3 - 0$$

$$T_f = T_l \frac{t_3}{t_2 - t_3}$$

where, t_3 and t_2 are the time segments.

From the measurement of indicated and brake power we can directly calculate f_p from the measurement of indicated and brake power,

$$f_p = i_p - b_p$$

Indicated Power

Following methods are used for the measurement of an indicated power

1. Using the indicator diagram.
2. By adding two measured quantities *viz* brake power and friction power.

Brake Power

Brake power measurement involves the determination of the torque and angular speed of the engine output shaft. The torque measuring device is called dynamometer.

Brake Power for Proxy Brake Dynamometer

$$b_p = 2\pi NT$$

N = Speed

Torque $T = wl$

$w \rightarrow$ Weight applied at distance for proxy brake.

Brake Power for Rope Brake Dynamometer

$$b_p = \pi DN(w - s)$$

where, D = Brake drum diameter

w = Weight

s = Spring scale reading.

Heat Balance

Energy supplied to an engine is the heat value of fuel consumed. Some past transfer into work by engine while other part loss.

Heat Balance Diagram for a SI Engine

The heat balance diagram for a SI engine is shown in figure below

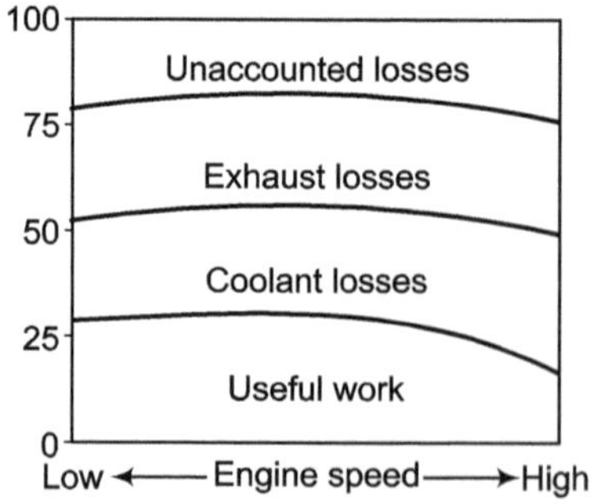

SI engine heat balance diagram

Heat Balance Diagrams for a Typical CI Engine

The heat balance diagrams for a CI engine is shown in figure below

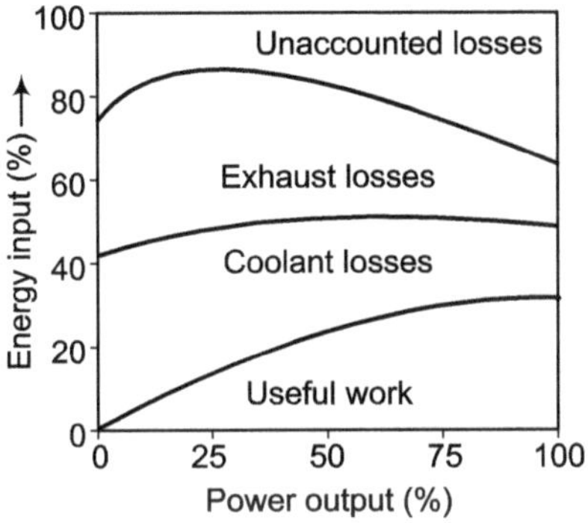

CI engine heat balance diagram

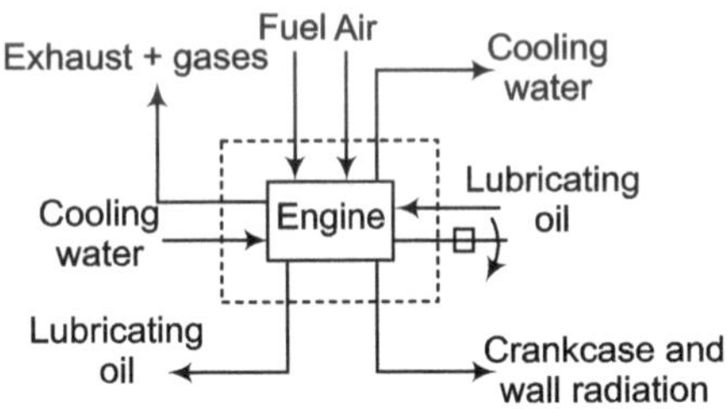

External heat balance

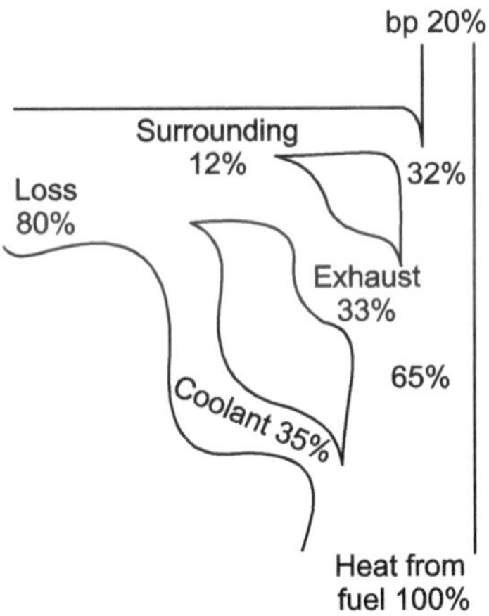

Sankey diagram

Supercharging

Supercharging is a method for improving engine power output. The supercharger used to increase the pressure, temperature and density of air supplied to an internal combustion engine. *Following points to be noted in supercharging*

- Supercharging increase the power output of the engine. It does not increase the fuel consumption per brake kW hour.

- The purpose of supercharging an engine is to raise the density of the air charge before it enters the cylinder. So, more air and fuel per cycle entered in tank and this can be effectively burnt during the combustion process to raise the engine power output to a higher value.
- Increase pressure and temperature as a result of supercharging may lead to detonation. Therefore the fuel used must have better anti-knock characteristics.

Types of Superchargers

There are two main types of superchargers defined according to the method of compression : positive displacement and dynamic compressors.

- Positive displacement pumps deliver a nearly fixed volume of air per revolution at all speeds.
- Dynamic compressors rely on accelerating the air to high speed and then exchanging that velocity for pressure by diffusing or slowing it down.

Turbocharging

In turbocharging the supercharger is being driven by a gas turbine which uses the energy in the exhaust gases. There is no mechanical linkages between the engine and supercharger.

Terminology and Definitions

Delivery Ratio

On the basis of mass, it is ratio of mass of fresh air delivered to the cylinder divided to a reference mass.

$$R_{del} = \frac{M_{del}}{M_{ref}}$$

On the basis of volume $R_{del} = \frac{V_{del}}{V_{ref}}$

where, V_{ref} = Total effective cylinder volume

M_{ef} = Reference mass

M_{del} = Mass of fresh air delivered to the cylinder

V_{del} = Volume of fresh air delivered to the cylinder

V_{ref} = Reference volume.

Trapping Efficiency

It is defined as the ratio of the amount of charge retained in the cylinder to the total charge delivered to the engine.

$$\eta_{trap} = \frac{V_{ret}}{V_{del}}$$

where, V_{ret} = Amount of charge retained in cylinder

V_{del} = Total charge delivered to engine.

Relative Cylinder Charge

It is measure of the success of filling the cylinder irrespective of the composition of charge.

$$C_{ret} = \frac{V_{ch}}{V_{ref}} = \frac{V_{ret} + V_{res}}{V_{ref}}$$

V_{ret} = Volume of air or mixture retained after flushing out the product of combustion.

V_{res} = Volume of residue gas retained in cylinder after flushing out the product of combustion.

Scavenging Efficiency

It is the ratio of volume of air retained to the volume of residue of gas retained *i.e.*,

$$\eta_{sc} = \frac{V_{ret}}{V_{ch}} = \frac{V_{ret}}{V_{ret} + V_{res}}$$

$$R_{del} = \frac{C_{rel}\,\eta_{sc}}{\eta_{trap}}$$

where, P_{del} = Delivery ratio

C_{rel} = Relative cylinder charge

η_{sc} = Scavenging efficiency

η_{trap} = Trapping efficiency

Charge Efficiency

It is the ratio of useful fresh charge to the displacement volume *i.e.*,

$$\eta_{ch} = \frac{V_{ret}}{V_{ref}}$$

$$\Rightarrow \qquad \eta_{ch} = R_{del}\,\eta_{trap}$$

11

Material Science *and* Production Engineering

Structure of Materials

The structure of a material may be divided into four levels-atomic structure, atomic arrangement, micro structure and macro structure. Although the main thrust of the material science is to understand and control the micro structure and macro structure.

Crystal Structure of Materials

Crystal structure is defined as the arrangement of the crystal within metals. When metals solidify from molten state, the atoms arrange themselves into various orderly configuration called **crystal**.

There are seven basic crystal structures, they are as given below

S.No.	Crystal system	Relation between primitives	Interface angles	Examples
1.	Cubic	$a = b = c$	$\alpha = \beta = \gamma = 90°$	Fe, Al, Cu
2.	Tetragonal	$a = b \neq c$	$\alpha = \beta = \gamma = 90°$	Sn, SO_2
3.	Orthogonal	$a \neq b \neq c$	$\alpha = \beta = \gamma = 90°$	KNO_3, $BaSO_4$
4.	Hexagonal	$a = b \neq c$	$\alpha = \beta = 90°$; $\gamma = 120°$	SiO_2, AgCl, Zn
5.	Rhombohedral	$a = b = c$	$\alpha = \beta = \gamma \neq 90°$	$CaSO_4$, $CaCO_3$
6.	Monoclinic	$a \neq b \neq c$	$\alpha = \beta = 90° \neq \gamma$	$FeSO_4$, $NaSO_4$
7.	Triclinic	$a \neq b \neq c$	$\alpha \neq \beta \neq \gamma \neq 90°$	$CuSO_4$, $K_2Cr_2O_7$

There relatively simple crystal structures are found for most of the common metals i.e., SSC, BCC and FCC are discussed below

Simple Cubic Cell (SCC)

In this arrangement, atoms are present at each corner of the cell and their centres are coincide with each corner.

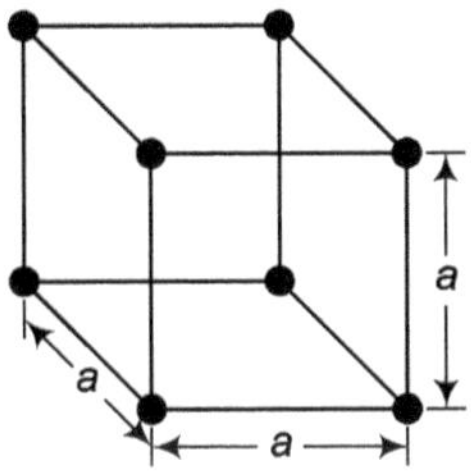

Simple cubic cell structure

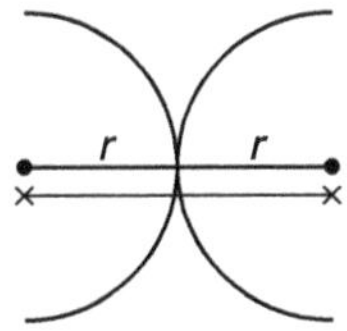

Atomic radius of SCC

So, $\frac{1}{8}$th part of the atom is present in the atom.

$$r + r = a$$

where, r = Atomic radius

a = Lattice constant.

The total number of atoms present in crystal structure,

$$n = \frac{1}{8} \times 8 = 1$$

- Atomic Packing Factor (APF) $= \dfrac{\text{Volume of atoms in a cell}}{\text{Volume of unit cell}}$

$$= \frac{4\pi a^3}{8 \times 3 \times a^3}$$

$$= \frac{\pi}{6} = \frac{3.14}{6}$$

$$= 0.52$$

Percentage APF = 52%

Percentage of voids = 100 − 52 = 48%

Key Points

- In crystal structure, the arrangement of the atoms in the crystal is called **crystalline structure.**
- The lattice structure of a particular metal is shown by a smallest group of atoms and known as a unit cell.

Body Centered Cubic (BCC) Structure

In this arrangement, an atom is present at the centre of the crystal and each corner is a centre of an atom.

Total effective number of atoms present in the crystal

$$= 1 + 8 \times \frac{1}{8}$$

$$= 1 + 1 = 2$$

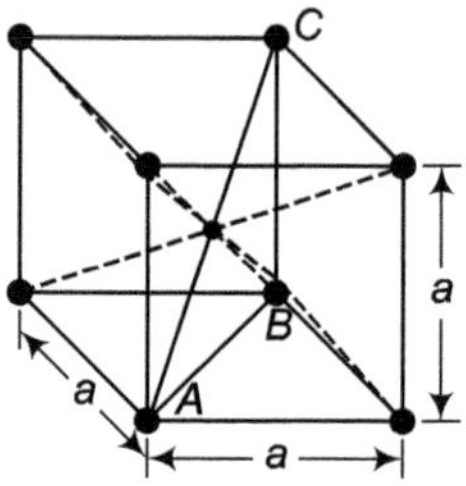

Body centred cubic structure

- Atomic Packing Factor (APF)

$$= \frac{\pi\sqrt{3}}{8} = \frac{3.14 \times 1.732}{8} = 0.68 \qquad \left(\because r = \frac{\sqrt{3}}{4}a\right)$$

- Percentage APF = 68%
- Percentage of voids = 100 − 68 = 32%

Face Centred Crystal (FCC)

In this arrangement, each face has an atom and corners are also occupied by atoms.

- Total effective number of atoms in cell

$$= 8 \times \frac{1}{8} + 6 \times \frac{1}{2} + 1 + 3 = 4$$

$$r = \frac{a}{2\sqrt{2}}$$

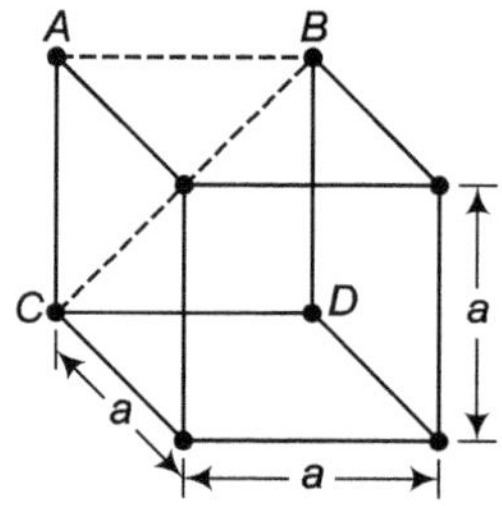

A FCC structure

- Atomic Packing Factor (APF) $= \dfrac{n \times \frac{4}{3}\pi r^3}{a^3}$

$$= \frac{16\pi}{16 \times 3\sqrt{2}}$$

$$= \frac{3.14}{3 \times 1.414}$$

$$= 0.74$$

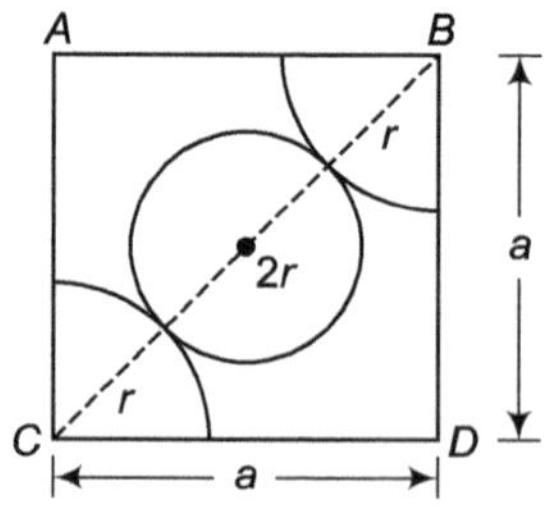

Atomic Packing Factor (APF) calculation

Percentage APF = 74%

Percentage of voids = 100 − 74 = 26%

Gibbs Phase Rule

The Gibbs Phase Rule describes the possible number of degree of freedom in a system at equilibrium, in terms of the number of separate phases and the number of chemical constituents in the system.

This is defined by the formula as

$$D = C - P + 2$$

where, D = degree of freedom,
C = number of components,
P = number of phases

Total number of variables in a system is given by

$$P(C - 1) + 2 = PC - P + 2$$

Key Points

- Coordination number of SCC crystal is 6.
- Coordination number of BCC crystal is 8.
- Coordination number of FCC crystal is 12.

Phase Diagram

It is also known as equilibrium diagram. The plots showing relations between phases in equilibrium *versus* composition, pressure and temperature are called **phase diagrams**. The composition is plotted on X-axis and temperature is plotted on Y-axis at any specified pressure. Pressure is generally taken atmospheric.

Eutectic Point

When a liquid phase changes into two different solid phases during cooling or two solid phases change into a single liquid phase during heating, this point is known as **eutectic point** and this temperature is known as **eutectic temperature**.

$$\underset{\text{(liquid phase)}}{L} \underset{\text{heating}}{\overset{\text{cooling}}{\rightleftharpoons}} \underset{\text{(solid phase)}}{\alpha} + \underset{\text{(solid phase)}}{\beta}$$

Eutectoid Point

When a solid phase changes into two solid phases during cooling and *vice-versa* that point is known as **eutectoid point** and temperature at this reaction occurs known as **eutectoid temperature.**

$$\underset{\text{(solid phase)}}{\gamma} \underset{\text{heating}}{\overset{\text{cooling}}{\rightleftharpoons}} \underset{\text{(solid phase)}}{\alpha} + \underset{\text{(solid phase)}}{\beta}$$

Peritectic Point

A binary system when solid and liquid phases changes solid phase on cooling and *vice-versa* on heating, then state of system is known as peritectic point.

$$\underset{\text{(solid Phase)}}{\beta} + \underset{\text{(liquid phase)}}{L} \underset{\text{heating}}{\overset{\text{cooling}}{\rightleftharpoons}} \underset{\text{(solid phase)}}{\alpha}$$

Peritectoid Point

If a binary phase diagram when two solid phases change to one solid phase, then state of system is known as peritectoid point.

$$\underset{\text{(solid phase)}}{\gamma} + \underset{\text{(solid phase)}}{\beta} \underset{\text{heating}}{\overset{\text{cooling}}{\rightleftharpoons}} \underset{\text{(solid phase)}}{\alpha}$$

Iron-Carbon Phase Diagram

Phase diagram of great technological importance is that of the iron- carbon system.

Allotropic Forms of Steel

The temperature 723°C is known as **Curie temperature**, below it steel shows magnetic properties and above it steel becomes non-magnetic. Diagram, the carbon percentage is plotted on *X*-axis and temperature plotted on *Y*-axis. The melting point of iron is about 1539°C. The melting temperature of iron varies with increasing carbon percentage. The iron has carbon upto 2% known as steel. The cast iron is a form of having 2 to 4.5% carbon. The iron has carbon upto 6.67% known as **pig iron**.

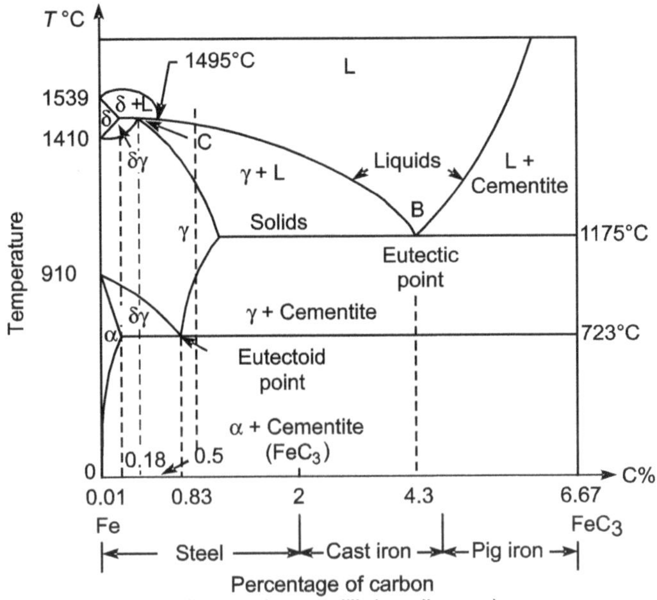

(Iron-carbon equilibrium diagram)

Eutectoid Point

Eutectoid reaction in this diagram occurs when temperature 1°C and carbon is 0.83%. At this point a solid form iron (γ) changes into two solid forms α-iron and cementite (Fe_3C).

$$\underset{\text{(FCC)}}{\text{(irons)}} \xrightleftharpoons{\text{Cooling}} \underset{\text{(BCC)}}{\alpha\text{-iron}} + \text{cementite } (Fe_3C)$$

A eutectoidal mixture of ferrite (α-iron) and cementite is known as pearlite. This is a microconstituent. The fraction of ferrite in eutectoid steel is 88%.

Eutectic Point

In this diagram, the location of this point is at 1175°C temperature and 4.3% at the carbon. At this point eutectic reaction occurs. In this reaction, a liquid phase changes two solid phases γ-iron (austenite) and cementite (FeC).

Peritectic reaction occurs at 1495°C and at this point carbon composition is 0.18%. This temperature is known as peritectic temperature.

Cast Iron

Cast iron is iron or a ferrous alloy which has been heated until it liquefies, and is then poured into a mould to solidify. It is usually made from pig iron. The alloy constituents affect its colour when fractured: white cast iron has carbide impurities which allow cracks to pass straight through. Grey cast iron has graphitic flakes which deflect a passing crack and initiate countless new cracks as the materials breaks.

Types of Cast Irons

Many types of cast irons can be defined below

Grey Cast Iron

It is also known as graphitic cast iron. In this type of cast iron graphite is in the forms of flake whose tips are sharp. Due to sharpness of tips, they becomes source of stress concentration. It has poor tensile strength.

White Cast Iron

White cast iron contains cementite phase. Due to presence of cementite, it is extremely hard besides being strong and brittle.

Mottled Cast Iron

It contains mixture of grey and white cast irons.

Malleable Cast Iron

It is produced when white cast iron is heated for a long time at 900°C and after that by slow cooling rate. Due to this process cementite decomposes to ferrite. Si is kept less than 1 %, so as to avoid graphitization.

Spheroidal Cast Iron

In this cast iron, graphites are present in the form of spherical module. Spheroidal cast iron is produced by adding Mg or Ce to molten iron. Content of Si is about 2.5% in it.

Inoculated Cast Iron

It is produced by inoculated molten pig iron by solution of silicon compounds such as calcium silicide.

Stress-Strain Diagram for Cast Iron

The properties of metal and alloys can be increased or decreased, according to our requirement by heating or cooling them into a controlled condition. This process is known as heat treatment.

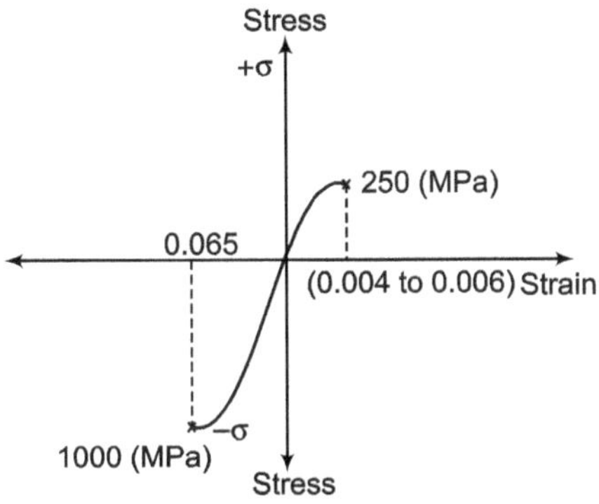

Stress-strain diagram for cast iron

Ferrous Alloy

Alloying elements	Symbols	Enhance mechanical properties
Tungsten	W	Imparts red hardness
Vanadium	V	Induces fine grain distribution
Chromium	Cr	Improves resistance to corrosion and oxidation
Molybdenum	Mo	Imparts hardneability
Phosphorus	P	Inhance machinability
Titanium	Ti	Increase abrasion resistance
Nickel	Ni	Improves tensile strength and toughness
Manganese	Mn	Imparts wear resistance

Alloy Mn-Steel	Composition percentage	Application
Silico-manganese	Mn 65%, Si 20%, Fe 15%	Railway crossing
Hadfield steel	Mn 13%, C 1.25%, rest Fe	Rock crushing machine, excavators
Mo-steel chlorimet V steel	Mo 33%, Ni 65%, Fe 1% other 1%	High speed cutting tool
Ferrovanadium	V 50%, Fe 50%	Spring, cutting tools

Material	Composition	Application
High carbon tool steel	C 0.8 to 1.3% Si 0.1 to 0.4%, Mn 0.15 to 0.4% rest Fe	Milling cutters turning tools for soft material
Low alloy carbon steel	C 1.2%, Mn 0.3 to 0.7% W. 2.75, Mo 0.4%, S 0.3%, rest Fe	High speed milling, drilling and turning tool
High speed steel (HSS)	W 18%, Cr 4%, V 1%, C 0.6 to 0.8%, rest Fe	Useful at high temperature operations

Non-ferrous Alloys

Alloys	Composition	Application
Mg alloy electron	Al 7.5%, Mn 0.3% Zn 2%, rest Mg	Parts of aircraft and automobiles
Al alloy duralumin	Cu 4%, Mg 0.5% Si 0.7% rest Al	Electrical cables
Cu alloy brass (Cu + Zn) muntz metal	Cu 60%, Zn 40%	Brass castings, stampings
Bronze (Cu + Sn) gun metal	Cu 88%, Zn 2%, Sn 10%	Steam pipe fitting
Bell metal	Cu 70%, Sn 30%	Casting of bells

Bearing Materials

Name	Composition	Application
Tin babbit	Sn 80%, Sb 12%, Cu 8%	High speed bearing bushes
Lead babbit	Pb 8.5% Sb 10%, Sn 5%	Railway wagon bearing

Common Non-ferrous Alloys

Alloy	Composition	Application
Invar	Ni 36%, Fe 64%	Precision measuring instrument
Nichrome	Ni 80%, Cr 20%	Thermocouple, strain gauge
Maganin	Cu 87%, Mn 13%	Resistor strain gauge
Canstantan	Cu 60%, Ni 40%	Electrically heated appliances

Key Points

- Recovery temperature is below 0.1 T_m.
- Cold working temperature is below recrystallization temperature.
- Hot working temperature is above recrystallization temperature.
- For pure metal recrystallization temperature is 0.3 T_m to 0.5 T_m.
- For alloys recrystallization temperature is 0.57 T_m to 0.7 T_m.
  where, T_m is melting point temperature in kelvin.

Metal Cutting *and* **Tools**

Cutting tools that are made from wrought high speed steels are shaped to the required geometries by machining. *The metal cutting and tools are explain as below*

Tool Geometry

A cutting tool has one or more sharp cutting edges. The cutting edges are used to separate a chip from the parent work material. A cutting tool has special geometry for a specific machining process.

The tools are classified as

(i) Single point cutting tool (ii) Multi-point cutting tool

A single point cutting tool has only one cutting edge and a multi-point cutting tool has one or more cutting edge.

Geometry of Single Point Cutting Tool

Some angles which are used to define the tool geometry are given below

Rake Angle

Rake angle is defined as angle between the face of the tool called the rake face and the normal to the cutting direction.

Side Relief and End Relief or Clearance Angle

The purpose of relief angles are increasing tool life and reducing the chances of tool breakage. **Clearance angle** is defined as the angle between the machined surface and the under side of the tool, called **flank face.**

If relief angle is too large, the tool may break the chip. If this angle is too small then, tool will rub the work piece and produce excessive heat and this will turn, cause premature dulling of the cutting tool.

For machining of hard material the relief angle should be small and for machining of softer or weaker material, it should be large. A smaller angle should be used for interrupted cuts or heavy feeds and a larger angle for semi-finish and finish cuts.

Lead Angle

This angle is determined by the tool holder which must be chosen for each particular job.

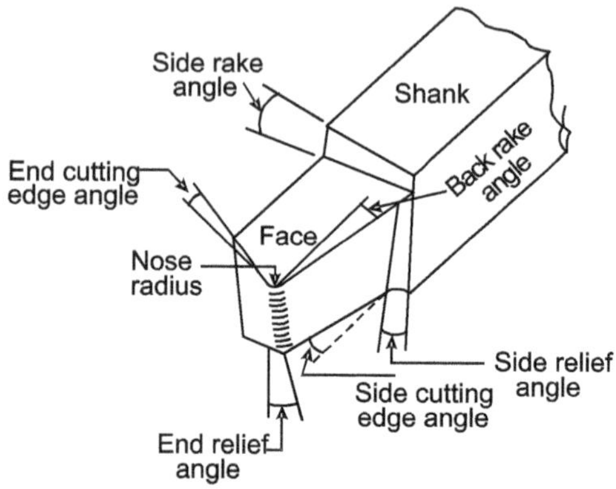

Nomenclature for right hand cutting tool

Positive lead angle performs two main functions

(i) Thins the chip

(ii) Protect the insert

The underformed chip thickness decreases when using positive lead angle-longitudinal force (in the direction of feed on work piece can be reduced by using positive lead angle tools.

Tool Nomenclatures

Tool nomenclatures are deperulson the following systems. Which are explain below

ASA system The tool angles in coordinate system are given in the following order

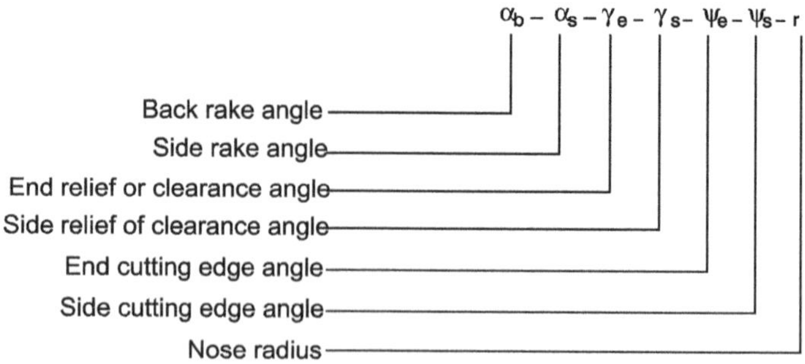

ASA system

ORS system The tool designation under ORS is given in the following order

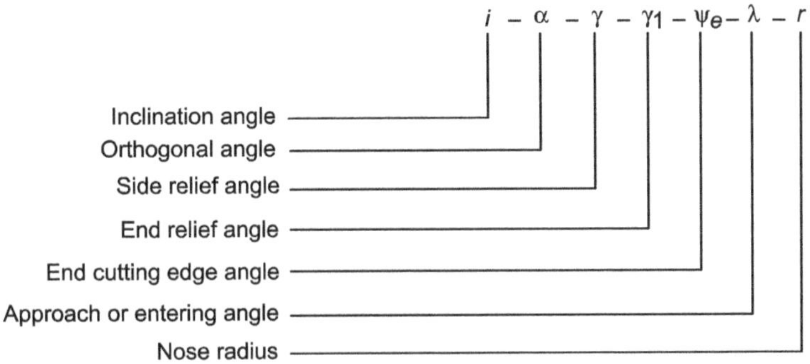

ORS System

- Relationship between these angles are given below

$$\tan \alpha = \tan \alpha_s \sin \lambda + \tan \alpha_b \cos \lambda$$
$$\tan \alpha_b = \cos \lambda \tan \alpha + \sin \lambda \tan i$$
$$\tan \alpha_s = \sin \lambda \tan \alpha - \cos \lambda \tan i$$
$$\tan i = \tan \alpha_b \sin \lambda - \tan \alpha_s \cos \lambda$$

The relationship between lead angle and side cutting edge angle is given by

$$\lambda = 90^\circ - \psi_s$$

Geometry of Turning Process

The mechanism of turning process as shown in the given figure

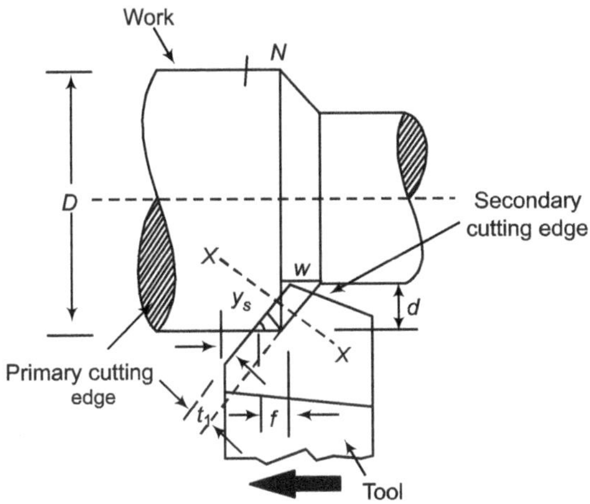

(a)

Geometry of turning process

From given Fig. (a), $t_1 = f \cos \psi_s$

$$\omega = d / \cos \psi_s$$

where, ψ = Side cutting edge angle

- Cutting speed is given by $v = \pi DN$

where, N = Number of jobs revolution per unit time
D = Initial diameter of the job
d = Depth of cut,
t_1 = Uncut chip thickness, f = Feed

- For orthogonal cutting, $\tan \alpha_b \cos \psi_s = \tan \alpha_s \sin \psi_s$
- The normal rake angle can be expressed in terms of other tool angles as

$$\tan \alpha = \frac{\tan \alpha_b \sin \psi_s + \tan \alpha_s \cos \psi_s}{\sqrt{1 + (\tan \alpha_b \cos \psi_s - \tan \alpha_s \sin \psi_s)^2}}$$

From given Fig. b, $F_f = F_t \cos \psi_s$

$$F_R = F_t \sin \psi_s$$

where, f_c = Cutting force; F_f = Feed force; F_R = Radial force

Power consumption in cutting operation is given by

$$W_c = F_c y$$

Mechanics of Orthogonal Metal Cutting

This analysis is based on Merchant's thin shear plane model, considering the minimum energy principle. This model is applicable at very high speeds which are generally practised in production.

Let, F_t = Thrust force (perpendicular to the job piece)
F_c = Cutting force (in the direction of cutting velocity)
F_s = Force along the shear force

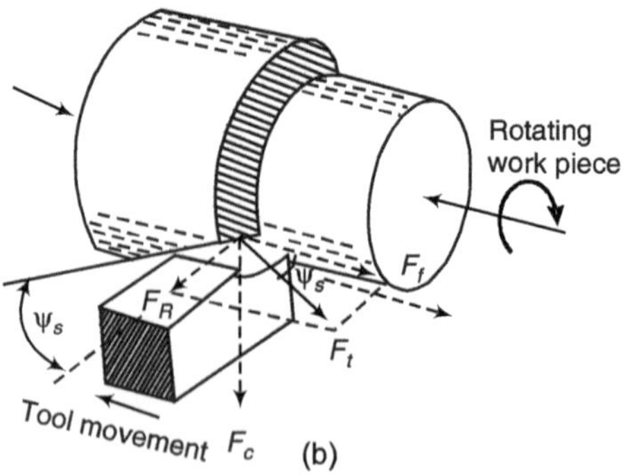

Orthogonal cutting

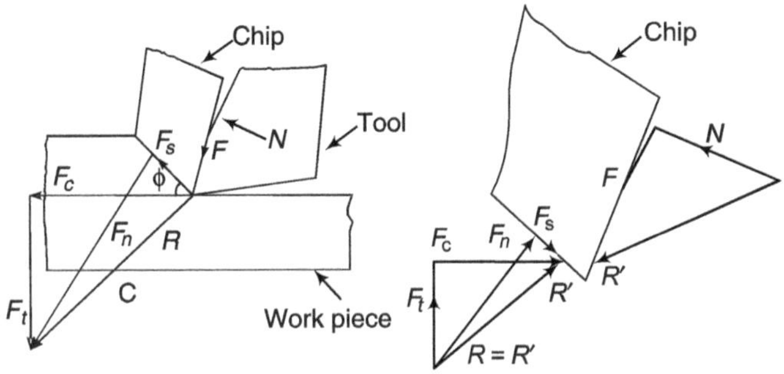

(c) (d)

Mechanics of orthogonal metal cutting

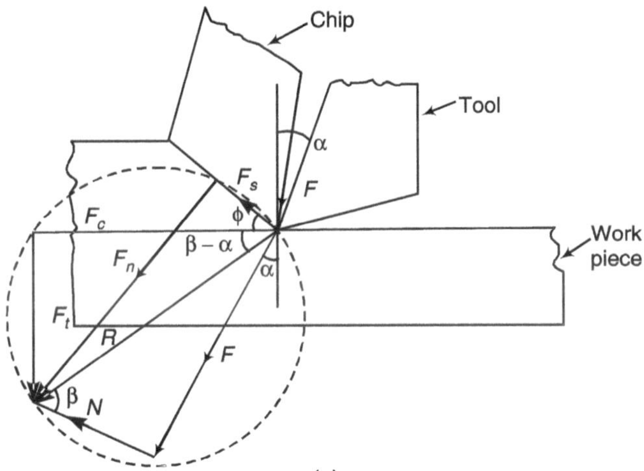

(e)

Parts of orthogonal metal cutting

F_n = Force normal to the shear force
N = Force acting normal to the rake face
F = Frictional force along the tool rake face.

- All the forces that are acting on chip can be represented with the help of a circle. This circle L is known as **Merchant force circle.**

From above Fig. (e),

$$F_s = F_c \cos\phi - F_t \sin\phi$$

$$F_n = F_c \sin\phi + F_t \cos\phi$$

$$= F_s \tan(\phi + \beta - \alpha)$$

$$F = F_c \sin\alpha + F_t \cos\alpha$$

$$N = F_c \cos\alpha - F_t \sin\alpha$$

- **Coefficient of friction** $\mu = \tan\beta = \dfrac{F}{N} = \dfrac{F_t + F_c \tan\alpha}{F_c - F_t \tan\alpha}$

  where, β = Friction angle $\quad \phi$ = Shear angle
- **The area of shear plane** A_s is given by

$$A_s = \frac{wt}{\sin\phi}$$

  where, w = Width of cut

  t = uncut chip thickness
- Shear force acting on shear plane = Shear stress × Area = $\tau A_s = \dfrac{\tau wt}{\sin\phi}$

  where, τ = Average shear stress in the shear plane
- Mean normal stress in the shear plane

$$\sigma = \frac{F_n}{A_s}$$

$\therefore$
$$F_n = \sigma \times A_s = \frac{\sigma\, wt}{\sin\phi}$$

$$F_s = R\cos(\theta + \beta - \alpha)$$

$$F_c = R\cos(\beta - \alpha) = \frac{F_s \cos(\beta - \alpha)}{\cos(\phi + \beta - \alpha)}$$

$$F_c = \frac{\tau wt \cos(\beta - \alpha)}{\cos(\phi + \beta - \alpha)}$$

and
$$F_t = \frac{\tau wt}{\sin\phi}\,\frac{\sin(\beta - \alpha)}{\cos(\phi + \beta - \alpha)}$$

Shear Angle Relationship

Merchant consider that τ would have the value of yield shear stress for the job material and that μ would have the usual for any dry sliding friction. For determining the value of ϕ, Merchant assumed the minimum energy principle applied in metal cutting so that the deformation process adjust itself to a minimum energy condition.

Hence,
$$\frac{dF_c}{d\phi} = \frac{\tau wt \cos(\beta - \alpha)\cos(2\phi + \beta - \alpha)}{\sin^2\phi\cos^2(\phi + \beta - \alpha)}$$

For minimum value,

Putting $$\frac{dF_c}{d\phi} = 0 2\phi + \beta - \alpha = \frac{\pi}{2}$$

$$\phi = \frac{\pi}{2} + \frac{\alpha}{2} - \frac{\beta}{2}$$

Chip Thickness Ratio

The ratio of uncut chip thickness to chip thickness is known as chip **thickness ratio** or cutting ratio and developed by *r*.

The chip thickness ratio diagram can be shown below

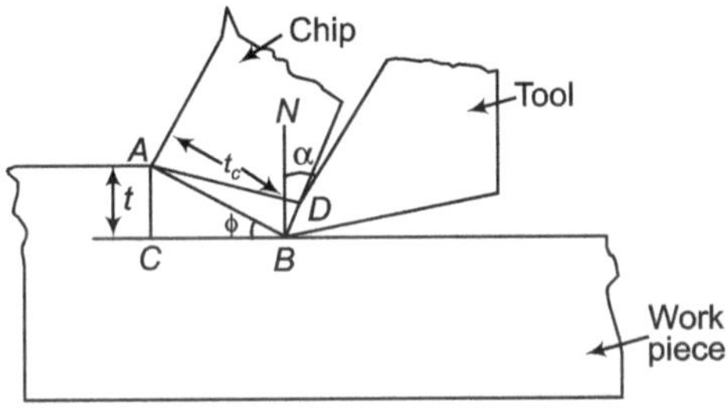

A schematic diagram to show chip thickness ratio

where, $AD = t_c$ = Chip thickness
$AC = t$ = Uncut chip thickness

$$r = \frac{t}{t_c} = \frac{\sin\phi}{\cos(\phi - \alpha)}$$

$$\Rightarrow \qquad \tan\phi = \frac{r\cos\alpha}{1 - r\sin\alpha}$$

- If the chip thickness is t_c and length is l_c, then $t_c = \frac{t\,l}{l_c}$

  where, t = Uncut chip thickness
  l = Length of uncut chip

- To obtain exact size of uncut chip length l, we may introduce a small cut parallel to the axis on the work piece so that the uncut chip size is given by

  $$l = \pi D$$

  where, D = diameter of the job before matching is done on it.

Velocity Ratio

Cutting velocity is more than the chip velocity. The mass of removal material is constant before and after machining.

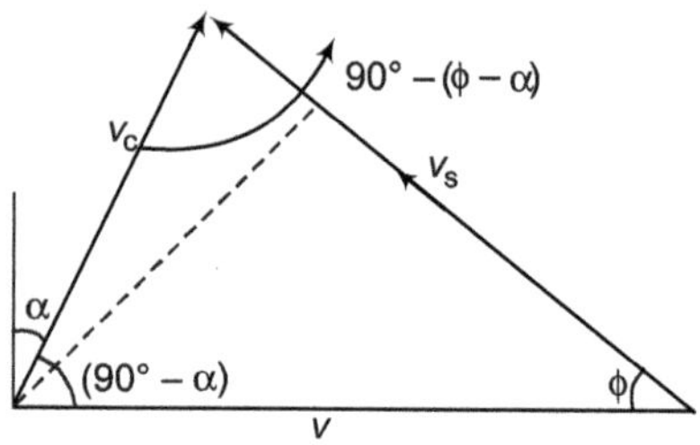

Velocity ratio diagram

From velocity triangle, $$v_t = v_c t_c$$

$$v_c = v\frac{t}{t_c} = \frac{v\sin\theta}{v\cos(\phi-\alpha)}$$

$$\frac{v_s}{\sin(90^\circ-\alpha)} = \frac{v}{\sin[90^\circ-(\phi-\alpha)]} = \frac{v_c}{\sin\phi}$$

$$\frac{v_s}{\cos\alpha} = \frac{v}{\cos(\phi-\alpha)} = \frac{v_c}{\sin\phi}$$

Shear strain is denoted by γ and expressed as

$$\gamma = \tan\phi + \cot(\phi-\alpha)$$

Tool Life

Tool life T is the cutting time at the end of which a given tool may be termed impossible on the basis of a selected tool failure criterion. Taylor established a relationship between tool life cutting velocity which is given by

$$vT^n = C$$

where, v = Cutting speed
T = Tool life in minutes
n = A exponent which depends on cutting condition
C = Constant.

- Depth of cut and feed rate also influence the tool life so that where, Taylor equation can be modified as

$$vT^n d^x f^y = \text{Constant}$$

where, d = Depth of cut
f = Feed rate (in mm/rev) in turning.

- Power is calculated by force multiplied by velocity in that direction

$$P_c = F_c v$$

$$P_c = \frac{F_c \times v}{60 \times 75 \times 1.36} \text{ kW} \qquad \text{(If } F_c \text{ is in kgf)}$$

If F_c is in Newton then,
$$P_c = \frac{F_c \times v}{1000} \text{ kW}$$

- Design power rating of electric motor of the main drive,

$$P_m = \frac{P_c}{n_m}$$

where, n_m = Efficiency of machine tool.

Machinability

The ease with which a given material may be machined under a given set of cutting conditions. *Factors affecting the machinability are*

Tool life Larger tool life, better machinability.

Surface finish Good finish, more machinable material.

Cutting forces Smaller cutting force, more machinable material.

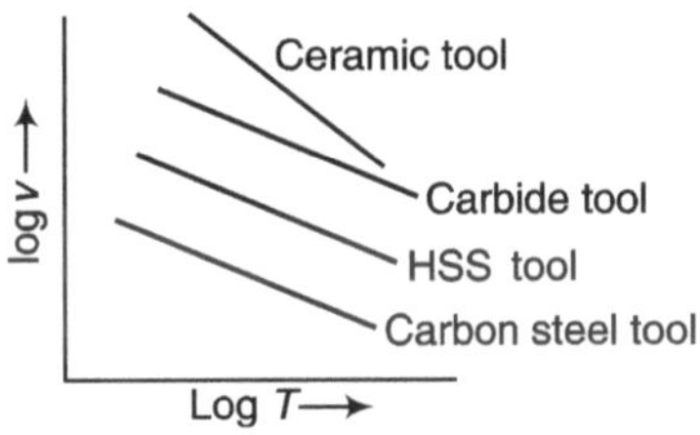

$$\text{Machinability index} = \frac{V_t}{V_s} \times 100$$

where, v_t = Cutting speed of metal for 1min tool etc.
v_s = Cutting speed of standard.

Variation of velocity with respect to time for different material.

Economics of Machining

The economic of machining means minimum cost is used during machining process. For total minimum cost,

$$V_{opt} = \frac{C}{\left[\left(\frac{C_e}{C_m} + T_c\right)\left(\frac{1}{n} - 1\right)\right]^n} \Rightarrow t_{opt} = \left[\left(\frac{C_e}{C_n} + T_c\right)\left(\frac{1}{n} - 1\right)\right]$$

where, C_m = Machine cost in ₹/time, T_c = Tool changing time
C = Total cost, C_e = Cost of tool/grind.

- For maximum production rate,

$$\text{Optimum speed, } v_{opt} = \frac{C}{\left[\left(\frac{1}{n}-1\right)T_c\right]^n}$$

$$\text{Optimum time, } T_{opt} = \left(\frac{1}{n}-1\right)T_c$$

- For maximum profit rate,

$$\text{Profit rate} = \frac{R - C_0}{T_n + T_h + \frac{T_m}{T} \times T_c}$$

R = Selling price, T_h = Idle time/component

$$T_m = \frac{l}{fN}$$

(v_{opt}) production rate > (v_{opt}) profit rate > (v_{opt}) minimum unit cost

Machining Time on Lathe in Turning

The machining time on lathe in turning depends on the following groups, which are described as

Cutting Speed

It is defined as distance travelled by cutting tool along circumference of the work piece in one minute. It is expressed in m/min. It is given by

$$v = \frac{\pi DN}{1000}$$

where, D = Diameter of job (in mm); N = rpm of the work piece.

Feed

It is the distance that tool bit advances along work piece for each revolution of the spindle (or work piece itself). It is expressed in terms of mm/rev. It is denoted by f.

Depth of Cut

This is the distance that tool bit moves into the work. It is denoted by t.

$$t = \frac{D-d}{2}$$

where, D = Original diameter of work piece

d = Diameter of work piece after machining.

Machining Time

If feed rate is f and work piece rotates at N rpm.
Then, distance travelled by tool in one revolution = f
and distance travelled by tool in N revolution = $f\,N$

Time taken by tool to travel this distance is one minute because rotational speed is given in rev/min.

$$\text{Speed of tool} = f\,N$$

If l is length of work piece then, time taken in turning = $\dfrac{1}{f\,N}$ min.

Material Removal Rate (MRR)

The quantity of material removed in unit time is known as **material removal rate.**

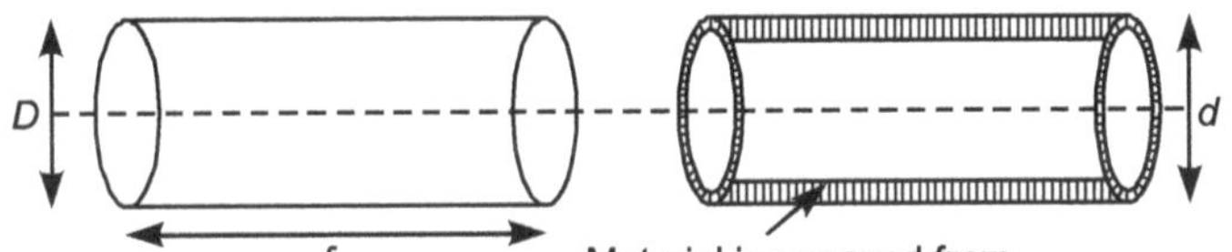

The unit of material removal rate is mm^3/min. Length of material removed in one minute from work piece feed = f mm. Material removed from work piece in one minute

$$= \frac{\pi}{4} D^2 f - \frac{\pi}{4} d^2 f = \frac{\pi}{4}(D^2 - d^2) f$$

$$= \pi \left(\frac{D+d}{2}\right)\left(\frac{D-d}{2}\right) f = \pi D_{av}\, t\, f$$

So, $$\text{MRR} = \pi D_{av}\, f\, t \text{ mm}^3$$

Operating Conditions in a Shaping Machine

Various conditions for operating in a shaping machine are explained ahead

Machining Time for in Plan or Slab Milling

The time for machining in a plan or slab milling can be describe by the following diagram

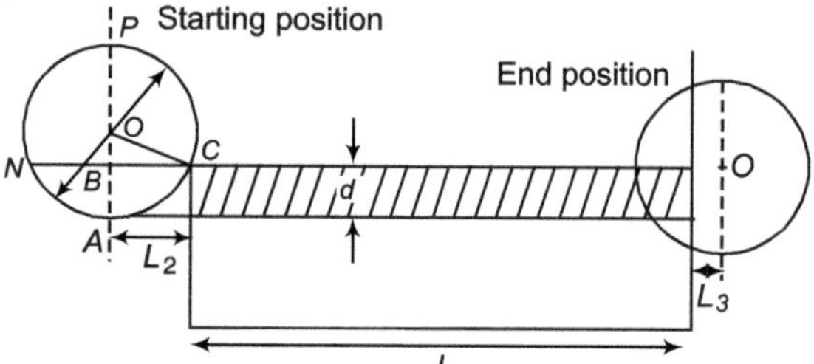

Slab milling

L_3 (over travel) approximately taken as 1 to 5 mm. $L_2 = BC$

By theorem of intersecting chord in a circle, $BN \times BC = PB \times AB$

where, AP = Diameter of cutter = D, AB = Depth of cut = d

Then, $$BN \times BC = (D - d) \times d$$

OB is perpendicular to CN.

Then, $$BN = BC \Rightarrow BC^2 = (D - d) \times d$$

$$L_2 = BC = \sqrt{d\,(D - d)} \text{ mm}$$

$$\text{Time for machining} = \frac{L_1 + \sqrt{d\,(D - d)} + L_3(5) \times n}{f}$$

where, n = Number of passes; f = Feed rate.

Machining Time for Face Milling

For symmetrical face milling, $L_2 = CM = OM - OC$

$$AC = BC = \frac{w}{2}$$

$$OC = \sqrt{OA^2 - AC^2} = \sqrt{\left(\frac{D}{2}\right)^2 - \left(\frac{w}{2}\right)^2} = \frac{1}{2}\sqrt{D^2 - w^2}$$

$$CM = OM - OC = \left(\frac{D}{2} - \frac{1}{2}\sqrt{D^2 - w^2}\right) \text{ mm}$$

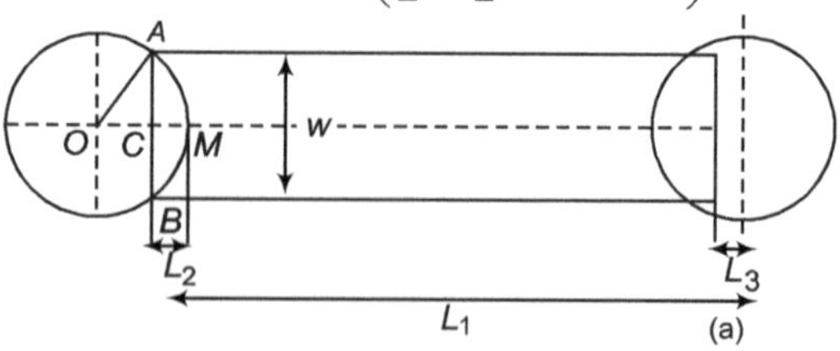

Face milling diagram

Here, the length approach is cleared by Fig. (b).

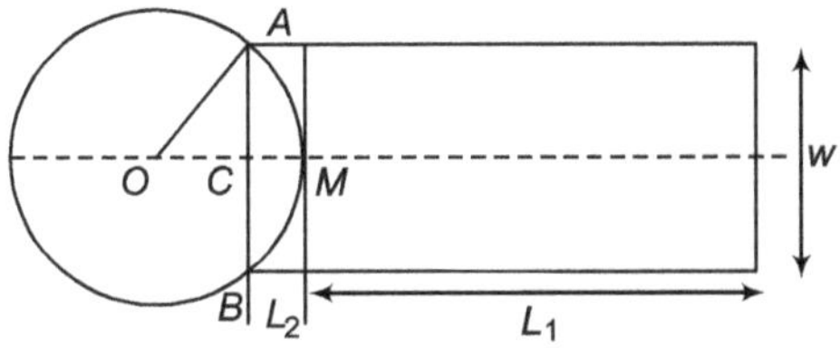

The condition before actual cutting (b)

$$CM = L_2$$

Grinding

The grinding process is defined as a machining process to produce high surface finish with tight tolerance. This process is used for machining of hard materials.

The basic purposes of grinding are

- Machining for hard materials
- Production of high quality surface with close tolerance.
- Removal of excess material from the surface of work piece.

Grinding Wheel

Grinding wheel contains **abrasive particles** bonding materials and voids. The abrasive particles are located at periphery of wheel acts like as cutting tool tips and remove the material from the surface of the job. The abrasives are used for making grinding wheel, they are hard substances. Some of them are aluminium oxides, ceramic, silicon carbide; zirconia, super abrasive (diamond), super abrasive (CBN), tungsten carbide etc. (CBN = Cubic Boron Nitride)

Properties of Abrasive Grains

Various types of properties of abrasive grains are defined as below

Friability

The ability to fracture of abrasive grain under pressure known as friability. By this ability, when cutting edges becomes dull, the grain breaks-off and exposes new cutting edges.

Hardness

Hardness of abrasive grain is defined as ability to scratch or penetrate the work piece during operation.

Size

Size of the screen through which the grain will pass.

Size of the screen

Bond

The bond material is used to hold abrasive particle together. The bonding materials are metal, resin/plastic, rubber shellac silicates, vitrified etc. Silicates are known as oxychlorides.

Grade

The grade of the wheel is a measure of ability to retain grit. A softer material is grinded by harder grade and *vice-versa*.

Structure

Structure is spacing between abrasive grains or the dense of the wheel. An open structure would be 12 or high while close structure would be 6 or so.

Wheel Specification

The schematic diagram of wheel specification can be shown below

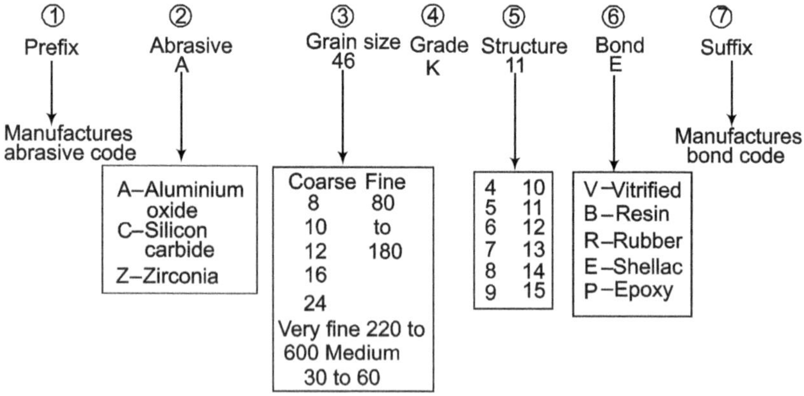

Specification of wheel

Note Coarse grits are used to aggressive stock removal and rough surface finish and finer grit are used to less stock removal but better surface finish.

Grinding Wheel Wear

Grinding wheel wear can be defined as the following groups.

Attrious Wear

The cutting edges become dull and developing a wear flat that similar to flank wear in cutting tool.

Dressing

It is a process of sharpening the abrasive elements of the wheel. It is done by star dressers, diamond dresser and dressing sticks.

Truing

It is the process of aligning the periphery of grinding wheel so that it runs concentric with its axis of rotation. Truing and dressing occurs simultaneously with regard conventional wheels. In super abrasive wheel, the two processes are accomplished separately with truing performed first.

Grinding Processes

- **Surface grinding** Used for grinding flat surfaces.
- **Cylindrical grinding** This process is used for grinding external and internal cylindrical surfaces.
- **Centred grinding** This is used for grinding axially symmetric surfaces, they are centred or centreless.
- **Centreless grinding** Centreless grinding is high speed, low cost operation. In this process, two wheels are used, first one is grinding wheel and second one is governing wheel.

Jigs

Jigs are used to hold work piece and guide to tool into a predetermined position on the work piece. Jigs size depend on the dimensions of the part that is designed to hold.

Fixture

A fixture is work holding device that is bolted or otherwise fastened to the machine. A fixture does not provide guiding of the processing tool.

Principle of Location

In this method, the work piece is located by means of six pins which restrict the work piece in nine of its degree of freedom.

Principle of Least Point

In this technique, only minimum locating point should be used to secure location of the work piece in anyone plane.

Principle of Extreme Position

Locating points should be taken as far away from another as possible on anyone work piece surface.

Principle of Mutually Perpendicular Planes

Locating points are in such a way that their positions are mutually perpendicular planes to each other.

Metal Forging

Forging can be defined as a method of shaping heated metal by compression. Forging can be done either at room or elevated temperature. Forging at the room temperature known as cold forging. In cold forging temperature of material is below its recrystallization temperature. If forging is carried out above recrystallization temperature then, it is known as hot forging.

Forging basically involves three different operations

Drawing

In this operation, the length of a job is increased by decreasing its cross-sectional area.

Upsetting

In this operation, the length of a metal is reduced while increasing the cross-sectional area.

Squeezing

This will lengthen the metal, but will not increase the cross-sectional area.

Operations of Smith Forging

Some basic operations of smith forging can be defined as

Swaging

It is accomplished by hammering the metal stock while it is held on anvil within anyone of various concave tools called swages.

Bending

It is accomplished by hammering the metal stock around a form.

Cogging

It is also known as drawing out. It is basically open die forging operation in which the thickness of a bar is reduced by successive forging steps at specific intervals. Because the contact area per stroke is small, a long section of bar can be reduced in thickness without requiring large forces or machinery.

Coining

Coining essentially a closed die forging process typically used in minting coins, medalions and jewellry. The slug is coined completely by closed die cavity. To produce fine details the pressure required can be high as five or six times the strength of material.

Heading

Heading is essentially an upsetting operation usually performed at the end of a round rod or wire in order to produce a larger cross-section. At the machine, this process is being done and known as headers, the head of bolt making by this process.

Piercing

Piercing is a process of identifying the surface of work piece with a punch in order to produce a cavity of an impression.

Rolling

It is a process of forging in which changing the cross-section of a long work piece is done by applying a compressive force through a set of rolls.

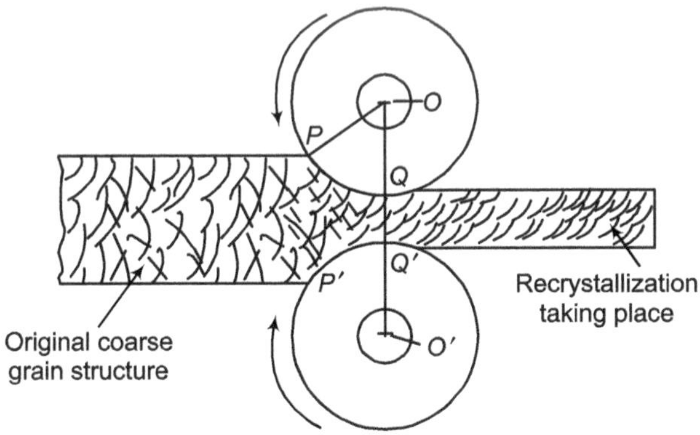

Rolling process diagram

Principle of Rolling

When stock enters in roll then, speed of stock is less than the speed of rolls (speed of rolls means peripheral speed). At a point between P and Q the stock speed is same as the roll peripheral speed. Most deformation takes place in thickness, although there is some increase in width. In all rolling process temperature uniformity is important.

In rolling,

$$\text{Volume enters in roll} = \text{Volume outs from the roll}$$
$$Q_1 = Q_2$$
$$A_1 v_1 = A_2 v_2$$

where, A_1 = area in front of roll (in m^2)
v_1 = velocity of metal before the roll (in m/s)
A_2 = area after rolling (in m^2)
v_2 = velocity of metal after rolling (in m/s).

Analysis of Rolling

A strip of initial thickness h_i enters the roll gap and after rolling the thickness of strip becomes h_f.' The peripheral speed of roll is v_r, When metal passes through roll gap then, velocity is increased due to volume constancy. At the exit of the roll gap, the velocity of strip is v_f. There is sliding occurs between roll gap and the strip because v_r remains constant along roll gap, where velocity of roll and strip are same known as neutral plane and that point is known as neutral point or no slip point.

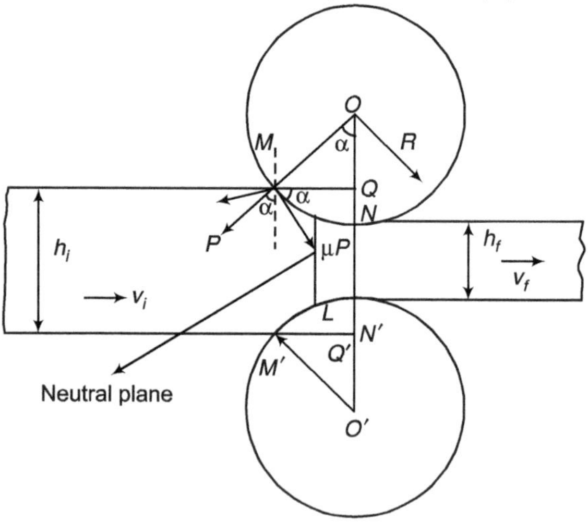

Analysis of rolling diagram

So,
$$v_i < v_r < v_f$$
$$\text{Forward slip} = \frac{v_f - v_r}{v_r}$$
$$\text{Backward slip} = \frac{v_r - v_i}{v_r}$$

The reduction could be achieved with a set of roll is designated as 'the angle of bite or angle of contact'. It is denoted by α in given figure.

Draft

Reduction in thickness known as draft.

$$\begin{aligned}\Delta h &= h_i - h_f \\ &= (ON - OQ) + (O'N' - O'Q') \\ &= (R - R\cos\alpha) + (R - R\cos\alpha) \\ &= 2R\,(1-\cos\alpha)\end{aligned}$$

Horizontal component of force $F_x = \mu P\cos\alpha - P\sin\alpha$

For rolling occurs $F_x \geq 0$

$$\Rightarrow \quad \mu P\cos\alpha \geq P\sin\alpha$$

$$\mu \geq \tan\alpha \;(\tan\alpha)_{max} = \mu$$

$$\Delta h_{max} = 2R\left[1 - \left(1 - \frac{1}{2}\mu^2 + \ldots\right)\right]$$

$$\Delta h_{max} = \mu^2 R \qquad \text{(neglect small terms)}$$

Roll Contact Length

It can be define by the diagram as shown below

Roll contact length AB is approximately equal to projected length AM.

$$BM = B'M' = \frac{\Delta h}{2} = \left(\frac{h_i - h_f}{2}\right)$$

$$AM \times MC = BM \times MD$$

$$AM^2 = \frac{\Delta h}{2} \times \left(2R - \frac{\Delta h}{2}\right)$$

$$L^2 = R\,\Delta h - \frac{\Delta h^2}{4}$$

$$L = \sqrt{R\Delta h - \frac{\Delta h^2}{4}} = \sqrt{R\Delta h}$$

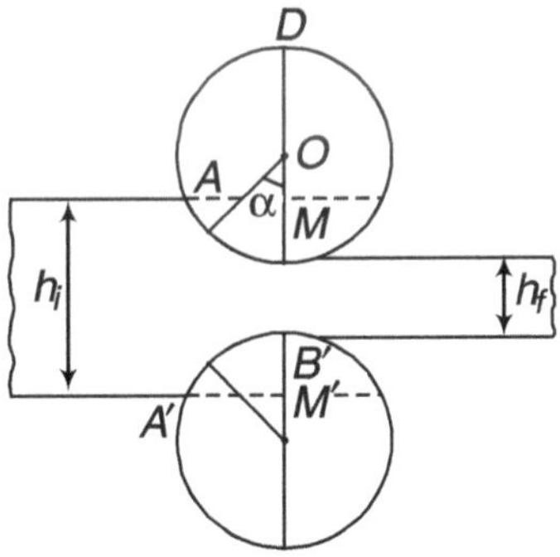

Roll contact length

Δh is small so $\frac{\Delta h^2}{4}$ is too small, we assume it as negligible.

Angle of Bite or Contact Angle

It is given as $\mu^2 R = \Delta h$

$$\mu^2 = \frac{\Delta h}{R}$$

$$\tan^2\alpha = \frac{\Delta h}{R} \qquad (\mu = \tan\alpha)$$

$$\tan\alpha = \sqrt{\frac{\Delta h}{R}}$$

$$= \sqrt{\frac{h_i - h_f}{R}}$$

Roll Force

Roll force is given by $F = P_{av} wL$

where, w = Width of strip

L = Contact length

P_{av} is given by $P_{av} = 2k\left(1 + \frac{\mu L}{2h_{av}}\right)$

So, $F = Lw\,2k\left(1 + \frac{\mu L}{2h_{av}}\right)$

$$h_{av} = \frac{h_i + h_f}{2}$$

Drawing

Wire drawing is an operation to produce wire of various sizes within certain specific tolerances. The process involves reducing diameter of rods or wires by passing them through a series of wire drawing dies with successive die having smaller diameter than the one preceding it. The maximum reduction in area of wire is less than 50%.

Defects in Wire Drawing

Some wire drawing defects are given as below

Bulge Formation

This occurs in front of die due to low reduction and high die angle.

Internal Cracks

It is also known as central burst or centre cracking.

Tendency of internal cracks

- increases with increasing die angle.
- increases with increasing amount of impurities.
- decreases the increasing drawing ratio and friction.

Surface Defects

This defect occurs on work piece, surface due to improper selection of drawing method and inadequate lubrication.

Seams

Seams are longitudinal scratches or fold in materials.

Drawing Force

The expression of drawing force under frictionless conditions is given by

$$F = Y_{av} A_f \ln\left(\frac{A_0}{A_f}\right)$$

where, A_0 = initial cross-sectional area of wire or rod A_f = final cross-sectional area Y_{av} = average true stress of the material in die gap.

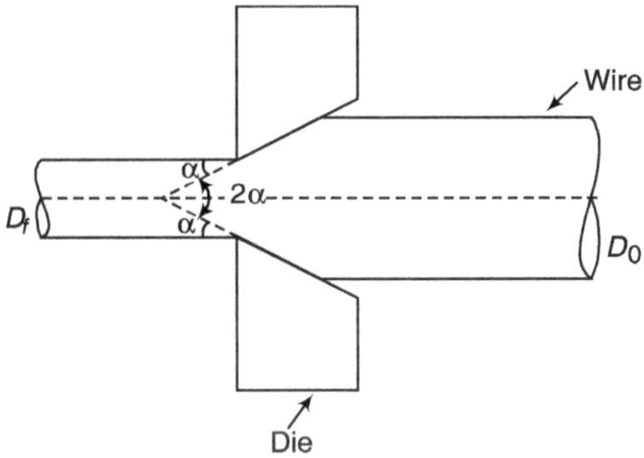

Drawing force diagram

Maximum reduction in diameter in drawing process,

$$R = 1 - \left(\frac{D_f}{D_0}\right)^2$$

or

$$R = 1 - \left(\frac{1}{1+B}\right)^{1/B}$$

where, D_f = Final diameter
D_0 = Initial diameter.

$$B = \frac{\mu}{\tan\alpha} = \mu \cot\alpha$$

where, μ = friction coefficient and α = semi-die angle.

Extrusion

Extrusion is similar to drawing process. Main difference between extrusion and drawing is that in extrusion process we applied compressive force but in drawing process, force in tensile to obtain reduction in diameter. Extrusion produces only compressive and shear forces in stock without any tensile force which makes high deformation possible without tearing the metal.

Extrusion Force

Extrusion force is given by $F = A_0 k \ln\left(\frac{A_0}{A_f}\right)$

where, k = Extursion constant; A_0 = Initial area
A_f = Final area of extrudent part.

Types of Extrusion

The following types of extrusions are given below

Direct Extrusion

It is also known as forward extrusion. In this process, a round billet is placed in a chamber (container) and forced through a die opening by the movement of extruded part in the direction of the applied force.

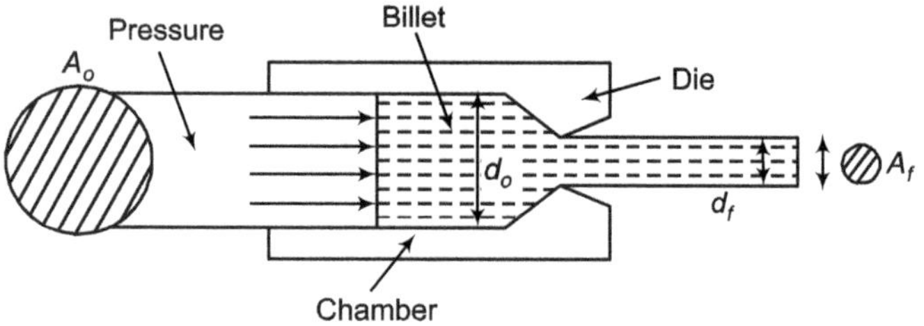

Direct extrusion

Indirect Extrusion

Indirect extrusion is a process in which metal confined in the cavity to flow in a direction opposite to that ram travel. There is no relative motion between billet and chamber wall so, the friction force is less than that for direct extrusion. Power required for process is also less than that for direct extrusion.

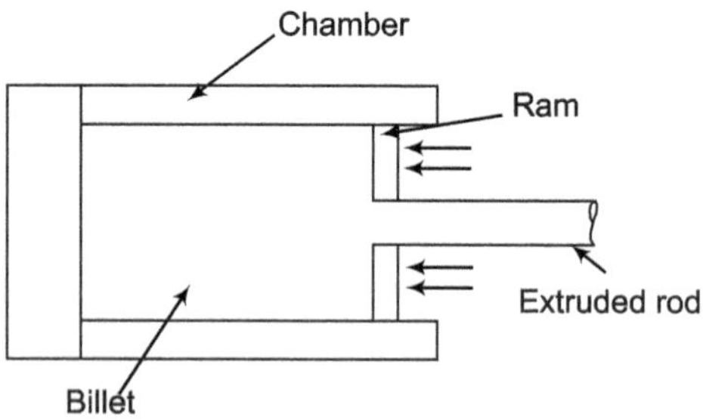

Indirect extrusion

Combined Extrusion

Combined extrusion uses a combination of forward and backward extrusion. The metal in chamber, flow in both up and down directions.

Hydrostatic

In hydrostatic extrusion, the pressure required for extrusion is supplied through an incompressible fluid medium surrounding the billet. There is no container wall friction. The high pressure in chamber transmits some of the fluid to die surface, where it significantly reduces friction and forces.

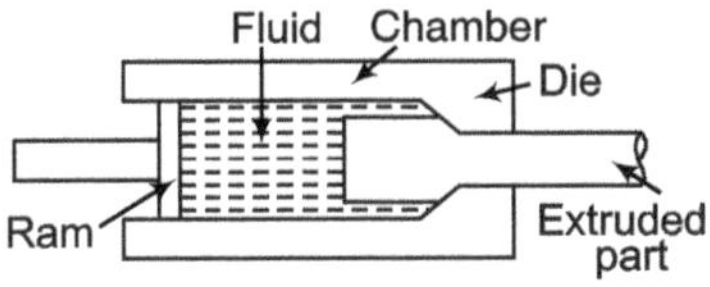

Hydrostatic

Impact Extrusion

It is similar to indirect extrusion. It is often included in the cold extrusion category. The punch descends rapidly on the blank (slug) which is extruded backward'. The thickness of the tubular section is a function of the clearance between die and punch. This is because of volume constancy.

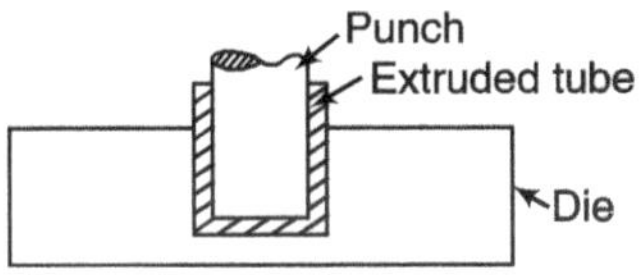

Impact extrusion

Extrusion Defects

Extrusion defects are classified in the following groups

Surface Cracking

If friction, extrusion temperature or speed is too high then, surface temperature rises significantly and this condition may cause surface cracking and tearing. These cracks are along the grain boundaries and usually caused by hot shortness.

Extrusion or Pipe Defect

The type of metal flow pattern tends to draw surface oxides and impurities towards the centre of billet, much like a funnel. This defect is known as pipe defect, extrusion defect or tail pipe or fish-tailing.

Internal Cracking

The centre of extruded part can develop cracks (these are known as centre burst, centre cracking arrow head fracture or chevron cracking) due to a state of hydrostatic tensile at the centre line of the deformation zone in the die.

Sheet Metal Working

Shearing

Shearing is defined as the process in which sheet is cut by subjected, it to shear stresses on a sheet metal part.

A burr is thin edge or ridge. Burr height increases with increasing in clearance and in ductility of sheet metal.

Burr formation also depends on tool sharpness. Dull tool edges greatly contribute to burr formation.

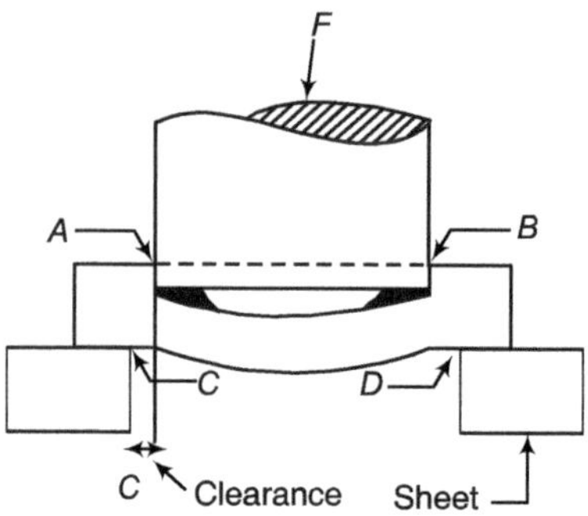

Shearing stress

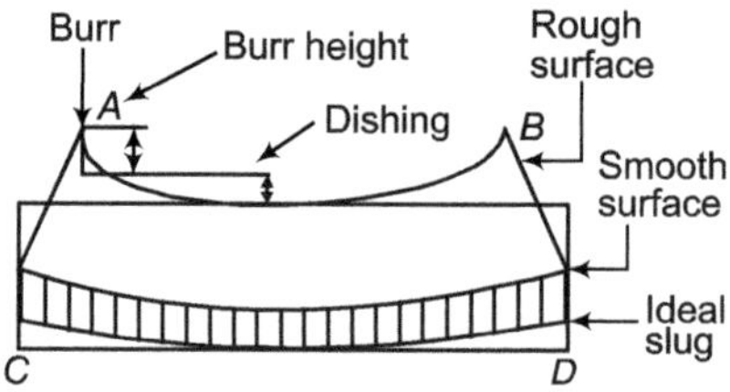

Sheet metal parts

Shearing Operations

The shearing operations are classified into the following manner as given below

- **Blanking** It is a metal fabricating process, during which a metal work piece is removed from a metal sheet when it is punched. In this process, slug is useful part and sheet metal is scrap.
- **Piercing** It is general term for cutting openings such as holes and slots in sheet metal part. In this process,' slug is scrap and sheet metal part is useful.
- **Slitting** Shearing operation can be carried out by means of a pair of circular blades similar to those that in a can opener, this process is called slitting.
- **Slotting** It is cutting of elongated holes or slots.
- **Shaving** It is a finishing or sizing process and very little material is removed from edges.
- **Preformating** Punching a number of similar hole in sheet. They are arranged in a regular pattern in a sheet.

- **Trimming** It is the cutting away of excess material in a flange or flash from a piece.
- **Parting** Shearing the sheet metal into two or more pieces is known as parting process.
- **Notching** Removing pieces (or various shapes) from the edge of a sheet metal is known as notching process.
- **Nibbing** It is a process in which making a contour or other shape by punching a series of overlapping round or square holes along the edge of part.
- **Lancing** Leaving a tab without removing of any material from the sheet metal is known as lancing process.
- **Clearance** In blanking process, cut out piece is desired part. In this case punch is made of smaller size and die is made of exact sizes as that of blank.

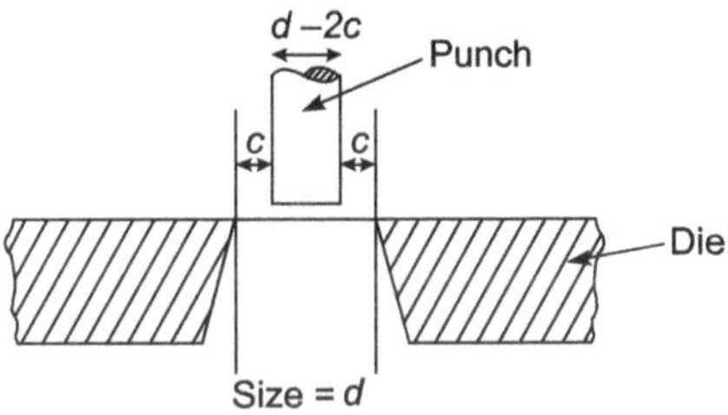

Punch-die system

In piercing process, the cut out part is waste and left out piece is required. In this case punch is made of exact size and die is made, bigger size. Cutting force in cutting shearing operation is given by

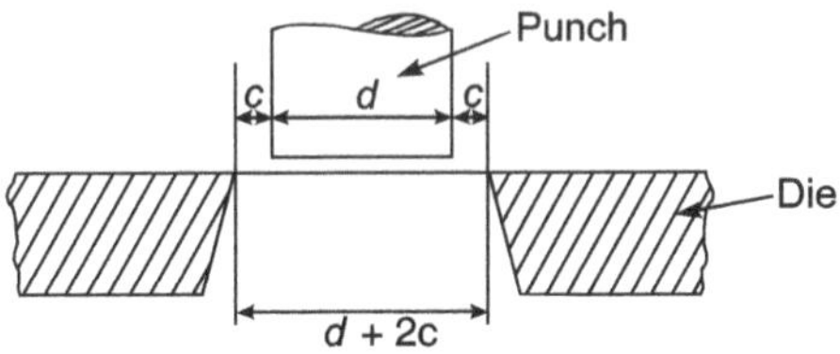

Punch-die system

$$F = \text{Shear area} \times \text{Shear stress of material}$$

$$= \pi D t \tau \quad \text{(for round hole)}$$

$$= L t \tau \quad \text{(for other contour)}$$

where, F = Blanking force (in newton); D = Diameter of punch (in mm)

t = Stock thickness (in mm)

τ = Shear strength of material (in N/mm^2)

L = Shear length (in mm)

The work done by the press $E = F \times p \times t$

where, E = Work done (in N-m)

p = Percentage penetration of punch into stock

Cutting force is reduced by providing shear in punch or die. The total work done remains same. If shear is provided on die or punch, then

Distance travelled by punch =
Percentage penetration × Stock thickness + Amount of shear = $pt + S$

Work done = Maximum shear force × Punch travel = $F_S \times (pt + S)$

Work done remains same in both process.

So, $$F \times p \times t = F_S \times (pt + S)$$

$$F_S = \frac{F \times p \times t}{pt + S}$$

If $S >> pt$,

Then, $$F_S = \frac{Fpt}{S}$$

Calculation for Shear Length in Various Cutting Shape

In this figure a rectangular shape slug is cut out.

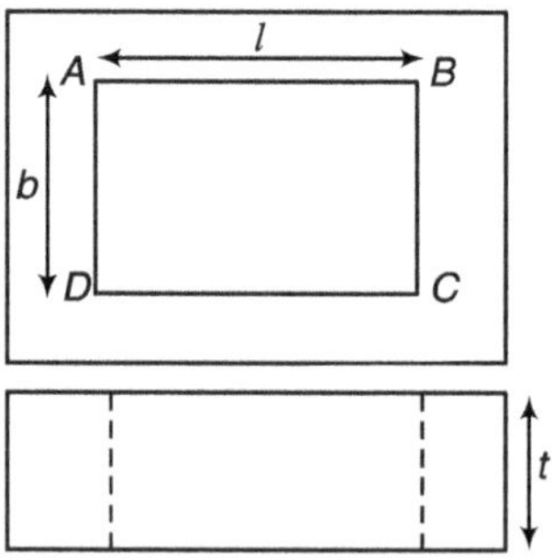

Shear length

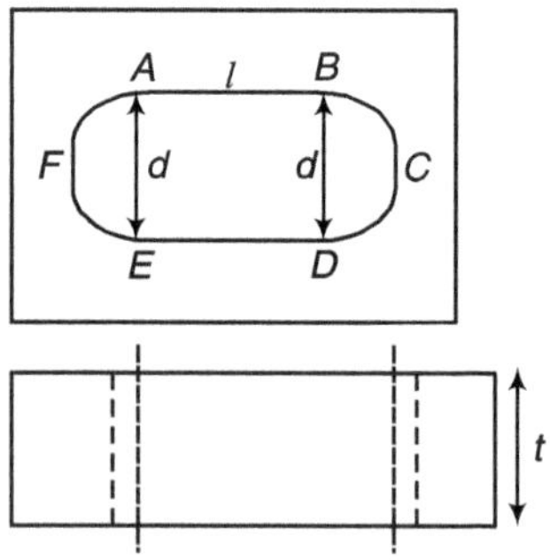

Shear length in various cutting shapes

Then, shear length = $AB + BC + CD + DA$

$$= 2\,(l + b)$$

Shear area $$= 2\,(b + l) \times t$$

$$= 2t\,(a + b)$$

In this figure,

Total shear length = $AB + BCD + DE + EFA$

$$= l + \frac{\pi}{2}d + l + \frac{\pi}{2}d = (2l + \pi d)$$

Shear area = $(2l + \pi d) \times t$

- When cut a square hole of side length a from a t thickness stock

$$\text{Shear area} = 4at$$

- In washer making, D = Outer diameter of washer
  d = Inner diameter of washer
  t = Stock thickness

$$\text{Shear area} = (\pi D + \pi d) \times t$$
$$= \pi t\,(D + d)$$

From above example, it is clear that first we should calculate perimeter of cutting lines and then the perimeter is multiplied by stock thickness for determining shear area.

Deep Drawing

Deep drawing is a sheet metal forming process in which a sheet metal blank is radially down into a forming die by the mechanical action of the punch. It is thus a shape transformation process with material retention.

Deep drawing are classified into the following groups

Ironing

It is the operation in which the thickness of the shell wall is reduced and its surface is made smoothen.

Redrawing

When the ratio of blank diameter and the final cup diameter is too large, the drawing operation is performed more than one stage. After the first one, the successive drawings are known as redrawing operation.

Reverse Drawing

In this operation or process, a bending is to be subjected on a drawn cup in the opposite direction to its original drawn direction. This helps in strain softening.

Defects in Deep Drawing

Defects in deep drawing divided into the following types

Earing

This is the formation of edges around the top of a drawn shell.

Buckling

When an uncontrolled deformation pattern perpendicular to the surface of sheet causes by compressive stress occurs then, it causes abend, kink or other wavy condition of the work piece, this defect is known as buckling. Buckling in the wall of the part is referred to a puckering and buckling in flange part is referred to as wrinkling.

Surface Scratch

This occurs on the surface of the draw cup.

Bulging

In this defect, the diameter of a cylindrical shell increases or expanding of the outer walls of any shell or box shape walls were previously straight.

Radius of Draw or Corner Radius of Die

The radius of die corner should be in a specified range in deep drawing processes. Too large radius causes edge wrinkling and too small radius causes thinning and tearing of side walls.

$$R_{cd} = 4t \quad \text{(normal)}$$

$$= 6t \text{ to } 8t \text{ (when blank holder is used)}$$

where, t = Thickness of blank.

Corner Radius of Punch

In ideal case, radius of punch corner should be equal to drawn cup radius. But it is not possible to find exact size in one drawing operation and hence, more redrawing operations are required to obtain final size.

Corner of punch radius should be in a range of $4t$ to $10t$.

where, t = Blank thickness

Draw Clearance

It should be 7 to 20% of blank thickness.

Blank Size

$$\frac{\pi}{4}D^2 = \pi dh + \frac{\pi}{4}d^2$$

$$D = \sqrt{d^2 + 4dh} \qquad \left(\text{when } \frac{d}{r} \geq 20\right)$$

$$= \sqrt{d^2 + 4\,dh - 0.5r} \qquad \left(\text{when } 15 \leq \frac{d}{r} \leq 20\right)$$

$$= \sqrt{d^2 + 4d\ h - 0.5r} \qquad \left(\text{when } 10 \leq \frac{d}{r} \leq 15\right)$$

$$= \sqrt{(d - 2r)^2 + 4\,d\,(h - r) + 2\pi r\,(d - 0.7r)} \qquad \left(\text{when } \frac{d}{r} \leq 10\right)$$

where, D = Blank diameter (in mm)
d = Outer diameter of shell (in mm)
h = Height of shell (in mm)
r = Corner radius of punch (in mm)

Bending

In this process straining of a metal sheet or plate is done, by moving it around a straight axis lying in the neutral plane.

Bend Allowance

Bend allowance is given by $BA = \frac{\pi\theta}{180^\circ}(R + kt)$

where, θ = Bend angle in degree
R = Inside bend radius (in mm)
t = Material thickness
$k = k$ factor $= \frac{t_1}{t}$
$= 0.33$ (when $R < 2t$)
$= 0.5$ (when $R > 2t$)
t_1 = Distance of neutral plane from inside face.
$L_t = A + B +$ BA (Bend allowance)
L_t = Total flat length
$L_t = A + B - BD$ (Bending deduction)

Bending Force

Bending force is given by $F_b = \dfrac{k\sigma_u L t^2}{w}$

where, k = Constant

σ_u = Ultimate tensile strength (in MPa)

t = Thickness of blank (in mm)

w = Width of die opening (in mm)

$k = 1.33$ for die opening of $8t$

$= 1.20$ for die opening of $16t$ for V-bending

$k = 2.66$ for die opening of $8t$

$= 2.4$ for die opening of $16t$ U-bending

$k = 0.33$ for edge bending or wiping bending

Single wiping or edge bending die

$$w = R_1 + R_2 + C$$

$$C = \text{Thickness of blank} = t$$

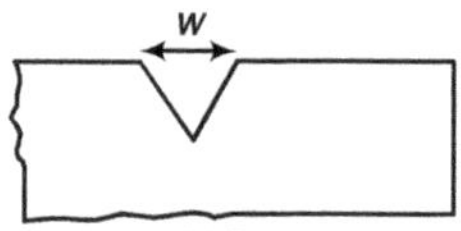

V-Die

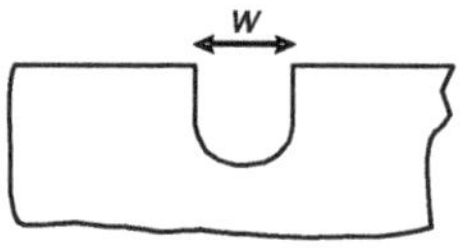

U-Die

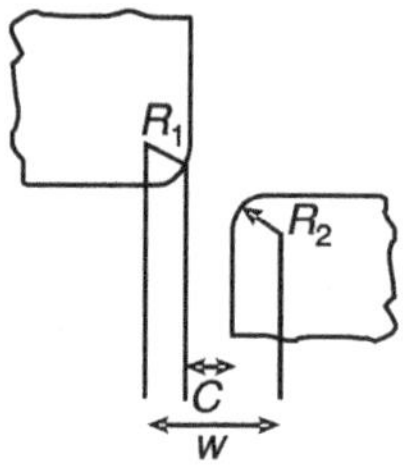

Edge bending die

Powder Metallurgy

Powder metallurgy is defined as making reliable ferrous and non-ferrous parts from powder of ferrous and non-ferrous material.

The basic steps of powder metallurgy are

- Powder production
- Blending and mixing elemental or alloy powders
- Compaction
- Sintering
- Finishing operation

Powder Production

The various methods used to power production. Some basic methods are defined below.

Atomization

In this process, molten metal is broken in form of metal powder with the help of water, air or inert gas spray.

Mechanical Comminution Method

In this process, metal powders are produced by mechanically, Mechanical operation like milling, lathe turning and chipping are to be done on raw materials. These methods produce hard and brittle powders.

Chemical Method

This process is included the production of powder by reduction of metal oxides, precipitation from and thermal decomposition.

Electrolytic Deposition

It involves the precipitation of metallic element at the cathode of an electrolytic cell. Copper powders are produced by this method.

Crushing

This process is used to produce hard and brittle material powder. Ceramic powders are produce by this method. Some metal and alloys are also produced by this method.

Agglomeration

This process is used to produce metal powder where constituents are physically mixed together with an organic binder. This process is used in the manufacture of NiAl, AlUi polyester powder.

Blending

The purpose of mixing is to provided homogeneous mixture and to incorporate the lubricant. The mixing is done in rotating containers called blenders.

Some lubricant is mixed with powder which reduces friction between particles and surfaces of tools along which the powder must slide during compaction.

Compacting

Compaction is the step in which blended powders are pressed into shape in dies. The presses are used driven by mechanically or hydraulically. The purpose of compaction are to obtain the required shape, density and particle to particle contact. After this process the parts are sufficiently strong to be further processed.

Sintering

Sintering is the process where green compacts are heated in controlled atmosphere furnace to a tempera- ture below the melting point. But sufficiently to allow fusion of individual particle.

Finishing Operations

The following are the various types of loads as given below

- **Coining** It is a press working operation on a sintered part to press detail into its surface.
- **Forging** In this process, desired shape is given to sintered part at elevated temperature.
- **Machining** Making threads, holes or slot on sintered part is called machining.
- **Sizing** In this process, the dimensional accuracy of sintered part is improved.
- **Heat Treating** This process is used to improve hardness of sintered part.
- **Stream Treating** In this process, sintered part obtained a oxide casting on its surface.
- **Plating** This operation is used to obtain spectic surface finish and colour.
- **Joining** In this process, two sintered parts are joined by typical joining method like welding, brazing etc.
- **Repressing** Repressing increases the density and improves the properties of a sintered part.
- **Impregnation** The powder metallurgy part has tiny capillary pores all over which can be filled by oil or other lubricant, this process is called impregnation.
- **Infiltration** The pores can also filled by a metal of lower melting point than sintered part by heating the sintered parts when metal is placed against it. It increases strength and density of sintered part. Sintered steel parts are infiltrated by copper or bronze.

Metal Casting

It is a metal forming process, in which metal is formed by allowing molten metal to solidify in a mould. *E* casting, we can make most intricate of shapes, both external and internal.

Casting process can be shown in a flow chart as

Pattern making
↓
Core making
↓
Moulding
↓
Melting and pouring
↓
Cooling and solidification
↓
Cleaning

Pattern Making

First of all, pattern is a replica or facsimile model of desired casting which produces a cavity. Molten metal solidifies in this cavity. The process of making such cavity by using pattern is known as pattern making.

Pattern materials are

1. Wood   2. Plastic   3. Plaster of Paris   4. Wax

Core Making

Cores are used to produce the internal features of part such as hole or any opening. Cores are placed in the cavity produced by pattern.

Moulding

Moulding is the process to prepare a mould for receiving mould metal. Mould must be strong enough to hold the weight metal. ***Moulding process*** *generally involves*

- Placing a moulding aggregate around a pattern
- Pattern is held with supporting frame

- Removal of pattern to produce cavity
- Setting the cores in cavity
- Closing of mould

Melting and Pouring

Melting is used to melt the metal, so that it can flow inside mould cavity. Molten metal is poured in pouring cup. This is known as sprue and horizontal portion of this cup and is known as runner. The points through which metal enters into cavity are known as gate.

Cooling and Solidification

We know, density of solid is more than liquid. Hence, when molten metal solidifies, the size of solid is less in comparison of molten metal. This process is known as shrinkage.

Cleaning

When metal is cooled and solidified, we withdraw it. Cleaning is done to remove the dust or other particles of moulding sand.

Design of Gating System

Bernoulli's theorem plays an important role in regulating the flow of metal in gating system. According to Bernoulli's theorem, 'the total energy head remains constant at any system'. It is given by

$$\frac{p}{\rho g} + \frac{v^2}{2g} + h = \text{constant}$$

where, p = Pressure (in N/m^2 or Pa)

v = Velocity of the liquid (in m/s)

h = Potential head (in m)

ρ = Density of the liquid (in kg/m^3)

$g = 9.8$ m/s^2 = Acceleration due to gravity

According to continuity law, 'mass of the metal flowing at any section in the would is constant'.

$$m = \rho_1 A_1 v_1 = \rho_2 A_2 v_2$$

But density remain constant.

So,

$$\rho_1 = \rho_2$$

$$A_1 v_1 = A_2 v_2$$

$$Q = A_1 v_1 = A_2 v_2$$

where, Q = Rate of flow of the metal (in m^3/s)

A = Area of cross-section of that section, where metal is flowing (in m^2)

v = Velocity of the metal (in m/s)

Mould Filling-up Time

Various possible mechanism of mould filling-up time are given as

Top Gate

Top gate can be shown in the given figure as

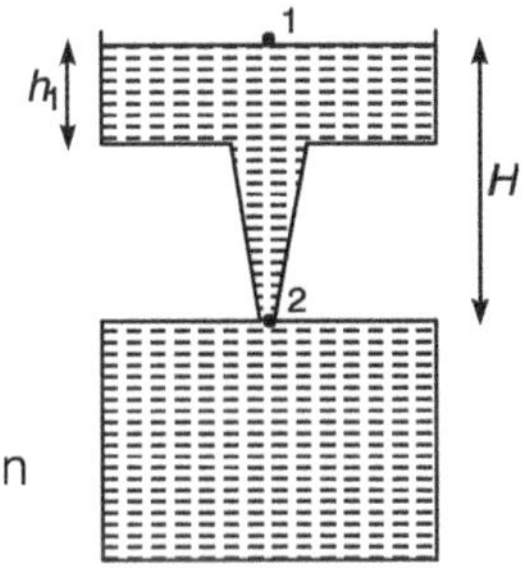

Top gate

Velocity of the liquid at the point 2 = $\sqrt{2gH}$

If the area of gate is A_g, then volume flow rate

$$= A_g \times \sqrt{2gH}$$

If the volume of mould cavity is V_m and it is filled in time t, then

$$A_g \times \sqrt{2gH} \times t = V_m$$

$$\therefore \quad t = \frac{V_m}{A_g \sqrt{2gH}}$$

Bottom Gate

It can be shown below in the figure

Let a dx thickness or height in mould cavity filled by gate in dt time. The height of strip from the bottom of the cavity is x. Applying Bernoulli's equation between points 1 and 2.

$$\frac{p_1}{\rho g} + \frac{v_1^2}{2g} + H = \frac{p_2}{\rho g} + \frac{v_2^2}{2g} + 0 \quad \text{...(i)}$$

v_1 can be neglected because pouring basin area is very large and p_1 is atmospheric pressure. So, Eq. (i) becomes

$$H = \frac{p_2}{\rho g} + \frac{v_2^2}{2g} \quad \text{...(ii)}$$

- Since, pressure on the 2-2′ plane will remain same. So, applying Bernoulli's equation between points 2 and 3

$$\frac{v_2^2}{2g} + \frac{p_2}{\rho g} = \frac{p_3}{\rho g} + x + \frac{v_3^2}{2g}$$

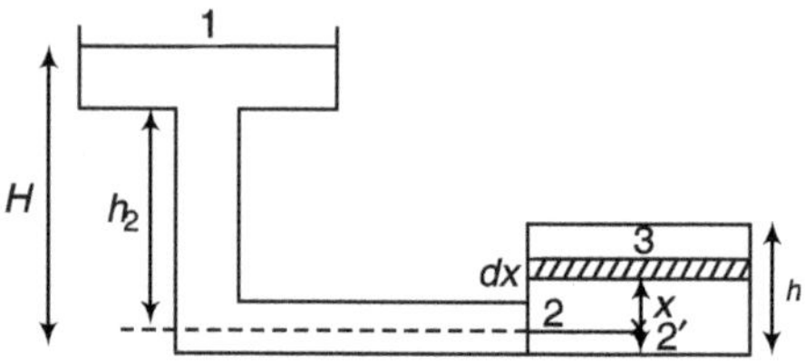

Bottom gate

But p_3 is atmospheric pressure and mould area is same for point 2 and point 3

So, $$v_2 = v_3 \frac{p_2}{\rho g} = x$$

Putting this value in Eq. (ii)

$$\Rightarrow \quad H = x + \frac{v_2^2}{2g} \quad \frac{v_2^2}{2g} = H - x$$

$$v_2 = v_g = \sqrt{2g(H - x)}$$

Applying continuity equation, $A_g \times V_g \times dt = A_m dx$

where, A_m = Area of mould, A_g = Area of gate

$$t = \sqrt{\frac{2}{g}} \frac{A_m}{A_g} [\sqrt{H} - \sqrt{H - h}]$$

Design of Sprue

Applying Bernoulli's equation between points 2 and 3, designing of sorue

$$\frac{p_2}{\rho g} + \frac{v_2^2}{2g} + h_2 = \frac{p_3}{\rho g} + \frac{v_3^2}{2g} + 0 \quad \ldots(i)$$

When the mould is empty then, air will be present there and that time the pressure at the point 3 will be equal to atmospheric pressure. If the area of sprue is constant throughout then,

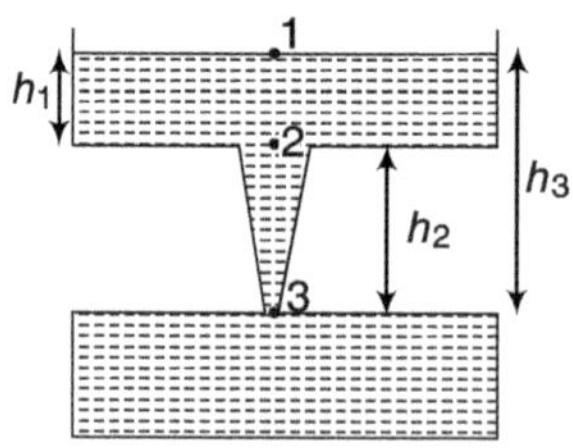

A schematic diagram of sprue

$$v_2 = v_3 \qquad \text{(From continuity equation)}$$

Then, Eq. (i) becomes $\dfrac{p_2}{\rho g} + h_2 = 0$

($p_3 = 0$, here atmospheric pressure is assumed zero)

$$\frac{p_2}{\rho g} = -h_2$$

It means that at point 2 in the sprue, there will be vacuum and it will capture atmosphere into liquid through sand voids. This will produce blow holes in the casted part. This phenomenon, is known as aspiration effect. So, avoid this phenomenon, the cross-section area of the sprue is changed to have uniform pressure (atmospheric pressure) throughout the sprue. Let again consider points 2 and 3 when pressure is uniform (at atmospheric pressure) throughout the sprue.

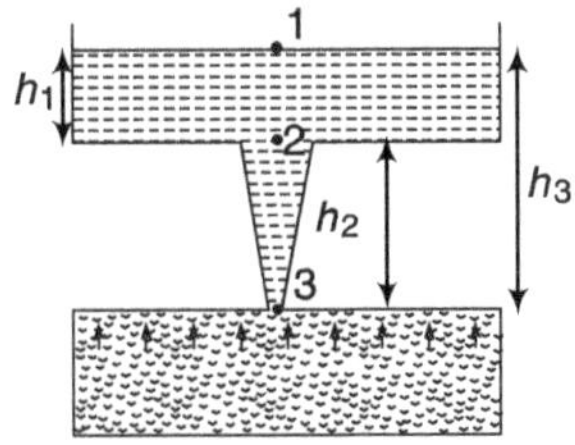

Sprue diagram (aspiration effect)

$$\frac{v_2^2}{2g} + h_2 = \frac{v_3^2}{2g} \qquad \ldots\text{(ii)}$$

From continuity equation $A_2 v_2 = A_3 v_3$

$$\frac{A_2}{A_3} = \frac{v_3}{v_2} \quad v_2 = \frac{A_3}{A_2} v_3$$

where, $\dfrac{A_3}{A_2} = R$

R is known as area ratio $\quad v_2 = R\, v_3$

Applying Bernoulli's equation between points 1 and 2,

$$\frac{p_1}{\rho g} + \frac{v_1^2}{2g} + h_3 = \frac{p_2}{\rho g} + \frac{v_3^2}{2g} + 0$$

$v_1 = 0$ (because basin has large cross-section area)

$p_1 = p_3 = 0$ (because they are at atmospheric pressure)

Design of Risers

The purpose of riser is to provide liquid metal to compensate liquid and shrinkage during phase change. The secondary purpose of riser is to show that whether casting is full or not. So, a riser is designed in such a way that it stays molten longer than the casting.

The solidification time both for casting and riser given as

$$t = k\left(\frac{V}{SA}\right)^2 \qquad \text{(Chvorinov rule)}$$

t = Solidification time (in second)
SA = Surface area (in m^2)
V = Volume (in m^3)
k = Mould constant

- The freezing ratio x of a mould is defined as the ratio of cooling characteristics of casting to the riser.

$$x = \frac{\left(\frac{SA}{V}\right)_{\text{casting}}}{\left(\frac{SA}{V}\right)_{\text{riser}}}$$

- In order to be able to feed the casting, the riser should solidify last and hence, its freezing should be greater than unity.
  So, we can write it as

$$\left(\frac{SA}{V}\right)_{\text{casting}} \geq \left(\frac{SA}{V}\right)_{\text{riser}}$$

- Based on the Chvorinovi's rule, Caine developed a relationship empirically for the freezing ratio of steel casting. The equation is

$$x = \frac{0.10}{y - 0.03} + 1$$

where, $y = \dfrac{\text{Riser volume}}{\text{Casting volume}}$

For aluminium casting, $x = \dfrac{0.10}{y - 0.06} + 1.08$

Sphere have minimum ratio of surface area to the volume. But it is not used as riser for casting because hottest metal being at centre of sphere and it is difficult to use for feeding the casting.

Generally, cylindrical shaped risers are used for casting purpose.

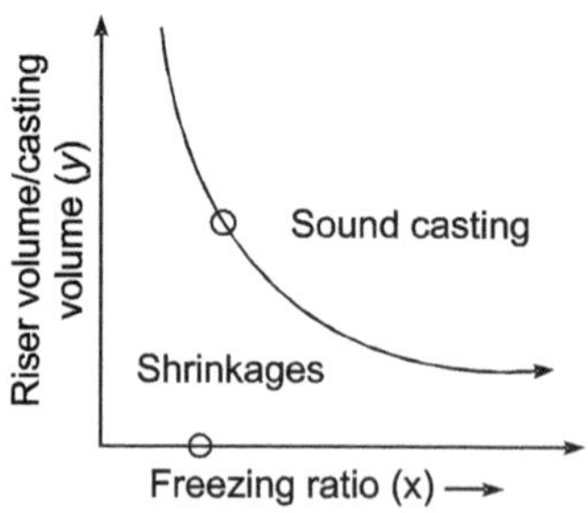

y-x graph

Design for Top Riser

The riser which is placed on the top of casting and this surface in contact with metal is called top riser.

Area of riser exposed to atmosphere

$$A = 2\pi rh + \pi r^2 \qquad \ldots\text{(i)}$$

where, r = radius of riser

h = height of riser

Volume of riser $\quad V = \pi r^2 h \; h = \dfrac{V}{\pi r^2}$

$$A = 2\pi r \frac{V}{\pi r^2} + \pi r^2$$

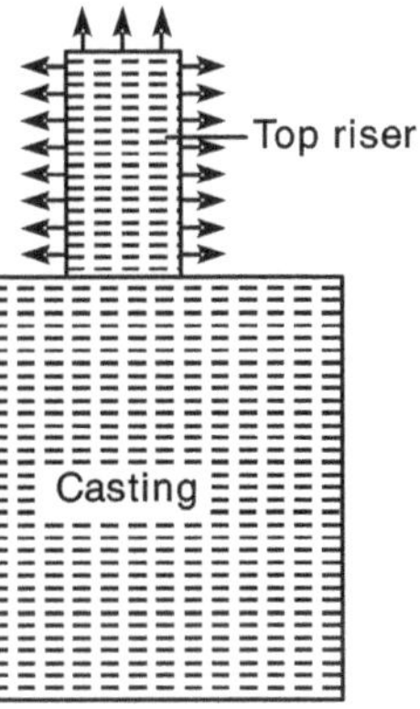

Top riser

Area of riser should be minimum,

So, $\quad \dfrac{dA}{dr} = 0 \; r = \left(\dfrac{V}{\pi}\right)^{1/3} \; r = h$

or $\quad \dfrac{h}{r} = 1 \quad \text{or} \quad \dfrac{h}{d} = \dfrac{1}{2}$

where, d = diameter of the riser

$$\left(\frac{V}{A}\right)_{\text{riser}} = \frac{\pi r^3}{2\pi r^2 + \pi r^2} = \frac{r}{3} = \frac{d}{6}$$

Design for Side Riser

Side riser which is placed in the side of casting and this surface in contact with metal is called side riser.

Volume of riser $V = \pi r^2 h$

Surface area exposed to atmosphere of the riser

$$A = 2\pi\, rh + \pi r^2 + \pi r^2$$

$$= 2\pi rh + 2\pi r^2$$

$$= 2\pi r \frac{V}{\pi r^2} + 2\pi r^2$$

$$= \frac{2V}{r} + 2\pi r^2$$

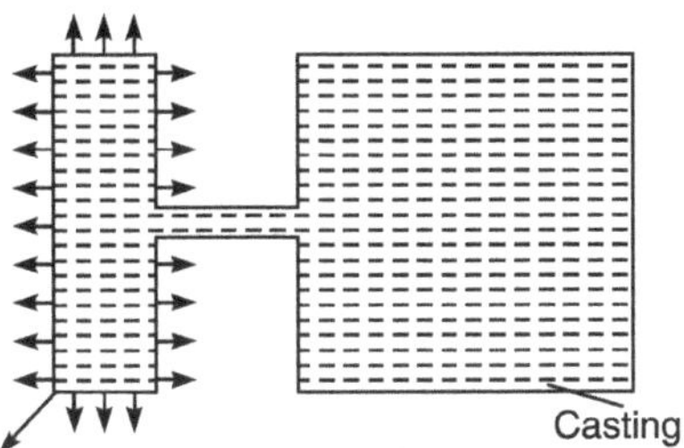

Side riser

- Area of riser which exposed to atmosphere should be minimum.

So, $$\frac{dA}{dr} = 0$$

$\Rightarrow$ $$r = \left(\frac{V}{2\pi}\right)^{1/3} \frac{h}{d} = 1$$

where, d = diameter of the riser $$\left(\frac{V}{A}\right)_{\text{riser}} = \frac{2\pi r^3}{4\pi r^2 + 2\pi r^2}$$

$$= \frac{r}{3} = \frac{d}{6}$$

- In both cases (top riser and side riser), volume to area ratio $\left(\frac{V}{A}\right)$ remains constant.

Welding

Welding is defined as a material joining process which produces coalescence of materials by heating them to suitable temperatures with or without the application of pressure or by the application of pressure alone and with or without the use of filler material.

Oxy-Fuel Gas Welding

Oxy-Fuel gas Welding (OFW) is a welding process that uses a fuel gas combined with O_2 to produce a flame. This flame is the source of heat that is used to melt the metals at the joint. If fuel gas is acetylene (C_2H_2) then, it is known as oxy-acetylene gas welding. In this process, heat is generated in two pair of chemical reaction. The first reaction occurs at inner core of the flame is

$$C_2H_2 + O_2 \longrightarrow 2CO + H_2 + \text{Heat}$$

This produces one-third of the total heat generated in the flame. Secondary reaction is

$$2CO + H_2 + \frac{3}{2} O_2 \longrightarrow 2CO_2 + H_2O + \text{Heat}$$

Secondary reaction produce about 2/3rd of the total heat. The temperature developed in flame about 3200°C as the result of these reactions.

Types of Flame

Flame can be classified into the following types

Neutral Flame

In neutral flame, the ratio of oxygen and acetylene is 1 : 1.

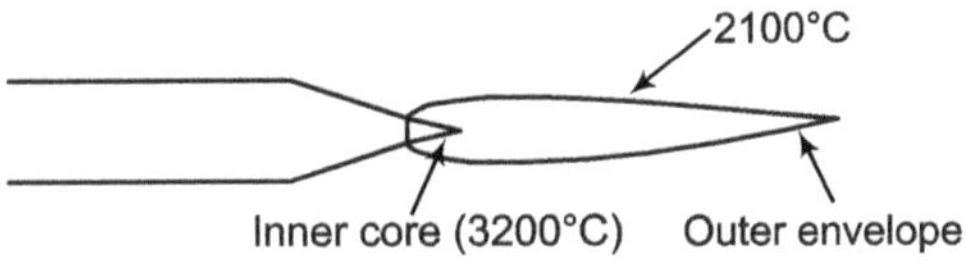

Neutral flame

Oxidising Flame

In this flame, the quantity of oxygen is greater than acetylene. This is harmful, especially for steel because it oxidises the steel. This is useful in welding of copper and copper based alloy.

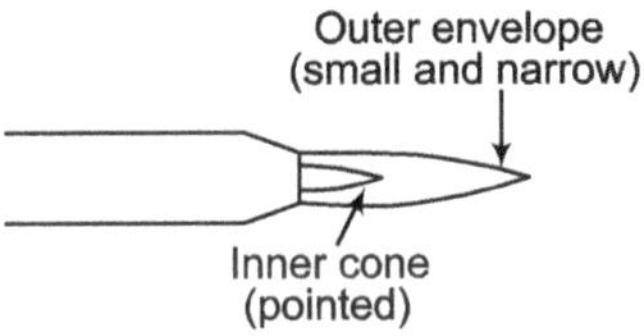

Oxidising flame

Carburising Flame

In this flame, the quantity of the oxygen is less than the acetylene. It is suitable for application requiring low heat such as brazing, soldering and flame hardening

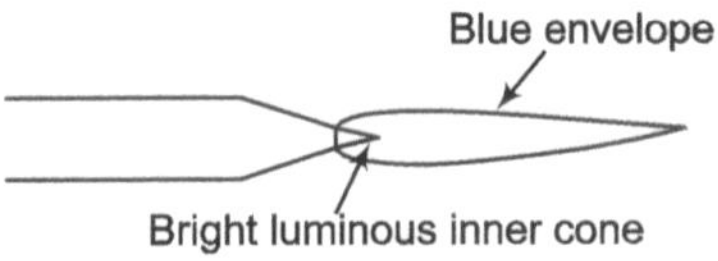

Carburising flame

Arc Welding

In this process, heat is obtained from electrical energy. Electric arc is generated by touching the tip of a coated electrode against the work piece and the withdrawing it quickly to a sufficient distance to maintain the arc. In this welding method, the temperature is generated about 6500°C.

Straight Polarity

In straight polarity, workpiece is positive terminal and electrode is connected with negative terminal respectively. It is preferred for sheet metals because it produces shallow penetration and joints for very wide gaps.

Reverse Polarity

In reverse polarity, electrode is positive terminal and work piece is connected with negative terminal. It is used for welding of thick sheet because it provides deep penetration. DC is used for sheet welding because it produces steady arc. AC is used for welding of thick section and using large electrode diameter at maximum current.

Shielded Metal Arc Welding

According to our requirement, we use AC or DC supply or straight or reverse polarity. Generally, it is used in construction, in ship building, on pipe lines and for maintenance work.

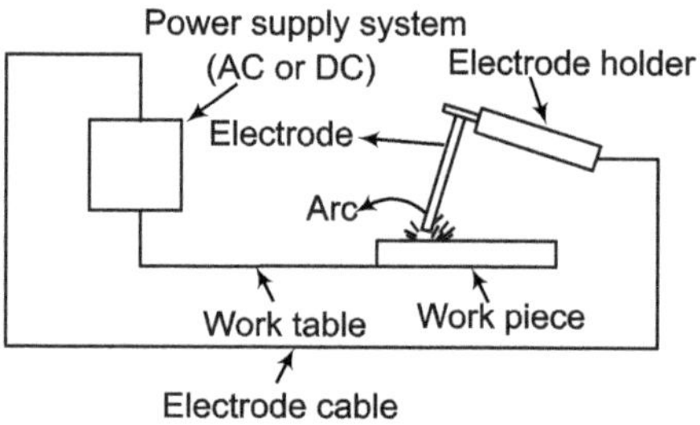

Shielded metal arc welding

Submerged Arc Welding

In this welding process, arc is shielded by a granular flux, consisting of lime, silica, manganese oxide, calcium fluoride and other compounds. The flux is fed into the weld zone by gravity flow through nozzle. The flux layer completely covers the molten metal, it prevents spatter and spark which reduce radiation and fume characteristic of the SAW process. It provides very high welding productivity depositing 4-10 times the amount of weld metal per hour as SAW. By this method we can weld steel, stainless steel sheet, plate often at high speed (5 m/min) rate.

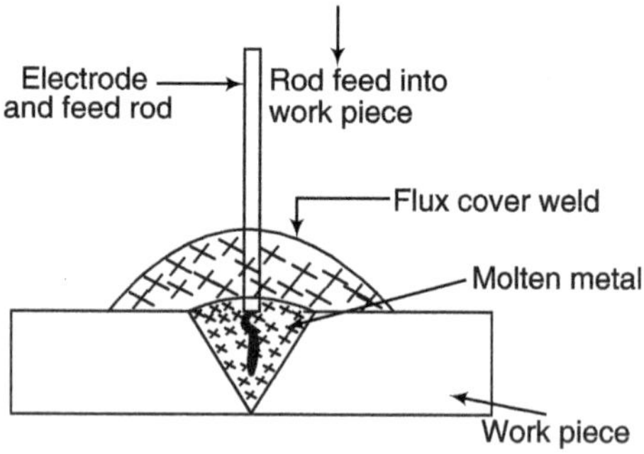

Submerged arc welding

Gas Metal Arc Welding

It is also known as Metal Inert Gas (MIG) welding. The weld area is shielded by an effectively inert atmosphere of argon, helium, CO_2 and various other gas mixtures. The consumable bare wire is fed automatically through a nozzle into weld area or arc.

Deoxidizers are usually present in the electrode metal itself which also prevent oxidation of molten weld puddle.

Metal transfer in this process by four modes

1. Spray transfer 2. Globular 3. Short circuiting 4. Pulsed spray

Spray Transfer

In this metal transfer mode, small droplets of metal from electrode transferred to the weld area at a rate of several hundred droplets per second. Transfer is spatter free and very stable. High DC current and voltages of large diameter electrodes are used.

Globular Transfer

In this transfer, carbon dioxides rich gases are utilized and globules propelled by the forces of the electric arc transfer the metal with considerable spatter. High welding current are used, so high weld penetration occurs. Thick sections are joined by this method.

Short Circuiting

The metal transfer in individual is more than 50 per second as electrode tip touch the weld pool and short circuiting occurs. Low current and voltage are generally used with CO_2 like rich gases. The small diameter used in this method.

Pulsed Spray

The average current required is reduced by using a pulsed arc which superimpose high amplitude pulse onto a low, steady current. This process can be used in all welding positions.

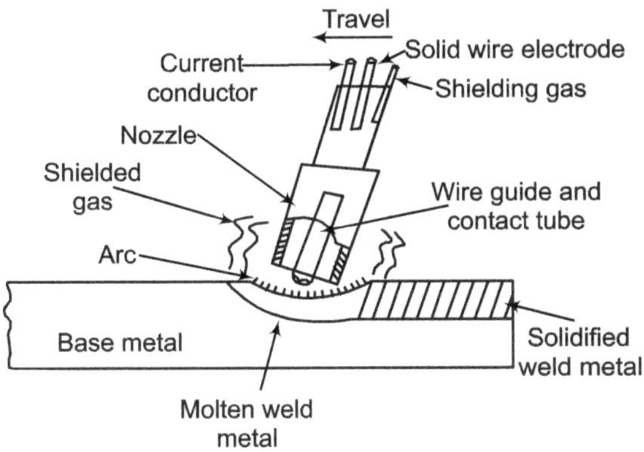

Pulsed and pray

Fluxed Cored Arc Welding

It is similar to gas metal arc welding except that the electrode is tubular in shape and is filled with flux (flux-cored). This is used for steel, stainless steel and nickel alloys.

Electro Gas Welding

Electro gas welding is used primarily for welding the edges of sections vertically in one pass with pieces placed edge to edge (butt joint). The weld metal is deposited into cavity between the two pieces to be joined. This space is enclosed by two copper dam to prevent molten slag to run-off from cavity. Electrode used in this welding is consumable. This is used for welding of steels, titanium and aluminium alloys. The weld thickness range is from 12 mm to 75 mm.

Non-consumable Electrode

The non-consumable electrode basically are used in the following arc welding process which defined as

Gas Tungsten Arc Welding

It is also known as Tungsten Inert Gas (TIG) welding. The filler metal is supplied from a filler wire because, tungsten electrode is not consumed in this operation, a constant and stable arc gap is maintained at constant current level.

The shielding gas used in this process is helium or argon or mixture of the both gases. Depending on the metals to be welded, the power supply is either DC at 200 A or AC at 500 A. AC current is removed oxides and improved weld quality.

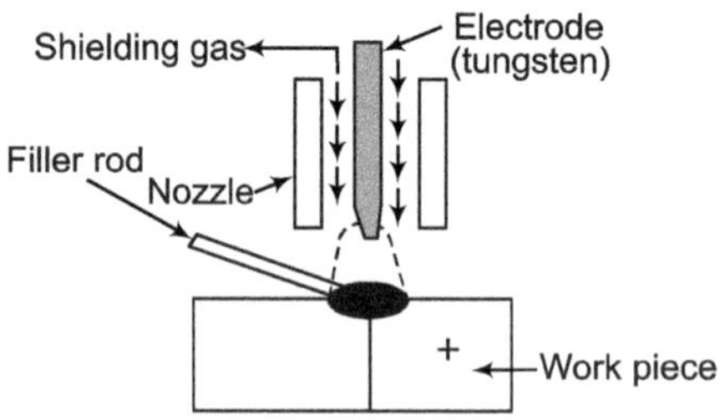

Gas tungsten arc welding

Electro Slag Welding

It is similar like electro gas welding. The main difference is that the arc is started between electrode tip and the bottom of the part to be welded. Flux is added then is melted by the heat of the arc. When molten slag touches the tip of electrode, the arc is extinguished. Then, heat is produced by electrical resistance of the molten slag. It is used for welding of thick section, ranging from 50 mm to more than 900 mm. Current required 600 A at 40-50 V, although higher currents are used for thick plates. Welding speed is about 12 to 36 m/min.

Atomic Hydrogen Welding

In this welding process, arc in a shielding atmosphere of hydrogen. The arc between two tungsten or carbon electrodes. The hydrogen gas also cools the electrodes.

Plasma Arc Welding

It is similar to gas tungsten arc welding. A concentrated plasma arc is produced and aimed at the weld area. A plasma is ionised hot gas, composed of nearly equal number of electron and ions. The plasma is initiated between the tungsten electrode and the orifice by low current pilot arc. Operating current is usually 100 A but they can be higher value for some special applications.

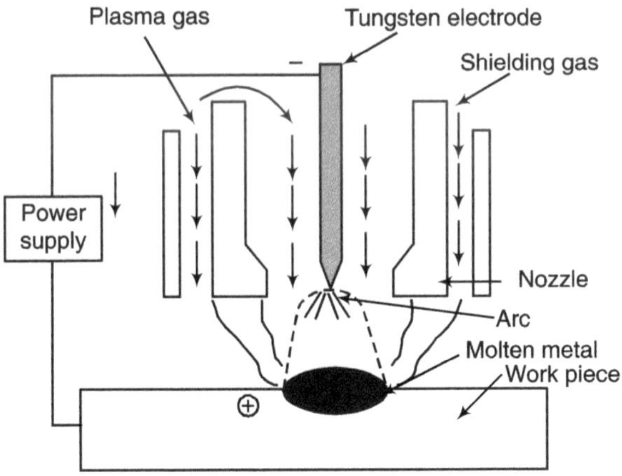

Plasma arc welding

There are two processes of plasma arc welding. In transferred arc method, the work piece is the part of electrical circuit. The arc transfers from the electrode to work piece. In non-transferred method work piece is not a part of electrical circuit and arc produce between tungsten electrode and nozzle and heat is carried to work piece by the plasma gas.

Thermit Welding

In this welding process, heat is produced by exothermic reaction between metal oxides and metallic reducing element. This heat is utilized in the welding. The most common mixture of materials used in welding of steel and cast iron is finely divided particles of Fe_3O_4, Al_2O_3, iron and aluminium. This non-explosive mixture produces maximum theoretical temperature of 3200°C is less than one minute. However, it reaches only about 2200°C.

The reaction is started by applying a magnesium fuse to special compounds of peroxides, chromates or chlorates known as oxidizing agent at an ignition temperature of about 1200°C. It is suitable welding and repairing large forgings and castings. It can also be used to join rails. pipe and thick steel structural sections.

Electron Beam Welding

In this welding process, heat is generated by high velocity narrow-beam electrons. When they strikes to work piece, their kinetic energies change into the heat. The process requires special equipment to focus the beam on the work piece in a vacuum. In higher vacuum beam penetrate more depth greater than the depth to width ratio. This is used to weld thin foil to thick sections. EBM process have small heat affected zone. By this method 150 mm thick section can be weld.

Laser Beam Welding

In this welding process, a high power laser beam as the source of heat is used to make a weld. The beam is focused on small area and it has high energy density so, it has more penetrating power. By this method aluminium, ferrous metals, copper, super alloy' and refractory metal can be weld.

Non-conventional Welding

The non-conventional welding process are used in solid state welding. Some welding methods are defined as

Cold Welding

Cold welding is a solid state welding, this method uses pressure at room temperature to produce coalescence of metals with substantial deformation at the weld. The pressure is applied by roller or die. The both metal should be ductile in nature (most cases).

Diffusion Welding

It is a welding process in which the strength of the joint results primarily from diffusion (movement of atoms across the interface) and secondarily from plastic deformation of the faying surfaces. The temperature involves about 0.5 T_m, where T_m is melting temperature of welding material in kelvin.

Explosion Welding

Explosion welding is a solid state welding process in which coalescence is affected by high movement of the metal to be joined produced by a controlled detonation.

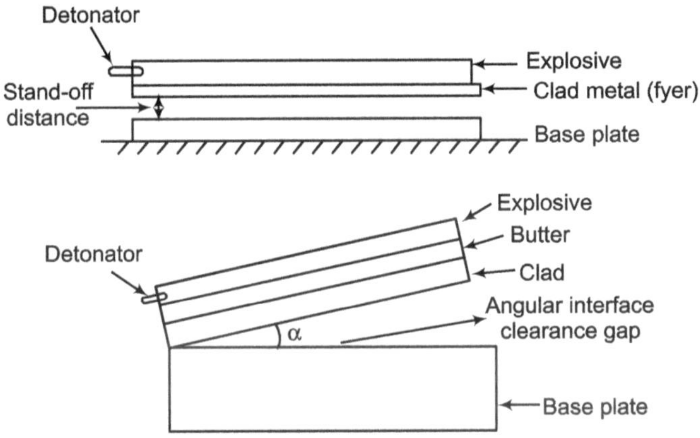

Explosion welding

Resistance Welding

In this welding process, heat is generated by means of electrical resistance across the two components to be joined.

$$H = I^2Rt$$

where, H = Heat generated (in joules)

R = Resistance (in Ω)

I = Current (in A)

t = Time of current flow.

Resistance Spot Welding

In this welding process, the tips of two opposing solid cylindrical electrodes touch a lap joint of two sheet metals and resistance heating produces a spot weld. Obtained a strong bond in the weld nugget, pressure is applied until the current is turned off.

High Frequency Resistance Welding

High frequency resistance welding, is similar like seam welding but supply is replaced by"a high frequency current (up to 450 kHz).

Resistance Projection Welding

In this welding process, high resistance at the joint is developed by embossing one or more projections on one of the surface to be welded. These projections are converted into nugget when a current is passing through material due to more heat generated at the points of projection.

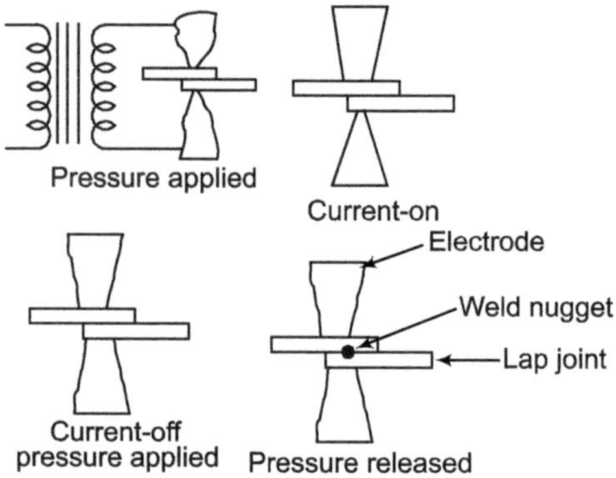

Resistance spot welding

Resistance Seam Welding

It is similar to resistance spot welding but electrode is replaced by roller electrodes. In AC power supply, the electrically conducting rollers produce a spot weld whenever the current reaches a maximum level in AC cycle.

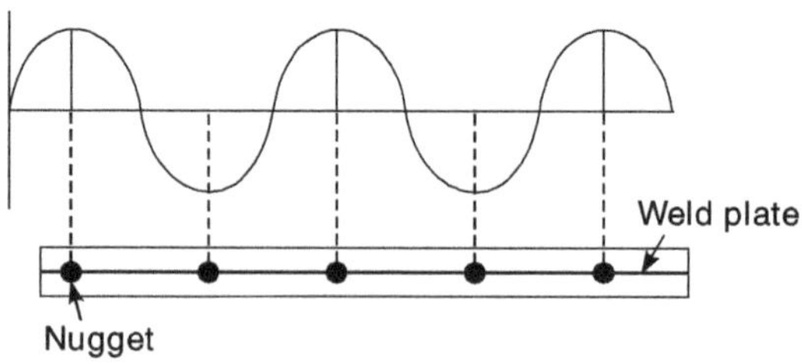

Resistance seam welding

Flash Welding

Flash welding is also known as flash butt welding. Heat is generated from the arc as end of the two member begin to make contact and develop an electrical resistance at the joint.

Stud Welding

Stud welding is also known as stud arc welding. Stud which is a small part of threaded rod or hanger serves as one of the electrode while being joined to another component which is usually a flat plate. A ceramic ring is placed around the joint to prevent from oxidation to weld zone.

Percussion Welding

In this welding process, current is supplied by a capacitor. Capacitor is discharged within 1 to 10 ms to develop localized high heat at joint. This process is used in electronic industries.

Welding Defects

After welding some defects are found which make the strength of joint weak. Such defects are called welding defects. Some welding defects are given below.

Porosity

In this defect, air bubbles (or gases) are present in the weld zone. The distribution of air bubbles in weld zone is random. Porosity caused by gases release during melting of the weld area but trapped during solidification, chemical reaction during welding or by contaminants.

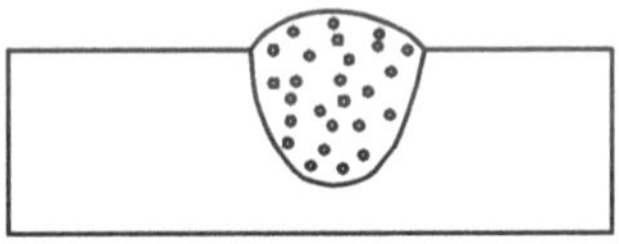

Porosity

Slag Inclusions

Slag inclusions are compound such as oxides, fluxes and electrode contain meterials that are trapped in the weld zone.

Incomplete Fusion

When gap is not totally filled by molten metal, it is known as incomplete fusion.

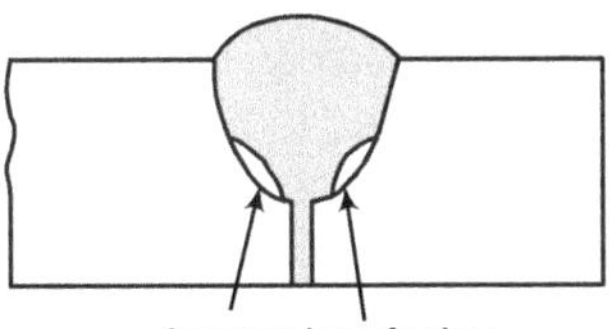

Incomplete fusion

Incomplete Penetration

Incomplete penetration occurs when the depth of the welded joint is insufficient.

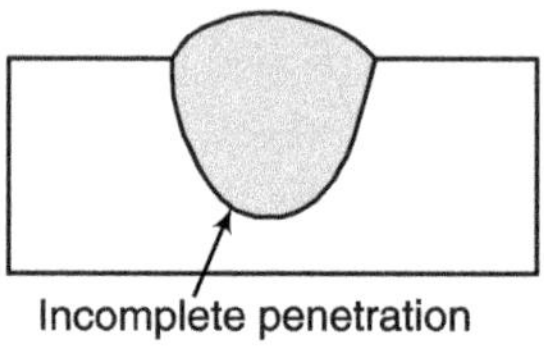

Under Filling

It occurs when the joint is not filled with the proper amount of molten metal.

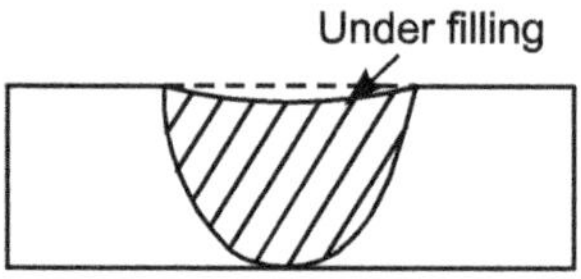

Under filling

Under Cutting

It occurs when base metal melts away from the weld zone and the consequent generation of a groove is in the shape of sharp recess or no tch. It reduces the fatigue strength of the joint.

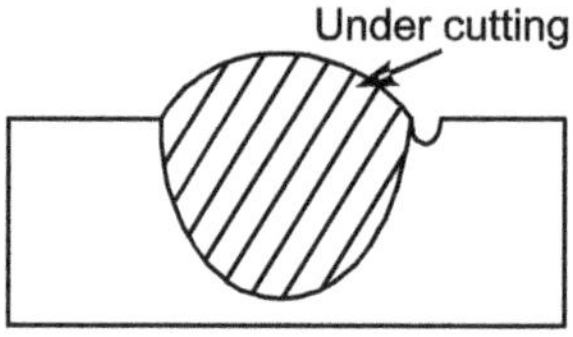

Under cutting

Inclusions

These can occur when several passes are made along a V-joint when joining thick plate using flux cored or flux coated rods and the slag covering a run is not totally removed every pass before following pass.

Lamellar Tears

This is mainly a problem with low quality steel. It occurs in plate that has a low ductility in the thickness direction which caused by non-metallic inclusions (sulphides, oxides) that have been elongated during the rolling process. Lamellar tearing can occur in both fillet and butt welds but the most affected joints are T and corner joints where the fusion boundary is parallel to the rolling plane.

Cracks

Cracks may occur in various locations and directions in the weld area. The typical types of cracks are longitudinal, transverse, crater, under bead and toe cracks. When joint is at elevated temperature then a crack occurs and it is known as hot crack. When it occurs after solidifying weld metal, it is known as cold crack.

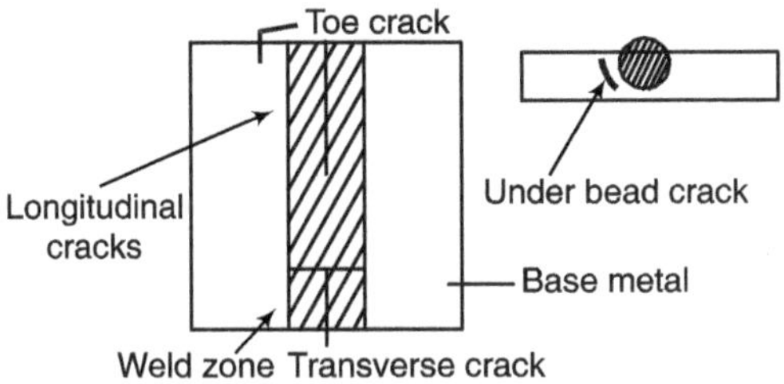

Crack in Welded joints

Electro Chemical Machining (ECM)

In ECM process, material removal takes place, that is reverse of electrolysis. This is high current, low voltage process. The tool material used in electro chemical machining are brass, bronze, copper and aluminium. The gap between electrode and work piece is about 0.03 – 0.05 mm. The material removal rate in electrochemical machining is proportional to the current settings and the temperature of the electrolyte. In ECM process, temperature should be controlled very precisely because even temperature increase by 5°C, it produces over cut of 0.2 mm. According to Faraday's law,

Mass of material removal $\propto$ charge

$$m \propto q$$

$$m = Zq$$

where, m = Mass of material

q = Charge

Z = Electro chemical equivalent and $Z = \frac{E}{F}$

where, E = Chemical equivalent $= \frac{\text{Atomic weight}}{\text{Valency}}$

where, F = Faraday's constant = 96500 C

Then, $m = \frac{E}{F} q$ and $q = It$

where, I = Current passed through electrode

t = Time of current flow in electrode

$$m = \frac{EIt}{F} \quad \frac{m}{t} = \frac{EI}{F}$$

$\frac{m}{t}$ is mass removed in unit time.

$$\text{Metal Removal Rate (MRR)} = \frac{m}{\rho t} = \frac{EI}{\rho F}$$

where, ρ = Material density $MRR = \frac{EI}{\rho F} = sI$

where, s = Specific material removal rate $= \frac{E}{\rho F}$

Electrode feed rate is given by $= s \times J$

where, J = Current density $= \frac{I}{A}$

$$I = \frac{V}{R}\, R = \rho\, \frac{l}{A}$$

where, V = Voltage

ρ = Specific resistance or electrical resistivity

$$s_1 = \frac{I}{A} = \frac{V}{RA} = \frac{V}{\rho \frac{l}{A} A} = \frac{V}{\rho l}$$

where, l = Gap between work piece and tool

When any alloy is given like as (alloy having different elements)

	A	B	C
Atomic weight	x	y	z
Valency	m	n	Q
Percentage	p	q	r

where, $p + q + r = 100$

Then,
$$\frac{1}{E_{alloy}} = \frac{p \times m}{100 \times x} + \frac{q \times n}{100 \times y} + \frac{r \times Q}{100 \times z}$$

where, E_{alloy} = Chemical equivalent weight

If only A and B are given then, $p + q = 100$

$$\frac{1}{E_{alloy}} = \frac{p \times m}{100 \times x} + \frac{q \times n}{100 \times y}$$

Application of ECM

It has main application in machining steam turbine blades and die sinking.

Electrode Discharge Machining (EDM)

EDM process is also called spark erosion machining where the material is removed by erosion of repeated spark.

V_d = Discharging voltage

V_0 = Charging voltage

$$V_d = V_0\,(1 - e^{-t/RC})$$

Frequency of spark $f = \frac{1}{t}$

Energy released per spark $E = \frac{1}{2} CV_d^2$

Average value of power delivered

$$P_{av} = \frac{E}{t_1 + t_2}, t_1 = \text{Charging time}$$

$$t_2 = \text{Discharging time}$$

$$t_1 >> t_2$$

Optimum value of discharge voltage

$$\frac{V_d}{V_0} = 0.72$$

$$V_d = 0.72\ V_0$$

$\Rightarrow V_d$ is 72% of charging voltage.

The major application of EDM is in die sinking.

Ultrasonic Machining (USM)

In ultrasonic machining tool is made up of ductile material between tool and workpiece there is slowy of abrasives. Non-conducting brittle materials can be machined by this process like glasses and ceramics.

MRR ECM > EDM>USM

Total wear USM>EDM>ECM

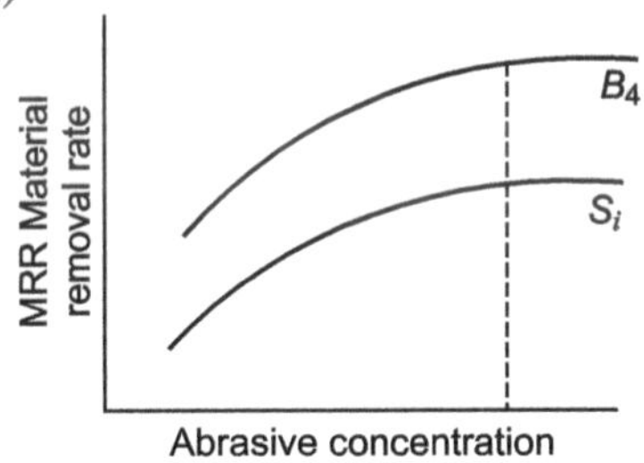

Abrasive Jet Machining (AJM)

For machining very fine holes of about 0.002 mm size abrasive jet machining is used. Abrasive's and air are mixed thrown over the surface at a very high velocity. A crack will be developed as result. By increasing stand of *f* distance. MRR will initially increase but beyond certain value MRR decreases.

Advantages of (AJM)

- Low cost and ease of operation.
- No heat generation

Disadvantages of (AJM)

- Poor surface finish.
- Very low MRR

Laser Beam Machining (LBM)

In this method focusing of laser on workpiece result into a very high surface temperature resulting in metal removal due to melting and evaporation. It is used to drill micro holes cutting very narrow slot in very hard material like diamond. For making wire drawing dies, heat affected zone in this process is very less. It is costly method and is used only when it is not feasible to machine with other process.

Electron Beam Machining (EBM)

It is process of machining material with the use of very high velocity beam of electrons. The workpiece is held in vacuum chamber and electron beam focused on it when strikes the surface of materials, it results into generation of very high temperature leading to metal removal due to melting and evaporation.

The setup consists of electron gun that is the source for electron. A high DC power source, electromagnetic focusing lens, enclosed vacuum chamber and deflecting coils. It is used for drilling very fine holes cutting contours and very narrow slots.

Metrology

Metrology is defined by the International Bureau of Weights and Measures (BIPM) as "the science of measurement embracing both experimental and theoretical determinations at any level of uncertainly in any field of science and technology."

Accuracy *and* Precision

Accuracy is defined as closeness to the exact value and precision is repeatively. Accuracy is related to mean value of the parts produced and precision is related to the process dispersion.

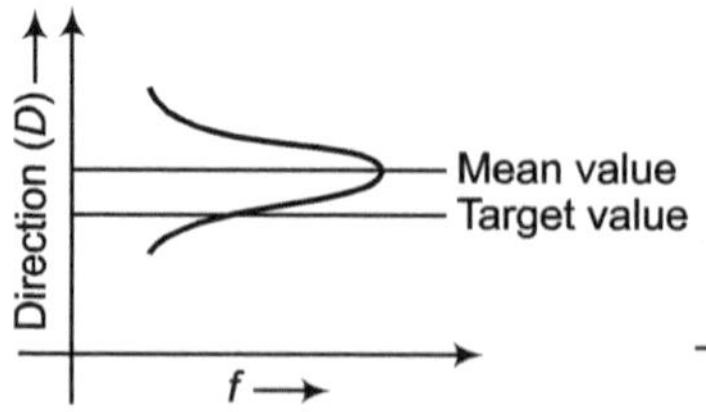

Accurate but not precise

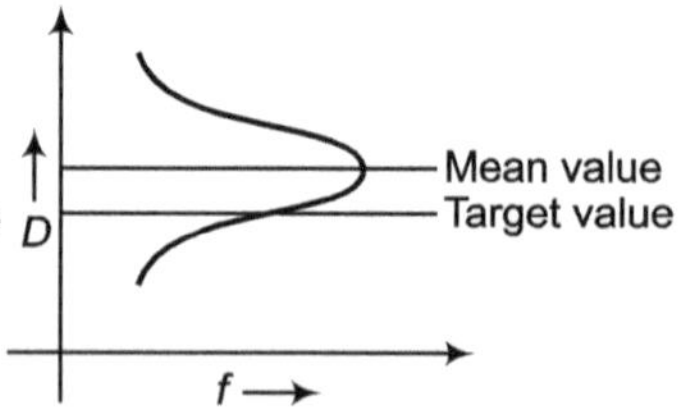

Accurate but not precise

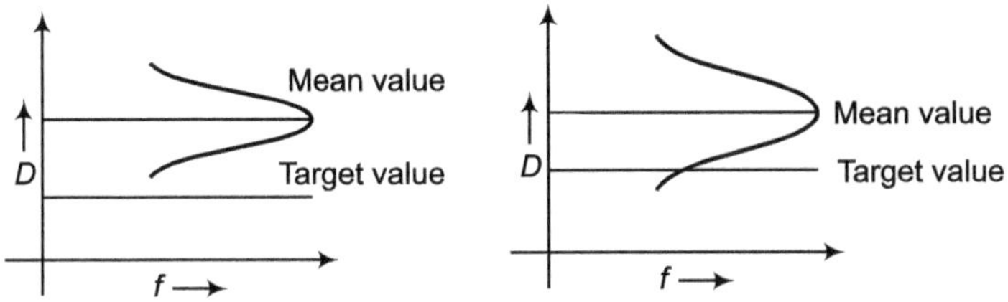

Neither precise nor accurate

Selective Assembly

The process of achieving fully interchangeable system is called selective assembly.

Fully interchangeable system are those system which does not require any inspection after machining. Selected number hole and shaft have desired tolerance individually so that hole and shaft is selected randomly from hole and shaft lot and assemble can made.

Limits, Fits and Tolerances

Assembly parts (either hole or shaft) should come within certain range of dimensions. This acceptable variation is called tolerance larger size is called upper limit and lower size is called lower limit.

Distance from the basic size where the tolerance zone is situated is called fundamental deviation.

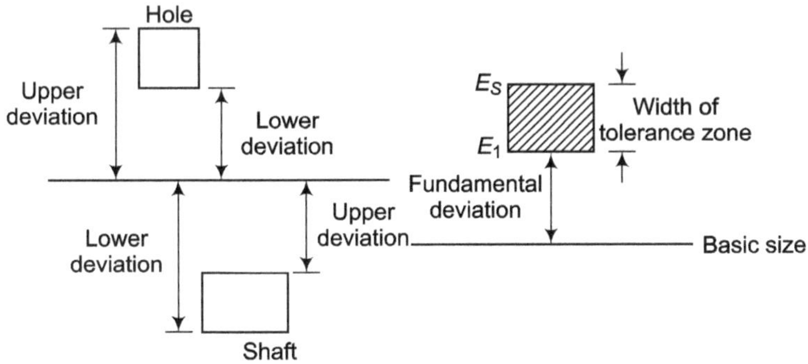

Hole Basis System and Shaft Basis System

If *H* type of hole appears, the system is called hole basis system and if *h* type of shaft is present it is called shaft basis system.

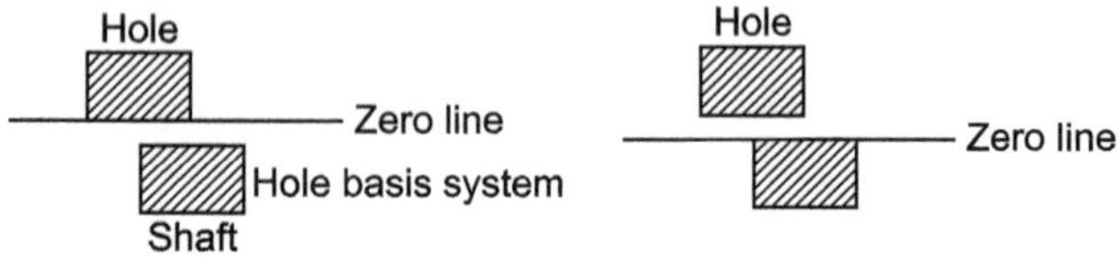

Shaft basis system

Maximum and Minimum Material Limit

Large size shaft and small size hole will have more material. So, these limits are called maximum material limit. Lower size shaft and large size hole will have less material. So, these limits are called minimum material limit.

Broadly there are three types of fits

Clearance Fit

If lower limit of hole is larger than the upper limit of shaft. It is considered as clearance fit.

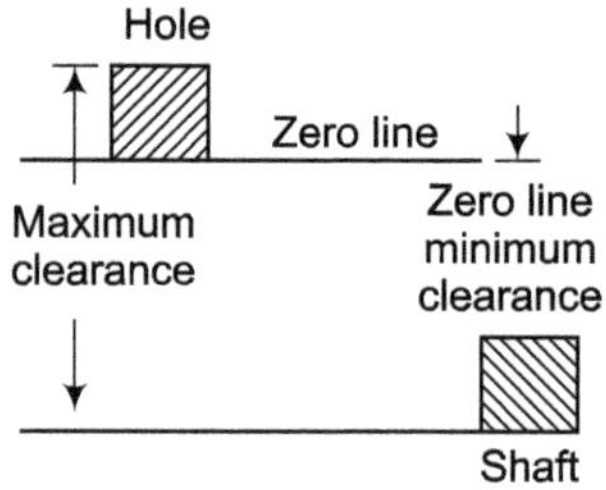

Clearance fit

Transition Fit

This type of fit appears when there is overlap in the tolerance zones. Physically, it indicates that when a part is selected randomly from hole and a shaft lot. Some of the assemblies can be made without the application of force and for some of the assemblies force will be required.

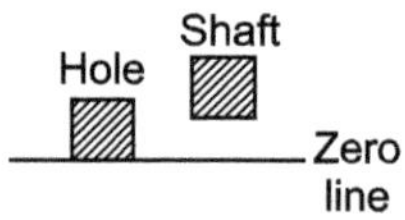

Transition fit

Interference Fit

If maximum size of hole is smaller than the minimum size of shaft then force has to be applied to make the assembly. Such fits are called interference fits

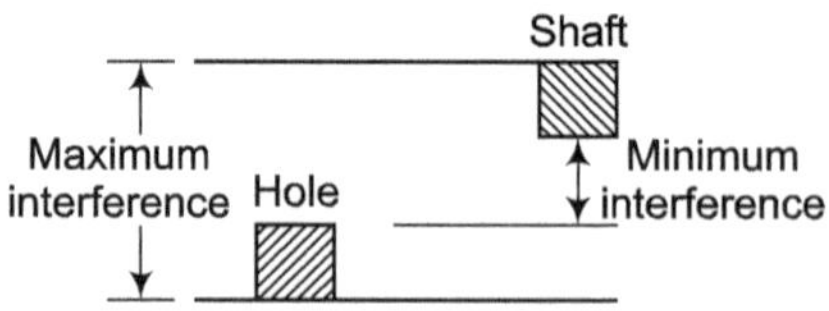

Interference fit

Allowance

It is defined as the difference between maximum material limit of hole and shaft. Depending upon the type of fit either it is equal minimum clearance or maximum interference.

Formula for Fundamental Deviation

$$D = \sqrt{D_{max} \times D_{min}}$$

$$i = 0.45\, D^{1/3} + 0.001\, D$$

Five steps preferred series

$IT01 - IT1 \rightarrow$ Empirical formula

$IT1 - IT15 \rightarrow$ Exact geometric mean

$IT5 - IT16 \rightarrow$ Preferred series

Width of Tolerance

IT 01	IT 0	IT1	IT 2
0.3 + 0.008D	0.5 + 0.012 D	0.8 + 0.02 D	ar
IT_3/ar^2	IT_4/ar^3	$IT_5 / 7i = ar^4$	$IT_6 / 10i$
$IT7 / 16i$	$IT8 / 25i$	$IT9 / 40i$	$IT\ 10 / 64i$
$IT11 / 100i$	$IT12 / 160i$	$IT\ /13\ 250i$	$IT14\ \ 400i$
$IT15 / 640i$	$IT\ 16 / 1000i$		

Design of Limit Gauges

Gauges (GO and NOGO) are designed to check both holes and shafts that wheather they come within their tolerance limit or not.

The gauges used for shafts are called ring gauges and gauges used for checking holes are plug gauges.

$$\text{Gauge tolerance} = \frac{1}{10}\,(\text{work tolerance})$$

$$\text{Wear tolerance} = \frac{1}{10}\,(\text{GT}) = \frac{1}{1000}\,\text{work tolerance}$$

Taylor Principle

- GO and NOGO gauges must be designed to check size as well as shape and has to be in the full form *i.e.*, full working length of holes or shaft.
- GO gauges are designed at maximum material limit condition and NOGO gauges are designed at minimum material limit condition.

Sine Bar

It is used for indirect measurement of an angle of an machined surface. High carbon and high chromium material are used for sine bar as it gives good surface finish.

Types of Error

Errors are classified into the following categories

System Error (systematic error)

These errors in the instrument follow a particular pattern. These errors can be neutralized by calibrating the system.

Short period Error

These errors may appear due to change in environmental conditions like humidity temperature variations in electric instruments, fluctuations recorded while conducting tensile test on USM. These errors can be taken care of by neglecting the irrelevant data.

Erratic Error

These errors appear due to incompleteness of any link. Such error does not follow any pattern. These magnitude is also uncertain so these errors have to be eliminated by inspecting the instrument before starting the experiment.

Comparator

Comparators are used for measuring the dimensions from a reference value.

Pneumatic Comparator

Pneumatic comparator work on the principle that of an air jet. Upon decreasing the stand off distance, the pressure on the back ground of jet will increase. This pressure is called back pressure and can be directly related to the measurement

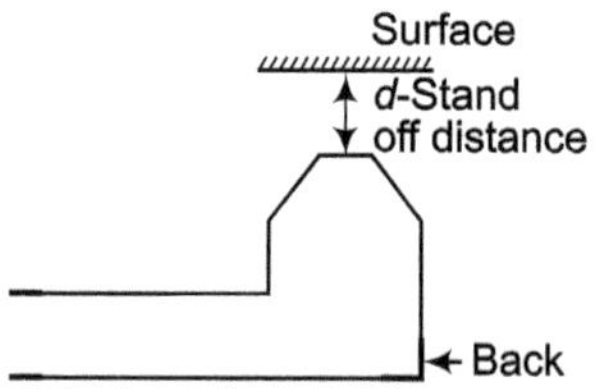

Pneumatic comparator

These comparators can be further divided into three groups

1. Flow type 2. Differential type 3. Back pressure type

$$\frac{p}{P} = A - b\left(\frac{M}{c}\right), \; p = \text{back pressure}$$

P = total pressure B/P and M graph

$$0.6 = 1.1 - \frac{b}{c} M_{max}$$

$$0.8 = 1.1 - \frac{b}{c} M_{min}$$

$$\frac{dP}{dM} = \frac{0.4P}{M_{av}}$$

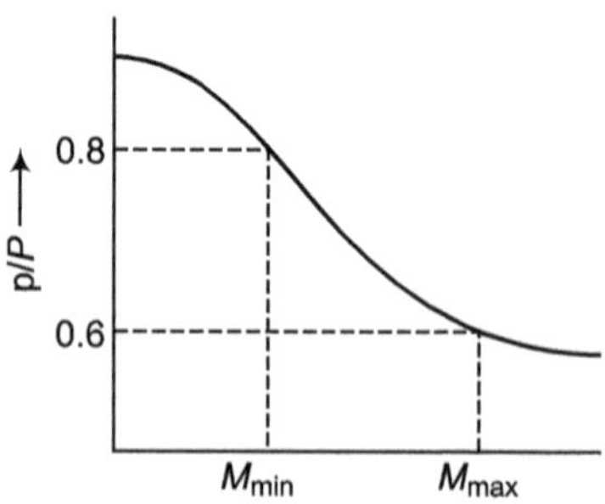

B/*P* and *M* graph

$$\text{Magnification} = \frac{dR}{dl} = \frac{dR}{dP} \times \frac{dp}{dM} \times \frac{dM}{dl}$$

$$\frac{dR}{dP} = \text{Indication sensitivity}$$

$$\frac{dp}{dM} = \text{Pneumatic sensitivity}$$

$$\frac{dM}{dl} = \text{Measuring head sensitivity}$$

$$A = \text{area of control orifice} = \left(\frac{\pi}{4} d_c^2\right)$$

$$M = \text{Measuring head area} = Td_m l$$

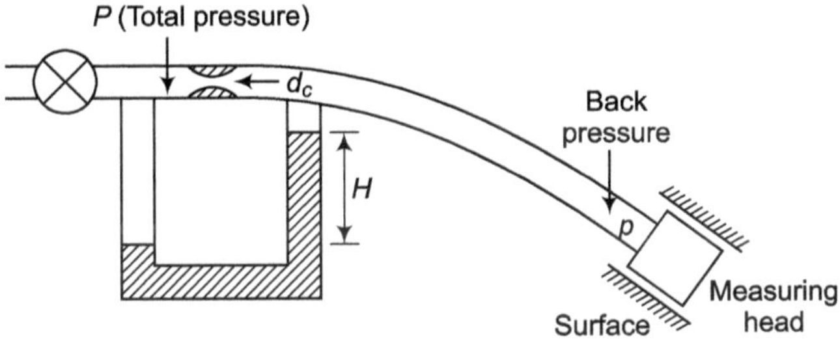

Back pressure type comparator

Surface Finish

A machined surface is having 2 types of irregularities *i.e.*, roughness and waviness. Small wavelength fluctuation are called roughness (primary texture) and large wavelength deviations are called waviness (secondary texture). Waviness is being produced by machine vibrations, chatter, errors, in guide ways etc. Roughness appears due to improper selection of cutting fluid development of temperature on robe face.

Lay and Flow

Predominant surface patterns produced by feed marks is called lay and surface irregularities present which are random is known as flow.

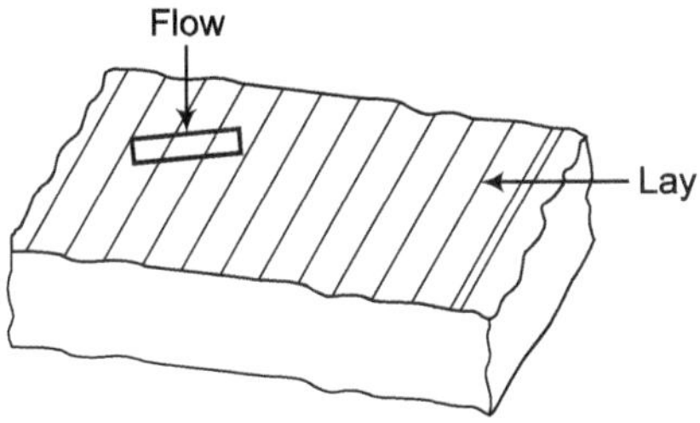

Lay and flow

12

Industrial Engineering

Forecasting

Forecasting is the first major activity in the planning. It involves careful study of past data and present scenario. Forecasting is the projection of past into future while prediction is the judgement in management after taking all available information into account.

We can define forecasting based on time duration as

- Short term forecasting for 1 to 3 months
- Intermediate term forecasting for 3 to 12 months
- Long term forecasting for more than 1year

Classification of Forecasting Method

Forecasting method can be divided into two parts

- Quantitative method
- Qualitative method

Quantitative Method

It is objective in nature and employ numerical information as the basis of making forecast.

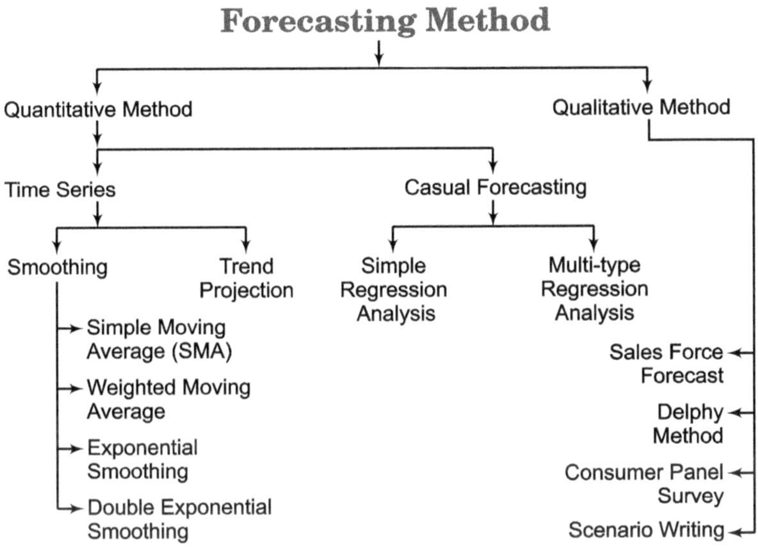

Classification of Forecasting Method

Quantitative method can be further classified in Time series and Casual method

Time Series Method

It identifies the historical pattern of demand for the product and project.

It has four components

(i) Trend (Gradual shift)

(ii) Cyclic (Recurrent upward or downward in time series)

(iii) Seasonal (Periodic fluctuation)

(iv) Irregular

Time series method further divided into Smoothing and Projection methods

Smoothing Method

Smoothing method is used to smooth out the random fluctuation caused by the irregular component of time series.

There are following types of smoothing method

- **Simple Moving Average Method** A Moving average is obtained by summing and averaging the values from a given number of periods respectively, each time, deleting the oldest value and adding a new value. The forecast for $(t+1)$ th period is given by

$$F_{t+1} = \left(\frac{1}{n}\right) \sum_{t=i+1-n}^{t} D_i$$

where, D_i = Actual demand for the i_{th} period.

n = Number of periods included in each average.

- **Weighted Moving Average** This approach is based on the principle that more weight age should be given to relatively newer data.

  Forecast for $(t+1)$ period

  $$F_{t+1} = \sum_{i=t+1-n}^{t} w_i D_i$$

  where, w_i = Relative weight of data for the i_{th} period

  $$\sum_{i=t+1-n}^{t} w_i = 1$$

- **Exponential Smoothing** In this method, the weightage of the data diminishes exponentially as the data become older.

  $$F_t = F_{t-1} + \alpha (D_{t-1} - F_{t-1}) \quad \text{or} \quad F_{t+1} = F_t + \alpha (D_t - F_t)$$

  F_t is also given by

  $$F_t = \alpha D_{t-1} + \alpha (1-\alpha) D_{t-2} + \alpha (1-\alpha)^2 D_{t-3} + \alpha (1-\alpha)^3 D_{t-4} + \ldots$$

  where, F_t = One period ahead forecast made at time t

  D_t = Actual demand for period t

  α = Smoothing constant $(0 \le \alpha \le 1)$

  $$= \frac{2}{N+1} \qquad (N = \text{Number of time periods})$$

Key Points

- Large value of α is selected for fluctuating demand while lower valve of α is selected for smooth demand.
- In exponential smoothing method of forecasting, the forecast for higher values of the smoothing constant will be more sensitive to changing patterns in demand.
- In exponential smoothing method of demand forecasting, demand for the more recent data is given more weightage.

- **Double Exponential Smoothing** Double exponential method is used when there is a trend in data, the raw data sequence of observations is represented by x_t beginning at time $t = 0$. We use S_t to represent the smooth value for time t and b_t is our best estimate of the trend at time t.

$$S_0 = x_0$$
$$S_t = \alpha\, x_t + (1-\alpha)\,(S_{t-1} + b_{t-1})$$
$$b_t = \beta\,(S_t - S_{t-1}) + (1-\beta)\, b_{t-1}$$
$$F_{t+m} = S_t + mb_t$$

where, α = Data smoothing factor ($0 < \alpha < 1$)

β = Trend smoothing factor

$$b_0 = \frac{x_{n-1} - x_0}{n-1}$$

$\Rightarrow$ $F_1 = S_0 + b_0,$ for ($n > 1$)

Trend Projection in Forecasting

In the trend projection method of forecasting, the value of time series exhibits a long term linear trend.

$$y_t = a + bx$$
$$\Sigma y = Na + b\Sigma x,$$
$$\Sigma xy = a\Sigma x + b\Sigma x^2,$$

where, y_t = The trend value, a = Intercept of trend line

b = Slope of trend line, x = Independent variable

Alternatively, $$b = \frac{\Sigma xy - N\,\bar{x}\,\bar{y}}{\Sigma x^2 - N\,(\bar{x})^2}$$

Casual Forecasting Method

In casual method, we try to establish a cause and effect relationship between changes in the sales level for the product and set up relevant explanatory variable.

It can be divided into simple regression analysis and multiple regression analysis

Forecasting using Simple Regression Analysis

$$y_e = a + bx$$

where, y_e = Estimated value of dependent variable

$\Sigma y = na + b\,\Sigma x$

a = y intercept of regression line

$\Sigma xy = a\,\Sigma x + b\Sigma x^2$

b = Slope of the regression line

x = Independent variable

Forecasting using Multiple Regression Analysis

$$y_e = a + b_1x_1 + b_2x_2$$
$$\Sigma y = na + b_1\Sigma x_1 + b_2\,\Sigma x_2$$
$$\Sigma\,x_1y = a\Sigma\,x_1 + b_1\,\Sigma\,x_1^2 + b_2\,\Sigma\,x_1x_2$$
$$\Sigma\,x_2y = a\,\Sigma\,x_2 + b_2\,\Sigma\,x_1x_2 + b_2\,\Sigma\,x_2^2$$

Qualitative Method

Qualitative forecasting is based on person judgement and involve qualities like intuition and experience.

It can be divided into following types

Delphi Method

It is long range forecast based on group consumer by panel of outside expert. It is extremely useful for forecast of new technology or new product and past data are absent. However, accuracy and reliability of forecast is relatively poor and may be time consuming process.

Sales Force Forecast

It is combined estimation at the district, regional and national level to obtain overall forecast by all members comprising sales force of a company. It is closest to consumer.

Consumer Panel Survey (Marketing Research)

It is based on subjective judgements of subjects involves about possible purchases.

Scenario Writing

It consists of developing a conceptual scenario of the future based on a well defined set of assumptions.

Responsiveness and Stability in Forecasting

Responsiveness indicates that forecast as calculated have a fluctuating and swinging pattern whereas the stability means that the forecast show a labelled or flat character.

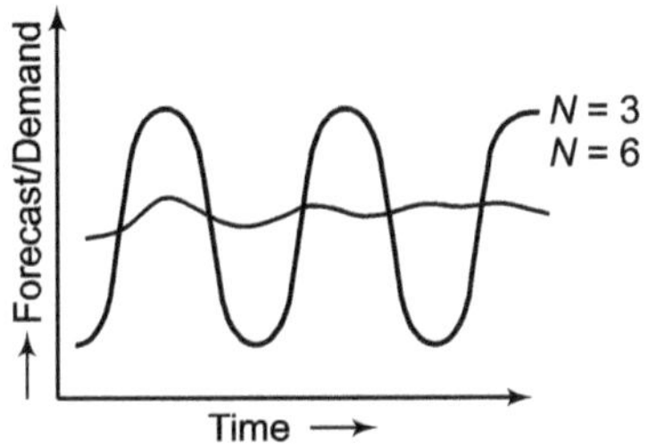

Demand *versus* time

As value of N increases the forecast become stable and lower value of N result in forecast being more responsive.

Measuring Forecast Error

A forecast error is the difference between the actual or real and the predicted or forecast value of time series or any other phenomenon of interest. *There are two methods which are given below*

Mean Absolute Deviation (MAD)

In this method, we calculate as the average of absolute value of difference between actual and forecasted values. The negative sign in this difference is ignored as overestimate as well as underestimate are both off-target and thus desirable.

$$\text{MAD} = \frac{\sum_{t=1}^{n} |D_t - F_t|}{n}$$

where, D_t = Actual value of demand for period t

F_t = Forecasted demand for period t

n = Number of periods considered for calculating error.

Mean Sum of Square Error (MSE)

Average of all squares of all errors in forecast is termed as mean sum of square method. It is written as

$$\text{MSE} = \frac{\sum_{t=1}^{n} (D_t - F_t)^2}{n}$$

Bias

It is a measure of over estimation or under estimation. If the bias value is positive, then it is under estimation and if bias is negative, then it is over estimation.

$$\text{Bias} = \frac{\sum_{t=1}^{n} (D_t - F_t)}{n}$$

Tracking Signal

It is ratio of bias till n periods to mean absolute deviation till n periods. It is used to identify those items which do not keep pace with either positive or negative bias or trend.

$$T_S = \frac{\sum_{t=1}^{n} (D_t - F_t)}{(\text{MAD})_n} = \frac{(\text{Bias})_n}{(\text{MAD})_n}$$

Inventory

Inventory is defined as the list of movable goods which helps directly or indirectly in production of goods for sale. We can also defined inventory as a comprehensive list of movable items which are required for manufacturing the products and to maintain the plant facilities in working conditions.

It can be divided in two parts

Direct Inventories

The inventories which play a direct role in manufacturing of a product and become an integral part of the finished product are called direct inventories. *e.g.,* raw materials, purchases part and finished goods.

Indirect Inventories

The inventories which helps the raw material to get converted into products but not integral part of finished product is called indirect inventories.

e.g., tool and supplies (material used in running the plant but do not go into the product) are indirect inventories.

Inventory Control

It means making the desired item of required quality and in required quantity available to various departments when needed.

Determining Inventory Control

The amount of inventory, a company should carry is determined by five basic variables

(a) Order quantity (b) Reorder point (c) Lead time

(d) Safety stock (e) Buffer stock

Order Quantity

It is the volume of stock at which order is placed or total quantity of buy or sell order.

Reorder Point

It is time between initiating the order and receiving the required quantity.
Reorder point = Minimum inventory + Procurement time × Consumption rate

Lead Time

The time gap between placing of an order and its actual arrival in the inventory is known as **lead time**. It consist of requisition time and procurement time.

It has two components

Administrative Lead Time

From initiation of procurement action until the placing of an order.

Delivery Lead Time

From placing of an order until the delivery of the ordered material.

Safety Stock

If the maximum inventory would be equal to the order quantity Q and minimum inventory would be zero.

$$\text{Average inventory in this case} = \frac{Q}{2}$$

$$\text{Safety stock} = k\sqrt{\text{Average consumption during lead time}}$$

k = A factor based on acceptable frequency of stock out in a given number of years.

Buffer Stock

For an average demand during average lead time the additional stock termed as **buffer stock**.

Buffer stock = Average demand × Average lead time

- When no stock outs are desired.
- Buffer stock = Maximum Demand During Lead Time (DDLT) – Average Demand During Lead Time (DDLT)
- When demand rate varies about the average demand during a constant Lead Time (LT) period

Reorder Level (ROL) = Average (DDLT) × LT + BS

Inventory Models

- **Static Inventory Model** (Only one order can be placed to meet demand. Repeated order are impossible or too expensive.)
- **Dynamic Inventory Model** (Repeated order can be placed to replenish stock.)
  - **Deterministic Model** (Assume demand as well as lead time of an item are deterministic.)
  - **Probabilistic Model** (It amount the variation in demand and lead time.)

Classification of inventory model

Inventory Cost

The costs that are affected by firm's decision to maintain particular level of inventory are called cost associated with inventories or relevant inventory cost.

Total Inventory Costs (TIC)

TIC = Purchase cost + Total Variable Cost (TVC) of managing the inventory

TIC = Purchase cost + Inventory cost + Ordering cost + Shortage cost

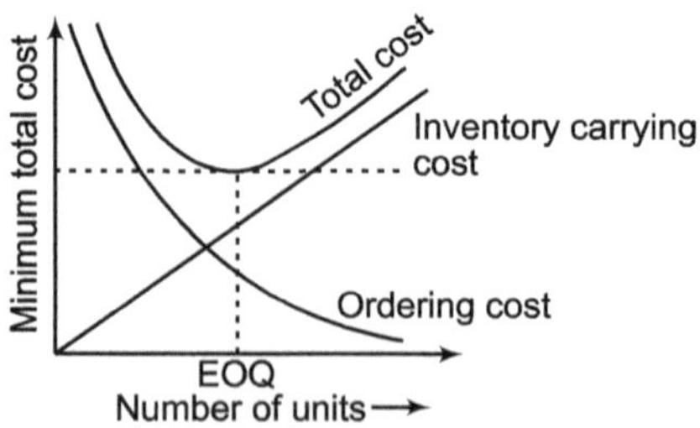

Inventory cost curve

Purchase Cost

It is defined as the cost of purchasing a unit of an item.

$$\text{Purchase cost} = \text{Price per unit} \times \text{Demand per unit time}$$
$$= C_u \times D$$

where, C_u = Unit cost
D = Annual demand

Ordering Cost

It is defined as the cost of placing an order from a vendor. This represents the expenses involved in placing an order with the outside supplier. This includes the costs involved in processing and ordering for purchase, expediting over the orders, receiving the consignment and inspection.

$$\text{Annual ordering cost} = \frac{C_o}{Q} \times D$$

where,

Q = Produced purchased or supplied throughout the entire time period (one year) or order quantity

C_o = Cost of placing an order

D = Annual demand

Carrying Cost

Carrying or holding costs are the costs incurred maintaining the stores in the firm. It is proportional to the amount of inventory and the time over which it is held.

$$\text{Annual carrying cost } C_{ui} = C_u \times i \times \frac{a}{2}$$

where, C_u = Unit purchase cost,
i = Interest rate

Shortage Cost

When an item cannot be supplied on consumer's demand, the penalty cost for running out of stock is called **shortage cost** or **stock out cost**.

Shortage cost = Cost of being short one unit in inventory
× Average number of unit short in the inventory.

Economic Order Quantity (EOQ)

Economic order quantity is the order quantity that minimizes total inventory holding costs and ordering costs. It is one of the oldest classical production scheduling models.

Assumptions of EOQ Model

- The ordering cost is constant
- Lead time is fixed
- Replenishment is made instantaneously
- The rate of demand is known
- Purchase price of the item is constant
- Only one product is involved.

EOQ When Stock Replenishment is Instantaneous

Economic Order Quantity (EOQ) is the order quantity that minimizes total inventory holding costs and ordering costs. It is one of the oldest classical production scheduling models.

It is defined as the quantity which will minimise the total variable cost of managing the inventory.

$$\text{TVC} = \frac{C_o}{Q} \times D + \frac{Q}{2} \times C_u \times i$$

$$\text{EOQ}\ (Q^\circ) = \sqrt{\frac{2C_o D}{C_u \times i}}$$

where, C_o = Cost of placing an order, C_u = Unit purchase cost

i = Interest rate, D = Annual consumption of the product.

- **Optimum number of order placed per year**

$$\therefore \quad n^o = \frac{D}{Q^\circ} = \sqrt{\frac{D\,C_u \times i}{2\,C_o}}$$

where, Q° = Economic order quantity

- **Optimum length of time between order** (Economic review period)

$$t^o = \frac{T}{n^o} = T\sqrt{\frac{2C_o}{D\,C_u \times i}} = \sqrt{\frac{2\,C_o}{D\,C_u \times i}} \qquad \text{(since } T = 1 \text{ year)}$$

- **Total minimum inventory production cost** $= \frac{1}{2} Q^o\, C_u \times i + \frac{D}{Q^o} C_o$

$$TC^\circ = \sqrt{2D\,C_u \times i \times C_o}$$

where, TC° = Total optimum inventory cost

- **Total cost** is calculated by

$$TC = C_u \times i \ \times Q + \frac{C_o}{Q} \times D + \frac{Q}{2} \times C_u \times i$$

- **Model sensitivity** $= \frac{TC}{TC^\circ}$

where, TC = Total cost

EOQ When Replenishment is not Instantaneous

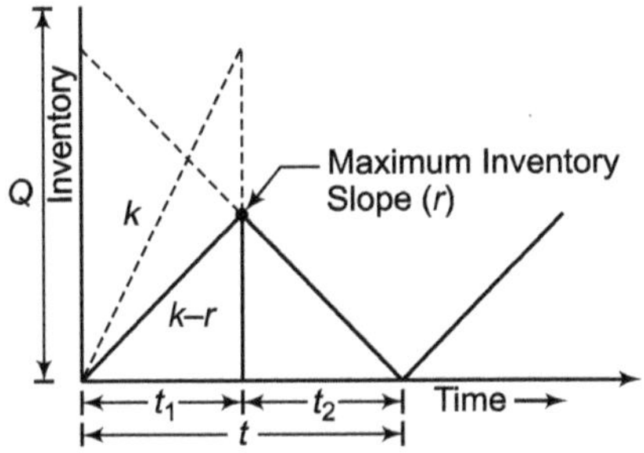

EOQ *versus* time graph

Let, k = Daily production rate or rate of replenishment

r = Depletion (demand) rate or daily consumption rate

- Economic Order Quantity (Q^o)

$$Q^o = \sqrt{\frac{2\,C_oD}{\sqrt{\left(1-\frac{r}{k}\right)C_u i}}}$$

- Optimum number of production runs per year

$$n^o = \frac{D}{Q^o} = \sqrt{\frac{DC_u \times i}{2C_o}\left(1-\frac{r}{k}\right)}$$

- Optimum length of each lot size production run

$$t_1^o = \frac{Q^o}{k} = \sqrt{\frac{2\,DC_o}{C_u \times i \times k\,(k-1)}}$$

- Total minimum production inventory cost

$$TC^o = \frac{C_oD}{Q^o} + \frac{1}{2}Q^o\left(1-\frac{r}{k}\right)C_u \times i = \sqrt{2DC_oC_u \times i\left(1-\frac{r}{k}\right)}$$

EOQ When Shortage are Permitted and Instantaneous Production and Variable Order Cycle Time

Let C_1 = Holding cost per unit inventory for a unit

C_2 = Shortage cost 0per unit of time per quantity

Q_1 = Amount which goes into inventory

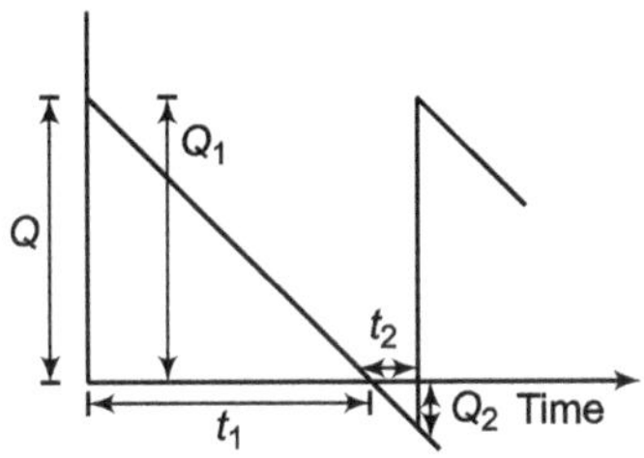

EOQ *versus* time graph

Q_2 = Amount which immediately taken to satisfy past orders or unfilled demand

C_o = Ordering cost

- EOQ $(Q^o) = \sqrt{\frac{2C_oD}{C_1}}\sqrt{\frac{C_1+C_2}{C_2}} \Rightarrow Q^o{}_1 = \left(\frac{C_2}{C_1+C_2}\right)Q^o$
- Time between receipts of orders (when to order)

$$t^o = \frac{Q^o}{D} = \sqrt{\frac{2C_o}{DC_1}}\sqrt{\frac{C_1+C_2}{C_2}}$$

- Total optimum inventory cost

$$TC^o = \sqrt{2DC_oC_1}\sqrt{\frac{C_2}{C_1+C_2}}$$

- Maximum inventory level $Q^o - Q_1^o = Q^o\left(1-\frac{C_1}{C_1+C_2}\right)$

$$= \sqrt{\frac{2C_o\,D}{C_1}}\sqrt{\frac{C_2}{C_1+C_2}}$$

Probabilistic Model

The ideal inventory models such as wilson and production models, do not account for the rise and uncertainty in their formulation. In reality these situations rarely occur. The major variations are in the demand rate and lead time. Both the demand rate and lead time may vary from cycle to cycle and their variations are probabilistic in nature. In the probabilistic models, the inventory decisions are based on the probability distribution of demand and/or that of the lead time.

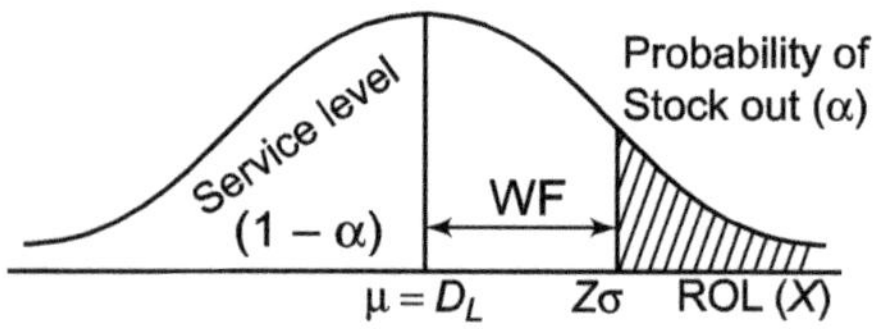

Probabilistic Model

Reserve stock level for normally distribute Demand During Lead Time (DDLT).

$$\text{Reserve Stock (RS)} = \text{ROL} - D_{LT} = X - \mu$$

$$Z = (X-\mu)/\sigma = \frac{\text{RS}}{\sigma}$$

where, X = Mean demand + Reserve stock

σ = Standard deviation

ROL = Reorder Level

One Period Problem

Marginal analysis approach

MP = Marginal Profit from each unit sold

ML = Marginal Loss from each unit that could not be sold

$P(r \geq \phi)$ = Probability that is demand for ϕ units or more for an optimum order quantity,

Expected marginal profit ≥ Expected marginal loss

$$P(r \geq \phi) \geq \frac{\text{ML}}{\text{MP} + \text{ML}}$$

If C = Cost per unit; P = Selling price per unit
S = Salvage per unit

$$P(r \geq \phi) \geq \frac{(C - S)}{(P - C) + (C - S)}$$

Linear Programming

A linear programming is defined as it is the optimisation of a linear function of variables subject to constrain of linear inequalities.

A Linear Programming Problem (LPP) *consists of three components namely*

(i) Decision Variables (Activities)
(ii) The Objective (Goal)
(iii) The Constraints (Restrictions)

Decision Variables (Activities)

It refer to the activities that are competing one another for sharing the resource available.

Objective (Goal)

The linear programming problem must have an objective which should be clearly identifiable and measurable in quantitative terms.

The Constraints (Restrictions)

There are always certain limitations (or constraints) on the use of resources, such as labour, space, raw material, money etc, that limit the degree on which an objective can be achieved.

Basic Assumptions of LPP

- **Certainty** All parameters, availability of resources, profit consumption of resources must be fixed.
- **Divisibility** (or continuity) Non-negative value of decision variable can positive frictional value.
- **Proportionality** Decision variable directly proportional to value of variable.
- **Additivity** The value of objective function for a given value of decision variables and the total sum of resources used must be equal to the sum of contributions earned from each variable and sum of resources used by each decision respectively.

LPP-Graphical Solutions

LPP involving two decision variables can easily solved by graphical method.

Basic Solutions

Given a system of m simultaneous linear equations in n unknowns ($m < n$). A is an $m \times n$ matrix of rank m, let B be any $m \times n$ submatrix formed by m linearly independent columns of A. Then,

A solution obtained by setting $(n - m)$ variables not associated with the columns of B equal to zero and solving the resulting system is called a basic solution to the given system of equations.

$$A_x = b, x^T \varepsilon R^n$$

The m variables, which may be all different from zero are called basic variables. The $m \times m$ non-singular submatrix B is called a basis matrix with the columns of B as basis vectors.

$$n - m \rightarrow \text{Non-basic variable}$$
$$m \rightarrow \text{Basic variable}$$

Degenerate Solution

A basic solution to the system is called degenerate if one or more of basic variable **vanish** (zero).

Basic Feasible Solution

A feasible solution an LPP which is also a basic solution to the problem is called a **basic feasible solution** to the LPP.

Simplex Method

When the decision variables are more than two, graphical method becomes inadequate and linear programming problem is solved by simplex method. Simplex method is defined as an algebraic procedure that through a series of repetitive operations, progressively approaches an optimal solution.

Conditions in Simplex Method

The two fundamental conditions on which the simplex method is based are

Conditions of Feasibility

It assume that if the initial solution is basic feasible, then during computation only basic feasible solutions will be obtained.

Condition of Optimality

It guarantees that only better solutions will be encountered.

Steps in the Simplex Algorithm

Let us assume that the existence of an initial basic feasible solution

Let $x = (x_1 \; x_2 \ldots x_n)$ be a feasible solution so that $x_j \geq 0$ and $m < n$

$$z = c_1 x_1 + c_2 x_2 + a_{i1} x_1 + a_{i2} x_2 + \ldots$$

$$\geq b_i \; x_j \geq 0 \left(\frac{i = 1, 2 \ldots m}{j = 1, 2 \ldots n} \right)$$

$$a_{i1} x_1 + a_{i2} x_2 + \ldots \leq b_i$$

- Check whether the objective function the given LPP is to be maximised or minimised. If it is to be minimised, then we convert it into a problem of maximising it by using the results.
- Minimum $z = -$ Maximum $(-z)$
- Check whether all b_i $(i = 1, 2, \ldots m)$ are non-negative if one of b_i is negative, then multiply the corresponding inequation of the constraint by -1, so as to get all b_i $(i = 1, 2 \ldots m)$ non-negative.
- Convert all inequation of constraint into equation by introducing slack and for surplus variables.
- Obtain an initial basic feasible solution to the problem in form of $x_B = B^{-1} b$ and put it in the first column of simple table.
- Compute the net evaluation $z_j - c_j (j = 1, 2 \ldots n)$ by using the relation. $z_j - c_j = c_B y_j - c_j$, where $y_i = B^{-1} a$

Examining the Sign $z_j - c_j$

If there are more than one negative $z_j - c_j$, then close the most negative of them.

- Compute the ratios $\left\{\frac{x_{Bi}}{y_{ir}}, y_{ir} > 0, i = 1, 2, 3..m\right\}$ and choose minimum of them. Let the minimum of these ratio be $\frac{X_{Bk}}{Y_{Kr}}$, then the vector y_R will level the basis y_B. The column element y_{kr}, which is in the kth row and the rth column is known as the **leading element** of the table.
- Convert the leading element to unity by dividing its row by the leading element itself and all other elements in its column to zeros by making use of the relations.

$$\hat{\mathbf{y}}_{ij} = y_{ij} - \frac{y_{kj}}{y_{kr}} y_{ir}, i = 1, 2 \ldots m + 1, \; i \neq k$$

$$\hat{\mathbf{y}}_{kj} = \frac{y_{kj}}{y_{kr}}, j = 0, 1, 2 \ldots n$$

- Repeat the procedure until get an optimum solution.

Key Points

- **Infinite Solutions** When a non-basic variable in an optimum solution has a zero value for Δj row, then the solution is not unique.
- **Unbounded Solution** When all replacement ratio are either infinite (or) negative, then the solution terminate. This indicate the problem has unbound solution.
- **Infeasible Solution** It is identified in the simplex table when optimality condition is satisfied and there is a positive value for artificial variable in the basic solution.

Duality in Linear Programming

Associated with every linear programming problem there always exist another linear programming problem which is based upon the same data and having the same solution. The original problem is called the **primal problem** while the associated one is called its **dual problem.**

Standard Primal Problem

Maximise $z = c_1 x_1 + c_2 x_2 \ldots + c_n x_n$

Subject to constraints

$$a_{i1}, x_1 + a_{i2} x_2 + \ldots + a_{in} x_n = b_i \quad \Rightarrow \quad x_j \geq 0 \begin{pmatrix} i = 1, 2 \ldots m \\ j = 1, 2 \ldots n \end{pmatrix}$$

Dual Problem

Minimize $z^* = b_1 w_1 + b_2 w_2 \ldots\ldots + b_m w_m$

Subject to the constraints

$$a_{1j} w_1 + a_{2j} w_2 + \ldots\ldots + a_{mj} w_m \geq c_j \; (j = 1, 2, \ldots n)$$

$$w_i = (i = 1, 2 \ldots\ldots m) \text{ unrestricted}$$

Minimization and Maximization Conditions in Dual Problem

Standard Primal Objective	Dual
	objective constraints variables
Maximisation	Minimisation ≥ unrestricted
Minimisation	Maximisation ≤ unrestricted

Transportation *and* **Assignment Model**

Transportation Problem (TP) is one of the subclass of LPP in which the objective is to transport various quantities of a single homogenous commodity that are initially stored at various sources orgins to the different destinations in such a way that the total transportation is a minimum.

Mathematically,

Minimise, $$z = \sum_{i=1}^{m} \sum_{j=1}^{n} x_{ij} c_{ij}$$

Subject to the constraints

$$\sum_{j=1}^{n} x_{ij} = a_i \; , i = 1, 2 \ldots m$$

$$\sum_{j=1}^{m} x_{ij} = b_j \; , j = 1, 2 \ldots n$$

$$x_{ij} \geq 0 \text{ for all } i \text{ and } j$$

Existence of Feasible Solution

A necessary and sufficient condition for the existence of a feasible solution to the general transportation problem is that

Total supply = Total demand

$$\sum_{i=1}^{m} a_i = \sum_{j=1}^{n} b_j = \lambda$$

Existence of Basic Feasible Solution

The number of basic variables of the general transportation problem at any stage of feasible solution must be $(m + n - 1)$. Now degenerate basic feasible solution (a feasible solution) involving exactly $(m + n - 1)$ positive variables is known as **non-degenerate basic feasible solution** otherwise it is said to be **degenerate basic feasible.** These allocation should be independent positions in case of non-degenerate basic feasible solutions.

Key Points

- **Optimum Solution** A feasible solution is said to be optimal, if it minimises the total transportation cost.
- **Unbalance TP** If total supply is not equal to total demand, then it balance with dummy source or destination.

Solution of a Transportation Problem

***Step* 1** Formulate problem as LPP.

***Step* 2** Set up transportation table.

***Step* 3** Examine total supply = total demand if not then introduce a dummy row or column having all its cost element as zero.

***Step* 4** Find an initial basic feasible solution that must satisfy all the supply and demand conditions.

***Step* 5** Examine the solution obtained in Step 4 for optimality *i.e.*, examine whether an improved transportation schedule with lower cost is possible.

***Step* 6** If the solution is not optimum modify the shipping schedule by including that unoccupied cell whose inclusion may result in an improved solution.

***Step* 7** Repeat step 4 until no further improvement is possible.

Finding an Initial Basic Feasible Solutions

There are three methods as given below

1. North west corner method
2. Least cost method
3. Vogel's approximation method (or Penalty method)

North West Corner Method

***Step* 1** Select the north west corner cell of the table and allocate as possible as first row is exhausted or the destination of the first column is satisfied.

$$x_{11} = \min(a_1, b_1)$$

***Step* 2** If $b_1 > a_1$, we move down vertically to the second row and make the second allocation of magnitude $x_{21} = \min(a_2\ b_1 - x_{11})$ in the cell 2, 1 if $b_1 < a_1$, we move right horizontally to the second column and make the second allocation of magnitude $x_{12} = \min(a_1 - x_{11}, b_2)$ in the cell (1, 2) If $b_1 = a_1$, there is a tie for the second allocation.

$$x_{12} = \min(a_1 - a_1, b_1) = 0 \text{ is the cell } (1, 2)$$
$$x_{21} = \min(a_2, b_1 - b_1) = 0 \text{ in the cell } (2, 1)$$

***Step* 3** Repeat Steps 1 and 2 moving down towards the lower/right corner of the transportation table until all the rim requirements are satisfied.

Least-Cost Method or Matrix Minima Method

***Step* 1** Determine the smallest cost in the cost matrix of the table

$$x_{ij} = \min(a_i, b_j) \text{ in the cell } (i, j)$$

***Step* 2** If $x_{ij} = a_i$ cross of the *i*th row of the table and decreases b_j by a_i go to Step 1 If $x_{ij} = b_j$ cross off the *j*th column of the transportation table and decreases a_i by b_j go to Step 3 If $x_{ij} = a_i = b_j$ cross off either the *i*th row or *j*th column but not both.

***Step* 3** Repeat Steps 1 and 2 for the resulting transportation table until all the rim requirements are satisfied whenever the minimum cost is not unique make an arbitrary choice among the minima.

Vogel's Approximation Method (VAM)

VAM takes into account not only the last cost c_{ij} but also the costs that just exceed c_{ij}.

The step of the method are given below

***Step* 1** For the row identify the smallest and next to smaller cost and determine difference between them. Display them along side the transportation table by enclosing them in parenthesis against the respective rows. Similarly compute the difference for each column.

***Step* 2** Identify row or column with largest difference use any arbitrary tie-breaking choice. Allocate the maximum feasible amount $x_{ij} = \min(a_i, b_j)$ in the (i, j)th cell and cross off either the *i* th row or the *j* th column in the usual manner.

***Step* 3** Recompute the column and row differences for the reduced transportation table and go to Step 2. Repeat the procedure until all the rim requirements are satisfied.

Test for Optimality in Transportation Problems

Step 1 Determine the net evaluations for non-basic variables (unoccupied cells) (u, v method).

Step 2 Choosing that net evaluation which may improve the current basic feasible solution.

Step 3 Determining the current occupied cell which leave the basis and repeating step 1 through step 2 ultimate an optimum solution is attained.

Assignment Problem

The assignment problem is a special case of the transportation problem in which the objective is to assign a number of resources to the equal number of activities at a minimum cost (or maximum profit).

Assignment problem is complete degenerate form of transportation problem. That means exactly one occupied cell in each row and each column of the transportation table *i.e.*, only n occupied cells in place of the required $(n + n - 1) = (2n - 1)$

Let x_{ij} denote the assignment of the ith resource to the jth activity such that

$$x_{ij} = \begin{cases} 1, & \text{if resource } i \text{ is assigned to activity } j \\ 0, & \text{otherwise} \end{cases}$$

then the mathematical formulation of the assignment problem is minimise

$$z = \sum_{i=1}^{n} \sum_{j=1}^{n} C_{ij} x_{ij}$$

Subject to the constraints

$$\sum_{i=1}^{n} x_{ij} = 1 \quad \text{and} \quad \sum_{j=1}^{n} x_{ij} = 1, x_{ij} = 0 \text{ or } 1$$

Hungarian Assignment Method (Reduced Matrix Method)

***Step* 1** Determine the cost table from the given problem.

***Step* 2** In case of number of resources is not equal to number of destinations, add a dummy source or dummy destinations so that cost table becomes a square matrix. The cost entries of dummy sources/destinations are always zero.

***Step* 3** Locate the smallest element in each row of the given cost matrix and then subtract the same from each element of that row.

***Step* 4** In the reduced matrix obtained in Step 3. Locate the smallest element of each column and then subtract the same from each element of that row.

***Step* 5** (a) Examine the rows successively until a row with single zero is found. (□) this zero and cross all other zero. Do it same for all rows.

(b) Repeat the procedure for each column.

(c) In case of two zero in same row or column choose arbitrary any one of these zero and cross off other zeros.

(d) Repeat (a) through (c) above successively until the chain of assigning (□) or cross (x) ends.

***Step* 6** If number of assignments (□) is equal to n an optimum solution is reached. If number of (□) $< n$ then go to next step.

***Step* 7** Draw the minimum number of horizontal or vertical lines to cover all the zeros of the reduced matrix.

***Step* 8** Developed the new revised cost matrix.

***Step* 9** Go to Step 6 and repeat the procedure until an optimum solution is attained.

Break Even Analysis (BEA)

BEA is also known as **cost volume** profit analysis and deals with inter relationship between total revenue cost and profit in order to determine the potential effects of decisions necessary to achieve or maintain various levels of financial structure of a company.

Fixed Cost

The cost which do not vary with increase or decrease of its production and sale activity for a given period. They are constant upto specific volume or range of volume. *e.g.*, rent

e.g., rent

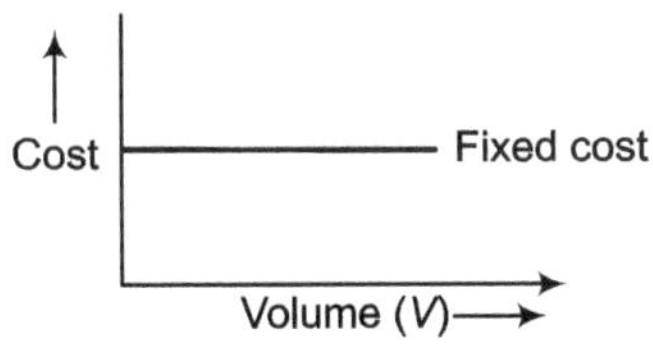

Fixed cost

Variable Cost

Costs that are incurred by the enterprise which increase with the rise in the level of production and decrease with its falls *e.g.,*electricity, water, power taxes.

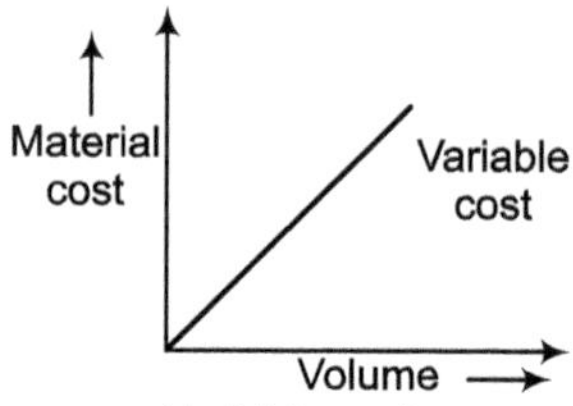

Variable cost

Semi Variable Cost

Costs which are partly fixed and partly variable such as maintenance costs of machines and equipment depreciation etc., these however can be broken up into variable and fixed portions.

Key Points

- Increase in fixed cost (possible owing to the purchase of a new machine) increases the total costs and thus shift BEP towards the right hand side, so profit will be decreased for the same units of output.
- Increase in variable costs and so in total cost possible owing to an increase in labour cost would shift the BEP towards the right hand side. This involves a decrease in profit for the same units of output.
- If the price of an article rises a new sales revenue will be drawn with a greater slope, this shift BEP towards the left hand side and thus increase the company profit for the same units of output.

Assumptions of BEA

- Constant selling price, linear relationship between sales volume and cost.
- Production match with sales.
- Only quantity can affect the cost, not any other factor.

  F = Fixed cost
  V = Variable cost per unit
  S = Sales price per unit
  P = Profit
  Q = Quantity produces and sold
  (BEP) Q = Break even quantity (no loss, no profit)
  (BEP) Q^*, (in units)

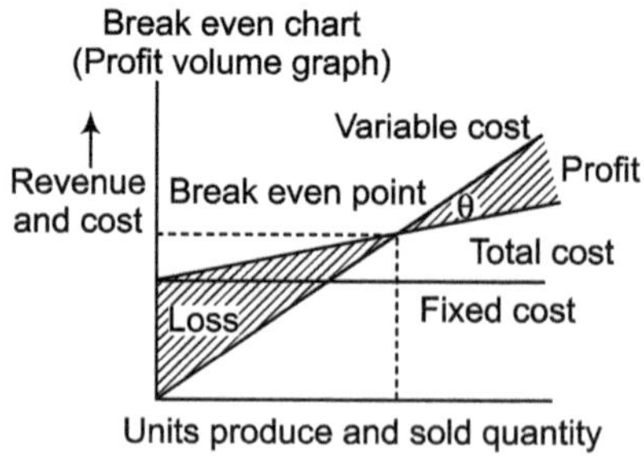

Break even graph

$$= \frac{F}{S - V}$$

$$Q^* \text{ (in ₹)} = \frac{F}{(S - V)} \times S = \frac{F}{1 - \frac{V}{S}}$$

Contribution Margin

The contribution margin is given as

$$= \text{Total sales} - \text{Total variable cost}$$
$$= \text{Fixed cost} + \text{Profit}$$

Contribution Margin Ratio or *P*/*V* Ratio

The contribution margin ratio is given as

$$= \frac{\text{Contribution margin per unit}}{\text{Selling price per unit}} \times 100$$

$$= \frac{S - V}{S} \times 100$$

Margin of Safety (MOS)

It is the excess of actual sales over break even sales.

$$\text{Margin of Safety (MOS)} = \text{Actual sales} - \text{Break even sales}$$

$$\text{Margin of Safety (MOS) ratio} = \frac{\text{Actual sales} - \text{Break even sales}}{\text{Actual sales}} \times 100$$

(i) If Q^* is break even quantity (at no profit) at the break even point

$$\text{Total revenue} = \text{Total cost} + \text{Variable cost}$$

$$Q^* S = F + Q^* V$$

(ii) If P is profit if sell Q quantity.

$$\text{Fixed cost} + \text{Variable cost} = \text{Sales} - \text{Profit}$$

$$F + VQ = SQ - P$$

$$Q \text{ (units)} = \frac{F + P}{S - V} \text{ in terms of ₹}$$

$$Q \text{ (₹)} = S \times \left(\frac{F + P}{S - V}\right)$$

where, S = Sales price

F = Fixed cost

V = Variable cost

Q = Quantity at BEP

Angle of Incidence (θ)

It is defined as the angle between the lines of total cost and total revenue. If θ will large, the profit will high rate.

Note θ large with high margin of safety marks the extremely favourable business position. Profit volume ratio can also be defined as

$$= \frac{\text{Fixed cost } + \text{Profit}}{\text{Quantity at profit}}$$

Disadvantages of Break Even Chart

- It is used for short run analysis and added difficulty when a company produces a variety of products.
- In actual total cost do not vary in direct proportion to unit produced.

Line Balancing

Assembly line is a special case of product layout in which the operations pertain to assembly of different parts at few station line (product) layout is useful for high volume, single type of manufacturing activity.

Work Element (i)

The part of total job content in the line is called work element.

$$1 \leq i \leq N$$

Work Stations (w)

The location on the assembly line where a combination of few work elements are performed is called work stations since minimum number of work stations (w) cannot be less than 1.

Total Work Content (T_{wc})

It is equal to the sum of processing time of each operation.

$$T_{wc} = \sum_{i=1}^{N} T_{iN}$$

where, T_{iN} = Processing line of operations.

Station Time (T_{Si})

It is equal to sum of processing time of operation on a particular work station.

$$T_{Si} = \sum_{n_1}^{n_2} T_{iN} \quad \text{or} \quad T_{wc} = \Sigma\, T_{Si}$$

Cycle Time

It is rate of production. $\quad T_C \geq \max \{T_{Si}\}$

Delay or Idle Time at Station

The difference between cycle time and station time is called the delay or idle time at station.

$$T_{dS} = T_C - T_{Si}$$

Balance Delay or Balancing Less (d)

It is the measure of the inefficiency of the line.

$$d = \left(\frac{nT_c - T_{wc}}{nT_c}\right) \times 100$$

$$\text{Line efficiency } L = \frac{\sum_{i=1}^{n} T_{Si}}{n(T_c)}$$

where, T_{Si} = Station time at station i, n = Total number of stations

T_C = Total cycle time.

Smoothness Index (SI)

$$\text{SI} = \sqrt{\sum_{i=1}^{n} [(T_{Si})_{\max} - T_{Si}]^2}$$

where, $(T_{Si})_{\max}$ = Maximum station time.

Note If smoothness index is low, it means line balancing is good.

Theoretically Minimum Number of Work Stations ($n_{\min}$)

The minimum number of work stations is defined as the ratio of total work content to the total cycle time.

$$n_{\min} = \frac{T_{wc}}{T_c}$$

Actual production rate $PR_{(\text{actual})} = \dfrac{1}{T_{C\,(\text{actual})}}$

where, T_{actual} = Actual exceed cycle time.

Largest Candidate Rule for Line Balancing

Step 1 Firstly enlist the elemental time value (T_C) in descending order.

Step 2 Select feasible element placement at the station, starting at the top of list. A feasible element is one that satisfies precedence requirement and does not cause the sum of elemental time value (T_e) at the station to exceed cycle time (T_C).

Step 3 Continue the process of assigning the work element to the station as in Step 2 until further elements can be added without exceeding cycle time (T_C).

Step 4 Repeat Step 2 and 3 for other station until all the elements are assigned.

PERT *and* **CPM**

Network scheduling is a technique used for planning and scheduling large projects in the fields of construction, maintenance, fabrication, purchasing, computer system installation etc.

There are two basic techniques of planning and control that utilise a network to complete a predetermined project or schedule

1. Program Evaluation and Review Technique (PERT)
2. Critical Path Method (CPM)

Basic Components Used in Planning and Control

The basic components used in program evaluation and Review Technique (PERT) can be given as follows

Activity

It lies between two events, called the preceding and succeeding. An effort is a task or item of work to be done that consumes time, efforts, money or other resources.

Key Points

- Predecessor Activity An activity which must be completed before one or more other activities start is known as **predecessor activity**.
- Successor Activity An activity which started immediately after one or more of other activities are completed is known as **successor activity**.
- **Dummy Activity** An activity which does not consume either any resource and time is known as **dummy activity**. A dummy activity is depicted by dotted line in the network diagram. It is used to maintain the logical sequence.

Event

An event represent the start or completion of some activity and as such it consumes no time. It has no time duration and does not consume any resources. So, an event is a node.

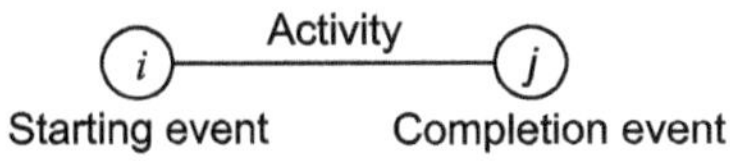

Activity-event flow diagram

CPM (Critical Path Method)

An activity in a network diagram is said to be critical, if the delay in its start will further delay the project completion time. In this activity times are known with certainity.

Prepare Project Scheduling

The project scheduling can be done in the following steps

(i) Time schedule for each activity.

(ii) Earlier and latest start time as well as earlier finish and latest finish of each activity.

(iii) Float for each activity.

(iv) Critical activities and critical path for the network.

Earliest and Latest Time

- Earliest Start Time (EST)
  It is earliest event time of tail and event.
- Earliest Finish Time (EFT)
  It is the sum of earliest start time and activity time.
  EFT = EST + activity time
- Earliest Event Time (EET)
  It is maximum of earliest finish times of all activities ending into that event.
- Latest Start Time (LST)
  It is the latest possible time by which an activity can start without delaying the date of completion of the project.
- Latest Finish Time (LFT)
  It is latest possible time by which the activity must be completed so that the scheduled date for the completion of the project may not be delayed. It is calculated by moving backward *i.e.*, from last event to first event of the network diagram.
  LFT = LST + Duration of activity
- Latest Event Time (LET)
  It is minimum of the latest start time of all activities originating from that event.

Float or Slack of an Activity and Event

Float or slack means a margin of extra time over and above its duration which a non-critical activity can consume without delaying the date of completion of the project. In other words, the float of an activity is the amount of time by which it is possible to delay its completion time without affecting the total project completion time.

Event Float

It is difference between its latest time (L_i) and its earliest time (ε_i).

$$\text{Event float} = L_i - \varepsilon_i$$

Activity Float

It is the float in the activity time estimates.

Total Float

The total float of an activity represents the amount of time by which an activity can be delayed without delay in the project completion date.

$$\begin{aligned} \text{Total float } (TF_{ij}) &= L_j - (E_i + t_{ij}) = LF_{ij} - EF_{ij} \\ &= (L_j - t_{ij}) - E_i \\ &= LS_{ij} - ES_{ij} \end{aligned}$$

Free Float

Free float is that portion of the total float within which an activity can be manipulated without affecting the float of subsequent activities.

$$\begin{aligned} FF_{ij} &= (E_j - E_i) - t_{ij} \\ &= E_j - (E_1 + t_{ij}) \\ &= \min\{ES_{ij}\} - EF_{ij} \ (i < j) \end{aligned}$$

Independent Float

It is that portion of total float within which an activity can be delayed for start without affecting floats of the preceding activities.

$$\begin{aligned} IF_{ij} &= (E_j - L_i) - t_{ij} \\ &= FF_{ij} - (\text{Slack of event } i) \end{aligned}$$

e.g.,

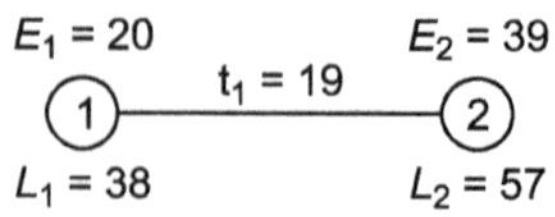

$$\begin{aligned} \text{Total float } &= L_2 - (E_1 + t_{12}) = 57 - (20 + 19) = 18 \\ \text{Free float } &= E_2 - E_1 - t_{12} = 0 \\ \text{Independent float } &= E_2 - (L_1 + t_{12}) = -18 \end{aligned}$$

Interfering Float

It can be defined as that part of the total float which causes a reduction in the float of the successor activities. It can be defined as the difference between the latest finish time of the activity under consideration and earliest start time of the following activity or zero whichever is larger.

PERT

The Program (or Project) Evaluation and Review Technique is commonly abbreviated (PERT)

It is used when the activity times are not known with certainity, for *e.g.,* in research and development. In PERT analysis, a beta distribution is assumed because it is unimodal, has non-negative end points and is approximate symmetric.

Expected time $$t_e = \frac{t_o + 4t_m + t_p}{6}$$

t_o = Optimistic time *i.e.,* shortest possible time to complete the activity if all goes well.

t_p = Pessimistic time *i.e.*, longest time that an activity could take if everything goes wrong.

t_m = Most likely time *i.e.*, normal time of an activity would take.

$$\text{Standard deviation } (\sigma) = \frac{t_p - t_o}{6}$$

$$\text{Variance } (\sigma^2) = \left(\frac{t_p - t_o}{6}\right)^2$$

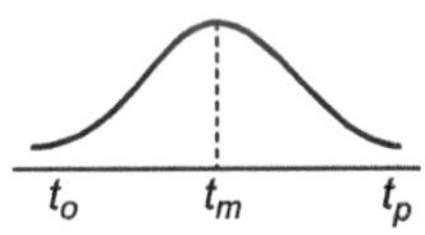

Most likely time segment

Probability of Meeting the Schedule Time

With PERT, it is possible to determine the probability of completing a contract on schedule.

Probability of completing the project by scheduled time (T_s) is given by

$$\text{Prob}\left(Z < \frac{T_s - T_e}{\sigma_e}\right)$$

Standard Normal Variation (SNV) is given by

$$Z = \frac{T_s - T_e}{\sigma_e}$$

where, T_e = Expected completion time of the project

T_s = Scheduled time of the project

σ_e = Number of standard deviation.

The scheduled time lies from the expected time.

$$Z = \frac{T_s - T_e}{\sigma_e}$$

Crashing

For reducing time of activity, more resources and funds need. For instance a point will be reached beyond which no further reduction in time will be possible irrespective of resources spent.

$$\text{Cost Slope (CS)} = \frac{\text{Crash cost} - \text{Normal cost}}{\text{Normal time} - \text{Crash time}}$$

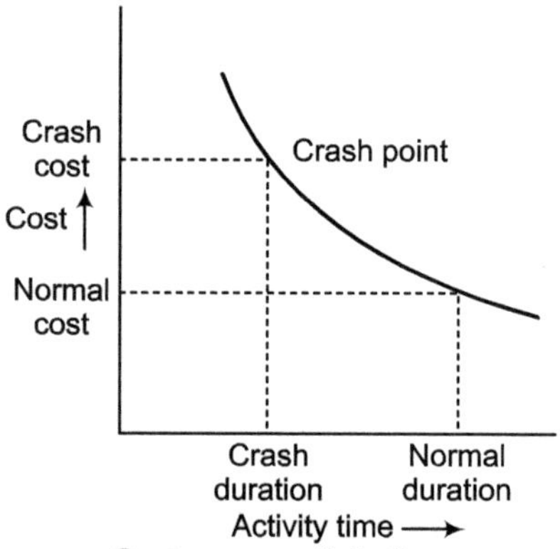

Cost *versus* activity time

Difference between CPM and PERT

CPM	PERT
Activity oriented	Event oriented
Deterministic approach	Probabilistic approach
One time estimate	Three time estimate (a) Optimistic (b) Most likely (c) Pessimistic
Planning device	Control device

Quality Control

It can be defined as the entire collection of activities which ensures that the operation will produce the optimum quality products at minimum cost.

Quality

Quality in business, engineering and manufacturing has a pragmatic interpretation as the non-inferiority or superiority of something. Quality is also defined as the fitness for use at the most economic level.

Types of Quality

The types of quality are given below

1. Quality of design
2. Quality of conformance
3. Quality of performance

Control

Control can be defined as a process by means of which we observe the actual performance and compare it with some standard.

Quality Characteristics

- Technological
- Psychological
- Time oriented
- Contractual
- Ethical

Key Points

- Quality control should not be confused with inspection.
- Inspection means checking of materials, product or components of product at various stages with reference to certain predetermined factors and detecting and sorting out the faulty or defective items.
- Quality control is a broad term. It involves inspection at particular stage but rise inspection does not mean quality control.

Cost of Quality

Cost of quality is a means to quantify the total cost of quality-related efforts and deficiencies.

It can be defined in four categories

- Cost of prevention
- Cost of appraisal
- Cost of internal failures
- Cost of external failures

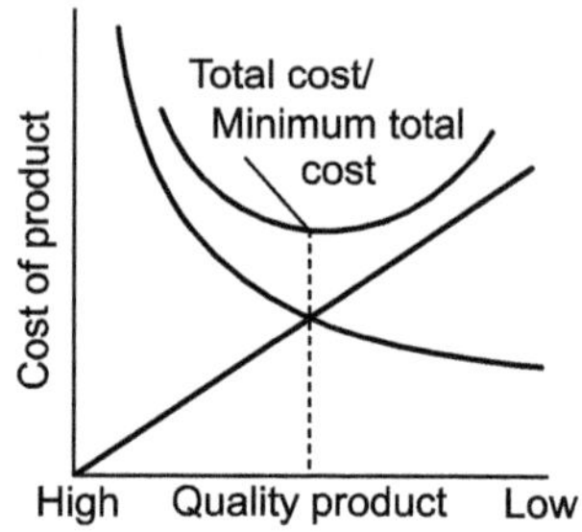

Cost of quality curve

Statistical Quality Control (SQC)

It is defined as the quality control system where statistical techniques are used to control, improve and maintain quality.

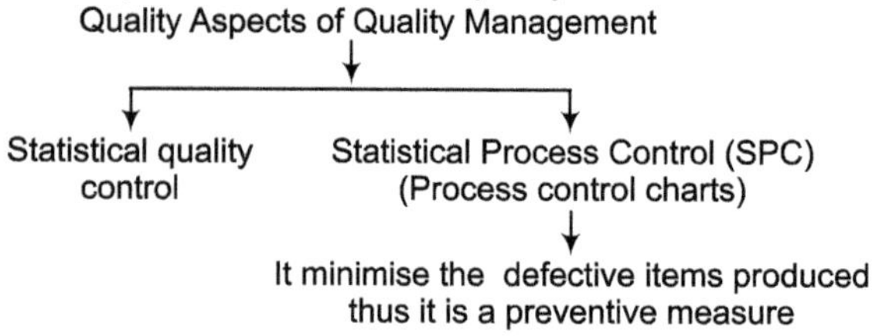

Classification of quality aspects of quality management

Total Quality Management (TQM)

TQM refers to total involvement of staff in an organisation together, which includes suppliers, distributors and even customers in bringing about quality satisfaction by promoting quality culture.

TQM is control of all transformation processes of an organisation to best satisfy customer's needs in the most economical manner.

Just In Time (JIT)

It is a new concept in manufacturing where material is pulled through the process by the next operation, when the problem occurs the process is shut down.

Key Points

- **KANBAN** is the Japanese name to JIT, actually it means in Japanese language 'A market used to control the sequence of jobs through a sequential process.'
- The purpose of JIT is to produce and deliver finished goods just in time to be sold.

Zero Defect Concept

Zero defect concept seek the volumentary participation of workers in understanding personal responsibility for quality of task they perform.

Quality Assurance

It deals with the question of assuring the desired quality, reliability service and other aspects in manufacturing product through scientific techniques.

Control Charts for Variables

A control chart is based upon measurement of quality characteristics are called as **control charts for variables**. Control charts for variables are often found to be more economical means of controlling quality than control charts based on attributes.

There are two types of variations

1. Variation due to assignable cause
2. Variation due to chance cause chance variations (Random Variations) $\overline{X}$ and R chart or $\overline{X}$ and σ charts are used in combination for the control process.

$\overline{X}$ Chart

It shows the centering of the process *i.e.*, it shows the variation in the average of samples.

R Chart

It shows the uniformity or consistency of the process *i.e.*, it shows the variations in the ranges of samples.

Calculation for $\overline{X}$ and R Charts

For X-chart, centre line (control limit) $\overline{X} = \dfrac{\sum_{i=1}^{N} \overline{X}_i}{N}$

where, $\overline{X}$ = Average of average
N = Number of subgroups

Upper Control Limit (UCL)

$$\text{UCL} = \overline{\overline{X}} + 3\sigma_X$$

Lower Control Limit (LCL)

$$\text{LCL} = \overline{\overline{X}} - 3\sigma_X \quad \Rightarrow \quad \sigma_X = \frac{\sigma}{\sqrt{n}}$$

where, σ_x = Standard mean of sample mean
σ = Process standard deviation
n = Sample size.

For ***R*** chart, $$\overline{R} = \frac{\sum_{i=1}^{N} R_i}{N}$$

$$\text{UCL} = D_4 \overline{R} = D_2 \sigma$$

$$\text{LCL} = D_3 \overline{R} = D_1 \sigma$$

where, $\sigma = \dfrac{\overline{R}}{D_2}$ If $n < 7$, then $D_3 = 0$

Value of D_3 and D_4 depend on the sample size.

$\overline{X}$ for *R* chart

S. No	$\overline{X}$	*R* (Maximum-Minimum)
1	$\bar{x}_1$	R_1
2	$\bar{x}_2$	R_2
....	..	...
N	$\bar{x}_N$	R_N

Control Units

For plotting control charts generally $\pm 3\sigma$ limits are selected. Therefore, such control charts are known as **3 σ control charts**.

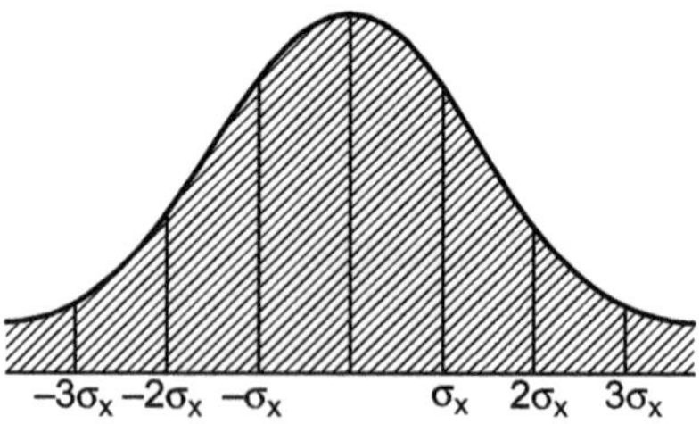

3σ control chart

σ-Chart

The larger the subgroup size the narrower the control limits on $\overline{X}$ chart and the easier is to detect small variations. Generally speaking, the larger the subgroup size, the more desirable it is to use standard deviation rather range as measure of subgroup of dispersion.

$$\sigma_6 = \frac{\sigma}{\sqrt{2n}} \quad \text{(when } n > 25\text{)}$$

$$\sigma_6 = [2\,(n-1) - 2n\,C_2]^{\frac{1}{2}} \frac{\sigma'}{\sqrt{2n}} \quad \text{(when } n \le 25\text{)}$$

where, n = Sample size

The control limits are given by

$$\text{UCL} = \bar{\sigma} + 3\sigma_6 = B_4\bar{\sigma}$$

$$\text{LCL} = \bar{\sigma} - 3\sigma_6 = B_3\bar{\sigma}$$

Process Capability

Process capability may be defined as the minimum spread of a specific measurement variation which will include 99.7% of the measurement from the given process.

$$\text{Process capability} = 6\,\sigma$$

$$(X_{max} - X_{min}) > 6\sigma$$

where, X_{max} = Upper specification limit, X_{min} = Lower specification limit

In this type, situation processes are always in control.

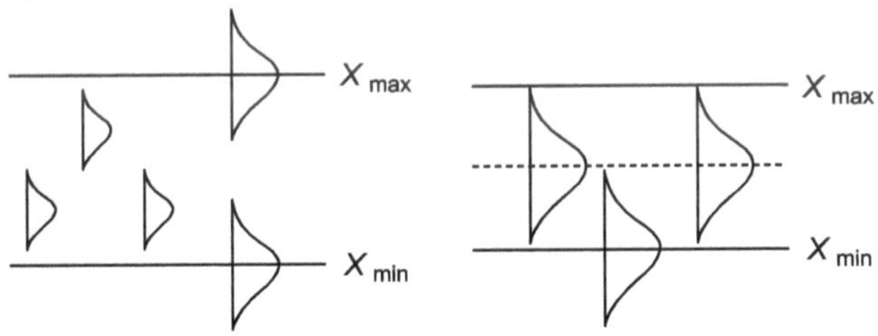

Upper and lower specification limit

$$(X_{max} - X_{min}) < 6\,\sigma$$

In this type of situation, defective parts will always be there. Therefore tolerance should be increase.

$$(X_{max} - X_{min}) = 6\,\sigma$$

In this case, a little change in centering will cause many components to fall outside the specification limits. So, it is advisable increase tolerances if there are tighter than really necessary.

On chart $\pm\,3\sigma$ – active limit

$\pm\,2\sigma$ – warning limit.

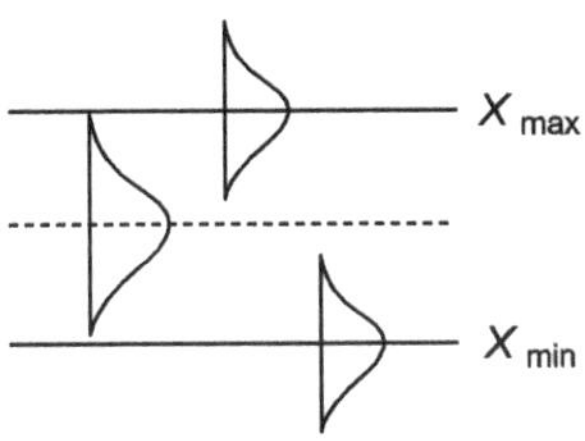

Increase in tolerances

Control Chart for Attributes

There are following chart use for attributes

P Chart

It is also known as fraction defective chart. This is made when the sample size is varying.

$$\text{Centre Line (CL)} = \overline{P} = \sum_{i=1}^{N} P_i$$

$$\text{Average sample size } (\overline{n}) = \frac{\sum_{i=1}^{N} n_i}{N}$$

$$\text{UCL} = P + 3\sigma = \overline{P} + 3\sqrt{\frac{\overline{P}(1-\overline{P})}{n}}$$

$$\text{LCL} = \overline{P} - 3\sigma = \overline{P} - 3\sqrt{\frac{\overline{P}(1-\overline{P})}{n}}$$

where, n = Sample size.

nP Chart

In the case of subgroup, size is constant nP chart is used for actual number of defectives.

$$\overline{P} = \frac{\sum_{i=1}^{N} P_i}{N} = \frac{\Sigma n\overline{P}}{\Sigma n}$$

$$\text{UCL} = n\overline{P} + 3\sigma_{np} = n\overline{P} + 3\sqrt{n\overline{P}(1-\overline{P})}$$

$$\text{LCL} = n\overline{P} - 3\sigma' nP$$

$$= n\overline{P} - 3\sqrt{n\overline{P}(1-\overline{P})}$$

Evaluation of *P*

Sample number	Sample size	Number of defective	$P = \frac{d}{n}$
1	n	d_1	$P_1 = \frac{d_1}{n_1}$
2	n	d_2	....
3	n	d_3	....
....	.....	....	
N	n	d_N	$P_N = \frac{d_N}{n}$

C Chart

It is based on Poisson distribution, *C* chart is the control chart for defects. It is made for number of defects which are present in a sample.

$$\bar{C} = \frac{\sum_{i=1}^{N} C_i}{N} = \frac{\text{Number of defects in all samples}}{\text{Total number of samples}}$$

$$\text{UCL} = \bar{C} + 3\sqrt{\bar{C}}, \quad \text{LCL} = \bar{C} - 3\sqrt{\bar{C}}$$

Number of defects when exceed a particular limit they make an item defective.

U Chart

When the sample size varies from sample to sample, it is necessary to use *U* chart.

Type of Charts

S.No.	Chart	Process observation	Process observations relationships	Process observations type	Size of shift to detect
1.	$\bar{x}$ and *R* chart	Quality characteristic measurement within one subgroup	Independent	Variables	Large ($\geq 1.5\,\sigma$)
2.	*P*-chart	Fraction nonconforming within one subgroup	Independent	Attributes	Large ($\geq 1.5\,\sigma$)
3.	*nP*-chart	Number nonconforming withing one subgroup	Independent	Attributes	Large ($\geq 1.5\,\sigma$)
4.	*C*-chart	Number of nonconformances within one subgroup	Independent	Attributes	Large ($\geq 1.5\,\sigma$)
5.	*U*-chart	Nonconformances per unit within one subgroup	Independent	Attributes	Large ($\geq 1.5\,\sigma$)

$$\text{Centre line } \bar{U} = \frac{C}{N} = \frac{\text{Number of defects in sample}}{\text{Number of units in a sample}}$$

$$\bar{n} = \frac{\sum_{i=1}^{N} n_i}{N} \Rightarrow \text{UCL} = \bar{U} + 3\sqrt{\frac{\bar{U}}{\bar{n}}}$$

$$\text{LCL} = \bar{U} - 3\sqrt{\frac{\bar{U}}{\bar{n}}}$$

Acceptance Sampling Process

It is a process of evaluating a portion of the product/material in a lot for the purpose of accepting or rejecting a lot as either conforming or not conforming to a quality specifications.

The sampling method can be classified as

- Simple random sampling
- Stratified sampling
- Systematic sampling
- Cluster sampling
- Two stage sampling

Operating Characteristic Curve (OCC)

The OC curve for an attribute sampling plan is a graph of fraction defective in a lot against the probability of acceptance.

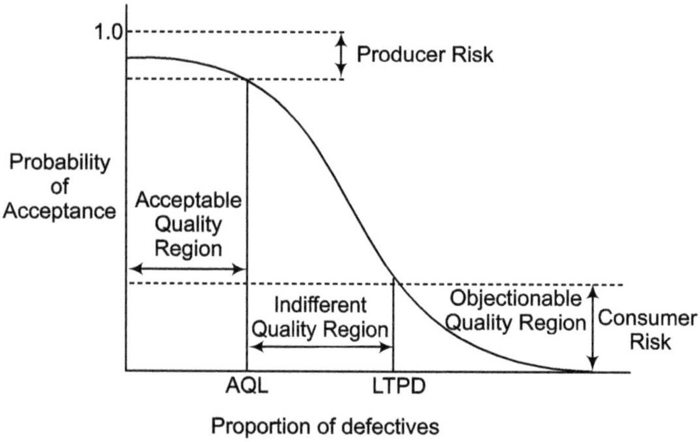

Operating Characteristic Curve

Acceptance Quality Level (AQL)

It represent the maximum proportion of defective which the consumer find definitely acceptable.

Lot Tolerance Per cent Defective (LTPD)

It represent a bad quality lot submitted by the producer and such a lot should be rejected most of the times by the sampling plan.

$$\text{Probability of acceptance } P_a = \frac{e^{-np}(np)^x}{x!}$$

where, P = Fraction defective, x = Number of defective item
n = Sample size

Average Outgoing Quality (AOQ)

It represent the average percentage defective in the outgoing products after inspection including all accepted and all rejected lots which have been 100% inspected and defective replaced by non-defectives.

Queueing Theory

Queues (waiting line) stands for a number of customers waiting to be serviced. The Queue does not include the customer being serviced. The process or system that performs the services to the customer is called service facility. Queueing theory is used in service oriented organisation.

Elements of a Queueing System

1. Input or arrival process
   (i) Size of queue
   (ii) Pattern of arrivals
   (iii) Customer's behaviour
2. Queue discipline
3. Service mechanism
   (i) Single queue one server
   (ii) Single queue several server
   (iii) Several queue one server
   (iv) Several server
4. Capacity of the system

Operating Characteristics of a Queueing System

- Expected number of customers in the system denoted by $[E(n)]$ or L. It is the average number of customers in the system, both waiting and in service.
- Expected number of customers in the queue $[E(m)$ or $Lq)]$. It is the average number of customers waiting in queue.

  Here, $m = n - 1$, *i.e.*, excluding the customer being served

  or $L_q = L - 1$
- Expected waiting time in the system denoted by $E(v)$ or w is the average total time spent by a customer in the system. It is generally taken to be waiting time plus servicing time.
- Expected waiting time in queue denoted by $E(w)$ or w_q. It is average time spent by a customer in the queue before the comencement of the service.
- The server utilisation factor or (many period) denoted by $P = (\lambda/\mu)$. It is proportion of time that a service actually stands with a customer.

Here, λ = Average number of customer arriving per unit of time.

μ = Average number of customer completing service per unit of time.

The server utilisation factor is also known as traffic intensity or the clearing ratio.

Deterministic Queueing System

A queueing system wherein the customers arrives at regular intervals and service time for each customer is known and constant is known as **deterministic queueing system**.

(i) if $\lambda > \mu$, the waiting line (queue) shall be formed and will increase indefinitely, the service facility would always be busy and service system will eventually fail.

(ii) if $\lambda \leq \mu$, there shall be no queue and hence no waiting time, the proportion of time the service facility would be idle is $1 - \frac{\lambda}{\mu}$.

Probability Distribution in Queueing Systems

It is assumed that customers joining the queueing system arrive in a random manner (variable) and follow a poisson distribution or equivalently the inter-arrival times obey exponential distribution.

Queueing Model

$N(t)$ = Number of customers in queueing system at time t ($t \geq 0$)

S = Number of servers in queueing system

$P_n(t)$ = Probability of n units in queueing system

λ_n = Mean arrival rate (units/ unit time)

L_q = Average number of customers in the queue system

n = Mean number of units in the queueing system including the one being served

w_s = Average waiting time in the queue

w_q = Average time in the queue system

M/M/1 : (∞/FIFO)

Single service channel, poisson input, exponential service, no limit on the system capacity.

First in, first out

Steady state solution $P_n = \rho^n (1 - \rho)$, where $\rho = \frac{\lambda}{\mu} < 1$ and $n \geq 0$

- Inter arrival time $= \frac{1}{\lambda}$
- Traffic intensity factor or utilisation factor or channel efficiency $\rho = \frac{\lambda}{\mu}$
- Average number of customer in system

$$L = \frac{\rho}{1-\rho} = \frac{\lambda}{\mu - \lambda}$$

- Average number of customer of queue or average queue length

$$L_q = L - \rho = \frac{\rho}{1-\rho} - \rho = \frac{\lambda}{\mu - \lambda} - \frac{\lambda}{\mu}$$

- Waiting line in system

$$w = \frac{1}{\mu - \lambda} \qquad \left(w = \frac{L}{\lambda}\right)$$

where, μ = Average number of customer completing service per unit of time

λ = Average number of customer arriving per unit of time

- Average or expected waiting line in queue

$$w_q = \frac{1}{\mu - \lambda} - \frac{1}{\mu} \qquad \left(w_q = \frac{L_q}{\lambda}\right)$$

- Probability that the service facility is idle

$$P_0(t) = (1 - \rho)$$

- Probability of ***n*** customer in the service at time

$$P_n(t) = \rho^n \times P_0(t)$$

- Average length of non-empty queue

$$L_n = \frac{1}{1-\rho} = \frac{\mu}{\lambda - \mu}$$

- The fluctuation (variance) of queue length

$$v(n) = \frac{\rho}{(1-\rho)^2} = \frac{\lambda\mu}{(\mu - \lambda)^2}$$

- Probability of ***n*** arrival in time *t* $P(nt)$

$$P(nt) = \frac{e^{-\lambda} (\lambda t)^n}{n!}$$

- Probability that the waiting time in the queue is greater than or equal to t

$$P(w_q \geq t) = \frac{\lambda}{\mu} e^{-(\mu - \lambda)t}$$

where, w_q = Waiting time in queue

- Probability that waiting time in system is greater than or equal to t

$$P(w_s \geq t) = e^{-(\mu - \lambda)t}$$

where, w_s = Waiting time in system

- Probability that waiting time in system is less than or equal to t

$$P(w_s \leq t) = 1 - e^{-(\mu - \lambda)t}$$

(M/G/1): (∞ / FIFO)

G = Gesal service time
σ = Standard deviation for the service time
σ^2= Variance

$$L_q = \frac{\lambda^2\sigma^2 + \rho^2}{2(1-\rho)}$$

$$L = L_q + \rho$$

$$w_q = \frac{L_q}{\lambda}$$

$$w = w_q + \frac{1}{\mu}$$

where, L_q = Average queue length
w_q = Waiting time in the queue

(M/D/1) : (∞ / FIFO)

D = Deterministic or constant service times *i.e.*, standard deviation for service time is zero ($\sigma = 0$)

$$L_q = \frac{\rho^2}{2(1-\rho)} \text{ and } L = L_q + \rho$$

$$w_q = \frac{L_q}{i}$$

$$\Rightarrow \quad w = w_q + \frac{1}{\mu}$$

where, w_q = Expected waiting times

Work Study *and* **Value Engineering**

The method of investigation of work done and finding best most efficient way of using available resources *i.e.*, money and machinery is called **work study**.

It can be divided into two parts

Method Study

It find best possible manufacturing procedure involves least time and not cause fatigue to workers.

Work Measurements

It is the application of technique designed to establish the time for a qualified worker to carry out a specific job at a defined level of performance. It concern the total time allow to complete the job by that procedure.

Techniques for Carrying Out Work Measurement

There are following three techniques for carrying out work measurement

Stop Watch (Time Study Method)

Stop watch time study is defined as a work measurement technique for recording the times and rate of working for the element of a specified job, carried out under specified condition and for necessary for carrying out the job at a defined level of performance.

Number of cycles or observations to be taken while conducting stop watch time study is given by

$$N = \left[\frac{K \sqrt{N' (\Sigma x)^2 - (\Sigma x)^2}}{\Sigma x} \right]^2$$

where, N = Total number of cycles

N' = Number of observations already taken

X = Individual observation

K = Constant = 40, 95% confidence level $\pm$ 5%, accuracy

= 20, 95% confidence level $\pm$ 10% accuracy

Predetermined Motion Time Standard (PMTS)

In this time established for human motions is used to buildup the time for a job at a defined level of performance. In this system, the operations under study are divided into the basic motions. Then, the timing for these basic motions are individually computed and added for getting the basic operations time.

MTM → (Method Time Measurement) Most popular values in MTM are given in Time Measurement Unit (TMU)

$$1\,\text{TMU} = 10^{-5}\,\text{h}$$

Work Sampling or Ratio Delay

Work sampling is a method of finding the percentage occurrence of a certain activity by statistical sampling and random observations,

$$n = \frac{K^2 P (1 - P)}{(SP)^2}$$

where, n = Sample size

P = Percentage occuring of activity

S = Error in fraction

K = A factor depending upon the confidence level

= 1 for 68% confidence level

= 2 for 95% confidence level

= 3 for 99.7% confidence level

Comparison of Work Measurement Technique

Issue	Stop Time Study	PTSM	Work Sampling
Type of standard produced	Very accurate for repeated task	Extremly accurate for repeated task	Fairy accurate for long and short cycle operation
Advantage	Tight standard	Very tight standard	Loose standard
	Speedy application	Speedy application	Ease of application
	Generate detailed information	Generate detailed information	Employ's reaction is favourable and may be used for long cycle operation
Disadvantage	Chances of employee reaction	Specialised training is necessary	Lack detailed information
	Not suited for long cycle and mental work	Not suited for long cycle or mental operation	Need long sample period
	Chances of subjective leveling		Difficult to explain and justify

Ratio Delay

Ratio of percentage of the day the man working and the percentage time man is idle.

Performance Rating

Rating as the assessment of the workers rate of working relative to observer's concept of the rate corresponding to the standard pace.

Standard Performance

Standard performance is the rate of output which qualified work will naturally achieve without over exertion. It is denoted 100 on standard rating and performance scalar.

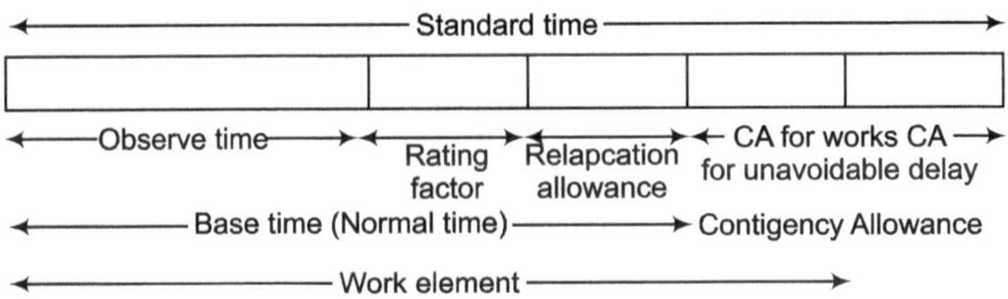

Normal time (Base time) = Elemental time × Rating factor

Standard time = Normal time + Allowances

Method Study's Criteria

There are 3 criteria use for method study

1. Economic criteria 2. Technical criteria 3. Human criteria

Process Chart Symbol

Operation	
Storage	
Delay or Temporary Storage	
Transportation	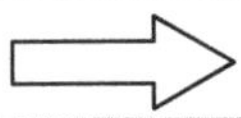
Inspection	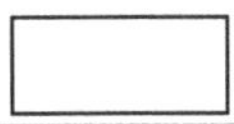
Combined Symbol	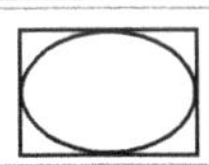

Recording Techniques

There are three types of recording technique that are allowable for carrying out work study

1. Charts 2. Diagrams 3. Photographic films

Charts

1. Outline process chart (Operation process chart)
2. Flow process chart
   (i) Man type chart, (ii) Material type chart
   (iii) Equipment type chart
3. Multiple activity chart
4. Two hand (Left hand Right hand) process chart

SIMO Chart (Simultaneous Motion Chart)

It is used to record (18 nos) agains the time scale which are performed by the operator or operator at work. SIMO chart is the micromotion form of operator process chart. Because SIMO charts are used primarily for operation of short duration often performed with extreme quickness. It becomes necessary to combine then with films. In this chart movement are recorded against time measured in wink.

$$1 \text{ wink} = \frac{1}{2000} \text{th minute.}$$

Trave Chart

It records the movement and or man between different departments.

Diagram

A diagram is used to define or describe the flow of a method involving step by step considerations with their process charts symbols. *Diagram can be divided into flow and string diagram*

Flow Diagram

It show the path of man material and components on a scale model of the factory.

String Diagram

It is a scale diagram on which movement in a given area and over a given period of time is plotted by means of a continuous thread.

Photographic Films

Photographic film is a sheet of coated with an emulsion containing light-sensitive silver halide salts with variable crystal sizes that determine the sensitivity; contrast and resolution of the film. It consists Cycle graph and Chromocycle graph which are defined below

Cycle Graph

It have been developed for studying the motion path of an operator.

Chromocycle Graph

Chromocycle graph have been appeared by using electric circuit with circuit breaker and with the bulb. Switch ON and OFF create the dotted line with pear shaped dotes by the bulb.

Value Engineering

It is the application of the concept of value analysis at the design or premanufacture stage of the component parts with a view to cut down the unnecessary costs, without impairing the function or utility of the product.

Value

It is the cost of proportionate to the function *i.e.,*

$$\text{Value} = \frac{\text{Function (or utility)}}{\text{Cost}}$$

Types of Values

1. Cost value Manufacturing component.
2. Use value *e.g.,* Tie chip, water, salt, medicine.
3. Esteem value *e.g.,* Jewellery, TV, Car.
4. Exchange value Product *i.e.,* exchangable with other one.

Value Analysis

Value analysis examines the design functions and cost of each and every component in order to produce it economically without decreasing its utility, function and reliability. Value analysis is a much faster cost reduction techniques.

Procedure of Value Analysis

The procedure of value analysis is given below

1. Blast : Identify product
   Collect information
   Define various function.
2. Create : Create and critically evaluate different alternatives.
3. Refine : Develop the best alternative and implement.

Sequencing

The selection of an appropriate order for a series of jobs to be done on a finite number of facilities is called sequencing. When the number of jobs and the number of machines are very few the problem can be solved by making mental picture. As the number of jobs and machines increasers the problem becomes complex.

Basic Terms Used in Sequencing

1. **Number of machines** The number of machines which are used in system.
2. **Processing order** It is the order in which machines are used or processed.
3. **Processing time** It indicates the time required by a job on each machine.
4. **Total elapsed time** It is the time interval between starting the first job and completing the last job including the idle time (if any) in a particular order by the given set of machines.
5. **Idle time on a machine** It is time for which a machine does not have a job to process *i.e.*, idle time from the end of job $(i-1)$ to the start of job i.
6. **No passing rule** It refers to the rule of maintaining the order in which jobs are to be processed on given machines.

Processing n Jobs through Two Machines

n Jobs are processing through two machines between different time interval which is shown in figure below

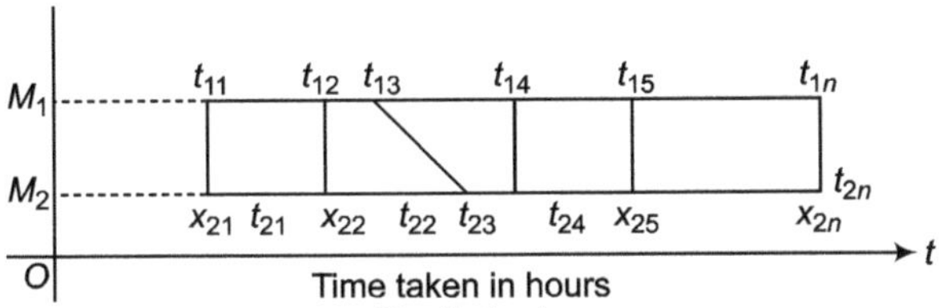

Time taken in hours
Gantt chart

Total elapsed time

Where, $$T = \sum_{j=1}^{n} t_{2j} + \sum_{j=1}^{n} x_{2j}$$

t_{2j} = Time required for processing jth job on machine M_2.
X_{2j} =Time for which M_2 remain idle after processing $(j - 1)$th job and before starting work on jth job.

Processing n Jobs through k machines

Job Number 1 2 3 ... n

Processing time on machine $M_1 t_{11} t_{12} t_{13} \dots t_{1n}$

$M_2 t_{21} t_{22} t_{23} \dots t_{2n}$

$M_k t_{k1} t_{k2} t_{k3} \dots t_{kn}$

An optimum solution to this problem can be obtained, if either or both of the following conditions hold

(a) min1 $t_{ij} \geq \max t_{ij}$ for $i = 1, 2, 3, \dots k - 1$ or (b) $\min t_{kj} \geq \max t_{ij}$

Processing 2 Jobs through k Machines

***Step* 1** Draw 2 perpendicular line, first represent job 1 and second job 2.

***Step* 2** Mark the processing time for job 1 and 2 according to the given order of machines.

***Step* 3** Constant various blocks starting from the origin by pairing the same machines until the end point.

***Step* 4** Draw a diagonal line (45°) in which horizontal segment line indicate the first job is under process while 2nd job is idle verticle segment of line show 2nd job under process while first is idle diagment segment of the line shows the both jobs are under process simultaneously.

***Step* 5** Choose the path on which diagonal movement is maximum.

***Step* 6** The total elapsed time is obtained by adding the idle time for either job to the processing time for that job.

ABC **Analysis** *and* **Material Requirement Planning** (MRP)

Always Better Control (ABC) analysis is a basic tool which helps the management to place their effort where the results would be useful to the greatest possible extent. This technqiues involve the classification of inventory items into 3 categories *A*, *B* and *C* in the decesending order of annual consumption and annual monetary value of each item.

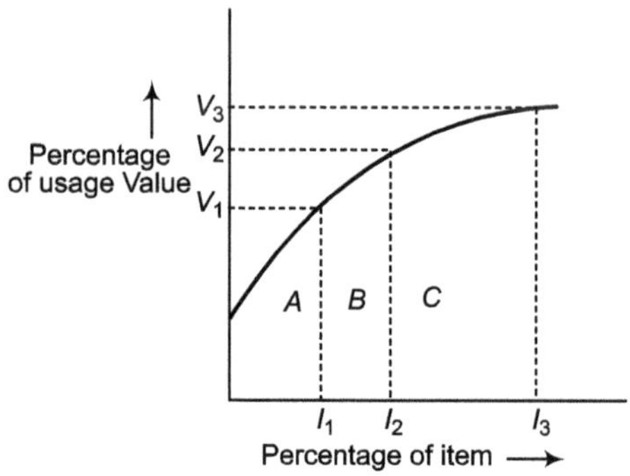

Curve between item and usage value

ABC analysis is based on Parito's law.

Category of Item in *ABC* Analysis

Category of item	Percentage annual usage items	Percentage of usage value (cost)
A	I_1	V_1
B	I_2	V_2
C	I_3	V_3

Category *A* items

- Most cost and valuable items.
- More careful and close control is needed.

Category *B* items

- Average consumption value.
- Lesser costly than *A*.
- Require lesser degree of control than *A*.

Category *C* items

- Lowest consumption value.
- Loose control of C item.

Material Requirement Planning (MRP)

MRP is a computational technique that converts the master schedule for end produce into a detailed schedule for raw materials and components used in the end products.
The detailed schedule identifies the quantities of each raw material and component items. It also falls when each item must be ordered and delivered so as to meet the master schedule for the final product.

The purpose of MRP is to ensure that materials and components are available in the right quantities and at the right time so that finished products can be completed during according to the master production schedule.

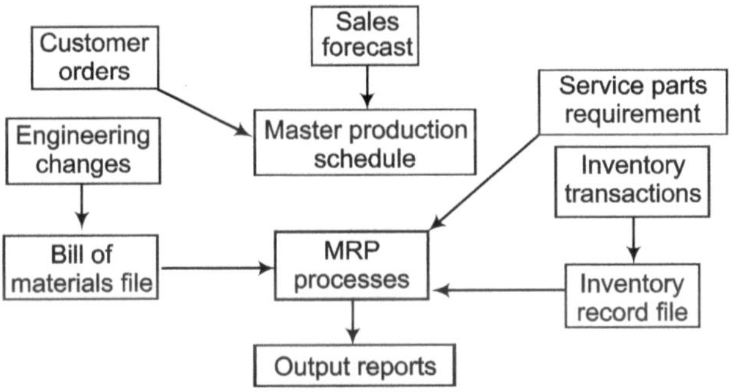

A system of MRP

There are three input to MRP

1. Master Production Schedule (MPS)
   Series of time phase quantites for each item.
2. Bill of material file which define the product structure.
3. Inventory record file or item master file.

Manufacturing Resource Planning (MRP II)

- MRP (II) is a total company system in which functional group interact commonly and formally and make joint decisions. Finally the system is user transparent users at all levels understand and accept the logic and realism of system and need not work outside the formal system.
- It includes strategic financial planning as well as production planning through the use of simulation capabilities.

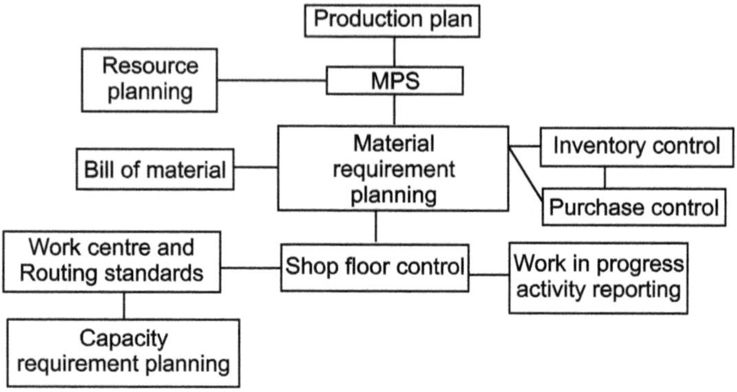

A system of MRP II

Plant Layout *and* Production Planning and Control (PPC)

Plant layout is an area of arranging fascilities such as equipment, department, section etc., inside plant or work place, it is one of the most critical strategic decisions.

There are three basic methods (types) of layout

1. Product layout (line layout)
2. Process (functional) layout

There are three basic methods of clustering an organisation activities. Those three methods are given below

1. Product Layout
2. Process Layout
3. Mixed or Group layout

Product Layout

This is also called line layout.When the product is produced in mass, this type of layout is used. This type of layout is adopted in an industry whose production is continuous.

Process Layout

It is also known as functional layout. In this layout machines or processes of similar facilities are grouped together. In such a layout, the work usually moves though a series of departments and rarely from one work station to another within a single department.

Mixed or Group layout

Mixed or grouped layout is a combination of product and process type layouts. In this type of layout advantages of both product and process is obtained. The selection of product type and process type layout depends upon the initial fixed cost and the variable cost of producing parts. *Some of the characteristics of grouped type layout are listed below*

Cellular or Group Layout

- Large variety of products in small volumes.
- Intercell transfer part should ideally be zero.
- Cellular of process & product layout.

Production Planning and Control (PPC)

PPC *has many functions to perform*

- Planning phase
- Action phase
- Control phase

Type of Layout

Factors	Fixed	**Product** (Line)	**Cellular** (Group technology)	**Process** (Functional)
Types of operations	Ship building large scale project construction project	Continuous and repetitive	Small to medium batch	Job or small batch
Arrangement of facilities	Facilities to a fixed product/ project	Placed along the line of product flow	Similar parts are grouped in part family	Grouped of speciality
Cost of layout	Moderate to low	Moderate to low	Moderate to high	Moderate to low
Material handeling	Moderate	Less	Less	High
Material travel	Variable path	Fixed path	Fixed path	Variable path
Utilisation of facilities	Moderate	Very high	High	Low
Operating facilities	General Purpose	Special purpose	Special purpose	General purpose
Employee skill	Unskilled/Spilled	Unskilled	Multiskilled	Skilled
$\frac{\text{Production of quality}}{\text{Number of production}} = \frac{Q}{P}$	Normally one	Large	Moderate	Small

Steps in PPC

The various steps of PPC are given below

- **Loadings** Assignments of work to manpower machinery etc., without specifying when the work is to be done.
- **Scheduling** Time phase of loading sequence of work.
- **Dispatching** Execute planning function properly implementation of plans.
- **Routing** Flow of work in plant, what, where and how work done.
- **Progress reporting** It can be interpreted by comparison with present level of performance.

  (i) Bar chart
  (ii) Curve chart
  (iii) Mechanical chart
  (iv) Gantt chart
- **Expediting** Taking action if the progress reporting indicates a deviation of the plan from the originally set target.
- **Replanning** Replanning of whole affairs become essential in case expediting fails to bring deviate plant to its actual path.

13

Element of Computation

Numerical Control & Computer Numerical Control

NC is an acronym for Numerical Control and CNC is an acronym for Computer Numerical Control. NC machines performed limited functions and moments controlled by punched tape or punch cards. As the technology evolved. NC machines become CNC machines. CNC machines have far more capability than their predecessor.

Stepper Motor

The stepper motor is special type of synchronous motor which is designed for rotate through a specific angle for each electrical pulse received from the control unit.

Basic Length Unit (BLU)

In NC machines the displacement length per one pulse output from machine is defined as a basic length unit.

In the CNC computer each bit (binary digit) represent a BLU.

Bit = BLU

Part Programming

The programming is the sequence of instructions, which describe the work, which has to be done on a part, in the form required by a computer under the control of a numerical control computer program.

CNC Programming

- Coordinate system
- Units incremental or absolute positioning
- Coordinates *X*, *Y*, *Z*, *RX*, *RY*, *RZ*.
- Feed rate and spindle speed.
- Coolant control : on/off, flood, mist.
- Tool control : tool and tool parameters.

Important *G* Codes

Code	Meaning	Format
G00	Rapid transverse	N_G00X_Y_Z_
G01	Linear interpolation	N_G01 X_Y_Z_F_
G02	Circular interpolation (clockwise)	N_G02 X_Y_Z_R_F_
G02	Circular interpolation (anticlockwise)	N_G02 X_Y_Z_R_F_
G04	Dwell	N_G04 P_
G17	*XY* plane	
G18	*XZ* plane	
G19	*YZ* plane	
G20/G70	Inch unit	
G21/G71	Metric unit	
G28	Automatic return to reference point	
G40	Cutter compensation (Cancel)	
G41	Cutter compensation left	N_G41_D_
G42	Cutter compensation right	N_G42D_
G43	Tool length compensation	N_G43H_
G44	Tool length compensation	N_G44H
G49	Tool length compensation (cancel)	
G80	Cancel corved cycle	
G81	Drilling cycle	N_G81T_R_F
G90	Absolute positioning	
G91	Incremental positioning	
G92	Absolute present change the	N_G92 X_Y_Z_

Important *M* Codes

Code	Meaning
*M*00	Program stop
*M*01	Optional program stop
*M*03	Spindle on clockwise
*M*04	Spindle on counter clockwise
*M*05	Spindle stop
*M*06	Tool change
*M*08	Coolant on
*M*09	Coolant off
*M*10	Clamps on
*M*11	Clamps off
*M*02 or *M*30	Program stop, reset to start

Key Points

- Part programming is the task of preparing a program sheet from a drawing sheet.
- In part programming, all data is fed into the numerical control system using a standardized format.
- Programming is where all the machining data are compiled and where the data are translated into a language which can be understood by the control system of the machine tool.

APT Language

(Automatically Programmed Tools Language)

APT is a high level computer programming language used to generate instructions for numerically controlled machine tools.

The APT language consists of many different types of statements made up of the following valid letters, numerals and punctuations marks.

Letters : ABCDEFGHIJKLMNOPQRSTUVWXYZ

Numerals : 0123456789

(/) A slash divides a statement into to sections.
e.g., GO/PAST.

(,) A comma is used as a separator between the elements in a statement generally to the right of the slash.

(=) An equals is used for assigning an entity to a symbol name.
e.g., P_1 = POINT/25, 50, 30

The complete APT part program consists of the following four type of statements

1. Geometry
2. Motion
3. Post Precessor
4. Compilation Control

Robotics

Industrial Robot A robot is a reprogrammable, multifunctional manipulator designed to handle material, parts, tools or specialised devices through variable programmed motions for the performance of a variety of tasks.

Asimov's Three Law of Robotics

1. **Human Safety** A robot may not injur a human being or through inaction, allow a human being to come to harm.
2. **Robots are Slaves** A robot must obey orders given it by human beings except where such orders would conflict with the first law.
3. **Robot Survival** A robot must protect its own existence as long as such protection does not conflict with the first or second law.

All robots have following basic components

1. Manipulators
2. Feed back devices
3. End effectors
4. Controller
5. Power Supply

Wrist Assembly

Wrist assembly is attached to end of arm. End effector is attached to wrist assembly function of wrist assembly is to orient end effector.

Two or 3 degree of freedom is needed to wrist for orientation and work effectively.

Note A total of 6 degree of freedom is needed to locate a robot's hand at any point in its work force.

FMS (Flexible Manufacturing System)

FMS is a manufacturing system in which there is same amount of flexibility that allows the system to react in the case of changes.

There are two types of categories of flexibility.

1. Machine flexibility
2. Routing flexibility

Computer

A computer is an electronic machine that helps to process data. It is used to solve problems relating to almost all fields such as education, home. medicine, science and technology, research, designing, publishing, communication etc.
A computer is an information-processing and information-accessing tool. This means that a computer accepts some information or data from the outside world. It processes it to produce a new information.

Types of Computer

According to size and capacity these are following types of computers

Micro Computer

These computers are used by individual, thus also called PC or Personal Computer. These days P.Cs are largely used for domestic and official purposes etc.

Mini Computer

This type of computer is comparatively larger than that of micro computer. This is 5 to 50 times more powerful than that of a Micro Computer.

Main Frame Computer

These are large sized computers. By Time Sharing and Multi Tasking techniques many people rather more than 100 people can work at a time on different terminals of this computer.

Super Computer

These are very powerful computers and have more storage capacity. These are the most expensive and the fastest computer able to process most complex jobs with a very high speed.

Quantum Computer

The development of this type is in final stage. Probably Quantum Computers will be more advanced than that of human brain. In Quantum Computer Q-Bit will be used in lieu of Binary Bits.

Programming Languages of Different Generation

Generation	Languages
1st generation (1940-52)	FORTRAN-i
2nd generation (1952-64)	FORTRAN-ii, ALGOL-60, COBOL
3nd generation (1964-71)	PL/I, ALGOL-W, ALGOL-68, SIMULA-67, APL, SNOBOL, 4 BASIC
4nd generation (1971- —)	CLUE, ALFARD, UCLID, Reformed MODULA, EDA, ORACLE
5nd generation (For future)	Artificial Intelligence Languages.

Key Points

- A computer is a data processing machine.
- Hardware comprises of the physical units of a computer system.
- Both hardware and software together make a computer system functional.
- An operating system is an interface between the user and the computer hardware.
- A database is a collection of interrelated data.
- A relational database is a collection of data items organized as a set of formally described tables from which data can be accessed or reassembled in many different ways without having to reorganize the database tables.

Appendix

Symbols *and* Units

K Absolute temperature degree Kelvin

C Temperature, degree Celsius

F Temperature, degree Fahrenheit

ρ Density, (kg/m^3)

C_p Specific heat capacity, (kJ/kg.K)

C_p/C_V Specific heat capacity, ratio, dimensi- onless

μ Viscosity, $N\text{-}s/m^2 \times 10^6$(For $N\text{-}s/m^2$ $(= kg/m.s)$ multiply tabulated values by 10^{-6})

K Thermal conductivity, $W/m.K \times 10^3$ (For $W/m.K$ multiply tabulated values by 10^{-3})

Pr Prandtl number, dimensionless

h Enthalpy, (kJ/kg)

V_s Sound velocity, (m/s)

T Absolute temperature, degree Kelvin

$\bar{h}$ Enthalpy, (kJ/kmol)

$\bar{u}$ Internal energy, (kJ/kmol)

$\bar{s}^o$ Absolute entropy at standard reference pressure, (kJ/kmol K)

p_s Pressure of water vapour at saturation. (N/m^2)

W_s Humidity ratio at saturation, mass of water vapour associated with unit mass of dry air

V_a Specific volume of day, (m^3/kg)

V_s Specific volume of saturated mixture, $(m^3/kg$ dry air)

h_a^a Specific enthalpy of day air(kJ/kg)

h_s Specific enthalpy of saturated mixture (kJ/kg dry air)

S_s Specific entropy of saturated mixture, (J /K· kg dry air)

t Thermodynamic temperature, degree C

T Thermodynamic temperature, K

pV RT,(kJ/kg)

u_0 Specific internal energy at zero pressure, (kJ/kg)

h_0 Specific enthalpy at zero pressure, (kJ/kg)

s_1 Specific entropy of semiperfect vapour at 0.1 MN/m^2, (kJ /kg.K)

ψ_1 Specific Helmholtz free energy of semiperfect vapour at 0.1 MN/m^2, (kJ /kg)

ς_1 Specific Gibbs free energy of semiperfect vapour at 0.1 MN/m^2, (kJ /kg)

p_r Relative pressure, pressure of semiperfect vapour at zero entropy, (TN/m^2)

V_r Relative specific volume, specific volume of semiperfect vapour at zero entropy, (mm^3/kg)

C_{p0} Specific heat capacity at constant pressure for zero pressure, $(kJ/kg \cdot K)$

C_{V0}	Specific heat capacity at constant volume for zero pressure, $(kJ / kg \cdot K)$	T_{sat}	Saturation temperature, °C S = Entropy, (kJ/kg.K)
k	C_{p0}/C_{V0} = Isentropic exponent $-(\partial \log p / \partial \log V)_s$	V	Specific volume, (m^3/kg)
T	Temperature, °C h = enthalpy, (kJ/kg)	p	Pressure, bar and μPa
		U	Internal energy.(kJ/kg)

Mechanical Engineering Glossary

- **Bearing** The part of a machine within which a rotating or sliding shaft is held.
- **Bearing Type Connection** A connection made using snug-tight connection where the load is transferred by bearing of bolts against plate (inside the bolt hole).
- **Buckling Strength** Force or moment which a member can withstand without buckling.
- **Block Shear** A fracture type shear failure where fracture may occur on either the tension plane or the shear plane followed by yielding on the other plane.
- **Braced Frame** A frame that has resistance to lateral loads supplied by some types of bracing system.
- **Brittle Fracture** Abrupt fracture with little or no prior ductile deformation.
- **Buckling** Instability of the compression member characterised by their deflection in their own plane.
- **Buckling Load** The load at which a straight compression member assumes a deflected position.
- **Bell Crank** A pivoting double lever used to change the direction of applied motion.
- **Boss** A cylindrical projection, as on a casting or a forging. Usually provides a contact surface around a hole.
- **Broach** To finish the inside of a hole to a shape other than round, as in a keyway (*n*). The tool for the process, which has serrated edges and is pushed or pulled through the hole to produce the required shape.
- **Burnish** To smooth or polish by a rolling or sliding tool under pressure.
- **Bushing** A smooth walled bearing (AKA a plain bearing). Also a tool guide in a jig or fixture.
- **Cam** A mechanical device consisting of an eccentric or multiply curved wheel mounted on a rotating shaft, used to produce variable or reciprocating motion in another engaged or contacted part (cam follower).

- **Casting** Any object made by pouring molten metal into a mold.
- **Chamfer** A flat surface made by cutting off the edge or corner of a object (bevel *v*). This process of creating a chafer.
- **Clevis** A U-shaped piece with holes into which a link is inserted and through which a pin or bolt is run. It is used as a fastening device which allows rotational motion.
- **Collar** A cylindrical feature on a part fitted on a shaft used to prevent sliding (axial) movement.
- **Collet** A cone shaped sleeve used for holding circular or rod like pieces in a lathe or other machine.
- **Core** To form the hollow part of a casting, using a solid form placed in the mold (*n*) .The solid form used in the coring process, often made of wood, sand, or metal.
- **Counterbore** A cylindrical flat-bottomed hole, which enlarges the diameter of an existing pilot hole.
- **Countersunk** A conical depression added to an existing hole to accommodate and the conic head of a fastener recessing it below the surface of a face.
- **Coupling** A device used to connect two shafts together at their ends for the purpose of transmitting power may be used to account for minor misalignment or for mitigating shock loads.
- **Die** One of a pair of hardened metal plates or impressing or forming desired shape. Also, a tool for cutting external threads.
- **Face** To machine a flat surface perpendicular to the axis of rotation of a piece.
- **Fillet** A rounded surface filling the internal angle between two intersection surfaces.
- **Fit** The class of contact between two machined surfaces, based upon their respective specified size tolerances (clearance, transitional, interference).
- **Fixture** A device used to hold a workpiece while manufacturing operations are performed upon that workpiece.
- **Flange** A projecting rim or edge for fastening, stiffening or positioning.
- **Gage** A device used for determining the accuracy of specified manufactured parts by direct comparison.
- **Gage Blocks** Precision machined steel blocks having two flats, parallel surfaces whose separation distance is fabricated to a guaranteed accuracy of a few million of an inch;
- **Gear Hobbing** A special form of manufacturing that cuts gear tooth geometries. It is the major industrial process for cutting involute form spur gears.

- **Geneva Cam** A device to turn constant rotational motion into intermittent rotational motion.
- **Gusset** (Plate) A triangular metal piece used to strengthen a joint.
- **Hasp** A metal fastener with a slotted, hinged part that fits over a loop and is secured by a pin, bolt, or padlock.
- **Idler** A mechanism used to regulate the tension in belt or chain. Or, a gear used between a driver and follower gear to maintain the direction of rotation.
- **Jig** A special device used to guide a cutting tool (drill jig) or to hold material in the correct position for cutting or fitting together (as in welding or brazing).
- **Journal** The part of a shaft that rotates within a bearing.
- **Kerf** A channel or groove cut by a saw or other tool.
- **Key** A small block or wedge inserted between a shaft and hub to prevent circumferential movement.
- **Keyseat** A slot or groove cut in a shaft to fit a key. A key rests in a keyseat.
- **Keyway** A slot cut into a hub to fit a key. A key slides in a keyway.
- **Knurl** To roughen a turned surface, as in a handle or a knob.
- **Lug** Projection on (typically) a cast or forged part to provide support or allow mounting or the attachment of another component.
- **Neck** To cut a groove around a shaft, usually toward the end or at a change in diameter. A portion of reduced diameter between the ends of a shaft.
- **Pad** A rectangular or irregular projection, as on a casting or a forging. Usually provides a contact surface around a set of holes.
- **Pawl** A device used to prevent a toothed wheel (ratchet) from rotating backwards, or a device that stops, locks, or releases a mechanism.
- **Pillow Block** A bearing housing which typically mounts to a single planar face. May be split or un-split to accommodate insertion /removal of the bearing.
- **Pinion** A plain gear, often the smallest gear in a gearset, often the driving gear. May be used in conjunction with a gear rack (rack and pinion, see below).
- **Planetary Gears** A gearset characterised by one or more planet gear(s) rotating around a sun gear.
- **Rack (w/pinion gear)** A toothed bar acting on (or acted upon), by a gear (pinion).
- **Ratchet** A mechanical device used to permit motion in one direction only.
- **Relief** A groove or cut on a part used to facilitate machining.
- **Retaining Ring** A tool steel ring used in conjunction with a shaft groove or internal groove to located or control position of a component.

- **Rocker Arm** A pivoted arm-like lever used to transfer the application direction of a linear force.
- **Round** A rounded external intersection between two surfaces. Compare to Fillet.
- **Scotch Yoke** Mechanism used to convert rotational motion to linear motion.
- **Sheave** A grooved wheel used to accommodate a belt for the transmission of power. Sometimes referred to as a pulley sheave.
- **Shim** A thin strip of metal inserted between two surfaces to adjust for fit. The process of inserting shims.
- **Shoulder** A plane surface on a shaft, normal to the axis, produced by a change in diameter.
- **Spline** A cylindrical pattern of keyways. May be external (L) or internal (R).
- **Spotface** A round machine surface around a hole on a casting or forging, usually to provide a contact surface for a fastener or other mating component, the process used to create that feature.
- **Standoffs** A mounting designed to position objects a predetermined distance above or away from the surface upon which they are mounted.
- **Tap** To cut internal machine threads in a hole, the tool used to create that feature.
- **Undercut** A cut having inward sloping sides, to cut leaving an overhanging edge.
- **Yoke** A clamp or vise that holds a machine part in place or controls its movement or that holds two such parts together. A crosshead of relatively thick cross section, that secures two or more components so that they move together.
- **Compressive Stress** (or compression) is the stress state caused by an applied load that acts to reduce the length of the material (compression member) in the axis of the applied load, in other words stress state caused by squeezing the material.
- **Tensile Stress** is the stress state caused by an applied load that tends to elongate the material in the axis of the applied load, in other words the stress caused by pulling the material.
- **Shear Stress** is the stress state caused by the combined energy of a pair of opposing forces acting along parallel lines of action through the material.
- **Yield Strength** is the lowest stress that produces a permanent deformation in a material.
- **Compressive Strength** is a limit state of **compressive stress** that leads to failure in the manner of ductile failure (infinite theoretical yield) or brittle.

- **Tensile Strength** or ultimate tensile strength is a limit state of tensile stress that leads to tensile failure in the manner of ductile failure.
- **Fatigue Strength** is a measure of the strength of a material or a component under cyclic loading, and is usually more difficult to assess than the static strength measures.
- **Strain** or reduced deformation is a mathematical term that expresses the trend of the deformation change among the material field.
- **Deflection** is a term to describe the magnitude to which a structural element bends under a load.
- **Elasticity** is the ability of a material to return to its previous shape after stress is released.
- **Plasticity** or plastic deformation is the opposite of elastic deformation and is defined as unrecoverable strain.
- **Thermodynamics** is a branch of natural science concerned with heat and its relation to energy and work.
- **Equilibrium Thermodynamics** In thermodynamic equilibrium, a system's properties are, by definition, unchanging in time.
- **Non-equilibrium Thermodynamics** is a branch of thermodynamics that deals with systems that are not in thermodynamic equilibrium; it is also called thermodynamics of irreversible processes.
- **Zeroth law of Thermodynamics** If two systems are each in thermal equilibrium with a third, they are also in thermal equilibrium with each other.
- **First law of Thermodynamics** The increase in internal energy of a closed system is equal to the difference of the heat supplied to the system and the work done by it: $\Delta U = Q - W$
- **Second law of Thermodynamics** Heat cannot spontaneously flow from a colder location to a hotter location.
- **The Internal Combustion Engine** is an engine in which the combustion of a fuel occurs with an oxidiser in a combustion chamber that is an integral part of the working fluid flow circuit.
- **A Two-Stroke** Two-cycle, or two-cycle engine is a type of internal combustion engine which completes a power cycle in only one crankshaft revolution and with two strokes, or up and down movements, of the piston in comparison to a "four-stroke engine", which uses four strokes to do so.
- **A Four-Stroke Engine** (Also known as four-cycle) is an internal combustion engine in which the piston completes four separate strokes-intake, compression, power, and exhaust-during two separate revolutions of the engine's crankshaft, and one single thermodynamic cycle.

- **The Six-Stroke Engine** is a type of internal combustion engine based on the four-stroke engine, but with additional complexity intended to make it more efficient and reduce emissions.
- **A Diesel Engine** also, known as a compression-ignition engine and is an internal combustion engine that uses the heat of compression to initiate ignition to burn the fuel that has been injected into the combustion chamber.
- **A Petrol Engine** known as a gasoline engine in North America and is an internal combustion engine with spark-ignition, designed to run on petrol (gasoline) and similar volatile fuels.

Frontiers of Research in Mechanical Engineering

- **Micro Electro-Mechanical Systems** (MEMS) Micro-scale mechanical components such as springs, gears, fluidic and heat transfer devices are fabricated from a variety of substrate materials such as silicon, glass and polymers like SU8.
- **Friction Stir Welding** (FSW) This innovative steady state (non-fusion) welding technique joins materials previously un-weldable, including several aluminum alloys. It plays an important role in the future construction of airplanes, potentially replacing rivets.
- **Composites** or composite materials are a combination of materials which provide different physical characteristics than either material separately.
- **Mechatronics** is the synergistic combination of mechanical engineering, electronic engineering and software engineering. The purpose of this interdisciplinary engineering field is the study of automation from an engineering perspective and serves the purposes of controlling advanced hybrid systems.
- **Nanotechnology** At the smallest scales, mechanical engineering becomes nanotechnology -one speculative goal of which is to create a molecular assembler to build molecules and materials *via* mechanosynthesis.
- **Biomechanics** is the application of mechanical principles to biological systems, such as humans, animals, plants, organs and cells.
- **Computational Fluid Dynamics** Computational fluid dynamics, usually abbreviated as CFD, is a branch of fluid mechanics that uses numerical methods and algorithms to solve and analyse problems that involve fluid flows.

Thermodynamics Relationships

- Entropy $S = -k_B \sum_i p_i \ln p_i$
- Enthalpy $H = U + pV$
- Gibbs free energy $G = H - TS$
- Helmholtz free energy $F = U - TS$
- Grand potential $\phi_G = U - TS - \mu N$
- General heat/thermal capacity $C = \partial Q / \partial T$
- Heat capacity (isobaric) $C_p = \partial Q / \partial T$
- Specific heat capacity (isobaric) $C_{mp} = \partial^2 Q / \partial m \partial T$
- Molar specific heat capacity (isobaric) $C_{np} = \partial^2 Q / \partial n \partial T$
- Heat capacity (isochoric/volumetric) $C_V = \partial Q / \partial T$
- Specific heat capacity (isochoric) $C_{mV} = \partial^2 Q / \partial m \partial T$
- Molar specific heat capacity (isochoric) $C_{nV} = \partial^2 Q / \partial n \partial T$
- Specific latent heat $L = \partial Q / \partial m$
- Ratio of isobaric to isochoric heat capacity, heat capacity ratio $\gamma = C_p / C_V = C_p / C_V = C_{mp} / C_{mV}$

Phase Transitions

- Adiabatic transition $\Delta Q = 0$, $\Delta U = W$
- Isothermal transition $\Delta U = 0$, $\Delta W = -\Delta Q$
- Isothermal transition $W = KTN \ln (V_2/V_1)$
- Isobaric transition $p_1 = p_2$, $p =$ constant
  $\Delta W = p\Delta V$, $\Delta U = \Delta Q + p\delta V$
- Isochoric transition $V_1 = V_2$, $V =$ constant
  $\Delta W = 0$, $\Delta Q = \Delta U$
- Adiabatic expansion $p_1 V_1^{\gamma} = p_2 V_2^{\gamma}$
  $T_1 V_1^{\gamma - 1} = T_2 V_2^{\gamma - 1}$
- Free expansion $\Delta U = 0$
- Work done by an expanding gas $\Delta w = \int_{V_1}^{V_1} p dv$

  Net work done in cyclic processes $\Delta w = \oint_{\text{cycle}} pdV$

Maxwell's Relations

Physical situation	Nomenclature	Equations
Thermodynamic potentials as functions of their natural variables	$U(S,V)$ = Internal energy	$\left(\frac{\partial T}{\partial V}\right)_S = -\left(\frac{\partial p}{\partial S}\right)_V = \frac{\partial^2 U}{\partial S \partial V}$
	H (S, p) = Enthalpy	$\left(\frac{\partial T}{\partial p}\right)_S = +\left(\frac{\partial V}{\partial S}\right)_p = \frac{\partial^2 H}{\partial S \partial p}$
	F (T, V) = Hemholtz free energy	$+\left(\frac{\partial S}{\partial V}\right)_T = \left(\frac{\partial p}{\partial T}\right)_V = \frac{\partial^2 F}{\partial T \partial V}$
	G (T, p) = Gibbs free energy	$-\left(\frac{\partial S}{\partial p}\right)_T = +\left(\frac{\partial V}{\partial T}\right)_p = \frac{\partial^2 G}{\partial T \partial p}$

Thermal Properties of Matter

- Joule-Thomson Coefficient $\mu_{JT} = \left(\frac{\partial T}{\partial p}\right)_H$
- Compressibiility (constant temperature) $K_T = -\frac{1}{V}\left(\frac{\partial V}{\partial p}\right)_{T,N}$
- Coefficient of thermal expansion (constant pressure) $\alpha_p = \frac{1}{V}\left(\frac{\partial V}{\partial pT}\right)_p$
- Heat capacity (constant pressure)

  $$C_p = \left(\frac{\partial Q_{rev}}{\partial T}\right)_p = \left(\frac{\partial U}{\partial T}\right)_p + p\left(\frac{\partial V}{\partial T}\right)_p = \left(\frac{\partial H}{\partial T}\right)_p = T\left(\frac{\partial S}{\partial T}\right)_p$$

- Heat capacity (constant volume) $C_V = \left(\frac{\partial Q_{rev}}{\partial T}\right)_V = \left(\frac{\partial U}{\partial T}\right)_V = T\left(\frac{\partial S}{\partial T}\right)_V$

www.ingramcontent.com/pod-product-compliance
Ingram Content Group UK Ltd.
Pitfield, Milton Keynes, MK11 3LW, UK
UKHW021337190726
13845UKWH00027B/605

9 789350 945841